Lecture Notes in Computer Science 2950
Edited by G. Goos, J. Hartmanis, and J. van Leeuwen

Springer-Verlag Berlin Heidelberg GmbH

Nataša Jonoska   Gheorghe Păun
Grzegorz Rozenberg (Eds.)

Aspects of Molecular Computing

Essays Dedicated to Tom Head
on the Occasion of His 70th Birthday

Springer

Volume Editors

Nataša Jonoska
University of South Florida, Department of Mathematics
4202 e. Fowler Av. PHY114, Tampa, FL 33620-5700, USA
E-mail: jonoska@math.usf.edu

Gheorghe Păun
Institute of Mathematics of the Romanian Academy
P.O. Box 1-764, 70700 Bucharest, Romania
and
Rovira i Virgili University
Pl. Imperial Tàrraco 1, 43005 Tarragona, Spain
E-mail: gp@astor.urv.es

Grzegorz Rozenberg
Leiden University, Leiden Institute for Advanced Computer Science
Niels Bohrweg 1, 2333 CA Leiden, The Netherlands
and
University of Colorado, Department of Computer Science
Boulder, CO 80309-0347, USA
E-mail: rozenber@liacs.nl

Cataloging-in-Publication Data applied for

A catalog record for this book is available from the Library of Congress.

Bibliographic information published by Die Deutsche Bibliothek
Die Deutsche Bibliothek lists this publication in the Deutsche Nationalbibliografie; detailed bibliographic data is available in the Internet at <http://dnb.ddb.de>.

CR Subject Classification (1998): F.1, J.3, G.2, F.4.3

ISSN 0302-9743
DOI 10.1007/978-3-540-24635-0

Springer-Verlag is a part of Springer Science+Business Media

springeronline.com

MyCopy version of the original edition 2004

Typesetting: Camera-ready by author, data conversion by Boller Mediendesign
Printed on acid-free paper SPIN: 10976616 06/3142 5 4 3 2 1 0
www.springer.com/mycopy

Thomas J. Head

Preface

Molecular Computing is a fast-emerging area of Natural Computing. On the one hand, it is concerned with the use of (bio)molecules for the purpose of computing while on the other hand it tries to understand the computational nature of the molecular processes going on in living cells.

The paper "Molecular computation of solutions to combinatorial problems" by L. Adleman, which describes a laboratory experiment and was published in *Science* in November 1994, was an important milestone for the area of molecular computing, as it provided a "proof-of-principle" that one can indeed perform computations in a biolab using biomolecules and their processing using biomolecular operations. However, research concerning the computational nature of biomolecular operations dates back to before 1994. In particular, a pioneering work concerning the mathematical theory of biooperations is the paper "Formal language theory and DNA: an analysis of the generative capacity of specific recombinant behaviors," authored by Tom Head in 1987, which appeared in the *Bulletin of Mathematical Biology.* The paper uses the framework of formal language theory to formulate and investigate the computational effects of biomolecular operations carried out by restriction enzymes. This paper has influenced research in both molecular computing and formal language theory. In molecular computing it has led to a clear computational understanding of important biomolecular operations occurring in nature, and it has also stimulated the design of a number of laboratory experiments utilizing Tom's ideas for the purpose of human-designed DNA computing. In formal language theory it has led to a novel, interesting and challenging research area, originally called "splicing systems" and then renamed "H systems" in honor of Tom ("H" stands for "Head"). Many papers stimulated by the pioneering ideas presented by Tom in his seminal paper were written by researchers from all over the world.

Adleman's paper was a great event in Tom's life: it has confirmed his conviction that biooperations can be used for the purpose of computing, but more importantly it has stimulated his interest in experimental research. One can safely say that since 1994 most of Tom's research has been focused on the design of experiments related to DNA computing. Also on this research path he remained highly innovative and original, combining his great talent for modeling with a passsion for experimental biology. A good manifestation of this line of Tom's research is aqueous computing – a really elegant but also experimentally feasible model of molecular computing invented by him.

An example of the recognition of Tom's research within the molecular computing community is the "DNA Scientist of the Year" award that Tom received in 2003.

Tom's multidisciplinary talents and interests become even more evident when one realizes that his original training and passion was mathematics, and in particular algebra. He moved from there to formal language theory. It is also important to keep in mind that his work on formal models for biology started long before his 1987 paper, as he was a very active and productive researcher in the area of Lindenmayer systems that model the development of simple multicellular organisms, not on the molecular but rather on the cellular level.

With this volume, presenting many aspects of research in (or stimulated by) molecular computing, we celebrate a scientist who has been a source of inspiration to many researchers, and to us a mentor, a scientific collaborator, and a warm and caring friend.

HAPPY BIRTHDAY, Tom!

November 2003

Nataša Jonoska
Gheorghe Păun
Grzegorz Rozenberg

Table of Contents

Solving Graph Problems by P Systems with Restricted Elementary Active Membranes

Artiom Alhazov[1,2], Carlos Martín-Vide[2], and Linqiang Pan[2,3]

[1] Institute of Mathematics and Computer Science
Academy of Science of Moldova
Str. Academiei 5, Chişinău, MD 2028, Moldova
artiom@math.md
[2] Research Group on Mathematical Linguistics
Rovira i Virgili University
Pl. Imperial Tàrraco 1, 43005 Tarragona, Spain
aa2.doc@estudiants.urv.es,
cmv@astor.urv.es, lp@fll.urv.es
[3] Department of Control Science and Engineering
Huazhong University of Science and Technology
Wuhan 430074, Hubei, People's Republic of China
lqpan@mail.hust.edu.cn

Abstract. P systems are parallel molecular computing models based on processing multisets of objects in cell-like membrane structures. In this paper we give membrane algorithms to solve the vertex cover problem and the clique problem in linear time with respect to the number of vertices and edges of the graph by recognizing P systems with active membranes using 2-division. Also, the linear time solution of the vertex cover problem is given using P systems with active membranes using 2-division and linear resources.

1 Introduction

The P systems are a class of distributed parallel computing devices of a biochemical type, introduced in [2], which can be seen as a general computing architecture where various types of objects can be processed by various operations. It comes from the observation that certain processes which take place in the complex structure of living organisms can be considered as computations. For a motivation and detailed description of various P system models we refer the reader to [2],[4].

In [3] Păun considered P systems where the number of membranes increases during a computation by using some division rules, which are called P systems with active membranes. These systems model the natural division of cells.

In [6] Pérez-Jiménez et al. solve the satisfiability problem in linear time with respect to the number of variables and clauses of propositional formula by recognizing P systems with active membranes using 2-division. Thus the vertex cover problem and the clique problem belonging to the class of **NP** problems can also

N. Jonoska et al. (Eds.): Molecular Computing (Head Festschrift), LNCS 2950, pp. 1–22, 2004.

be solved in a polynomial time by recognizing P systems with active membranes using 2-division. One can get solutions for the vertex cover problem and the clique problem by reducing these problems to the satisfiability problem in order to apply those P systems which solve satisfiability problem in linear time. In this paper, we do not apply the cumbersome and time consuming (polynomial time) reduction, but we directly give membrane algorithms to solve the vertex cover problem and the clique problem in linear time with respect to the number of vertices and edges of graph by recognizing P systems with restricted elementary active membranes. In this solution, the structure of P systems is uniform, but we need polynomial resources to describe P systems. Using linear resources to describe P systems, we also give a membrane algorithm to solve the vertex cover problem in linear time by semi-uniform P systems.

The paper is organized as follows: in Section 2 we introduce P systems with active membranes; in Section 3 we define the notion of recognizing P system; in Section 4 the polynomial complexity class in computing with membranes, $\mathbf{PMC}_{\mathcal{F}}$, is recalled. Section 5 gives a membrane algorithm to solve the vertex cover problem in linear time with respect to the number of vertices and edges of graph by recognizing P systems with restricted elementary active membranes, and give a formal verification. In Section 6, we give a membrane solution of the vertex cover problem by semi-uniform P systems with restricted elementary active membranes and linear resources. In Section 7, the clique problem is solved in linear time with respect to the number of vertices and edges of the graph by recognizing P systems with restricted elementary active membranes. Finally, Section 8 contains some discussion.

2 P Systems with Active Membranes

In this section we describe P systems with active membranes due to [3], where more details can also be found.

A *membrane structure* is represented by a Venn diagram and is identified by a string of correctly matching parentheses, with a unique external pair of parentheses; this external pair of parentheses corresponds to the external membrane, called the *skin*. A membrane without any other membrane inside is said to be *elementary*. For instance, the structure in Figure 1 contains 8 membranes; membranes 3, 5, 6 and 8 are elementary. The string of parentheses identifying this structure is

$$\mu = [_1[_2[_5\ ]_5[_6\ ]_6]_2[_3\ ]_3[_4[_7[_8\ ]_8]_7]_4]_1.$$

All membranes are labeled; here we have used the numbers from 1 to 8. We say that the number of membranes is the *degree* of the membrane structure, while the height of the tree associated in the usual way with the structure is its *depth*. In the example above we have a membrane structure of degree 8 and of depth 4.

In what follows, the membranes can be marked with + or −, and this is interpreted as an "electrical charge", or with 0, and this means "neutral charge". We will write $[_i\ ]_i^+, [_i\ ]_i^-, [_i\ ]_i^0$ in the three cases, respectively.

The membranes delimit *regions*, precisely identified by the membranes (the region of a membrane is delimited by the membrane and all membranes placed immediately inside it, if any such a membrane exists). In these regions we place *objects*, which are represented by symbols of an alphabet. Several copies of the same object can be present in a region, so we work with *multisets* of objects. A multiset over an alphabet V is represented by a string over V: the number of occurrences of a symbol $a \in V$ in a string $x \in V^*$ (V^* is the set of all strings over V; the empty string is denoted by λ) is denoted by $|x|_a$ and it represents the multiplicity of the object a in the multiset represented by x.

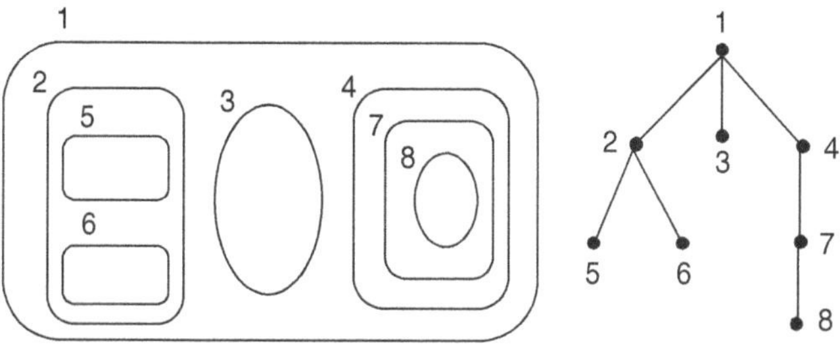

Figure 1. A membrane structure and its associated tree

A *P system with active membranes and 2-division* is a construct

$$\Pi = (O, H, \mu, w_1, \ldots, w_m, R),$$

where:

(i) $m \geq 1$ (the initial degree of the system);
(ii) O is the alphabet of *objects*;
(iii) H is a finite set of *labels* for membranes;
(iv) μ is a *membrane structure*, consisting of m membranes, labelled (not necessarily in a one-to-one manner) with elements of H;
(v) $w_1, \ldots, w_m$ are strings over O, describing the *multisets of objects* placed in the m regions of μ;
(vi) R is a finite set of *developmental rules*, of the following forms:
 (a) $[_h a \to v]_h^{\alpha}$,
 for $h \in H, \alpha \in \{+, -, 0\}, a \in O, v \in O^*$
 (object evolution rules, associated with membranes and depending on the label and the charge of the membranes, but not directly involving the membranes, in the sense that the membranes are neither taking part in the application of these rules nor are they modified by them);
 (b) $a[_h\ ]_h^{\alpha_1} \to [_h b]_h^{\alpha_2}$,
 for $h \in H, \alpha_1, \alpha_2 \in \{+, -, 0\}, a, b \in O$
 (communication rules; an object is introduced in the membrane, possibly modified during this process; also the polarization of the membrane can be modified, but not its label);

(c) $[_h a\,]_h^{\alpha_1} \rightarrow [_h\,]_h^{\alpha_2} b,$
for $h \in H, \alpha_1, \alpha_2 \in \{+,-,0\}, a, b \in O$
(communication rules; an object is sent out of the membrane, possibly modified during this process; also the polarization of the membrane can be modified, but not its label);

(d) $[_h a\,]_h^{\alpha} \rightarrow b,$
for $h \in H, \alpha \in \{+,-,0\}, a, b \in O$
(dissolving rules; in reaction with an object, a membrane can be dissolved, while the object specified in the rule can be modified);

(e) $[_h a\,]_h^{\alpha_1} \rightarrow [_h b\,]_h^{\alpha_2} [_h c\,]_h^{\alpha_3},$
for $h \in H, \alpha_1, \alpha_2, \alpha_3 \in \{+,-,0\}, a, b, c \in O$
(division rules for elementary membranes; in reaction with an object, the membrane is divided into two membranes with the same label, possibly of different polarizations; the object specified in the rule is replaced in the two new membranes by possibly new objects);

(f) $[_{h_0} [_{h_1}\,]_{h_1}^{\alpha_1} \dots [_{h_k}\,]_{h_k}^{\alpha_1} [_{h_{k+1}}\,]_{h_{k+1}}^{\alpha_2} \dots [_{h_n}\,]_{h_n}^{\alpha_2}]_{h_0}^{\alpha_0}$
$\rightarrow [_{h_0} [_{h_1}\,]_{h_1}^{\alpha_3} \dots [_{h_k}\,]_{h_k}^{\alpha_3}]_{h_0}^{\alpha_5} [_{h_0} [_{h_{k+1}}\,]_{h_{k+1}}^{\alpha_4} \dots [_{h_n}\,]_{h_n}^{\alpha_4}]_{h_0}^{\alpha_6},$
for $k \geq 1, n > k, h_i \in H, 0 \leq i \leq n$, and $\alpha_0, \dots, \alpha_6 \in \{+,-,0\}$ with $\{\alpha_1, \alpha_2\} = \{+,-\}$; if the membrane with the label h_0 contains other membranes than those with the labels $h_1, \dots, h_n$ specified above, then they must have neutral charges in order to make this rule applicable; these membranes are duplicated and then are part of the contents of both new copies of the membrane h_0
(division of non-elementary membranes; this is possible only if a membrane contains two immediately lower membranes of opposite polarization, $+$ and $-$; the membranes of opposite polarizations are separated in the two new membranes, but their polarization can change; always, all membranes of opposite polarizations are separated by applying this rule).

For a detailed description of using these rules we refer to [3]. Here we only mention that all the above rules are applied in parallel, but at one step, a membrane h can be subject of only one rule of types (b)-(f).

A *P system with restricted active membranes* is a P system with active membranes where the rules are of types (a), (b), (c), (e), and (f) only (i.e., a P system with active membranes not using membrane dissolving rules of type (d)).

A *P system with restricted elementary active membranes* is a P system with active membranes where the rules are of types (a), (b), (c), and (e) only (i.e., a P system with active membranes not using membrane dissolving rules of type (d) and the (f) type rules for non-elementary membrane division).

3 Recognizing P Systems

In this section, we introduce the notion of *recognizing P systems* following [5]. First of all, we consider P systems with input and without external output.

Definition 31 *A P system with input is a tuple (Π, Σ, i_Π), where:*

- *Π is a P system, with working alphabet Γ and initial multisets $w_1, \cdots, w_p$ (associated with membranes labelled by $1, \cdots, p$, respectively).*
- *Σ is an (input) alphabet strictly contained in Γ.*
- *$w_1, \cdots, w_p$ are multisets over $\Gamma - \Sigma$.*
- *i_Π is the label of a distinguished membrane (of input).*

If $w \in M(\Sigma)$ is a multiset over Σ, then the initial configuration of (Π, Σ, i_Π) with input w is (μ_0, M_0), where $\mu_0 = \mu$, $M_0(j) = w_j$, for each $j \neq i_\Pi$, and $M_0(i_\Pi) = w_{i_\Pi} \cup w$.

The computations of a P system with input $w \in M(\Sigma)$ are defined in a natural way. Note that the initial configuration must be the initial configuration of the system associated with the input multiset $w \in M(\Sigma)$.

In the case of P systems with input and with external output (where we can imagine that the internal processes are unknown, and we only obtain the information that the system sends out to the environment), the concept of computation is introduced in a similar way but with a slight variant. In the configurations, we will not work directly with the membrane structure μ but with another structure associated with it including, in some sense, the environment.

Definition 32 *Let $\mu = (V(\mu), E(\mu))$ be a membrane structure. The membrane structure with environment associated with μ is the rooted tree $Ext(\mu)$ such that: (a) the root of the tree is a new node that we will denote env; (b) the set of nodes is $V(\mu) \cup \{env\}$; (c) the set of edges is $E(\mu) \cup \{\{env, skin\}\}$. The node env is called the environment of the structure μ.*

Note that we have only included a new node representing the environment which is only connected with the skin, while the original membrane structure remains unchanged. In this way, every configuration of the system informs about the environment and its content.

Now we introduce recognizing P systems as devices able to accept or reject multisets considered as input.

Definition 33 *A recognizing P system is a P system with input, (Π, Σ, i_Π), and with external output such that:*

1. *The working alphabet contains two distinguished elements yes, no.*
2. *All computations of the system halt.*
3. *If C is a computation of Π, then either the object yes or the object no (but not both) have to be sent out to the environment, and only in the last step of the computation.*

Definition 34 *We say that C is an accepting computation (respectively, rejecting computation) if the object yes (respectively, no) appears in the environment associated with the corresponding halting configuration of C.*

4 The Polynomial Complexity Class $\mathbf{PMC}_{\mathcal{F}}$

Many practical problems are presumably intractable for conventional (electronic) computers, because all known algorithms solving these problems spend exponential time. P systems have an inherent parallelism and hence the capability to solve hard problems in feasible (polynomial or linear) time.

To understand what it means that a problem can be solved in polynomial time by membrane systems, it is necessary to recall some complexity measure for P systems as described in [5].

A decision problem will be solved by a family of recognizing P systems in such a way that given an instance of the problem it is necessary to fix the concrete P system (with a suitable input multiset) that will process it. The next definition (*polynomial encoding*) captures this idea.

Definition 41 *Let L be a language, $\mathcal{F}$ a class of P systems with input and $\mathbf{\Pi} = (\Pi(t))_{t \in \mathbb{N}}$ a family of P systems of type $\mathcal{F}$. A polynomial encoding of L in $\mathbf{\Pi}$ is a pair (g, h) of polynomial time computable functions, $g : L \to \cup_{t \in \mathbb{N}} I_{\Pi(t)}$ and $h : L \to \mathbb{N}$, such that for every $u \in L$ we have $g(u) \in I_{\Pi(h(u))}$.*

Now we define what it means solving a decision problem by a family of recognizing P systems in time bounded by a given function.

Definition 42 *Let $\mathcal{F}$ be a class of recognizing P systems, $f : \mathbb{N} \to \mathbb{N}$ a total computable function, and $X = (I_X, \theta_X)$ a decision problem. We say that X belongs to $\mathbf{MC}_{\mathcal{F}}(f)$ if there exists a family, $\mathbf{\Pi} = (\Pi(t))_{t \in \mathbb{N}}$, of P systems such that:*

- $\mathbf{\Pi}$ *is $\mathcal{F}$-consistent: that is, $\Pi(t) \in \mathcal{F}$ for all $t \in \mathbb{N}$.*
- $\mathbf{\Pi}$ *is polynomially uniform: that is, there exists a deterministic Turing machine that constructs $\Pi(t)$ in polynomial time from $t \in \mathbb{N}$.*
- *There exists a polynomial encoding (g, h) from I_X to $\mathbf{\Pi}$ verifying:*
  - $\mathbf{\Pi}$ *is f-bounded, with regard to (g, h); that is, for each $u \in I_X$, all computations of $\Pi(h(u))$ with input $g(u)$ halt in at most $f(|u|)$ steps.*
  - $\mathbf{\Pi}$ *is X-sound, with regard to (g, h); that is, for each $u \in I_X$, if there exists an accepting computation of $\Pi(h(u))$ with input $g(u)$, then $\theta_X(u) = 1$.*
  - $\mathbf{\Pi}$ *is X-complete, with regard to (g, h); that is, for each $u \in I_X$, if $\theta_X(u) = 1$, then every computation of $\Pi(h(u))$ with input $g(u)$ is an accepting computation.*

A polynomial encoding (g, h) from I_X to $\mathbf{\Pi}$ provides a *size* function, h, that gives us the set of instances of X processed through the same P system, and an *input* function, g, supplying the input multiset to be processed for the P system.

Note 1. In the above definition we have imposed a *confluence* property in the following sense: for every input $u \in I_X$, either every computation of $\Pi(h(u))$ with input $g(u)$ is an accepting computation or every computation of $\Pi(h(u))$ with input $g(u)$ is a rejecting computation.

Definition 43 *The polynomial complexity class associated with a collection of recognizing P systems, $\mathcal{F}$, is defined as follows:*

$$\mathbf{PMC}_{\mathcal{F}} = \bigcup_{f polynomial} \mathbf{MC}_{\mathcal{F}}(f).$$

In particular, the union of all classes $\mathbf{MC}_{\mathcal{F}}(f)$, for f a linear function, is denoted by $\mathbf{LMC}_{\mathcal{F}}$.

Note 2. If a decision problem belongs to $\mathbf{PMC}_{\mathcal{F}}$, then we say that it is solvable in *polynomial time* by a family of P systems which are constructed in *polynomial time* starting from the size of the instances of the problem.

We say that a family of P systems $\mathcal{F}$ is *polynomially semi-uniform*, if there exists a deterministic Turing machine that constructs a P system in polynomial time from each instance of the problem [4]. We denote the corresponding complexity classes by $\mathbf{LMC}^{S}_{\mathcal{F}}$ and $\mathbf{PMC}^{S}_{\mathcal{F}}$.

5 Solving the Vertex Cover Problem

Given a graph G with n vertices and m edges, a *vertex cover* of G is a subset of the vertices such that every edge of G is adjacent to one of those vertices; the *vertex cover problem* (denoted by VCP) asks whether or not there exists a vertex cover for G of size k, where k is a given integer less than or equal to n. The vertex cover problem is an **NP**-complete problem [1].

Next we construct a family of recognizing P systems with active membranes using 2-division solving the vertex cover problem in linear time.

For that, first we consider the (size) function h defined on I_{VCP} by $h(G) = ((m+n)(m+n+1)/2) + m$, with $G = (V, E)$, where $V = \{v_1, v_2, \cdots, v_n\}$, $E = \{e_1, e_2, \cdots, e_m\}$. The function h is polynomial time computable since the function $\langle m, n\rangle = ((m+n)(m+n+1)/2) + m$ is primitive recursive and bijective from $\mathbb{N}^2$ onto $\mathbb{N}$. Also, the inverse function of h is polynomial.

For each $(m, n) \in \mathbb{N}^2$ we consider the P system $(\Pi(\langle m, n\rangle), \Sigma(m, n), i(m, n))$, where $\Sigma(m, n) = \{e_{x,(i,j)} \mid 1 \leq x \leq m, 1 \leq i, j \leq n$, the two vertices of edge e_x are v_i and $v_j\}$, $i(m, n) = 2$ and $\Pi(\langle m, n\rangle) = (\Gamma(m, n), \{1, 2\}, [_1[_2\ ]_2]_1, w_1, w_2, R)$, with $\Gamma(m, n)$ defined as follows:

$$\begin{aligned}\Gamma(m, n) = {} & \Sigma(m, n) \cup \{d_i \mid 1 \leq i \leq 3n + 2m + 3\} \\ & \cup \{r_{i,j} \mid 0 \leq i \leq m, 1 \leq j \leq 2n\} \cup \{c_{i,j} \mid 0 \leq i \leq k, 1 \leq j \leq 4n - 1\} \\ & \cup \{f_i \mid 1 \leq i \leq m + 2\} \cup \{t, \lambda, yes, no\}.\end{aligned}$$

The initial content of each membrane is: $w_1 = \emptyset$ and $w_2 = \{d_1, c_{0,1}\}$. The set of rules, R, is given by (we also give explanations about the use of these rules during the computations):

1. $[_2 d_i]_2^0 \rightarrow [_2 d_i]_2^+ [_2 d_i]_2^-$, $1 \leq i \leq n$.
   By using a rule of type 1, a membrane with label 2 is divided into two membranes with the same label, but with different polarizations. These rules allow us to duplicate, in one step, the total number of membranes with label 2.
2. $[_2 e_{x,(1,j)} \rightarrow r_{x,1}]_2^+$, $1 \leq x \leq m$, $2 \leq j \leq n$.
   $[_2 e_{x,(1,j)} \rightarrow \lambda]_2^-$, $1 \leq x \leq m$, $2 \leq j \leq n$.
   The rules of type 2 try to implement a process allowing membranes with label 2 to encode whether vertex v_1 appears in a subset of vertices, in such a way that if vertex v_1 appears in a subset of vertices, then the objects $e_{x,(1,j)}$ encoding edges e_x which are adjacent to vertex v_1 will evolve to objects $r_{x,1}$ in the corresponding membranes with label 2; otherwise, the objects $e_{x,(1,j)}$ will disappear.
3. $[_2 e_{x,(i,j)} \rightarrow e_{x,(i-1,j)}]_2^+$, $1 \leq x \leq m$, $2 \leq i \leq n$, $i \leq j \leq n$.
   $[_2 e_{x,(i,j)} \rightarrow e_{x,(i-1,j)}]_2^-$, $1 \leq x \leq m$, $2 \leq i \leq n$, $i \leq j \leq n$.
   The evolving process described previously is always made with respect to the vertex v_1. Hence, the rules of type 3 take charge of making a cyclic path through all the vertices to get that, initially, the first vertex is v_1, then v_2, and so on.
4. $[_2 c_{i,j} \rightarrow c_{i+1,j}]_2^+$, $0 \leq i \leq k-1$, $1 \leq j \leq 2n-1$.
   $[_2 c_{k,j} \rightarrow \lambda]_2^+$, $1 \leq j \leq 2n-1$.
   The rules of type 4 supply counters in the membranes with label 2, in such a way that we increase the first subscript of $c_{i,j}$, when the membrane has not more than k vertices; when the membrane has more than k vertices, then the counter-object $c_{k,j}$ will disappear. So when the process of generating all subsets of vertices is finished, the object $c_{k,j}$ (note the first subscript is k) will appear only in the membrane with label 2 encoding a subset with cardinality exactly k.
5. $[_2 d_i]_2^+ \rightarrow [_2\ ]_2^0 d_i$, $1 \leq i \leq n$.
   $[_2 d_i]_2^- \rightarrow [_2\ ]_2^0 d_i$, $1 \leq i \leq n$.
   $d_i [_2\ ]_2^0 \rightarrow [_2 d_{i+1}]_2^0$, $1 \leq i \leq n-1$.
   The rules of type 5 are used as controllers of the generating process of all subsets of vertices and the listing of adjacent edges: the objects d are sent to the membrane with label 1 at the same time the listing of adjacent edges and the counting of vertices are made, and they come back to the membranes with label 2 to start the division of these membranes.
6. $[_2 r_{i,j} \rightarrow r_{i,j+1}]_2^0$, $1 \leq i \leq m$, $1 \leq j \leq 2n-1$.
   The use of objects r in the rules of types 12, 13, and 14 makes necessary to perform a "rotation" of these objects. This is the mission of the rules of type 6.
7. $[_2 c_{i,j} \rightarrow c_{i,j+1}]_2^0$, $0 \leq i \leq k$, $1 \leq j \leq 2n-1$.
   The second subscript of $c_{i,j}$ (also in the rule of type 9) is used to control when the process of checking whether a subset of vertices with cardinality k is a vertex cover will start.

8. $[_1 d_i \to d_{i+1}]_1^0$, $n \le i \le 3n-2$.
   Through the counter-objects d, the rules of type 8 control the rotation of the elements $r_{i,j}$ in the membranes with label 2.
9. $[_2 c_{k,j} \to c_{k,j+1}]_2^0$, $2n \le j \le 4n-3$.
   $[_2 c_{k,4n-2} \to c_{k,4n-1} f_1]_2^0$.
   When the generating process of all subsets of vertices and the listing of adjacent edges is finished, the second subscript of all objects $c_{i,j}$ $(1 \le i \le k)$ is $2n$. After that, the second subscript of only objects $c_{k,j}$ (note that the first subscript is k) will increase. The object $c_{k,4n-2}$ evolves to $c_{k,4n-1}f_1$. The object $c_{k,4n-1}$ will change the polarization of the membrane to negative (using the rule of type 10). The object f_1 is a new counter.
10. $[_2 c_{k,4n-1}]_2^0 \to [_2\ ]_2^- c_{k,4n-1}$.
    $[_1 d_{3n-1} \to d_{3n}]_1^0$.
    The application of these rules will show that the system is ready to check which edges are covered by the vertices in a subset with cardinality exactly k encoded by membrane with label 2.
11. $[_1 d_i \to d_{i+1}]_1^0$, $3n \le i \le 3n+2m+2$.
    The rules of type 11 supply counters in the membrane with label 1, through objects d, in such a way that if the objects d_{3n+2m} appear, then they show the end of the checking of vertex cover. The objects d_i, with $3n+2m+1 \le i \le 3n+2m+3$, will control the final stage of the computation.
12. $[_2 r_{1,2n}]_2^- \to [_2\ ]_2^+ r_{1,2n}$.
    The applicability of the rule of type 12 encodes the fact that the vertices encoded by a membrane with label 2 cover the edge e_1 represented by the object $r_{1,2n}$, through a change in the sign of its polarization.
13. $[_2 r_{i,2n} \to r_{i-1,2n}]_2^+$, $1 \le i \le m$.
    In the copies of membrane 2 with positive polarization, hence only those where we have found $r_{1,2n}$ in the previous step, we decrease the first subscripts of all objects $r_{i,2n}$ from those membranes. Thus, if the second edge was covered by the vertices from a membrane, that is $r_{2,2n}$ was present in a membrane, then $r_{2,2n}$ becomes $r_{1,2n}$ – hence the rule of type 12 can be applied again. Note the important fact that passing from $r_{2,2n}$ to $r_{1,2n}$ is possible only in membranes where we already had $r_{1,2n}$, hence we check whether the second edge is covered only after knowing that the first edge was covered.
14. $r_{1,2n}[_2\ ]_2^+ \to [_2 r_{0,2n}]_2^-$.
    At the same time as the use of rule of type 13, the object $r_{1,2n}$ from the skin membrane returns to membranes with label 2, changed to $r_{0,2n}$ (which will never evolve again), and returning the polarization of the membrane to negative (this makes possible the use of the rule of type 12).
15. $[_2 f_i \to f_{i+1}]_2^+$, $1 \le i \le m$.
    The presence of objects f_i (with $2 \le i \le m+1$) in the membranes with label 2 shows that the corresponding subsets of vertices covering every edge of $\{e_1, \cdots, e_{i-1}\}$ are being determined.

16. $[_2 f_{m+1}]_2^- \rightarrow [_2 \ ]_2^- f_{m+1}$.
The rule of type 16 sends to the skin membrane the objects f_{m+1} appearing in the membranes with label 2.
17. $[_1 f_{m+1} \rightarrow f_{m+2} t]_1^0$.
By using the rule of type 17 the objects f_{m+1} in the skin evolve to objects $f_{m+2}t$. The objects t in the skin are produced simultaneously with the appearance of the objects $d_{3n+2m+2}$ in the skin, and they will show that there exists some subset of vertices which is a vertex cover with cardinality k.
18. $[_1 t]_1^0 \rightarrow [_1 \ ]_1^+ t$.
The rule of type 18 sends out of the system an object t changing the polarization of the skin to positive, then objects t remaining in the skin are not able to evolve. Hence, an object f_{m+2} can exit the skin producing an object *yes*. This object is then sent out to the environment through the rule of type 19, telling us that there exists a vertex cover with cardinality k, and the computation halts.
19. $[_1 f_{m+2}]_1^+ \rightarrow [_1 \ ]_1^- yes$.
The applicability of the rule of type 19 changes the polarization in the skin membrane to negative in order that the objects f_{m+2} remaining in it are not able to continue evolving.
20. $[_1 d_{3n+2m+3}]_1^0 \rightarrow [_1 \ ]_1^+ no$.
By the rule of type 20 the object $d_{3n+2m+3}$ only evolves when the skin has neutral charge (this is the case when there does not exist any vertex cover with cardinality k). Then the system will evolve sending out to the environment an object *no* and changing the polarization of the skin to positive, in order that objects $d_{3n+2m+3}$ remaining in the skin do not evolve.

From the previous explanation of the use of rules, one can easily see how these P systems work. It is easy to prove that the designed P systems are deterministic.

Now, we prove that the family $\mathbf{\Pi} = (\Pi(t))_{t \in \mathbf{N}}$ solves the vertex cover problem in linear time.

The above description of the evolution rules is computable in an uniform way. So, the family $\mathbf{\Pi} = (\Pi(t))_{t \in \mathbf{N}}$ is *polynomially uniform* because:

- The total number of objects is $2nm + 4nk + 9n + 4m - k + 8 \in O(n^3)$.
- The number of membranes is 2.
- The cardinality of the initial multisets is 2.
- The total number of evolution rules is $O(n^3)$.
- The maximal length of a rule (the number of symbols necessary to write a rule, both its left and right sides, the membranes, and the polarizations of membranes involved in the rule) is 13.

We consider the (input) function $g : I_{VCP} \rightarrow \cup_{t \in \mathbf{N}} I_{\Pi(t)}$, defined as follows:

$$g(G) = \{e_{x,(i,j)} | 1 \leq x \leq m, 1 \leq i, j \leq n,$$
$$\text{the two vertices of edge } e_x \text{ are } v_i \text{ and } v_j\}.$$

Then g is a polynomial time computable function. Moreover, the pair (g, h) is a polynomial encoding from VCP to $\mathbf{\Pi}$ since for each $G \in I_{VCP}$ we have $g(G) \in I_{\Pi(h(G))}$.

We will denote by $\mathcal{C} = (\mathcal{C}^i)_{0 \leq i \leq s}$ the computation of the P system $\Pi(h(G))$ with input $g(G)$. That is, $\mathcal{C}^i$ is the configuration obtained after i steps of the computation $\mathcal{C}$.

The execution of the P system $\Pi(h(G))$ with input $g(G)$ can be structured in four stages: a stage of *generation* of all subsets of vertices and *counting* the cardinality of subsets of vertices; a stage of *synchronization*; a stage of *checking* whether there is some subset with cardinality k which is a vertex cover; and a stage of *output*.

The *generating* and *counting stages* are controlled by the objects d_i, with $1 \leq i \leq n$.

- The presence in the skin of one object d_i, with $1 \leq i \leq n$, will show that all possible subsets associated with $\{v_1, \cdots, v_i\}$ have been generated.
- The objects $c_{i,j}$ with $0 \leq i \leq k$, $1 \leq j \leq 2n$ are used to count the number of vertices in membranes with label 2. When this stage ends, the object $c_{k,j}$ (note the first subscript is k) will appear only in the membrane with label 2 encoding a subset with cardinality exactly k.
- In this stage, we simultaneously encode in every internal membrane all the edges being covered by the subset of vertices represented by the membrane (through the objects $r_{i,j}$).
- The object d_1 appears in the skin after the execution of 2 steps. From the appearance of d_i in the skin to the appearance of d_{i+1}, with $1 \leq i \leq n-1$, 3 steps have been executed.
- This stage ends when the object d_n appears in the skin.

Hence, the total number of steps in the generating and counting stages is $3n-1$.

The *synchronization stage* has the goal of unifying the second subscripts of the objects $r_{i,j}$, to make them equal to $2n$.

- This stage starts with the evolution of the object d_n in the skin.
- In every step of this stage the object d_i, with $n \leq i \leq 3n-1$, in the skin evolves to d_{i+1}.
- In every step of this stage, the second subscript of objects $c_{k,j}$ (note that the first subscript is k) will increase. The object $c_{k,4n-2}$ evolves to $c_{k,4n-1} f_1$.
- This stage ends as soon as the object d_{3n} appears in the skin, that is the moment when the membrane with label 2 encoding a subset with cardinality k has negative charge (by using the first rule of type 10).

Therefore, the synchronization stage needs a total of $2n$ steps.

The *checking stage* has the goal to determine how many edges are covered in the membrane with label 2 encoding the subset of vertices with cardinality k. This stage is controlled by the objects f_i, with $1 \leq i \leq m+1$.

- The presence of an object f_i in a membrane with label 2 shows that the edges $e_1, \cdots, e_{i-1}$ are covered by the subset of vertices represented by such membrane.

- From every f_i (with $1 \leq i \leq m$) the object f_{i+1} is obtained in some membranes after the execution of 2 steps.
- The checking stage ends as soon as the object d_{3n+2m} appears in the skin.

Therefore, the total number of steps of this stage is $2m$.

The *output stage* starts immediately after the appearance of the object d_{3n+2m} in the skin and it is controlled by the objects f_{m+1} and f_{m+2}.

- To produce the output *yes* the object f_{m+1} must have been produced in some membrane with label 2 of the configuration $\mathcal{C}^{5n+2m-1}$. Then, after 4 steps the system returns *yes* to the environment, through the evolution of objects f_{m+2} present in the skin, and when it has positive charge.
- To produce the output *no*, no object f_{m+1} appears in any membrane with label 2 of the configuration $\mathcal{C}^{5n+2m-1}$. Then after 4 steps the system returns *no* to the environment, through the evolution of objects $d_{3n+2m+3}$ present in the skin, and when it has neutral charge.

Therefore, the total number of steps in the output stage is 4.

Let us see that the family $\mathbf{\Pi}$ is *linearly bounded*, with regard to (g, h). For that, it is enough to note that the time of the stages of the execution of $\Pi(h(G))$ with input $g(G)$ is: (a) generation stage, $3n-1$ steps; (b) synchronization stage, $2n$ steps; (c) checking stage, $2m$ steps; and (d) output stage, 4 steps. Hence, the total execution time of $\Pi(h(G))$ with input $g(G)$ is $5n + 2m + 3 \in O(n + m)$.

Now, let us see that the family $\mathbf{\Pi}$ is VCP-*sound* and VCP-*complete*, with respect to the polynomial encoding (g, h). For that it is sufficient to verify that the following results are true:

1. If s is a subset with cardinality k covering all edges, then at the end of the checking stage (that is, in the configuration $\mathcal{C}^{5n+2m-1}$), the object f_{m+1} appears in the membrane *associated with* s.
2. If s is not a subset with cardinality k covering all edges, then at the end of the checking stage (that is, in the configuration $\mathcal{C}^{5n+2m-1}$), the object f_{m+1} does not appear in the membrane *associated with* s.

Next we justify that the designed P systems $\Pi(t)$, with $t \in \mathbb{N}$, are recognizing devices.

By inductive processes it can be proved that the configuration $\mathcal{C}^{5n+2m-1}$ verifies the following properties:

(a) It has 2^n membranes with label 2. The membranes with label 2 encoding subsets of vertices with cardinality exactly k have negative charge. The other membranes with label 2 have neutral charge.
(b) The skin has neutral charge and its content is $d_{3n+2m}^{2^n}$.
(c) If the object f_{m+1} is present in a membrane with label 2, then the subset of vertices encoded by it is a vertex cover.

Proposition 51 *Suppose that there exist exactly q membranes with label 2 (with $1 \leq q \leq \binom{n}{k}$) of $\mathcal{C}^{5n+2m-1}$ containing some object f_{m+1}. Then, in the last step of the computation, the P system sends out the object yes to environment.*

Proof. Under the hypothesis of the proposition it is verified that:

(a) Structure of $\mathcal{C}^{5n+2m}$.
The rule $[_2 f_{m+1}]_2^- \rightarrow [_2\ ]_2^- f_{m+1}$ is applicable over the membranes with label 2 of $\mathcal{C}^{5n+2m-1}$ containing the object f_{m+1}, since their charge is negative. Hence, in the skin of $\mathcal{C}^{5n+2m-1}$ the objects f_{m+1}^q and $d_{3n+2m+1}^{2^n}$ must appear, because of the application of the rule $[_1 d_{3n+2m} \rightarrow d_{3n+2m+1}]_1^0$. The other membranes of $\mathcal{C}^{5n+2m-1}$ have no applicable rules.
(b) Structure of $\mathcal{C}^{5n+2m+1}$.
The objects f_{m+1} and $d_{3n+2m+1}$ in the skin evolve to $f_{m+2}t$ and $d_{3n+2m+2}$, respectively, because of the application of the rules $[_1 f_{m+1} \rightarrow f_{m+2}t]_1^0$ and $[_1 d_{3n+2m+1} \rightarrow d_{3n+2m+2}]_1^0$. The other membranes of $\mathcal{C}^{5n+2m}$ have no applicable rules.
(c) Structure of $\mathcal{C}^{5n+2m+2}$.
The rule $[_1 t]_1^0 \rightarrow [_1\ ]_1^+ t$ sends out of the system an object t and changes the polarization of the skin to positive (in order that the objects t remaining in the skin do not continue being expelled of the system). Moreover, the objects $d_{3n+2m+2}$ in the skin produce $d_{3n+2m+3}$ because of the application of the rule $[_1 d_{3n+2m+2} \rightarrow d_{3n+2m+3}]_1^0$. The other membranes of $\mathcal{C}^{5n+2m+1}$ have no applicable rules.
(d) Structure of $\mathcal{C}^{5n+2m+3}$.
Having in mind that the polarization of the skin in the configuration $\mathcal{C}^{5n+2m+2}$ is positive, an object f_{m+2} will be expelled from the system producing the object *yes*, because of the application of the rule $[_1 f_{m+2}]_1^+ \rightarrow [_1\ ]_1^- yes$. Moreover, the polarization of the skin changes to negative in order that no objects f_{m+2} remaining in the skin is able to continue evolving. The other membranes of $\mathcal{C}^{5n+2m+2}$ have no applicable rules.

Finally, in the configuration $\mathcal{C}^{5n+2m+3}$ there is no applicable rule, so, it is a halting configuration. As the system has expelled the object *yes* to the environment in the last step, $\mathcal{C}$ is an accepting computation.

In a similar way we prove the following result.

Proposition 52 *Let us suppose that no membrane with label 2 of* $\mathcal{C}^{5n+2m-1}$ *contains an object* f_{m+1}*. Then, in the last step of the computation, the P system sends out the object no to environment.*

From the above propositions we deduce the soundness and completeness of the family $\mathbf{\Pi}$, with respect to the vertex cover problem.

Proposition 53 (Soundness) *Let G be a graph. If the computation of* $\Pi(h(G))$ *with input* $g(G)$ *is an accepting computation, then G has a vertex cover with cardinality k.*

Proof. Let G be a graph. Let us suppose that the computation of $\Pi(h(G))$ with input $g(G)$ is an accepting computation. From Propositions 51 and 52 we deduce that some membranes with label 2 of $\mathcal{C}^{5n+2m-1}$ contain the object f_{m+1}. Hence some membrane with label 2 of $\mathcal{C}^{5n+2m-1}$ encodes a subset of vertices with cardinality k covering all edges of the graph. Therefore, G has a vertex cover with cardinality k.

Proposition 54 (Completeness) *Let G be a graph. If it has a vertex cover with cardinality k, then the computation of $\Pi(h(G))$ with input $g(G)$ is an accepting computation.*

Proof. Let G be a graph, which has a vertex cover with cardinality k. Then some membrane with label 2 of the configuration $\mathcal{C}^{5n+2m-1}$ (encoding a relevant subset of vertices for G) contains an object f_{m+1} . From Proposition 52 it follows that $\mathcal{C}^{5n+2m-1}$ is an accepting configuration. That is, the computation of $\Pi(h(G))$ with input $g(G)$ is an accepting computation.

The main result follows from all above.

Theorem 1. *The vertex cover problem can be solved in linear time with respect to the number of vertices and edges of the graph by recognizing P systems with restricted elementary active membranes.*

6 Solving the Vertex Cover Problem

In this section we will solve the vertex cover problem in linear time $(n + 2k + 2m + 8)$ by P systems with restricted elementary active membranes. Unlike the solution in the previous section, the solution is semi-uniform, but the resources, such as the number of objects, rules, and symbols in the description of the P systems is linear with respect to the number of vertices and the number of edges in the graph.

Theorem 2. *The vertex cover problem can be solved in linear time with respect to the number of vertices and edges of the graph by semi-uniform P systems with restricted elementary active membranes and linear resources.*

Proof. Let $G = (V, E)$ be a graph with n vertices and m edges, $V = \{v_1, v_2, \cdots, v_n\}$, $E = \{e_1, e_2, \cdots, e_m\}$. For $k = 1, 2, n-1$, and n, it is easy to check whether a given graph has a vertex cover of size k, so we can assume $3 \leq k \leq n-2$.

We construct the system

$$\Pi = (O, H, \mu, w_1, w_2, R),$$

where

$$O = \{a_i, d_i, t_i, t'_i, f_i \mid 1 \le i \le n\} \cup \{h_i, h'_i \mid 0 \le i \le k\} \cup \{e_i \mid 0 \le i \le m\}$$
$$\cup \{g_i \mid 1 \le i \le m+1\} \cup \{g'_i \mid 1 \le i \le m\} \cup \{a, b, p, q, u, yes\},$$
$$H = \{1, 2\},$$
$$\mu = [_1 [_2\ ]_2^0]_1^0,$$
$$w_1 = \lambda,$$
$$w_2 = a_1 a_2 \cdots a_n d_1,$$

and the set R contains the following rules

1. $[_2 a_i]_2^0 \to [_2 t_i]_2^0 [_2 f_i]_2^0,\ 1 \le i \le n.$
   The objects a_i correspond to vertices $v_i, 1 \le i \le n$. By using a rule as above, for i non-deterministically chosen, we produce the two objects t_i and f_i associated to vertex v_i, placed in two separate copies of membrane 2, where t_i means that vertex v_i appears in some subset of vertices, and f_i means that vertex v_i does not appear in some subset of vertices. Note that the charge remains the same for both membranes, namely neutral, hence the process can continue. In this way, in n steps we get all 2^n subsets of V, placed in 2^n separate copies of membrane 2. In turn, these 2^n copies of membrane 2 are within membrane 1 – the system always has only two levels of membranes. Note that in spite of the fact that in each step the object a_i is non-deterministically chosen, after n steps we get the same result, irrespective of which objects were used in each step.
2. $[_2 d_i \to d_{i+1}]_2^0,\ 1 \le i \le n-1.$
3. $[_2 d_n \to q q h_0]_2^0.$
   The objects d_i are counters. Initially, d_1 is placed in membrane 2. By division (when using rules of type 1), we introduce copies of the counter in each new membrane. In each step, in each membrane with label 2, we pass from d_i to d_{i+1}, thus "counting to n". In step n we introduce both q and h_0; the objects q will exit the membrane (at steps $n+1$, $n+2$), changing its polarization, the object h_0 is a new counter which will be used at the subsequent steps as shown below. Note that at step n we have also completed the generation of all subsets.
   Now we pass to the next phase of computation – counting the number of objects t_i $(1 \le i \le n)$ in each membrane with label 2, which corresponds to the cardinality of each subset; we will select out the subsets with cardinality exactly k.
4. $[_2 q]_2^0 \to [_2\ ]_2^- u.$
5. $[_2 q]_2^- \to [_2\ ]_2^0 u.$
6. $[_2 t_i \to a b t'_i]_2^-,\ 1 \le i \le n.$
7. $[_2 h_i \to h'_i]_2^0,\ 0 \le i \le k.$
8. $[_2 h'_i \to h_{i+1}]_2^+,\ 0 \le i \le k-1.$
9. $[_2 a]_2^0 \to [_2\ ]_2^+ u.$

10. $[_2 b]_2^+ \rightarrow [_2\ ]_2^0 u$.
At step $n+1$, h_0 evolves to h'_0, and at the same time one copy of q exits the membrane, changing its polarization to negative. At step $n+2$, t_i evolves to abt'_i, and at the same time the other copy of q exits the membrane, changing its polarization to neutral. At step $n+3$, one copy of a exits the membrane, changing its polarization to positive. At step $n+4$, h'_0 evolves to h_1, and at the same time one copy of b exits the membrane, returning its polarization to neutral (this makes possible the use of rules of types 7 and 9).
The rules of types 7, 8, 9, and 10 are applied as many times as possible (in one step rules of types 7 and 9, in the next one rules of types 8 and 10, and then we repeat the cycle). Clearly, at step $n+2+2k$, a membrane contains object h_k if and only if the cardinality of the corresponding subset is at least k. At step $n+3+2k$, in the membranes whose corresponding subsets have cardinality more than k, h_k evolves to h'_k, and one copy of a changes their polarization to positive. These membranes will no longer evolve, as no further rule can be applied to them. In the membranes whose corresponding subsets have cardinality exactly k, h_k evolves to h'_k, and their polarization remains neutral, because there is no copy of a which can be used. We will begin the next phase of computation with the rule of type 11 – checking whether a subset with cardinality k is a vertex cover.
11. $[_2 h'_k \rightarrow qqg_1]_2^0$.
12. $[_2 t'_i \rightarrow e_{i_1} \cdots e_{i_l}]_2^-$, $1 \leq i \leq n$, and vertex v_i is adjacent to edges $e_{i_1} \cdots e_{i_l}$.
At step $n+4+2k$, in the membranes with label 2 and polarization 0, h'_k evolves to qqg_1. At step $n+5+2k$, one copy of q exits the membrane, changing its polarization to negative. At step $n+6+2k$, in parallel t'_i $(1 \leq i \leq n)$ evolves to $e_{i_1} \cdots e_{i_l}$, and at same time the other copy of q exits the membrane, changing its polarization to neutral. After completing this step, if there is at least one membrane with label 2 which contains all symbols $e_1, \cdots, e_m$, then this means that the subset corresponding to that membrane is a vertex cover of cardinality k. Otherwise (if in no membrane we get all objects $e_1, \cdots, e_m$), there exists no vertex cover of cardinality k. In the following steps, we will "read" the answer, and send a suitable signal out of the system. This will be done by the following rules.
13. $[_2 g_i \rightarrow g'_i p]_2^0$, $1 \leq i \leq m$.
14. $[_2 e_1]_2^0 \rightarrow [_2\ ]_2^+ u$.
Object g_i evolves to $g'_i p$. At the same time for all subsets of cardinality k, we check whether or not e_1 is present in each corresponding membrane. If this is the case, then one copy of e_1 exits the membrane where it is present, evolving to u, and changing in this way the polarization of that membrane to positive (the other copies of e_1 will immediately evolve to e_0, which will never evolve again). The membranes which do not contain the object e_1 remain neutrally charged and they will no longer evolve, as no further rule can be applied to them.
15. $[_2 e_i \rightarrow e_{i-1}]_2^+$, $1 \leq i \leq m$.
16. $[_2 g'_i \rightarrow g_{i+1}]_2^+$, $1 \leq i \leq m$.

17. $[_2 p]_2^+ \rightarrow [_2\ ]_2^0 u$.
In the copies of membrane 2 corresponding to subsets of cardinality k with a positive polarization, hence only those where we have found e_1 in the previous step, we perform two operations in parallel: g_i' evolves to g_{i+1} (so we count the number of edges), and we decrease the subscripts of all objects e_j from that membrane. Thus, if e_2 is present in a membrane, then e_2 becomes e_1 – hence the rule of type 14 can be applied again. Note the important fact that passing from e_2 to e_1 is possible only in membranes where we already had e_1, hence we check whether edge e_2 appears only after knowing that edge e_1 appears. At the same time, the object p exits the membrane, evolving to u, and returning the polarization of the membrane to neutral (this makes possible the use of rules of types 13 and 14).
The rules of types 13, 14, 15, 16, and 17 are applied as many times as possible (in one step rules of types 13 and 14, in the next one rules of types 15, 16, and 17, and then we repeat the cycle). Clearly, if a membrane does not contain an object e_i, then that membrane will stop evolving at the time when e_i is supposed to reach the subscript 1. In this way, after $2m$ steps we can find whether there is a membrane which contains all objects $e_1, e_2, \cdots, e_m$. The membranes with this property, and only they, will get the object g_{m+1}.
18. $[_2 g_{m+1}]_2^0 \rightarrow [_2\ ]_2^0 yes$.
19. $[_1 yes]_1^0 \rightarrow [_1\ ]_1^+ yes$.
The object g_{m+1} exits the membrane which corresponds to a vertex cover with cardinality k, producing the object yes. This object is then sent to the environment, telling us that there exists a vertex cover with cardinality k, and the computation stops. Note that the application of rule 19 changes the polarization of the skin membrane to positive in order that the objects yes remaining in it are not able to continue exiting it.

From the previous explanation of the use of rules, one can easily see how this P system works. It is clear that we get the object yes if only if there exists a vertex cover of size k. The object yes exits the system at moment $n + 2k + 2m + 8$. If there exists no vertex cover of size k, then the computation stops earlier and we get no output, because no membrane was able to produce the object g_{m+1}, hence the object yes. Therefore, the family of membrane systems we have constructed is sound, confluent, and linearly efficient. To prove that the family is semi-uniform in the sense of Section 1, we have to show that for a given instance, the construction described in the proof can be done in polynomial time by a Turing machine. We omit the detailed construction due to the fact that it is straightforward although cumbersome as explained in the proof of Theorem 7.2.3 in [4].

So the vertex cover problem was decided in linear time $(n + 2k + 2m + 8)$ by P systems with restricted elementary active membranes, and this concludes the proof.

7 Solving the Clique Problem

Given a graph G with n vertices and m edges, a *clique* of G is a subset of the vertices such that every vertex is connected to every other vertex by an edge; the *clique problem* (denoted by CP) asks whether or not there exists a clique for G of size k, where k is a given integer less than or equal to n. The clique problem is an **NP**-complete problem [1].

For a graph $G = (V, E)$, we define the *complementary graph* $G' = (V, E')$, where $E' = \{(v_i, v_j) \notin E \mid v_i, v_j \in V\}$. In figure 2, an example of a graph with five vertices and five edges is given.

In this section we will solve the clique problem in linear time $(3n + 2k + 12)$ by recognizing P systems with restricted elementary active membranes.

Theorem 3. *The clique problem can be solved in linear time with respect to the number of vertices of the graph by recognizing P systems with restricted elementary active membranes.*

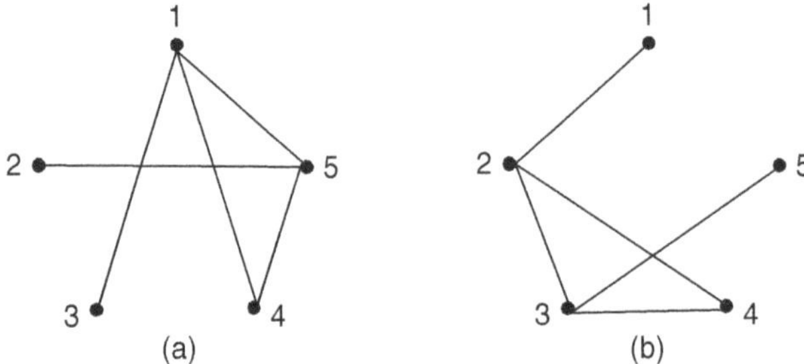

Figure 2. The original graph (a) and its complementary graph (b).

Proof. Let $G = (V, E)$ be a graph with n vertices and m edges, $V = \{v_1, v_2, \cdots, v_n\}$, $E = \{e_1, e_2, \cdots, e_m\}$. The instance of the problem is encoded by a (multi)set in the alphabet $\Sigma(\langle n, k\rangle) = \{p_{i,j,n+2k+5}, e_{i,j,n+2k+5} \mid 1 \le i, j \le n, i \neq j\}$. The object $p_{i,j,n+2k+5}$ stands for $(v_i, v_j) \in E$, while $e_{i,j,n+2k+5}$ represents $(v_i, v_j) \notin E$. For given values of n and k we construct a recognizing P system $(\Pi(\langle n, k\rangle), \Sigma(\langle n, k\rangle), i(\langle n, k\rangle))$ with $i(\langle n, k\rangle) = 2$

$$\Pi(\langle n, k\rangle) = (O(\langle n, k\rangle), H, \mu, w_1, w_2, R),$$

where

$$\begin{aligned}
O(\langle n, k\rangle) &= \{a_i, d_i, t_i, f_i, y_i, z_i \mid 1 \le i \le n\} \\
&\cup \{h_i, h'_i \mid 0 \le i \le k\} \cup \{g_i \mid 0 \le i \le n+3\} \\
&\cup \{e_{i,j,l} \mid 1 \le i \le n+2, 1 \le j \le n+2, 0 \le l \le n+2k+5\} \\
&\cup \{c_i \mid 0 \le i \le 3n+2k+11\} \cup \{a, b, g, p, q, u, yes, no\}, \\
H &= \{1, 2\}, \\
\mu &= [_1 [_2\ ]_2^0]_1^0, \\
w_1 &= c_{3n+2k+11}, \\
w_2 &= a_1 a_2 \cdots a_n d_1,
\end{aligned}$$

and the set R contains the following rules (we also give explanations about the use of these rules during the computations):

1. $[_2 a_i]_2^0 \rightarrow [_2 t_i]_2^0 [_2 f_i]_2^0,\ 1 \le i \le n.$
2. $[_2 d_i \rightarrow d_{i+1}]_2^0,\ 1 \le i \le n-1.$
3. $[_2 d_n \rightarrow q q h_0]_2^0.$
   We omit the explanations about the use of the rules of types 1, 2, and 3, because they are the same as the explanations in the proof of Theorem 2.
   Now we pass to the next phase of computation – counting the number of objects t_i $(1 \le i \le n)$ in each membrane with label 2, which corresponds to the cardinality of each subset.
4. $[_2 q]_2^0 \rightarrow [_2\ ]_2^- u.$
5. $[_2 q]_2^- \rightarrow [_2\ ]_2^0 u.$
6. $[_2 t_i \rightarrow ab]_2^-,\ 1 \le i \le n.$
7. $[_2 h_i \rightarrow h_i']_2^0,\ 0 \le i \le k.$
8. $[_2 h_i' \rightarrow h_{i+1}]_2^+,\ 0 \le i \le k-1.$
9. $[_2 a]_2^0 \rightarrow [_2\ ]_2^+ u.$
10. $[_2 b]_2^+ \rightarrow [_2\ ]_2^0 u.$
   The rules of types 4-10 are used in the same way as the corresponding rules from the proof of Theorem 2, the only difference is that at step $n+2$, t_i evolves to ab.
   The rules of types 7, 8, 9, and 10 are applied as many times as possible (in one step rules of types 7 and 9, in the next one rules of types 8 and 10, and then we repeat the cycle). Clearly, at step $n+2+2k$, a membrane contains object h_k if and only if the cardinality of the corresponding subset is at least k. At step $n+3+2k$, in the membrane whose corresponding subset has cardinality more than k, h_k evolves to h_k', and one copy of a changes its polarization to positive. This membrane will no longer evolve, as no further rule can be applied to it. In the membrane whose corresponding subset has cardinality exactly k, h_k evolves to h_k', and its polarization remains neutral, because there is no copy of a which can be used. We pass to the next phase of computation – checking whether a subset with cardinality k is a clique.
   (A subset A of vertices is a clique if and only if for each edge $(v_i, v_j) \in E'$, $x_i \in V-A$ or $x_j \in V-A$, i.e., $E' \subseteq V \times (V-A) \cup (V-A) \times V$. The process of checking whether a subset with cardinality k is a clique is based on this fact.)
11. $[_2 h_k' \rightarrow q q g]_2^0.$
12. $[_2 f_i \rightarrow y_i z_i]_2^-,\ 1 \le i \le n.$
13. $[_2 g \rightarrow g_0]_2^-,$
   $[_2 e_{i,j,l} \rightarrow e_{i,j,l-1}]_2^\alpha,\ 1 \le i,j \le n, 1 \le l \le n+2k+5, \alpha \in \{+,-,0\}.$
   At step $n+4+2k$, in the membranes with label 2 and polarization 0, h_k' evolves to qqg. At step $n+5+2k$, one copy of q exits the membrane, changing its polarization to negative. At step $n+6+2k$, in parallel, f_i $(1 \le i \le n)$ evolves to $y_i z_i$, g evolves to g_0 and an object $e_{i,j,0}$ appears for each $(v_i, v_j) \in$

E'. At same time the other copy of q exits the membrane, changing its polarization to neutral.

14. $[_2 y_i \to y_{i+1}]_2^0$, $1 \le i \le n-1$.
15. $[_2 z_i \to z_{i+1}]_2^0$, $1 \le i \le n-1$.
16. $[_2 e_{i,j,0} \to e_{s(i),s(j),0}]_2^0$, $1 \le i,j \le n$, where $s(t) = \min(t+1, n+2)$.
17. $[_2 g_i \to g_{i+1}]_2^0$, $0 \le i \le n-1$.
18. $[_2 z_n \to p]_2^0$.
19. $[_2 y_n]_2^0 \to [_2\ ]_2^+ u$.

At step $n+2k+7$, y_i, z_i $(1 \le i \le n-1)$ evolve to y_{i+1}, z_{i+1}, g_0 evolves to g_1, z_n evolves to p, $e_{i,j,0}$ $(1 \le i,j \le n)$ evolve to $e_{s(i),s(j),0}$, where $s(t) = \min(t+1, n+2)$; at same time, y_n exits the membrane where it appears, changing the polarization of that membrane to positive.

20. $[_2 e_{i,n+1,0} \to u]_2^+$, $1 \le i \le n+1$.
21. $[_2 e_{n+1,i,0} \to u]_2^+$, $1 \le i \le n+1$.
22. $[_2 p]_2^+ \to [_2\ ]_2^0 u$.

In the membranes with label 2 and positive polarization (i.e., the membranes where y_n appear in the last step, this means that vertex v_n does not belong to the corresponding subset), $e_{i,n+1,0}$ and $e_{n+1,i,0}$ $(1 \le i \le n+1)$ evolve to u (which will never evolve again); at the same time, object p exits the membrane, returning the polarization of the membrane to neutral (this makes possible the use of rules of types 14, 15, 16, 17, 18, and 19).

In the membranes with label 2 and neutral polarization (i.e., the membranes where y_n do not appear in the last step, this means that vertex v_n belongs to the corresponding subset), using rule of type 16, $e_{i,n+1,0}$ and $e_{n+1,i,0}$ $(1 \le i \le n+1)$ evolve to $e_{i+1,n+2,0}$ and $e_{n+2,i+1,0}$ (this means that the edges $e_{i,n}$ and $e_{n,i}$ do not belong to $V \times (V-A) \cup (V-A) \times V$, where A is the corresponding subset).

The rules of types 14, 15, 16, 17, 18, 19, 20, 21, and 22 are applied as many times as possible (in one step rules of types 14, 15, 16, 17, 18, and 19, in the next one rules of types 20, 21, and 22, and then we repeat the cycle). In this way, after $2n$ steps, a membrane will contain an object $e_{n+2,n+2,0}$ if and only if that membrane contains an edge which does not belong to $V \times (V-A) \cup (V-A) \times V$, where A is the corresponding subset. In the following steps, we will let the membranes corresponding to a positive answer send out an object.

23. $[_2 g_n \to g_{n+1} q]_2^0$.
24. $[_2 g_{n+1} \to g_{n+2}]_2^0$.
25. $[_2 g_{n+2} \to g_{n+3}]_2^-$.
26. $[_2 e_{n+2,n+2,0}]_2^- \to [_2\ ]_2^+ u$.
27. $[_2 g_{n+3}]_2^- \to [_2\ ]_2^- yes$.
28. $[_1 yes]_1^0 \to [_1\ ]_1^+ yes$.

At step $3n+2k+7$, g_n evolves to $g_{n+1}q$. At step $3n+2k+8$, g_{n+1} evolves to g_{n+2}, and at the same time, object q exits the membrane, changing the polarization to negative (using rule of type 4). At step $3n+2k+9$, in the membranes which contain object $e_{n+2,n+2,0}$, g_{n+2} evolve to g_{n+3}, and at the same time, $e_{n+2,n+2,0}$ exit these membranes, changing the polarization of these membranes to positive, they will never evolve again; in the membranes which do not contain object $e_{n+2,n+2,0}$, g_{n+2} evolve to g_{n+3}, the polarization of these membranes remains negative, in the next step they will produce the object *yes*. This object is then sent to the environment, telling us that there exists a clique with cardinality k, and the computation stops. The application of rule 28 changes the polarization of the skin membrane to positive in order that the objects *yes* remaining in it are not able to continue exiting it.

29. $[_1 c_i \to c_{i-1}]_1^0$, $1 \leq i \leq 3n + 2k + 11$.
30. $[_1 c_0]_1^0 \to [_1\ ]_1^0 no$.
    At step $3n + 2k + 11$, the skin membrane has object c_0, originating in $c_{3n+2k+11}$. Note that if there exists no vertex cover of size k, then at this moment the polarization of the skin membrane is neutral. In the next step, c_0 produces the object *no*, and the object *no* is sent to the environment, telling us that there exists no clique with cardinality k, and the computation stops.

From the previous explanation of the use of rules, one can easily see how this P system works. It is clear that we get the object *yes* if only if there exists a clique of size k. The object *yes* exits the system at moment $3n + 2k + 11$. If there exists no vertex cover of size k, then at step $3n + 2k + 12$ the system sends the object *no* to the environment. Therefore, the family of membrane systems we have constructed is sound, confluent, and linearly efficient. To prove that the family is uniform in the sense of Section 1, we have to show that for a given size, the construction of P systems described in the proof can be done in polynomial time by a Turing machine. Again, we omit the detailed construction.

So the clique problem was decided in linear time ($3n+2k+12$) by recognizing P systems with restricted elementary active membranes, and this concludes the proof.

8 Conclusions

We have shown that the vertex cover problem and the clique problem can be solved in linear time with respect to the number of vertices and edges of the graph by recognizing P systems with restricted elementary active membranes. It is also interesting to solve other related graph problems, in a “uniform” manner, by P systems which are the same, excepting a module specific to each problem.

The solution presented in Section 5 differs from the solution in Section 6 in the following sense: a family of recognizing P systems with active membranes is constructed, associated with the problem that is being solved, in such a way that all the instances of such problem that have the *same length* (according to a given polynomial-time computable criteria) are processed by the same P system (to which an appropriate input, that depends on the concrete instance, is supplied). On the contrary, in the solutions presented in Section 6, a single P system is associated with each one of the instances of the problem.

Let us denote by $\mathcal{AM}$ the class of recognizing P systems with active membranes and with 2-division. Then from Theorem 1 and Theorem 2 we have $VCP \in \mathbf{LMC}_{\mathcal{AM}} \subseteq \mathbf{PMC}_{\mathcal{AM}}$ and $CP \in \mathbf{LMC}_{\mathcal{AM}} \subseteq \mathbf{PMC}_{\mathcal{AM}}$. Because the class $\mathbf{PMC}_{\mathcal{AM}}$ is stable under polynomial time reduction, we have $\mathbf{NP} \subseteq \mathbf{PMC}_{\mathcal{AM}}$. Similarly, from Theorem 3 we have $VCP \in \mathbf{PMC}^S_{\mathcal{AM}}$ and $\mathbf{NP} \subseteq \mathbf{PMC}^S_{\mathcal{AM}}$, which can also be obtained from $VCP \in \mathbf{PMC}_{\mathcal{AM}}$, $\mathbf{NP} \subseteq \mathbf{PMC}_{\mathcal{AM}}$ and the fact $\mathbf{PMC}_{\mathcal{AM}} \subseteq \mathbf{PMC}^S_{\mathcal{AM}}$.

The following problems are open:

(1) Does $\mathbf{NP} = \mathbf{PMC}_{\mathcal{AM}}$ or $\mathbf{NP} = \mathbf{PMC}^S_{\mathcal{AM}}$ hold?

(2) Find a class $\mathcal{F}$ of P systems such that $\mathbf{NP} = \mathbf{PMC}_{\mathcal{F}}$ or $\mathbf{NP} = \mathbf{PMC}^S_{\mathcal{F}}$.

Acknowledgments. The work of the first author was supported by grant 2001CAJAL-BURV from Rovira i Virgili University; the work of the last author was supported by grant DGU-SB2001-0092 from Spanish Ministry for Education, Culture, and Sport, National Natural Science Foundation of China (Grant No. 60373089), and Huazhong University of Science and Technology Foundation.

References

1. M.R. Garey, D.J. Johnson, *Computers and Intractability. A Guide to the Theory of NP-Completeness.* W.H. Freeman, San Francisco, 1979.
2. Gh. Păun, Computing with Membranes, *Journal of Computer and System Sciences*, 61, 1 (2000), 108–143, and TUCS Research Report 208, 1998 (http://www.tucs.fi).
3. Gh. Păun, P Systems with Active Membranes: Attacking NP-Complete Problems, *Journal of Automata, Languages and Combinatorics*, 6, 1 (2001), 75–90.
4. Gh. Păun, *Membrane Computing: An Introduction*, Springer, Berlin, Heidelberg, 2002.
5. M.J. Pérez-Jiménez, A. Romero-Jiménez, F. Sancho-Caparrini, Complexity Classes in Cellular Computing with Membranes, *Proceedings of Brainstorming Week on Membrane Computing* (M. Cavaliere, C. Martín-Vide and Gh. Păun eds.), Tarragona, Spain, February, (2003) 270–278.
6. M.J. Pérez-Jiménez, A. Romero-Jiménez, F. Sancho-Caparrini, A Polynomial Complexity Class in P Systems Using Membrane Division, *Proceedings of the 5th Workshop on Descriptional Complexity of Formal Systems*, Budapest, Hungary, July 12–14, 2003.

Writing Information into DNA

Masanori Arita

Department of Computational Biology
Graduate School of Frontier Sciences
University of Tokyo
Kashiwanoha 5-1-5, 277-8561 Kashiwa, Japan
arita@k.u-tokyo.ac.jp

Abstract. The time is approaching when information can be written into DNA. This tutorial work surveys the methods for designing code words using DNA, and proposes a simple code that avoids unwanted hybridization in the presence of shift and concatenation of DNA words and their complements.

1 Introduction

As bio- and nano-technology advances, the demand for writing information into DNA increases. Areas of immediate application are:

- *DNA computation* which attempts to realize biological mathematics, i.e., solving mathematical problems by applying experimental methods in molecular biology [1]. Because a problem must be first encoded in DNA terms, the method of encoding is of crucial importance. Typically, a set of fixed-length oligonucleotides is used to denote logical variables or graph components.
- *DNA tag/antitag system* which designs fixed-length short oligonucleotide tags for identifying biomolecules (e.g., cDNA), used primarily for monitoring gene expressions [2,3,4].
- *DNA data storage* which advocates the use of bacterial DNA as a long-lasting high-density data storage, which can also be resistant to radiation [5].
- *DNA signature* which is important for registering a copyright of engineered bacterial and viral genomes. Steganography (an invisible signature hidden in other information) is useful for the exchange of engineered genomes among developers.

These fields are unlike conventional biotechnologies in that they attempt to *encode artificial information into DNA*. They can be referred to as 'encoding models'. Although various design strategies for DNA sequences have been proposed and some have been demonstrated, no standard code like the ASCII code exists for these models, presumably because data transfer in the form of DNA has not been a topic of research. In addition, requirements for DNA sequences differ for each encoding model.

In this tutorial work, the design of DNA words as information carriers is surveyed and a simple, general code for writing information into biopolymers is

N. Jonoska et al. (Eds.): Molecular Computing (Head Festschrift), LNCS 2950, pp. 23–35, 2004.

proposed. After this introduction, Section 2 introduces major constraints considered in the word design. In Section 3, three major design approaches are briefly overviewed and our approach is described in Section 4. Finally, Section 5 shows an exemplary construction of DNA code words using our method.

2 Requirements for a DNA Code

DNA sequences consist of four nucleotide bases (A: adenine, C: cytosine, G: guanine, and T: thymine), and are arrayed between chemically distinct terminals known as the 5'- and 3'-end. The double-helix DNA strands are formed by a sequence and its complement. The complementary strand, or *complement*, is obtained by the substitution of base A with base T, and base C with base G and vice versa, and reversing its direction. For example, the sequence `5'-AAGCGCTT-3'` is the complement of itself: $\begin{matrix} 5' - \texttt{AAGCGCTT} - 3' \\ 3' - \texttt{TTCGCGAA} - 5' \end{matrix}$. A non-complementary base in a double strand cannot form stable hydrogen bonds and is called a (base) *mismatch*. The stability of a DNA double helix depends on the number and distribution of base mismatches [6].

Now consider a set of DNA words for information interchange. Each word must be as distinct as possible so that no words will induce unwanted hybridization (*mishybridization*) regardless of their arrangement. At the same time, all words must be physico-chemically uniform (*concerted*) to guarantee an unbiased reaction in biological experiments.

In principle, there are two measures for evaluating the quality of designed DNA words: statistical thermodynamics and combinatorics. Although the thermodynamic method may yield a more accurate estimation, its computational cost is high. Therefore, since combinatorial estimations approximate the thermodynamic ones, the focus in this work is on the former method, described in terms of discrete constraints that DNA words should satisfy. In what follows, formal requirements for the DNA word set will be introduced.

2.1 Constraints on Sequences

DNA words are assumed to be of equal length. This assumption holds true in most encoding models. (Some models use oligonucleotides of different lengths for spacer- or marker sequences. As such modifications do not change the nature of the problem, their details are not discussed here.) The design problem posed by DNA words has much in common with the construction of classical error-correcting code words.

Let $x = x_1 x_2 \cdots x_n$ be a DNA word over four bases {A,C,G,T}. The *reverse* of x is denoted $x^R = x_n x_{n-1} \cdots x_1$, and the *complement* of x, obtained by replacing base A with T, and base C with G in x and vice versa, is denoted x^C. The *Hamming distance* $H(x, y)$ between two words $x = x_1 x_2 \ldots x_n$ and $y = y_1 y_2 \ldots y_n$ is the number of indices i such that $x_i \neq y_i$. For a set of DNA words S, S^{RC} is its complementation with reverse complement sequences, i.e., $\{x \mid x \in S \text{ or } (x^R)^C \in S\}$.

Hamming Constraints As in code theory, designed DNA words should keep a large Hamming distance between all word pairs. What makes the DNA code-design more complicated than the standard theory of error-correcting code is that we must consider not only $H(x, y)$ but also $H(x^C, y^R)$ to guarantee the mismatches in the hybridization with other words and their complements (Fig 1).

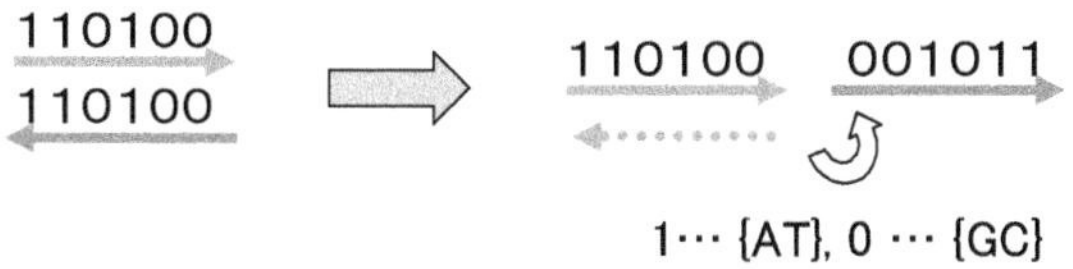

110100
110100
110100 001011
1··· {AT}, 0 ··· {GC}

Fig. 1. Binary Pattern of Hybridization. The complementary strand has a reverse pattern of {A,T} and {G,C} bases. A reverse complement of a DNA word corresponds to its complementary strand.

Comma-Free Constraints It is desirable for the designed words to be comma-free because DNA has no fixed reading frames. By definition, a code S is *comma-free* if the overlap of any two, not necessarily different, code words $x_1 x_2 \cdots x_n \in S$ and $y_1 y_2 \cdots y_n \in S$, (i.e., $x_{r+1} x_{r+2} \cdots x_n y_1 y_2 \cdots y_r$; $0 < r < n$) does not result in another code word in S [7,8]. The property by which any overlapping word differs from another word in at least d positions is called *comma-free with index d*. Thus, our DNA code should be comma-free with a high index. [1]

Note that comma-freeness is not replaced by introducing predefined 'spacer' words between code words. Such spacers may facilitate the decoding of words, but they do not contribute to the avoidance of mishybridization. Moreover, spacers lengthen the encoded DNA and lower its information content.

Energy Constraints In addition to the above constraints on mismatches, the melting temperatures of DNA words must be very similar to guarantee their concerted behavior *in vitro*. The most reliable estimation is the nearest neighbor approximation, where the temperature is computed from the frequency of 16 base dimers (from AA to TT) [12,6]. Arita and Kobayashi proposed its further approximation by grouping [GC] and [AT], where the temperature depends on the frequency of only 3 patterns ([GC][GC], [GC][AT] or [AT][GC], and [AT][AT]) [13]. *Dimer frequency* of a sequence x is the three tuple of integers, each describing the frequency of the above 3 patterns in this order. To integrate the terminal

[1] The idea of comma-freeness originated in the elucidation of DNA translation mechanism. Early on, DNA codons for 20 amino acids were thought to be encoded in the comma-free manner [9]. Incidentally, the number of comma-free code words of length 3 over 4 bases is at most 20. The systematic design of a comma-free code of index 1 was soon proposed [10,11].

bases, we assume as if x is cyclic in the computation of frequency. For example, AAGCGCTT and TACGGCAT exhibit close melting temperatures because they share the same dimer frequency $(3, 2, 3)$. Thus, all DNA code words should share the same dimer frequency to guarantee their concerted behavior.

Other Constraints Depending on the model used, there are constraints in terms of base mismatches. We focus on the first 2 constraints in this paper.

1. *Forbidden subwords* that correspond to restriction sites, simple repeats, or other biological signal sequences, should not appear anywhere in the designed words and their concatenations. This constraint arises when the encoding model uses pre-determined sequences such as genomic DNA or restriction sites for enzymes.
2. *Any subword of length k* should not appear more than once in the designed words. This constraint is imposed to ensure the avoidance of base pair nucleation that leads to mishybridization. The number k is usually ≥ 6.
3. *A secondary structure* that impedes expected hybridization of DNA words should not arise. To find an optimal structure for these words, the minimum free energy of the strand is computed by dynamic programming [14]. However, the requirement here is that the words do not form some structure. This constraint arises when temperature control is important in the encoding models.
4. *Only three bases*, A,C, and T, may be used in the word design. This constraint serves primarily to reduce the number of mismatches by biasing the base composition, and to eliminate G-stacking energy [15]. In RNA word design, this constraint is important because in RNA, both G-C pairs and G-U pairs (equivalent to G-T in DNA) form stably.

2.2 Data Storage Style

Because there is no standard DNA code, it may seem premature to discuss methods of aligning words or their storage, i.e., their data-addressing style. However, it is worth noting that the storage style depends on the word design; the immobilization technique, like DNA chips, has been popular partly because its weaker constraint on words alleviates design problems encountered in scaling up experiments.

Surface-Based Approach In the surface-based (or solid-phase) approach, DNA words are placed on a solid support (Fig 2). This method has two advantages: (1) since one strand of the double helix is immobilized, code words can be separated (washed out) from their complements, thereby reducing the risk of unexpected aggregation of words [16]; (2) since fluorescent labeling is effective, it is easier to recognize words, e.g., for information readout.

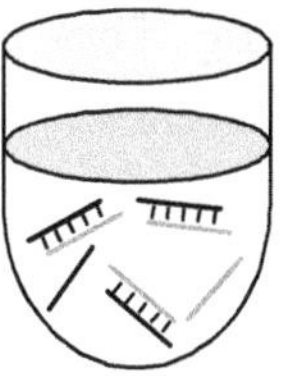

Fig. 2. The Surface-Based versus the Soluble Approach. While they are indistinguishable in solution, immobilization makes it easy to separate information words (gray) from their complements (black).

Soluble Approach Easier access to information on surfaces simultaneously limits the innate abilities of biomolecules. DNA fragments in solution acquire more flexibility as information carriers, and have been shown capable of simulating cellular automata [17]. Other advantages of the soluble approach are: (1) it opens the possibility of autonomous information processing [18]; (2) it is possible to introduce DNA words into microbes. The words can also be used for nano structure design.

Any systematic word design that avoids mishybridization should serve both approaches. Therefore, word constraints must extend to complements of code words. Our design problem is then summarized as follows.

> Problem: Given two integers l and d ($l > d > 0$), design a set S of length-l DNA words such that S^{RC} is comma-free with index d and for any two sequences $x, y \in S^{RC}$, $H(x, y) \geq d$ and $H(x^C, y^R) \geq d$. Moreover, all words in S^{RC} share the same dimer frequency.

3 Previous Works

Due to the different constraints, there is currently no standard method for designing DNA code words. In this section, three basic approaches are introduced: (1) the template-map strategy, (2) De Bruijn construction, and (3) the stochastic method.

3.1 Template-Map Strategy

This simple yet powerful construction was apparently first proposed by Condon's group [16]. Constraints on the DNA code are divided and separately assigned to two binary codes, e.g., one specifies the GC content (called *templates*), the other specifies mismatches between any word pairs (called *maps*). The product of two codes produces a quaternary code with the properties of both codes (Fig 3). Frutos et al. designed 108 words of length 8 where (1) each word has four GCs; (2) each pair of words, including reverse complements, differs in at least four bases [16]. Later, Li et al., who used the Hadamard code, generalized this construction to longer code words that have mismatches equal to or exceeding half their length [19]. They presented, as an example, the construction of 528 words of length 12 with 6 minimum mismatches.

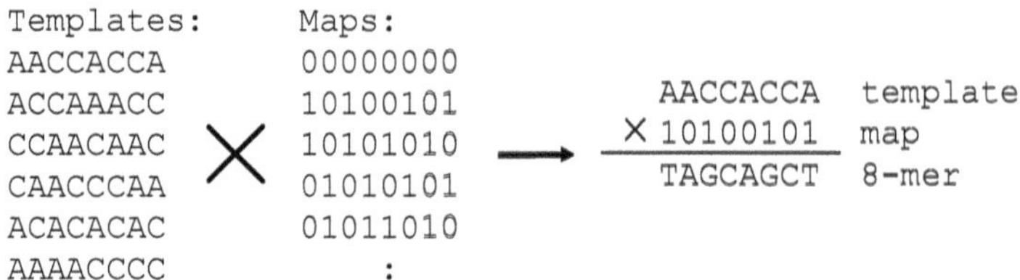

Fig. 3. Template-Map Strategy. In this figure, templates specify that the sequences contain 50% GCs and four mismatches between them and their complements. Maps are error-correcting code words and specify the choice between A and T, or G and C.

The drawback of this construction is twofold. First, the melting temperatures of the designed quaternary words may differ regardless of their uniform GC content. This property was analyzed in Li et al. and the predicted melting temperatures of the 528 words differed over 20 °C range [19]. The second problem is the comma-freeness. Although the design has been effectively demonstrated in the surface-based approach, scaling up to multiple words will be difficult due to mishybridization.

3.2 De Bruijn Construction

The longer a consecutive run of matched base pairs, the higher is the risk of mishybridization. The length-k subword constraint to avoid mishybridization is satisfied with a binary De Bruijn sequence of order k, a circular sequence of length 2^k in which each subword of length k occurs exactly once. [2] A linear time algorithm for the construction of a De Bruijn sequence is known [20]. Ben-Dor et al. showed an optimal choosing algorithm of oligonucleotide tags that satisfy the length-k subword constraint and also share similar melting temperatures [4].

One disadvantage is that the number of mismatches between words may be small, because the length-k constraint guarantees only one mismatch for each k-mer. Another disadvantage is again the comma-freeness.

3.3 Stochastic Method

The stochastic method is the most widely used approach in word design; there are as many types of design software as there are reported experiments.

[2] De Bruijn sequence can be quaternary. By using the binary version, however, we can easily satisfy the constraint that the subword does not occur in the complementary strand.

Deaton et al. used genetic algorithms to find code words of similar melting temperatures that satisfy the 'extended' Hamming constraint, i.e., a constraint where mismatches in the case of shift are also considered [21]. (The constraint they named the H-measure, is different from comma-freeness in that it considers mismatches between two words, not their overlaps.) Due to the complexity of the problem, they reported that genetic algorithms can be applied to code words of up to length 25 [22].

Landweber et al. used a random word-generation program to design two sets of 10 words of length 15 that satisfy the conditions (1) no more than five matches over a 20-nucleotide window in any concatenation between all 2^{10} combinations; (2) similar melting temperatures of 45 °C; (3) avoidance of secondary structures; and (4) no consecutive matches of more than 7 base pairs. [3] All of the strong constraints could be satisfied with only 3 bases [15]. Other groups that employed three-base words likewise used random word-generation for their word design [24,23].

Although no detailed analyses for such algorithms are available, the power of stochastic search is evident in the work of Tulpan et al., who could increase the number of code words designed by the template-map strategy [25]. However, they reported that the stochastic search failed to outperform the template-map strategy if searches were started from scratch. Therefore it is preferable to apply the stochastic method to enlarge already designed word sets.

4 Methods

4.1 Comma-Free Error-Correcting DNA Code

Among the different constraints on DNA code words, the most difficult to satisfy is comma-freeness; no systematic construction is known for a comma-free code of high index. The stochastic search is also not applicable because its computational cost is too high.

The comma-free property is, however, a necessary condition for the design of a general-purpose DNA code. This section presents the construction method for a comma-free error-correcting DNA code, and proposes a DNA code: 112 words of length 12 that mismatch at at least 4 positions in any mishybridization, share no more than 6 consecutive subsequences, and retain similar melting temperatures.

Basic Design For this design, we employed the method of Arita and Kobayashi [13]. It can systematically generate a set of words of length ℓ such that any of its members will have approximately $\ell/3$ mismatches with other words, their complements, and overlaps of their concatenations. They constructed sequences as a product of two types of binary words as in the template-map strategy, except that they used a single binary word, denoted T, as the template. Template T is

[3] The fourth condition is unnecessary if the first one is satisfied; presented here are all conditions considered in the original paper.

chosen so that its alignment with subsequent patterns always contains equal to or more than d mismatches.

$$T^R \quad TT^R \quad T^RT \quad TT \quad T^RT^R \tag{1}$$

The template specifies the GC positions of the designed words: [GC] corresponds to either 1's or 0's in the template. Since the pattern T^R specifies the AT/GC pattern of reverse complements, the mismatches between T and T^R guarantee the base mismatches between forward strands and reverse strands of designed DNAs. Other patterns from TT to T^RT^R are responsible for shifted patterns.

For the map words, any binary error-correcting code of minimum distance d or greater is used. Then, any pair of words in the resulting quaternary code induces at least d mismatches without block shift because of the error-correcting code, and with block shift or reversal because of the chosen template.

Comma-freeness is not the only advantage of their method. Because a single template is used to specify GC positions for all words, the GC arrangement of resulting code words is uniform, resulting in similar melting temperatures for all words in the nearest neighbor approximation [13].

Other Constraints In this subsection, methods to satisfy other practical constraints are introduced.

Forbidden subword
Since the error-correcting property of the map words is invariant under exchanging and 0-1 flipping columns of all words, this constraint can be easily satisfied.

Length-k subword
For the DNA words to satisfy this constraint, two additional conditions are necessary: First, the template should not share any length-k subword with patterns in (1). Second, the map words should not share any length-k subword among them.

The first condition can be imposed when the template is selected. To satisfy the second condition, the obvious transformation from word selection to the Max Clique Problem is used: the nodes correspond to the words, and the edges are linked only when two words do not share any length-k subword (without block shift). Note that the clique size is upper bounded by 2^k.

Secondary structure
Since all words are derived from the same template, in the absence of shifts, the number of mismatches can be the minimum distance of the error-correcting code words. Hybridization is therefore more likely to proceed without shifts. To avoid secondary structures, the minimum distance of the error-correcting code words is kept sufficiently large and base mismatches are as much distributed as possible. The latter constraint is already achieved by imposing the length-k subword constraint.

5 Results

5.1 DNA Code for the English Alphabet

Consider the design for the English alphabet using DNA. For each letter, one DNA word is required. One short error-correcting code is the nonlinear (12,144,4) code [26]. [4] Using a Max Clique Problem solver [5], 32, 56, and 104 words could be chosen that satisfied the length-6, -7, -8 subword constraint, respectively.

There are 74 template words of length 12 and of minimum distance 4; they are shown in the Appendix. Since 128 words cannot be derived from a single template under the subword constraint, two words, say S and T, were selected from the 74 templates such that both S and T induce more than 3 mismatches with any concatenation of 4 words T, S, T^R, and S^R (16 patterns), and each chosen word shares no more than length-5, -6, or -7 subword with the other and with their concatenations. Under the length-6 subword constraint, no template pair could satisfy all constraints. Under the length-7, and -8 subword constraints, 8 pairs were selected. (See the Appendix.) All pairs had the common dimer frequency. Under this condition, DNA words derived from these templates can be shown to share close melting temperatures.

Thus, we found 2 templates could be used simultaneously in the design of length-12 words. There were 8 candidate pairs. By combining one of 8 pairs with the 56 words in the Appendix, 112 words were obtained that satisfied the following conditions:

- They mismatched in at least 4 positions between any pair of words and their complements.
- The 4 mismatch was guaranteed under any shift and concatenation with themselves and their complements (comma-freeness of index 4).
- None shared a subword of length 7 or longer under any shift and concatenation.
- All words had close melting temperatures in the nearest neighbor approximation.
- Because all words were derived from only two templates, the occurrence of specific subsequences could be easily located. In addition, the avoidance of specific subsequences was also easy.

We consider that the 112 words serve as the standard code for the English alphabet. The number of words, 112, falls short of the 128 ASCII characters. However, some characters are usually unused. For example, HTML characters from to are not used. Therefore, the 112 words suffice for the DNA ASCII code. This is preferable to loosening the constraints to obtain 128 words.

[4] The notation (12,144,4) reads 'a length-12 code of 144 words with the minimum distance 4' (one error-correcting).

[5] http://rtm.science.unitn.it/intertools/

5.2 Discussion

The current status of information-encoding models was reviewed and the necessity and difficulty of constructing comma-free DNA code words was discussed. The proposed design method can provide 112 DNA words of length 12 and comma-free index 4. This result is superior to the current standard because it is the only work that considers arbitrary concatenation among words including their complementary strands.

In analyzing the encoding models, error and efficiency must be clearly distinguished. Error refers to the impairment of encoded information due to experimental missteps such as unexpected polymerization or excision. Efficiency refers to the processing speed, not the accuracy, of experiments.

Viewed in this light, the proposed DNA code effectively minimizes errors: First, the unexpected polymerization does not occur because all words satisfy the length-7 subword constraint. [6] Second, the site of possible excision under the application of enzymes is easily identified. Lastly, all words have uniform physico-chemical properties and their interaction is expected to be in concert. The efficiency, on the other hand, remains to be improved. It can be argued that 4 mismatches for words of length 12 are insufficient for avoiding unexpected secondary structures. Indispensable laboratory experiments are underway and confirmation of the applicability of the code presented here to any of the encoding models is awaited.

Regarding code size, it is likely that the number of words can be increased by a stochastic search.

Without systematic construction, however, the resulting code loses one good property, i.e., the easy location of specific subsequences under any concatenation.

The error-correcting comma-free property of the current DNA words opens a way to new biotechnologies. Important challenges include: 1. The design of a comma-free quaternary code of high indices; 2. Analysis of the distribution of mismatches in error-correcting code words; and 3. The development of criteria to avoid the formation of secondary structures.

Also important is the development of an experimental means to realize 'DNA signature'. Its presence may forestall and resolve lawsuits on the copyright of engineered genomes. Currently when a DNA message is introduced into a genome, no convenient method exists for the detection of its presence unless the message sequence is known. In the future, it should be possible to include English messages, not ACGTs, on the input window of DNA synthesizers.

References

1. L.M. Adleman, "Molecular Computation of Solutions to Combinatorial Problems," *Science* **266**(5187), 1021–1024 (1994).
2. S. Brenner and R.A. Lerner, "Encoded Combinatorial Chemistry," *Proc. Nation. Acad. Sci. USA* **89**(12), 5381–5383 (1992).

[6] The minimum length for primers to initiate polymerization is usually considered to be 8.

3. S. Brenner, S.R. Williams, E.H. Vermaas, T. Storck, K. Moon, C. McCollum, J.I. Mao, S. Luo, J.J. Kirchner, S. Eletr, R.B. DuBridge, T. Burcham and G. Albrecht, "In Vitro Cloning of Complex Mixtures of DNA on Microbeads: physical separation of differentially expressed cDNAs," *Proc. Nation. Acad. Sci. USA* **97**(4), 1665–1670 (2000).
4. A. Ben-Dor, R. Karp, B. Schwikowski and Z. Yakhini, "Universal DNA Tag Systems: a combinatorial design scheme," *J. Comput. Biol.* **7**(3-4), 503–519 (2000).
5. P.C. Wong, K-K. Wong and H. Foote, "Organic Data Memory Using the DNA Approach," *Comm. of ACM*, **46**(1), 95–98 (2003).
6. H.T. Allawi and J. SantaLucia Jr., "Nearest-neighbor Thermodynamics of Internal AC Mismatches in DNA: sequence dependence and pH effects," *Biochemistry*, **37**(26), 9435–9444 (1998).
7. S.W. Golomb, B. Gordon and L.R. Welch, "Comma-Free Codes," *Canadian J. of Math.* **10**, 202–209 (1958).
8. B. Tang, S.W. Golomb and R.L. Graham, "A New Result on Comma-Free Codes of Even Word-Length," *Canadian J. of Math.* **39**(3), 513–526 (1987).
9. H.F. Judson, *The Eighth Day of Creation: Makers of the Revolution in Biology,* Cold Spring Harbor Laboratory, (Original 1979; Expanded Edition 1996)
10. J.J. Stiffler, "Comma-Free Error-Correcting Codes," *IEEE Trans. on Inform. Theor.*, **IT-11**, 107–112 (1965).
11. J.J. Stiffler, *Theory of Synchronous Communication,* Prentice-Hall Inc., Englewood Cliffs, New Jersey, 1971.
12. K.J. Breslauer, R. Frank, H. Blocker and L.A. Marky, "Predicting DNA Duplex Stability from the Base Sequence," *Proc. Nation. Acad. Sci. USA* **83**(11), 3746–3750 (1986).
13. M. Arita and S. Kobayashi, "DNA Sequence Design Using Templates," *New Generation Comput.* **20**(3), 263–277 (2002). (Available as a sample paper at http://www.ohmsha.co.jp/ngc/index.htm.)
14. M. Zuker and P. Steigler, "Optimal Computer Folding of Large RNA Sequences Using Thermodynamics and Auxiliary Information," *Nucleic Acids Res.* **9**, 133–148 (1981).
15. D. Faulhammer, A.R. Cukras, R.J. Lipton and L.F. Landweber, "Molecular Computation: RNA Solutions to Chess Problems," *Proc. Nation. Acad. Sci. USA* **97**(4), 1385–1389 (2000).
16. A.G. Frutos, Q. Liu, A.J. Thiel, A.M. Sanner, A.E. Condon, L.M. Smith and R.M. Corn, "Demonstration of a Word Design Strategy for DNA Computing on Surfaces," *Nucleic Acids Res.* **25**(23), 4748–4757 (1997).
17. E. Winfree, X. Yang and N.C. Seeman, "Universal Computation Via Self-assembly of DNA: some theory and experiments," In *DNA Based Computers II, DIMACS Series in Discr. Math. and Theor. Comput. Sci.* **44**, 191–213 (1998).
18. Y. Benenson, T. Paz-Elizur, R. Adar, E. Keinan, Z. Livneh and E. Shapiro, "Programmable and Autonomous Computing Machine Made of Biomolecules," *Nature* **414**, 430–434 (2001).
19. M. Li, H.J. Lee, A.E. Condon and R.M. Corn, "DNA Word Design Strategy for Creating Sets of Non-interacting Oligonucleotides for DNA Microarrays," *Langmuir* **18**(3), 805–812 (2002).
20. K. Cattell, F. Ruskey, J. Sawada and M. Serra, "Fast Algorithms to Generate Necklaces, Unlabeled Necklaces, and Irreducible Polynomials over GF(2)," *J. Algorithms*, **37**, 267–282 (2000).

21. R. Deaton, R.C. Murphy, M. Garzon, D.R. Franceschetti and S.E. Stevens Jr., "Good Encodings for DNA-based Solution to Combinatorial Problems," In *DNA Based Computers II, DIMACS Series in Discr. Math. and Theor. Comput. Sci.* **44**, 247–258 (1998).
22. M. Garzon, P. Neathery, R. Deaton, D.R. Franceschetti, and S.E. Stevens Jr., "Encoding Genomes for DNA Computing," In *Proc. 3rd Annual Genet. Program. Conf.*, Morgan Kaufmann 684–690 (1998).
23. R.S. Braich, N. Chelyapov, C. Johnson, R.W. Rothemund and L. Adleman, "Solution of a 20-Variable 3-SAT Problem on a DNA Computer," *Science* **296**(5567), 499–502 (2002).
24. K. Komiya, K. Sakamoto, H. Gouzu, S. Yokoyama, M. Arita, A. Nishikawa and M. Hagiya, "Successive State Transitions with I/O Interface by Molecules," In *DNA Computing: 6th Intern. Workshop on DNA-Based Computers* (Leiden, The Netherlands), LNCS **2054**, 17–26 (2001).
25. D.C. Tulpan, H. Hoos and A. Condon, "Stochastic Local Search ALgorithms for DNA Word Design," In *Proc. 8th Intern. Meeting on DNA-Based Computers* (Sapporo, Japan), 311–323 (2002).
26. F.J. MacWilliams and N.J.A. Sloane, "The Theory of Error-Correcting Codes," New York, North-Holland, 2nd reprint (1983).

Appendix

110010100000	110001010000†	110000001010	110000000101	101100100000†	101001001000†
101000010001	101000000110†	100101000100†	100100011000	100100000011	100011000010
100010010100	100010001001	100001100001†	100000110010	100000101100†	011100000010
011010000100	011000110000†	011000001001	010110001000	010100100100	010100010001
010011000001	010010010010	010001101000	010001000110	010000100011†	010000011100
001110010000†	001101000001†	001100001100	001010101000†	001010000011	001001100010
001001010100†	001000100101	001000011010†	000110100010	000110000101	000101110000†
000101001010	000100101001†	000100010110	000011100100	000011011000	000010110001†
000010001110	000001010011	000001001101†	001101011111	001110101111	001111110101
001111111010	010011011111†	010110110111†	010111101110†	010111111001	011010111011†
011011100111	011011111100	011100111101†	011101101011	011101110110	011110011110†
011111001101	011111010011†	100011111101†	100101111011	100111001111†	100111110110
101001110111	101011011011	101011101110	101100111110	101101101101	101110010111
101110111001	101111011100†	101111100011	110001101111	110010111110†	110011110011†
110101010111	110101111100†	110110011101	110110101011†	110111011010†	110111100101
111001011101	111001111010	111010001111	111010110101	111011010110†	111011101001
111100011011	111100100111	111101001110†	111101110001	111110101100	111110110010†
000000000000†	111111111111†	000000111111	000011101011†	000101100111	000110011011†
000110111100	001001111001	001010011101	001010110110	001100110011†	001111000110†
010001110101†	010010101101†	010100001111†	010100111010	010111010100	011000010111
011000101110	011011001010†	011101011000†	011110100001	111111000000	111100010100†
111010011000†	111001100100	111001000011†	110110000110	110101100010	110101001001
110011001100	110000111001†	101110001010†	101101010010†	101011110000	101011000101†
101000101011	100111101000	100111010001	100100110101†	100010100111†	100001011110

(12,144,4) Code. Daggers indicate 56 words that satisfy the length-7-subword constraint.

101001100000 011001010000 101101110000 101100001000 011101101000 110011101000
001010011000 101110011000 111001011000 010110111000 001101000100 011101100100
001111010100 001110110100 111010001100 110010101100 101111000010 111001100010
010111100010 111100010010 011000001010 011010100110 100001110110 100100011110
111010010001 110110010001 100110101001 101110000101 111000100101 110101000011
110100100011

Templates of Length 12. When their reversals and 01-flips are included, the total number of words is 74.

000110011101 and 001010111100	000110011101 and 001111010100
001010111100 and 101110011000	001111010100 and 101110011000
010001100111 and 110000101011	010001100111 and 110101000011
110000101011 and 111001100010	110101000011 and 111001100010

Template Pairs Satisfying Minimum Distance 4 and Length-7-subword Constraint.

Balance Machines: Computing = Balancing

Joshua J. Arulanandham, Cristian S. Calude, and Michael J. Dinneen

Department of Computer Science
The University of Auckland
Auckland, New Zealand
{jaru003,cristian,mjd}@cs.auckland.ac.nz

Abstract. We propose a natural computational model called a *balance machine*. The computational model consists of components that resemble ordinary physical balances, each with an intrinsic property to automatically balance the weights on their left, right pans. If we start with certain fixed weights (representing *inputs*) on some of the pans, then the balance–like components would vigorously try to balance themselves by filling the rest of the pans with suitable weights (representing the *outputs*). This balancing act can be viewed as a computation. We will show that the model allows us to construct those primitive (hardware) components that serve as the building blocks of a general purpose (universal) digital computer: logic gates, memory cells (flip-flops), and transmission lines. One of the key features of the balance machine is its "bidirectional" operation: both a function and its (partial) inverse can be computed spontaneously using the same machine. Some practical applications of the model are discussed.

1 Computing as a "Balancing Feat"

A detailed account of the proposed model will be given in Section 2. In this section, we simply convey the essence of the model without delving into technical details. The computational model consists of components that resemble ordinary physical balances (see Figure 1), each with an intrinsic property to automatically balance the weights on their left, right pans. In other words, if we start with certain fixed weights (representing *inputs*) on some of the pans, then the balance–like components would vigorously try to balance themselves by filling the rest of the pans with suitable weights (representing the *outputs*). Roughly speaking, the proposed machine has a natural ability to load itself with (output) weights that "balance" the input. This balancing act can be viewed as a computation. There is just one rule that drives the whole computational process: *the left and right pans of the individual balances should be made equal.* Note that the machine is designed in such a way that the balancing act would happen automatically by virtue of physical laws (i.e., the machine is self-regulating).[1] One of our aims is to show that all computations can be ultimately expressed using one primitive operation:

[1] If the machine cannot (eventually) balance itself, it means that the particular instance does not have a solution.

N. Jonoska et al. (Eds.): Molecular Computing (Head Festschrift), LNCS 2950, pp. 36–48, 2004.

balancing. Armed with the *computing* = *balancing* intuition, we can see basic computational operations in a different light. In fact, an important result of this paper is that this sort of intuition suffices to conceptualize/implement *any* computation performed by a conventional computer.

Fig. 1. Physical balance.

The rest of the paper is organized as follows: Section 2 gives a brief introduction to the proposed computational model; Section 3 discusses a variety of examples showing how the model can be made to do basic computations; Section 4 is a brief note on the universality feature of the model; Section 5 reviews the notion of *bilateral computing* and discusses an application (solving the classic SAT problem); Section 6 concludes the paper.

2 The Balance Machine Model

At the core of the proposed natural computational model are components that resemble a physical balance. In ancient times, the shopkeeper at a grocery store would place a standard weight in the left pan and would try to load the right pan with a commodity whose weight equals that on the left pan, typically through repeated attempts. The physical balance of our model, though, has an intrinsic self-regulating mechanism: it can automatically load (without human intervention) the right pan with an object whose weight equals the one on the left pan. See Figure 2 for a possible implementation of the self-regulating mechanism.

In general, unlike the one in Figure 2, a balance machine may have more than just two pans. There are two types of pans: pans carrying *fixed* weights which remain unaltered by the computation and pans with *variable* (fluid) weights that are changed by activating the filler–spiller outlets. Some of the fixed weights represent the inputs, and some of the variable ones represent outputs. The inputs and outputs of balance machines are, by default, non–negative reals unless stated otherwise. The following steps comprise a typical computation by a given balance machine:

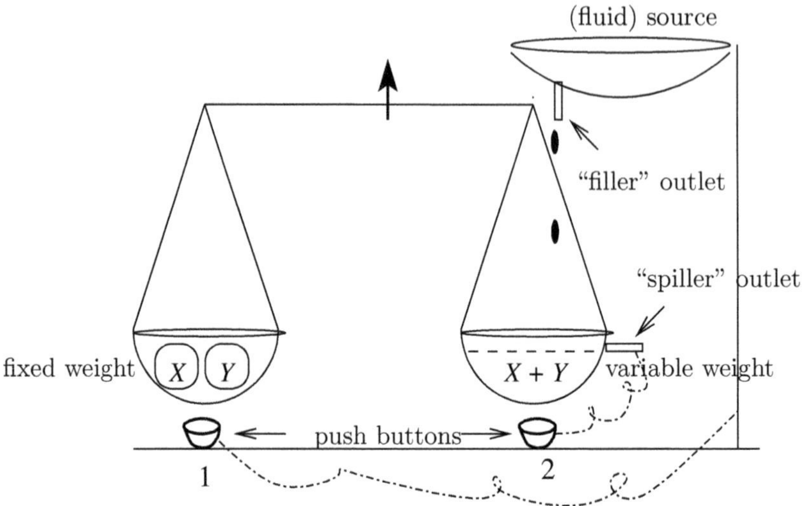

Fig. 2. A self-regulating balance. The *source* is assumed to have (an arbitrary amount of) a fluid–like substance. When activated, the *filler* outlet lets fluid from source into the right pan; the *spiller* outlet, on being activated, allows the right pan to spill some of its content. The balancing is achieved by the following mechanism: the spiller is activated if at some point the right pan becomes heavier than left (i.e., when push button (2) is pressed) to spill off the extra fluid; similarly, the filler is activated to add extra fluid to the right pan just when the left pan becomes heavier than right (i.e., when push button (1) is pressed). Thus, the balance machine can "add" (sum up) inputs x and y by balancing them with a suitable weight on its right: after being loaded with inputs, the pans would go up and down till it eventually finds itself balanced.

- Plug in the desired inputs by loading weights to pans (pans with variable weights can be left empty or assigned with arbitrary weights). This defines the initial configuration of the machine.
- Allow the machine to balance itself: the machine automatically adjusts the variable weights till left and right pans of the balance(s) become equal.
- Read output: weigh fluid collected in the output pans (say, with a standard weighting instrument).

See Figure 3 for a schematic representation of a machine that adds two quantities. We wish to point out that, to express computations more complex than addition, we would require a combination of balance machines such as the one in Figure 2. Section 3 gives examples of more complicated machines.

3 Computing with Balance Machines: Examples

In what follows, we give examples of a variety of balance machines that carry out a wide range of computing tasks—from the most basic arithmetic operations to

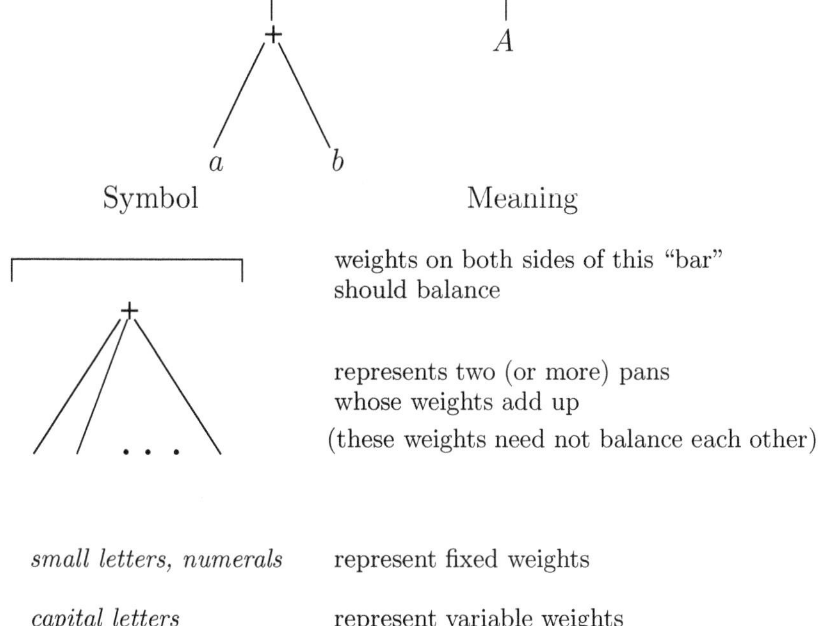

Fig. 3. Schematic representation of a simple balance machine that performs addition.

solving linear simultaneous equations. Balance machines that perform the operations *increment*, *decrement*, *addition*, and *subtraction* are shown in Figures 4, 5, 6, and 7, respectively. Legends accompanying the figures give details regarding how they work.

Balance machines that perform *multiplication by 2* and *division by 2* are shown in Figures 8, 9, respectively. Note that in these machines, one of the weights (or pans) takes the form of a balance machine.[2] This demonstrates that such a recursive arrangement is possible.

We now introduce another technique of constructing a balance machine: having a "common" weight shared by more than one machine. Another way of visualizing the same situation is to think of pans (belonging to two different machines) being placed one over the other. We use this idea to solve a simple instance of linear simultaneous equations. See Figures 10 and 11 which are self–explanatory.

An important property of balance machines is that they are *bilateral computing* devices. See [1], where we introduced the notion of bilateral computing. Typically, bilateral computing devices can compute both a function and its (partial) inverse using the same mechanism. For instance, the machines that increment and decrement (see Figures 4 and 5) share exactly the same mechanism, except

[2] The weight contributed by a balance machine is assumed to be simply the sum of the individual weights on each of its pans. The weight of the bar and the other parts is not taken into account.

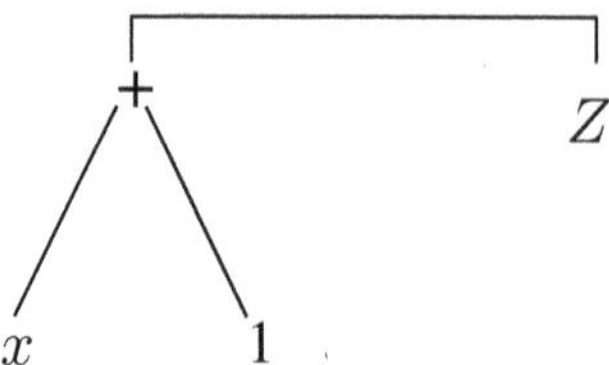

Fig. 4. Increment operation. Here x represents the input; Z represents the output. The machine computes *increment*(x). Both x and '1' are fixed weights clinging to the left side of the balance machine. The machine eventually loads into Z a suitable weight that would balance the combined weight of x and '1'. Thus, eventually $Z = x + 1$, i.e., Z represents *increment*(x).

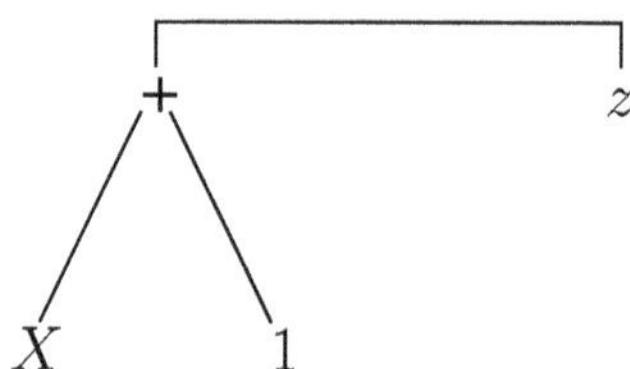

Fig. 5. Decrement operation. Here z represents the input; X represents the output. The machine computes *decrement*(z). The machine eventually loads into X a suitable weight, so that the combined weight of X and '1' would balance z. Thus, eventually $X + 1 = z$, i.e., X represents *decrement*(z).

for the fact that we have swapped the input and output pans. Also, compare machines that (i) add and subtract (see Figures 6 and 7) and (ii) multiply and divide by 2 (see Figures 8 and 9).

Though balance machines are basically analog computing machines, we can implement Boolean operations (AND, OR, NOT) using balance machines, provided we make the following assumption: *There are threshold units that return a desired value when the analog values (representing inputs and outputs) exceed a given threshold and some other value otherwise.* See Figures 12, 13, and 14 for balance machines that implement logical operations AND, OR, and NOT respectively. We represent $true$ inputs with the (analog) value 10 and $false$ inputs with 5; when the output exceeds a threshold value of 5 it is interpreted as $true$, and as $false$ otherwise. (Instead, we could have used the analog values 5 and 0 to represent $true$ and $false$; but, this would force the AND gate's output to a negative value for a certain input combination.) Tables 1, 2, and 3 give the truth tables (along with the actual input/output values of the balance machines).

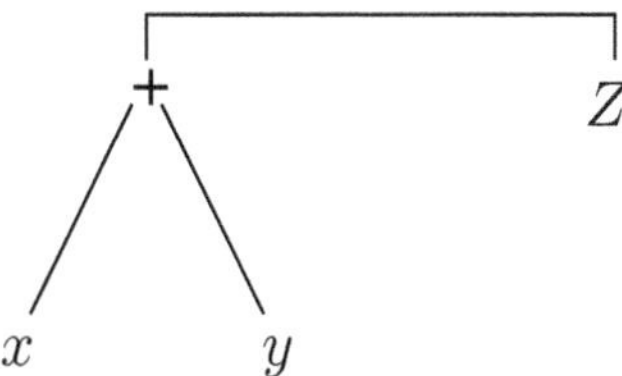

Fig. 6. Addition operation. The inputs are x and y; Z represents the output. The machine computes $x + y$. The machine loads into Z a suitable weight, so that the combined weight of x and y would balance Z. Thus, eventually $x+y = Z$, i.e., Z would represent $x + y$.

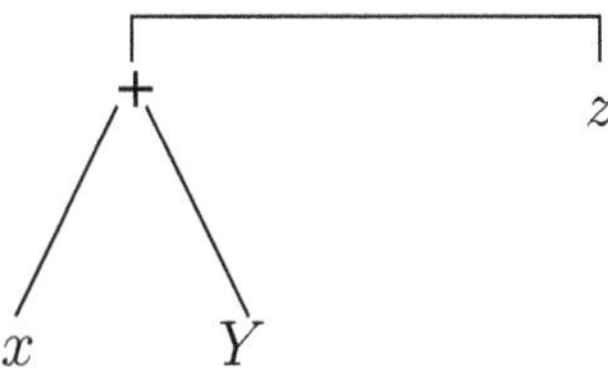

Fig. 7. Subtraction operation. Here z and x represent the inputs; Y represents the output. The machine computes $z - x$. The machine loads into Y a suitable weight, so that the combined weight of x and Y would balance z. Thus, eventually $x + Y = z$, i.e., Y would represent $z - x$.

4 Universality of Balance Machines

The balance machine model is capable of *universal discrete* computation, in the sense that it can simulate the computation of a practical, general purpose digital computer. We can show that the model allows us to construct those primitive (hardware) components that serve as the "building blocks" of a digital computer: logic gates, memory cells (flip-flops) and transmission lines.

1. **Logic gates**
   We can construct AND, OR, NOT gates using balance machines as shown in Section 3. Also, we could realize any given Boolean expression by connecting balance machines (primitives) together using "transmission lines" (discussed below).
2. **Memory cells**
   The weighting pans in the balance machine model can be viewed as "storage" areas. Also, a standard S–R flip-flop can be constructed by cross coupling two NOR gates, as shown in Figure 15. Table 4 gives its state table. They can be implemented with balance machines by replacing the OR, NOT gates in the diagram with their balance machine equivalents in a straightforward manner.

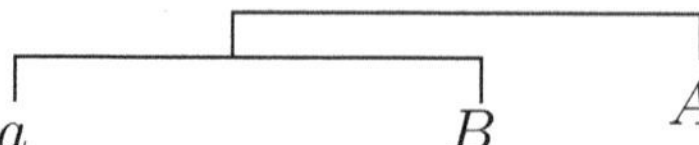

Fig. 8. Multiplication by 2. Here a represents the input; A represents the output. The machine computes $2a$. The combined weights of a and B should balance A: $a + B = A$; also, the individual weights a and B should balance each other: $a = B$. Therefore, eventually A will assume the weight $2a$.

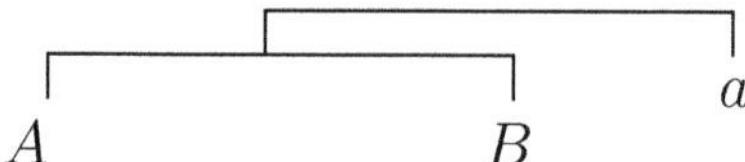

Fig. 9. Division by 2. The input is a; A represents the output. The machine "exactly" computes $a/2$. The combined weights of A and B should balance a: $A+B = a$; also, the individual weights A and B should balance each other: $A = B$. Therefore, eventually A will assume the weight $a/2$.

3. **Transmission lines**
   A balance machine like machine (2) of Figure 16 that does nothing but a "copy" operation (copying whatever is on left pan to right pan) would serve both as a transmission equipment, and as a *delay* element in some contexts. (The pans have been drawn as flat surfaces in the diagram.) Note that the left pan (of machine (2)) is of the fixed type (with no spiller–filler outlets) and the right pan is a variable one.

Note that the computational power of Boolean circuits is equivalent to that of a Finite State Machine (FSM) with *bounded* number of computational steps (see Theorem 3.1.2 of [4]).[3] But, balance machines are "machines with memory": using them we can build not just Boolean circuits, but also memory elements (flip-flops). Thus, the power of balance machines surpasses that of mere bounded FSM computations; to be precise, they can simulate any general *sequential circuit.* (A sequential circuit is a concrete machine constructed of gates and memory devices.) Since any finite state machine (with bounded or unbounded computations) can be realized as a sequential circuit using standard procedures (see [4]), one can conclude that balance machines have (at least) the computing power of *unbounded* FSM computations. Given the fact that any practical (general purpose) digital computer with only limited memory can be modeled by an FSM, we can in principle construct such a computer using balance machines. Note, however, that *notional* machines like Turing machines are more general than balance machines. Nevertheless, standard "physics–like" models in the litera-

[3] Also, according to Theorem 5.1 of [2] and Theorem 3.9.1 of [4], a Boolean circuit can simulate any T–step Turing machine computation.

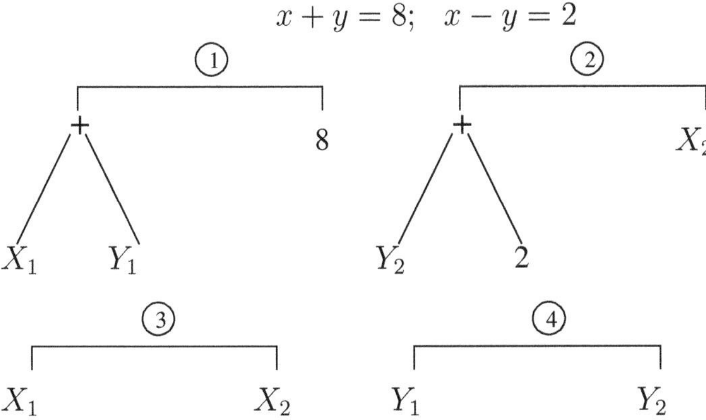

Fig. 10. Solving simultaneous linear equations. The constraints $X_1 = X_2$ and $Y_1 = Y_2$ will be taken care of by balance machines (3) and (4). Observe the sharing of pans between them. The individual machines work together as a single balance machine.

ture like the Billiard Ball Model[3] are universal only in this limited sense: there is actually no feature analogous to the infinite tape of the Turing machine.

5 Bilateral Computing

There is a fundamental asymmetry in the way we normally compute: while we are able to design circuits that can *multiply* quickly, we have relatively limited success in *factoring* numbers; we have fast digital circuits that can "combine" digital data using AND/OR operations and realize Boolean expressions, yet no fast circuits that determine the truth value assignment satisfying a Boolean expression. Why should computing be easy when done in one "direction", and not so when done in the other "direction"? In other words, why should *inverting* certain functions be hard, while *computing* them is quite easy? It may be because our computations have been based on rudimentary operations (like addition, multiplication, etc.) that force an explicit distinction between "combining" and "scrambling" data, i.e. *computing* and *inverting* a given function. On the other hand, a primitive operation like *balancing* does not do so. It is the same balance machine that does both addition and subtraction: all it has to do is to somehow balance the system by filling up the empty variable pan (representing output); whether the empty pan is on the right (addition) or the left (subtraction) of the balance does not particularly concern the balance machine! In the bilateral scheme of computing, there is no need to develop two distinct intuitions—one for addition and another for subtraction; there is no dichotomy between functions and their (partial) inverses. Thus, a bilateral computing system is one which can implement a function as well as (one of) its (partial) inverse(s), using the same intrinsic "mechanism" or "structure". See [1] where

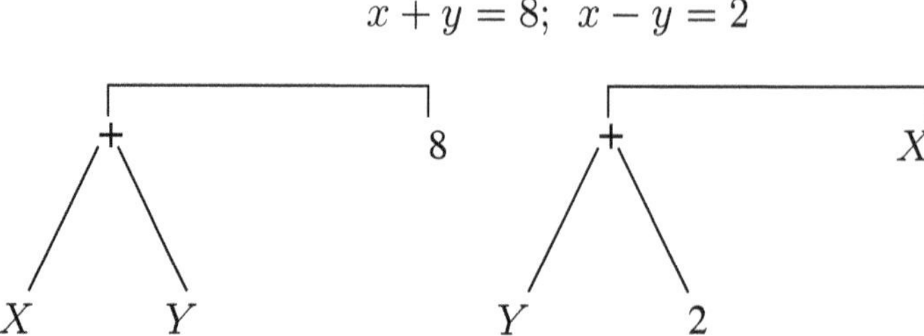

Fig. 11. Solving simultaneous linear equations (easier representation). This is a simpler representation of the balance machine shown in Figure 10. Machines (3) and (4) are not shown; instead, we have used the same (shared) variables for machines (1) and (2).

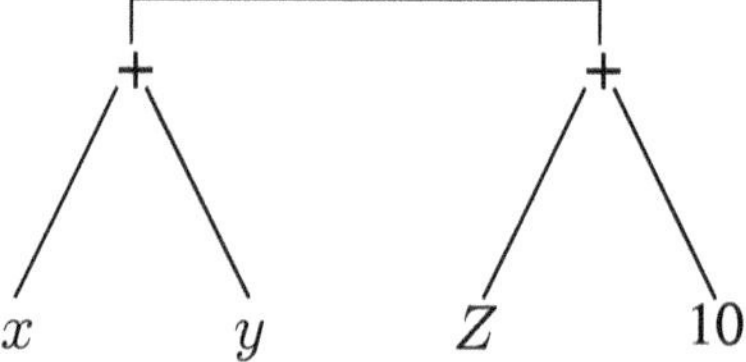

Fig. 12. AND logic operation. x and y are inputs to be ANDed; Z represents the output. The balance realizes the equality $x + y = Z + 10$.

we have developed bilateral computing systems based on fluid mechanics and have given a mathematical characterization of such systems.

We now show how the classic SAT problem can be solved under a bilateral computing scheme, using balance machines. For the time being, we make no claims regarding the time complexity of the approach since we have not analyzed the time characteristics of balance machines. However, we believe that it will not be exponential in terms of the number of variables (see also [1]). The main idea is this: first realize the given Boolean expression using gates made of balances; then, by setting the pan that represents the outcome of the Boolean expression to (the analog value representing) *true*, the balance machine can be made to automatically assume a set of values for its inputs that would "balance" it. In other words, by setting the output to be *true*, the inputs are forced to assume one of those possible truth assignments (if any) that generate a *true* output. The machine would never balance, when there is no such possible input assignment to the inputs (i.e., the formula is unsatisfiable). This is like operating a circuit realizing a Boolean expression in the "reverse direction": assigning the "output" first, and making the circuit produce the appropriate "inputs", rather than the other way round.

See Figure 17 where we illustrate the solution of a simple instance of SAT using a digital version of balance machine whose inputs/outputs are positive

Table 1. Truth table for AND.

x	y	Z
5 (*false*)	5 (*false*)	0 (*false*)
5 (*false*)	10 (*true*)	5 (*false*)
10 (*true*)	5 (*false*)	5 (*false*)
10 (*true*)	10 (*true*)	10 (*true*)

Table 2. Truth table for OR.

x	y	Z
5 (*false*)	5 (*false*)	5 (*false*)
5 (*false*)	10 (*true*)	10 (*true*)
10 (*true*)	5 (*false*)	10 (*true*)
10 (*true*)	10 (*true*)	15 (*true*)

Table 3. Truth table for NOT.

x	Y
5 (*false*)	10 (*true*)
10 (*true*)	5 (*false*)

Table 4. State table for S–R flip-flop.

S	R	Q	Q'
0	0	previous state of Q	previous state of Q'
0	1	0	1
1	0	1	0
1	1	undefined	undefined

integers (as opposed to reals). Note that these machines work based on the following assumptions:

1. Analog values 10 and 5 are used to represent *true* and *false* respectively.
2. The filler–spiller outlets let out fluid only in (discrete) "drops", each weighing 5 units.
3. The maximum weight a pan can hold is 10 units.

6 Conclusions

As said earlier, one of our aims has been to show that all computations can be ultimately expressed using one primitive operation: *balancing*. The main thrust of this paper is to introduce a natural intuition for computing by means of a generic model, and not on a detailed physical realization of the model. We have not analysed the time characteristics of the model, which might depend on how

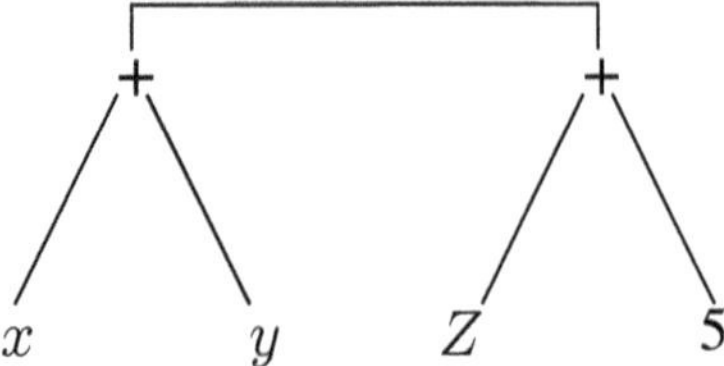

Fig. 13. OR logic operation. Here x and y are inputs to be ORed; Z represents the output. The balance realizes the equality $x + y = Z + 5$.

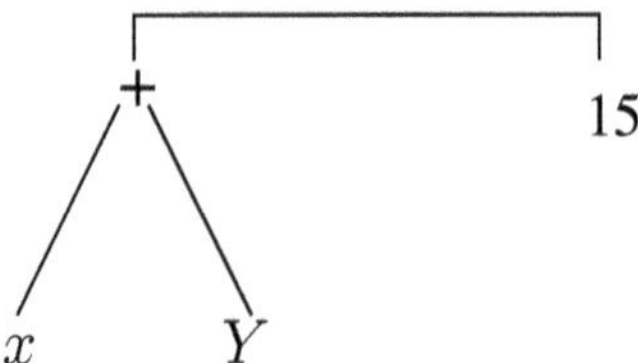

Fig. 14. NOT logic operation. Here x is the input to be negated; Y represents the output. The balance realizes the equality $x + Y = 15$.

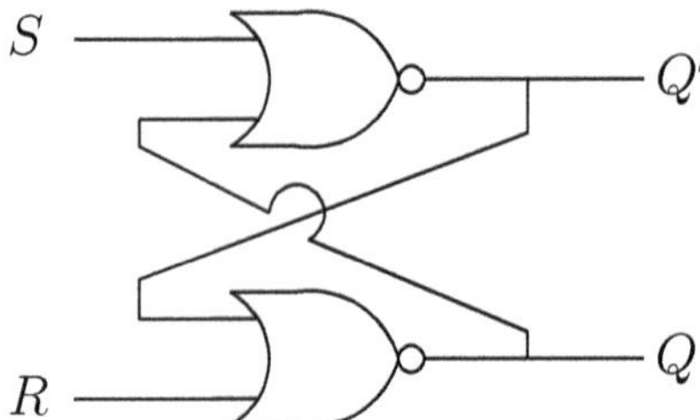

Fig. 15. S–R flip-flop constructed using cross coupled NOR gates.

we ultimately implement the model. Also, apart from showing with illustrative examples various possible (valid) ways of constructing balance machines, we have not detailed a formal "syntax" that governs them.

Finally, this note shows that one of the possible answers to the question "What does it mean to *compute*?" is: "To *balance* the inputs with suitable outputs (on a suitably designed balance machine)."

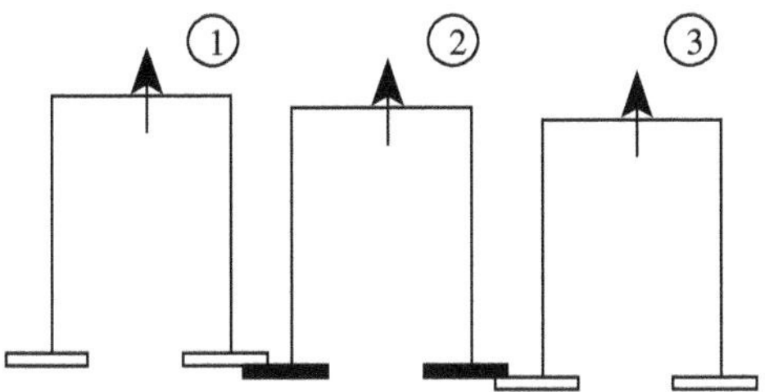

Fig. 16. Balance machine as a "transmission line". Balance machine (2) acts as a transmission line feeding the output of machine (1) into the input of machine (3).

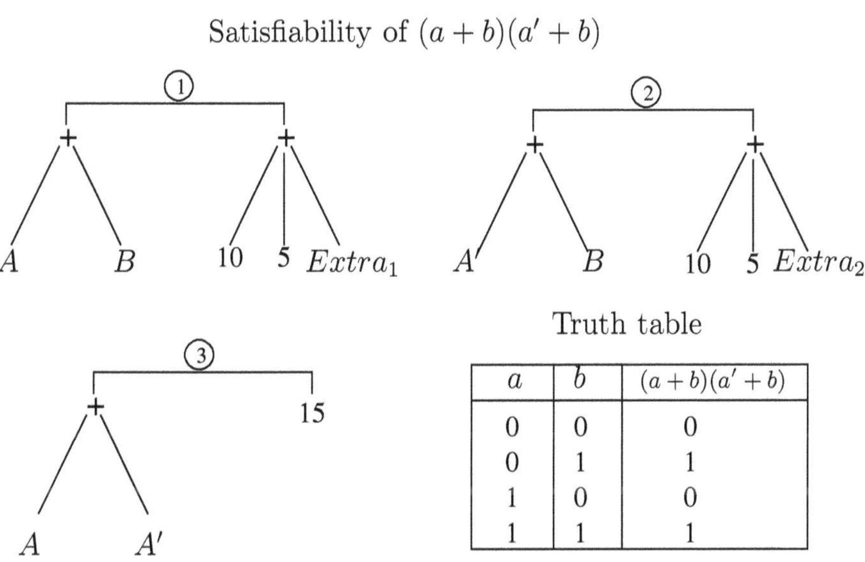

a	b	$(a + b)(a' + b)$
0	0	0
0	1	1
1	0	0
1	1	1

Fig. 17. Solving an instance of SAT: The satisfiability of the formula $(a + b)(a' + b)$ is verified. Machines (1), (2) and (3) work together sharing the variables A, B and A' between them. OR gates (labeled 1 and 2) realize $(a + b)$ and $(a' + b)$ respectively and the NOT gate (labeled 3) ensures that a and a' are "complementary". Note that the "output" of gates (1) and (2) are set to 10. Now, one has to observe the values eventually assumed by the variable weights A and B (that represent "inputs" of OR gate (1)). Given the assumptions already mentioned, one can easily verify that the machine *will balance*, assuming one of the two following settings: (i) $A = 5$, $B = 10$, ($Extra_1 = 0$, $Extra_2 = 5$) or (ii) $A = 10$, $B = 10$, ($Extra_1 = 5$, $Extra_2 = 0$). These are the *only* configurations that make the machine balanced. In situations when both the left pans of gate (1) assume 10, $Extra_1$ will automatically assume 5 to balance off the extra weight on the left side. ($Extra_2$ plays a similar role in gate (2).)

References

1. J.J. Arulanandham, C.S. Calude, M.J. Dinneen. Solving SAT with bilateral computing, *Romanian Journal of Information Science and Technology* (2003), to appear.
2. J.L. Balcázar, J.Díaz, J. Gabarró. *Structural Complexity I*, Springer–Verlag, Berlin, 1988.
3. E. Fredkin, T. Toffoli. Conservative logic, *Int'l J. Theoret. Phys.* 21 (1982), 219–253.
4. J.E. Savage. *Models of Computation*, Addison–Wesley, Reading, Mass., 1998.

Eilenberg P Systems with Symbol-Objects

Francesco Bernardini, Marian Gheorghe, and Mike Holcombe

Department of Computer Science
The University of Sheffield
Regent Court, Portobello Street, Sheffield, S1 4DP, UK
{F.Bernardini, M.Gheorghe, M.Holcombe}@dcs.shef.ac.uk

Abstract. A class of P systems, called EOP systems, with symbol objects processed by evolution rules distributed alongside the transitions of an Eilenberg machine, is introduced. A parallel variant of EOP systems, called EOPP systems, is also defined and the power of both EOP and EOPP systems is investigated in relationship with three parameters: number of membranes, states and set of distributed rules. It is proven that the family of Parikh sets of vectors of numbers generated by EOP systems with one membrane, one state and one single set of rules coincides with the family of Parikh sets of context-free languages. The hierarchy collapses when at least one membrane, two states and four sets of rules are used and in this case a characterization of the family of Parikh sets of vectors associated with $ET0L$ languages is obtained. It is also shown that every EOP system may be simulated by an EOPP system and EOPP systems may be used for solving NP-complete problems. In particular linear time solutions are provided for the SAT problem.

1 Introduction

P systems were introduced by Gh. Păun [12] as a computational model inspired by the structure and functioning of the cell. A central role in this context is played by membranes delimiting regions and allowing or preventing the transport of different molecules and chemicals among them. Different classes of P systems dealing with string objects or symbol objects, considering sets or multisets of elements leading to various families of languages were investigated [13] (an up-to-date bibliography of the whole area may be found at http://psystems.disco.unimib.it/). Because rewriting alone even in the context of a highly parallel environment of a membrane structure is not enough to lead to characterizations of recursively enumerable languages, various other features have been considered, such as a *priority relationship* over the set of rules, *permitting* or *forbidding* conditions associated with rules, *restrictions on the derivation mode*, the possibility to control the *membrane permeability* [7] etc (for more details see [13]). In general the most used priority relationship on the set of rewriting rules is a partial order relationship, well studied in the context of generative mechanisms with restrictions in derivation [5].

In [1] the priority relationship were replaced by a transition diagram associated with an Eilenberg machine giving birth to two classes of Eilenberg systems,

N. Jonoska et al. (Eds.): Molecular Computing (Head Festschrift), LNCS 2950, pp. 49–60, 2004.

a sequential version and a parallel one, called *EP systems* and *EPP systems*, respectively. In both variants, each transition has a specific set of evolution rules acting upon the string objects contained in different regions of the membrane system. The power of both variants working with string objects was investigated as well as the suitability of EPP systems to solve hard problems. In this paper multisets of symbol objects are considered and the corresponding of Eilenberg P systems are called *EOP systems* and *EOPP systems*. The definition and the behaviour of EOP and EOPP systems are very similar to those of EP and EPP systems, respectively. More precisely, the system will start in a given state and with an initial set of symbol objects. Given a state and a current multiset of symbol objects, in the case of EOP systems, the machine will evolve by applying rules associated with one of the transitions going out from the current state. The system will resume from the destination state of the current transition. In the parallel variant, instead of one state and a single multiset of symbol objects we may have a number of states, called *active states*, that are able to trigger outgoing transitions and such that each state hosts a different multiset of symbol objects; all the transitions emerging from every active state may be triggered once the rules associated with them may be applied; then the system will resume from the next states, which then become active states. EOP systems are models of cells evolving under various conditions when certain factors may inhibit some evolution rules or some catalysts may activate other rules. Both variants dealing with string objects and symbol objects have some similarities with the grammar systems controlled by graphs [4], replacing a one-level structure, which is the current sentential form, with a hierarchical structure defined by the membrane system. On the other hand, these variants of P systems may be viewed as Eilenberg machines [6] having sets of evolution rules as basic processing relationships. EP and EOP systems share some similar behaviour with Eilenberg machines based on distributed grammar systems [8].

Eilenberg machines, generally known under the name of X machines [6], have been initially used as a software specification language [9], further on intensively studied in connection with software testing [10]. Communicating X-machine systems were also considered [2] as a model of parallel and communicating processes.

In this paper it is investigated the power of EOP and EOPP systems in connection with three parameters: number of membranes, states and set of distributed rules. It is proven that the family of Parikh sets of vectors of numbers generated by EOP systems with one membrane, one state and one single set of rules coincides with the family of Parikh sets of context-free languages. The hierarchy collapses when at least one membrane, two states and four sets of rules are used and in this case a characterization of the family of Parikh sets of vectors associated with $ET0L$ languages is obtained. It is also shown that every EOP system may be simulated by an EOPP system and EOPP systems may be used for solving NP-complete problems. In particular linear time solutions are provided for the SAT problem. The last result relies heavily on similarities between EOPP systems and P systems with replicated rewriting [11] and EPP systems [1].

2 Definitions

Definition 1. *A* stream Eilenberg machine *is a tuple*

$$X = (\Sigma, \Gamma, Q, M, \Phi, F, I, T, m_0),$$

where:

- Σ *and* Γ *are finite sets called the* input *and the* output *alphabets, respectively;*
- Q *is the finite set of* states*;*
- M *is a (possibly infinite) set of* memory symbols*;*
- Φ *is a set of basic partial relations on* $\Sigma \times M \times M \times \Gamma^*$;
- F *is the* next state function $F : Q \times \Phi \to 2^Q$;
- I *and* T *are the sets of* initial *and* final *states;*
- m_0 *is the* initial memory *value.*

Definition 2. *An* EOP system *is a construct* $\Pi = (\mu, X)$*, where* μ *is a membrane structure consisting of* m *membranes, with the membranes and the regions labelled in a one to one manner with the elements* $1, \ldots, m$ *and an Eilenberg machine whose memory is defined by the regions* $1, \ldots, m$ *of* μ*. The Eilenberg machine is a system*

$$X = (V, Q, M_1, \ldots, M_m, \Phi, F, I),$$

having the following properties

- V *is the* alphabet *of the system;*
- Q, F *are as in Definition 1;*
- $M_1, \ldots, M_m$ *are finite multisets over* V *and represent the initial values occurring in the regions* $1, \ldots, m$ *of the system;*
- $\Phi = \{\phi_1, \ldots, \phi_p\}$, $\phi_i = (R_{i,1}, \ldots R_{i,m})$, $1 \leq i \leq p$ *and* $R_{i,j}$ *is a set of evolution rules (possibly empty) associated with region* j*, of the form* $X \to (u_1, tar_1) \ldots (u_h, tar_h)$*, with* X *a multiset over* V, $u_i \in V$, $tar_i \in \{here, out, in\}$, $1 \leq i \leq h$*; the indication here will be omitted.*
- $I = \{q_0\}$, $q_0 \in Q$ *is the initial state; all the states are final states (equivalent to* $Q = T$*).*

It may be observed that the set Σ and m_0 from Definition 1 are no longer used in the context of EOP systems. In fact, these concepts have been replaced by V and $M_1, \ldots, M_m$, respectively (all the symbols are output symbols, $\Gamma = V$).

A P system has m sets of evolution rules, each one associated with a region. An EOP system has the evolution rules distributed among p components ϕ_i, $1 \leq i \leq p$, each one containing m sets of evolution rules.

A computation in Π is defined as follows: it starts from the initial state q_0 and an initial configuration of the memory defined by $M_1, \ldots M_m$ and proceeds iteratively by applying in parallel rules in all regions, processing in each one all symbol objects that can be processed; in a given state q, each multiset that coincides with the left hand side of a rule is processed and the results are then distributed to various regions according to the target indications of that rule (for instance, when rewriting X by a rule $X \rightarrow (u_1, tar_1) \ldots (u_h, tar_h)$, the component of the multiset $u_1 \ldots u_h$ obtained will be send to the regions indicated by tar_i, $1 \leq i \leq h$ with the usual meaning in P systems (see [3], [13], [7])); the rules are from a component ϕ_i which is associated with one of the transitions emerging from the current state q and the resulting symbols constitute the new configuration of the membrane structure with the associated regions; the next state, belonging to $F(q, \phi_i)$, will be the target state of the selected transition. The result represent the number of symbols that are collected outside of the system at the end of a halting computation.

EOPP systems have the same underlying construct (μ, X), with the only difference that instead of one single membrane structure, it deals with a set of instances having the same organization (μ), but being distributed across the system. More precisely, these instances are associated with states called *active states;* these instances can divide up giving birth to more instances or collide into single elements depending on the current configuration of the active states and the general topology of the underlying machine. Initially only q_0 is an active state and the membrane configuration associated with q_0 is $M_1, \ldots, M_m$. All active states are processed in parallel in one step: all emerging transitions from these states are processed in parallel (and every single transition processes in parallel each string object in each region, if evolution rules match them).

Cell division: if q_j is one of the active states, $M_{j,1}, \ldots, M_{j,m}$ is its associated membrane configuration instance, and $\phi_{j,1}, \ldots, \phi_{j,t}$ are Φ's components associated with the emerging transitions from q_j, then the rules occurring in $\phi_{j,i}$, $1 \leq i \leq t$, are applied to the symbol objects from $M_{j,1}, \ldots, M_{j,m}$, the control passes onto $q_{j,1}, \ldots, q_{j,t}$, which are the target states of the transitions earlier nominated, with $M_{j,1,1}, \ldots, M_{j,m,1}, \ldots, M_{j,1,t}, \ldots, M_{j,m,t}$, their associated membrane configuration instances, obtained from $M_{j,1}, \ldots, M_{j,m}$, by applying rules of $\phi_{j,1}, \ldots, \phi_{j,t}$; the target states become active states, q is desactivated and $M_{j,1}, \ldots, M_{j,m}$ vanish. Only $\phi_{j,i}$ components that have rules matching the symbol objects of $M_{j,1}, \ldots, M_{j,m}$, are triggered and consequently only their target states become active and associated with memory instances $M_{j,1,i}, \ldots, M_{j,m,i}$. If none of $\phi_{j,i}$ is triggered, then in the next step q is desactivated and $M_{j,1}, \ldots, M_{j,m}$ vanish too. If some of $\phi_{j,i}$ are indicating the same component of Φ, then the corresponding memory configurations $M_{j,1,i}, \ldots, M_{j,m,i}$ are the same as well; this means that always identical transitions emerging from a state yield the same result.

Cell collision: if $\phi_1, \ldots, \phi_t$ enter the same state r and some or all of them emerge from active states, then the result associated with r is the union of membrane instances produced by those ϕ_i's emerging from active states and matching string objects from their membrane instances.

A computation of an EOP (EOPP) system halts when none of the rules associated with the transitions emerging from the current states (active states) may be applied.

If Π is an EOP system, then $Ps(\Pi)$ will denote the set of Parikh vectors of natural numbers computed as result of halting computations. The result of a halting computation in Π is the Parikh vector $\Psi_V(w) = (|w|_{a_1}, \ldots, |w|_{a_h})$, where w is the multiset formed by all the objects sent out of the system during the computation with $|w|_{a_i}$ denoting the number of a_i's occurring in w and $V = \{a_1, \ldots, a_n\}$.

The family of Parikh sets of vectors generated by EOP (EOPP) systems with at most m membranes, at most s states and using at most p sets of rules is denoted by $PsEOP_{m,s,p}(PsEOPP_{m,s,p})$. If one of these parameters is not bounded, then the corresponding subscript is replaced by $*$. The family of sets of vectors computed by P systems with non-cooperative rules in the basic model is denoted by $PsOP(ncoo)$ [13].

In what follows, we need the notion of an ET0L system, which is a construct $G = (V, T, w, P_1, \ldots, P_m)$, $m \geq 1$, where V is an alphabet, $T \subseteq V$, $w \in V^*$, and P_i, $1 \leq i \leq m$, are finite sets of rules (tables) of context-free rules over V of the form $a \to x$. In a derivation step, all the symbols present in the current sentential form are rewritten using one table. The language generated by G, denoted by $L(G)$, consists consists of all strings over T which can be generated in this way, starting from w. An ET0L system with only one table is called an E0L system. We denote by $E0L$ and $ET0L$ the families of languages generated by E0L systems and ET0L systems, respectively. Furthermore, we denote by by $PsE0L$ and $PsET0L$ the families of Parikh sets of vectors associated with languages in $ET0L$ and $E0L$, respectively. It is know that $PsCF \subset PsE0L \subset PsET0L \subset PsCS$. Details can be found in [14].

3 Computational Power of EP Systems and EPP Systems

We start by presenting some preliminary results concerning the hierarchy on the number of membranes and on the number of states.

Lemma 1. (i) $PsEOP_{1,1,1} = PsOP_1(ncoo) = PsCF$,
(ii) $PsEOP_{*,*,*} = PsEOP_{1,*,*}$,
(iii) $PsEOP_{1,*,*} = PsEOP_{1,2,*}$.

Proof. **(i)** EP systems with one membrane, one state and one set of rules are equivalent to P systems with non-cooperative rules in the basic model.

(ii) The hierarchy on the number of membranes collapses at level 1. The inclusion $PsEOP_{1,*,*} \subseteq PsEOP_{*,*,*}$ is obvious. For the opposite inclusion, the construction is nearly the same as those provided in [13] for the basic model of P systems. We associate to each symbol an index that represent the membrane where this object is placed and, when we move an object from a membrane to another one, we just change the corresponding index.

(iii) The hierarchy on the number of states collapses at level 2. The inclusion $PsEOP_{1,2,*} \subseteq PsEOP_{*,*,*}$ is obvious. On the other hand, consider an EP systems Π, with $Ps(\Pi) \in PsEOP_{1,s,p}$ for some $s \geq 3$, $p \geq 1$ (yet again, the cases $s = 1$ or $s = 2$ are not interesting at all), such that:

$$\Pi = ([_1\ ]_1, X),$$

where $X = (V, Q, M_1, \Phi, F, q_0)$, with $\Phi = \{\phi_1, \ldots, \phi_p\}$. We construct an EP systems Π', with $Ps(\Pi') \in PsEOP_{1,2,p+2}$ such that:

$$\Pi' = ([_1\ ]_1, X'),$$

where

$$X' = ((V \cup Q \cup \{\#, \epsilon\}), \{q, q'\}, M_1', \Phi', F', q),$$

with $q, q', \#, \epsilon \notin (V \cup Q)$ (i.e., $q, q', \#$ are new symbols that do not appear neither in V nor in Q),

$$M_1' = M_1' \cup \{q_0\}, \Phi' = \{\phi_1', \ldots, \phi_p', \phi_{p+1}, \phi_{p+2}\}.$$

For each $1 \leq j \leq p$, we have:

$$\phi_j' = (\phi_j \cup \{p \to q \,|\, F(\phi_j, p) = q\} \cup \{p \to \# \,|\, p \in Q\} \cup \{\# \to \#\}),$$

and $F'(\phi_p', q) = q$. Moreover, we have:

$$\phi_{p+1} = (\{p \to \epsilon \,|\, p \in Q\}),$$
$$\phi_{p+2} = (\{a \to \# \,|\, a \to v \in \phi_j, 1 \leq j \leq p\} \cup \{\# \to \#\}),$$

and $F' = (\phi_{p+1}, q) = q'$, $F' = (\phi_{p+2}, q') = q'$.
We have placed inside the skin membrane the initial state of the system Π. In general, we may suppose to have inside membrane 1 an object p that represent the current state of the state machine associated with the system Π. Thus, we apply the rules as in the system Π, by using some ϕ_j', and we change the state by using rule $p \to q$, if $F(\phi_j, p) = q$. At any moment, if we choose the wrong set of rules with respect to the current state (i.e., there does not exists any state q such that $F(\phi_j, p) = q$), then we are forced to apply a rule $p \to \#$, and, due to the rule $\# \to \#$, we generate an infinite computation. In order to finish a computation, we have to trigger on ϕ_{p+1}, which replaces the current state with ϵ and lead the system to the state q'. Here, if there are rules that can be still applied to the objects present inside, then an infinite computation is generated, as we can continue to use the rules inside membrane 1 ϕ_{p+2} forever.
It follows that $Ps(\Pi') = Ps(\Pi)$. □

Now, we are able to show the main result concerning the power of EP systems, which provides a characterization of the family of Parikh sets of vectors associated with $ET0L$ languages.

Theorem 1. $PsEOP_{1,2,*} = PsEOP_{1,2,4} = PsET0L$.

Proof. **(i)** $PsET0L \subseteq PsEOP_{1,2,4}$. According to Theorem 1.3 in [14], for each language $L \in ET0L$ there is an ET0L system that generates L and contains only two tables, that is, $G = (V, T, w, P_1, P_2)$. Therefore, we can construct an EP system

$$\Pi = ([_1\,]_1, X),$$

where

$$X' = ((V \cup T \cup \{\#, f, f'\}), \{q, q'\}, M_1, \Phi, F, q),$$

with $q, q', \#, f \notin (V \cup T)$ (i.e., $q, q', \#, f, f'$ are new symbols that do not appear in V or in T),

$$M_1 = w \cup \{f\}, \Phi = \{\phi_1, \phi_2, \phi_3, \phi_4\}.$$

We have:

$$\phi_1 = (P_1),$$
$$\phi_2 = (P_2),$$
$$\phi_3 = (\{f \to f'\}),$$
$$\phi_4 = (\{a \to \# \,|\, a \in (V - T)\} \cup \{a \to (a, out) \,|\, a \in T\}\{\# \to \#\}),$$

and $F = (\phi_1, q) = q$, $F = (\phi_2, q) = q$, $F = (\phi_3, q) = q'$, $F = (\phi_4, q') = q'$. The EP system Π works as follows. We start in the state q; here, we can use either ϕ_1 or ϕ_2 as many times as we want in order to simulate either the application of the rules in P_1 or the application of the rules in P_2. At any moment, we can decide to move from state q to state q' by triggering ϕ_3. In q', we use ϕ_4 in order to replace each non-terminal symbol with # and send out of the system each terminal symbol. In this way, if membrane 1 contains only terminal symbols, the computation halts successfully, otherwise we generate an infinite computation that yields no result. Thus, it is easy to see that $Ps(\Pi) = Ps(L(G))$. Furthermore, we have $Ps(Pi) \in PsEOP_{1,2,4}$.

(ii) Consider an EP systems Π, with $Ps(\Pi) \in PsEOP_{1,2,p}$ for some $s \geq 3$, $p \geq 1$ (yet again, the cases $s = 1$ or $s = 2$ are not interesting at all), such that

$$\Pi = ([_1\,]_1, X),$$

where $X = (V, Q, M_1, \Phi, F, q_0)$, with $\Phi = (\phi_1, \ldots, \phi_p)$. Thus, we construct an ET0L systems

$$G = ((V \cup Q \cup \cup\{\bar{a} \,|\, a V\} \cup \{\#\}), V, \bar{M_1} q_0, P_1, \ldots, P_p, P_{p+1}),$$

where $\bar{M_1} q_0$ denotes a string containing symbols $\bar{a}$ for every $a \in M_1$ and for each $1 \leq j \leq p$, P_j is a table that contains:

- a rule $\bar{a} \to v'$, for each rule $a \to v \in \phi_p$, with v' a string obtained from v as follows: it contains a symbol b if $(b, out) \subseteq v$, a symbol $\bar{b}$, if $b \subseteq v$ and there exists a rule $b \to u \in \phi_i$, $1 \leq i \leq p$, and λ (i.e., no symbol), for each $b \subseteq v$ such that there does not exist any rule $b \to u \in \phi_i$, $1 \leq i \leq p$;
- a rule $p \to q$, for each $p, q \in Q$ such that $F(\phi_j, p) = q$;
- a rule $p \to \#$, for each $p \in Q$.

Moreover, we have:

$$P_{p+1} = \{p \to \lambda \,|\, p \in Q\} \cup \{\bar{a} \to \#, a \to a \,|\, a \in V\}.$$

Now, it is easy to see that the ET0L system simulates the EP P system Π correctly. We start with a string $\bar{M_1}q_0$; we apply the tables $P_1, \ldots, P_p$ in order to simulate the application of the rules of Π an the corresponding transitions in the underlying state machine. At any moment, if we choose the wrong tables with respect to the current states, then we are forced to introduce in the string the non-terminal #, which cannot be replaced by any rule anymore. Finally, we have the P_{p+1} that is used to simulate the end of a successful computation in Π: if we use this table when there still exist some rules that can be applied to the symbols present in the current configuration, then we introduce the non-terminal # which cannot be removed from the string anymore; otherwise, we get a terminal string. □

As an E0L system is a an ET0L system with only one table, we get immediately the following result.

Corollary 1. $PsE0L \subseteq PsEOP_{1,2,3}$.

EPP systems exhibit a parallel behaviour not only inside of the membrane structure but also at the underlying machine level. Potentially, all transitions emerging from active states may be triggered in one step giving birth to new cells or colliding others. One problem addressed in this case is also related to the power of these mechanisms. In the sequel we will show that EPP systems are at least as powerful as EP systems.

Theorem 2. *If Π is an EP system with m membranes, s states and p sets of rules then there exists Π an EPP system with $m' \geq m$ membranes, $s' \geq s$ states and $p' \geq p$ rule transitions such that $Ps(\Pi) = Ps(\Pi)$.*

Proof. Let $\Pi = (\mu, X)$, be an EP system where μ is a membrane structure consisting of m membranes, and X an Eilenberg machine

$$X = (V, \Gamma, Q, M_1, \ldots M_m, \Phi, F, I),$$

where Q has s states and Φ contains p components. The following EPP system is built $\Pi' = (\mu, X')$, where

$$X' = (V', \Gamma, Q', M_1', \ldots M_m', \Phi', F', I),$$

with

- $V' = V \cup \{x\} \cup \{k \mid 1 \leq k \leq t\}$, where t is the maximum number of transitions going out from every state of X;
- $Q' = Q \cup \{q_{j,0} \mid q_j \in Q\} \cup \{q_{j,k} \mid q_j \in Q, 1 \leq k \leq t,\}$;
- $M_1' = M_1 \cup \{x\}; M_j' = M_j, 2 \leq j \leq m$;

- $\Phi' = \Phi \cup \{\phi_x, \phi_{1x}, \ldots, \phi_{tx}\}$, where
  - $\phi_x = (\{x \to k \mid 1 \leq k \leq t\}, \emptyset, \ldots, \emptyset)$,
  - $\phi_{kx} = (\{k \to x\}, \emptyset, \ldots, \emptyset), 1 \leq k \leq t$,
- for any $q_j \in Q$ if there are $1 \leq u \leq t$, transitions emerging from q_j and $F(q_j, \phi_{j,k}) = \{q_{j_k}\}$, $1 \leq k \leq u$ (not all $\phi_{j,k}$ are supposed to be distinct) then the following transitions are built in $\Pi\Pi$:
  $F'(q_j, \phi_x) = \{q_{j,0}\}$, $F'(q_{j,0}, \phi_{kx}) = \{q_{j,k}\}$, $1 \leq k \leq u$,
  $F'(q_{j,k}, \phi_{j,k}) = \{q_{j_k}\}$.

A computation in Π' proceeds as follows: at the beginning only the initial state is active and the memory configuration in this state is $M_1', \ldots, M_m'$. If the EP system Π is in a state q_j and the memory configuration is $M_{j,1}, \ldots, M_{j,m}$, then Π' must be in q_j as well. We will show that Π' has always only one active state. Indeed, if q_j is an active state in Π' and $M_{j,1}, \ldots, M_{j,m}$ are its associated membrane configuration, then in one step x from $M_{j,1}$ is changed by ϕ_x into k, a value between 1 and t; if u is the number of emerging transitions from q_j in Π, then $k > u$ implies that in the next step the current membrane configuration will vanish as no more continuation is then allowed from $q_{j,0}$; otherwise, when $1 \leq k \leq u$, only one transition may be triggered from $q_{j,0}$ and this is associated with ϕ_{kx} which restores x back into $M_{j,1}$ (the other transitions emerging from $q_{j,0}$ cannot be triggered). ϕ_{kx} leads the EPP system into $q_{j,k,1}$. From this state there are two transitions both associated with $\Phi_{j,k}$ that are triggered and consequently $M_{j,1}, \ldots, M_{j,m}$ are processed yielding $M_{j,1}', \ldots, M_{j,m}'$ and some symbol objects may be sent out of the system. In every step only one state is active. In this way Π and Π' compute the same objects, thus $Ps(\Pi) = Ps(\Pi')$. □

Note 1. From Theorem 2 it follows that if the EP system Π has m membranes, s states, p components of Φ, and the maximum number of transitions emerging from every state is t then the equivalent EPP system has $m' = m$ membranes, at most $s' = (1 + t)s$ states, and at most $p' = p + t + 1$ sets of rules.

4 Linear Solution to SAT Problem

SAT (satisfiability of propositional formulae in conjunctive normal form) is a well known NP-complete problem. This problem asks whether or not, for a given formula in the conjunctive normal form, there is a truth-assignment of the variables such that it becomes true. So far some methods to solve in polynomial or just linear time this problem have been indicated, but at the expense of using an exponential space of values. In [1] the problem has been solved in time linear in the number of variables and the number of clauses by using EPP systems. It is worth mentioning that in [1] the method used relied essentially on string objects dealt with by EPP systems. In the sequel we will show that SAT problem may be solved in linear Time by using EOPP systems.

Theorem 3. *The SAT problem can be solved in a time linear in the number of variables and the number of clauses by using an EOPP system.*

Proof. Let γ be a formula in conjunctive normal form with m clauses, $C_1, \ldots, C_m$, each one being a disjunction, and the variables used are $x_1, \ldots, x_n$. The following EOPP system, $\Pi = (\mu, X)$, may be then constructed:

$$\mu = [_1[_2 \ldots [_{m+1}]_{m+1} \ldots]_2]_1,$$
$$X = (V, Q, M_1, \ldots M_{m+1}, \Phi, F, I),$$

where:

- $V = \{(i_1, \ldots, i_k) \mid i_j \in \{0, 1\}, 1 \leq j \leq k, 1 \leq k \leq n\}$;
- $Q = \{q\}$;
- $M_1 = \ldots = M_m = \emptyset$, $M_{m+1} = \{(0), (1)\}$;
- $\Phi = \{\phi_0, \phi_1, \phi_2\}$;
  - $\phi_0 = (\emptyset, \ldots, \emptyset, \{(i_1, \ldots, i_k) \rightarrow (i_1, \ldots, i_k, 0) \mid$
    $i_j \in \{0, 1\}, 1 \leq j \leq k, 1 \leq k \leq n-1\})$,
  - $\phi_1 = (\emptyset, \ldots, \emptyset, \{(i_1, \ldots, i_k) \rightarrow (i_1, \ldots, i_k, 1) \mid$
    $i_j \in \{0, 1\}, 1 \leq j \leq k, 1 \leq k \leq n-1\})$,
  - $\phi_2 = (\{(i_1, \ldots, i_j = 1, \ldots, i_n) \rightarrow ((i_1, \ldots, i_j = 1, \ldots, i_n), out) \mid$
    x_j is present in $C_1, 1 \leq j \leq n\} \cup$
    $\{(i_1, \ldots, i_j = 0, \ldots, i_n) \rightarrow ((i_1, \ldots, i_j = 0, \ldots, i_n), out) \mid$
    $\neg x_j$ is present in $C_1, 1 \leq j \leq n\}$,
    $\ldots$,
    $\{(i_1, \ldots, i_j = 1, \ldots, i_n) \rightarrow ((i_1, \ldots, i_j = 1, \ldots, i_n), out) \mid$
    x_j is present in $C_m, 1 \leq j \leq n\} \cup$
    $\{(i_1, \ldots, i_j = 0, \ldots, i_n) \rightarrow ((i_1, \ldots, i_j = 0, \ldots, i_n), out) \mid$
    $\neg x_j$ is present in $C_m, 1 \leq j \leq n\}$,
    $\emptyset)$;
    where $(i_1, \ldots, i_j = 1, \ldots, i_n)$ and $(i_1, \ldots, i_j = 0, \ldots, i_n)$ denote elements of V having the j−th component equal to 1 and 0, respectively;
- $F(q, \phi_k) = \{q\}$, $0 \leq k \leq 2$;
- $I = \{q\}$.

The system Π starts from state q with $\emptyset, \ldots, \emptyset, \{(0), (1)\}$. By applying n times ϕ_0 and ϕ_1 in parallel one generates all truth values for the n variables in the form of 2^n symbols $(i_1, \ldots, i_n)$ with $i_j = 1$ or $i_j = 0$ indicating that variable x_j is either *true* or *false*. All these combinations are obtained in n steps in state q. In the next m steps ϕ_2 checks whether or not at least one truth-assignment satisfies all clause; this, if exists, will exit the system. The SAT problem is solved in this way in $n + m$ steps. □

There are some important similarities between the above theorem and Theorem 5 in [1]:

- the same membrane structure;
- the first m initial regions empty;
- the truth and false values introduced in parallel;

but also relevant distinct features:

- less states and a simpler definition of F in the above theorem;
- linear (in n) number of symbols in V and rules, in the case of Theorem 5 [1], but exponential number of corresponding components, in the above theorem.

5 Conclusions

In this paper two types of Eilenberg P systems, namely EOP systems and EOPP systems, have been introduced. They combine the control structure of an Eilenberg machine as a driven mechanism of the computation with a cell-like structure having a hierarchical organisation of the objects involved in the computational process. The computational power of EOP systems is investigated in respect of three parameters: number of membranes, number of states, and number of sets of rules.

It is proven that the family of Parikh sets of vectors of numbers generated by EOP systems with one membrane, one state and one single set of rules coincides with the family of Parikh sets of context-free languages. The hierarchy collapses when at least one membrane, two states and four sets of rules are used and in this case a characterization of the family of Parikh sets of vectors associated with ET0L languages is obtained. It is also shown that every EOP system may be simulated by an EOPP system.

EOPP systems represent the parallel counterpart of EOP systems, allowing not only the rules inside of the cell-like structure to develop in parallel, but also the transitions emerging from the same state. More than this, all states that are reached during the computation process as target states, may trigger in the next step all transitions emerging from them. It is shown that a general method to simulate an EOP system as an EOPP system may be stated. Apart from the fact that EOPP systems might describe interesting biological phenomena like cell division and collision, it is also an effective mechanism for solving NP-complete problems, like SAT, in linear time.

References

1. Bălănescu T, Gheorghe M, Holcombe M, Ipate, F (2003) Eilenberg P systems. In: Păun Gh, Rozenberg G, Salomaa A, Zandron C, (eds.), *Membrane Computing. International Workshop*, WMC-CdeA 02, Curtea de Arges, Romania, August 19-23, 2002. LNCS 2597, Springer, Berlin, 43–57
2. Bălănescu T, Cowling T, Georgescu H, Gheorghe M, Holcombe M, Vertan C (1997) Communicating stream X-machines systems are no more than X-machines, *J. Universal Comp. Sci.* 5:494–507
3. Calude C, Păun Gh (2000) *Computing with Cells and Atoms.* Taylor and Francis, London
4. Csuhaj-Varju E, Dassow J, Kelemen J, Păun Gh (1994) *Grammar Systems. A Grammatical Approach to Distribution and Cooperation.* Gordon & Breach, London
5. Dassow J, Păun Gh (1989) *Regulated Rewriting in Formal Language Theory.* Springer, Berlin
6. Eilenberg S (1974) *Automata, Languages and Machines.* Academic Press, New York
7. Ferretti C, Mauri G, Păun Gh, Zandron C (2001) On three variants of rewriting P systems. In: Martin-Vide C, Păun Gh (eds.), *Pre-proceedings of Workshop on Membrane Computing* (WMC-CdeA2001), Curtea de Argeş, Romania, August 2001, 63–76, and *Theor. Comp. Sci.* (to appear)

8. Gheorghe M (2000) Generalised stream X-machines and cooperating grammar systems, *Formal Aspects of Computing* 12:459-472
9. Holcombe M (1998) X-machines as a basis for dynamic system specification, *Software Engineering Journal* 3:69-76
10. Holcombe M, Ipate F (1998) *Correct Systems Building a Business Process Solution.* Springer, Berlin, Applied Computing Series
11. Krishna SN, Rama R (2001) P systems with replicated rewriting, *Journal of Automata, Languages and Combinatorics* 6:345-350
12. Păun Gh (1998) Computing with membranes, *Journal of Computer System Sciences* 61:108–143 (and Turku Center for Computer Science (1998) TUCS Report 208, http://www.tucs.fi)
13. Păun Gh (2002) *Membrane computing. An Introduction*, Springer, Berlin
14. Rozenberg G, Salomaa A (1980) *The Mathematical Theory of L Systems*, Academic Press. New York

Molecular Tiling and DNA Self-assembly

Alessandra Carbone[1] and Nadrian C. Seeman[2]

[1] Institut des Hautes Études Scientifiques
35, route de Chartres, 91440 Bures-sur-Yvette, France
carbone@ihes.fr
[2] Department of Chemistry, New York University
New York, NY 10003, USA
ned.seeman@nyu.edu

Abstract. We examine hypotheses coming from the physical world and address new mathematical issues on tiling. We hope to bring to the attention of mathematicians the way that chemists use tiling in nanotechnology, where the aim is to propose building blocks and experimental protocols suitable for the construction of 1D, 2D and 3D macromolecular assembly. We shall especially concentrate on DNA nanotechnology, which has been demonstrated in recent years to be the most effective programmable self-assembly system. Here, the controlled construction of supramolecular assemblies containing components of fixed sizes and shapes is the principal objective. We shall spell out the algorithmic properties and combinatorial constraints of "physical protocols", to bring the working hypotheses of chemists closer to a mathematical formulation.

1 Introduction to Molecular Self-assembly

Molecular self-assembly is the spontaneous organisation of molecules under thermodynamic equilibrium conditions into a structurally well-defined and rather stable arrangement through a number of non-covalent interactions [5,26,52]. It should not be forgotten that periodic self-assemblies of molecules lead to crystals in one, two or three dimensions; we often do not understand the interactions between the constituents of a crystal, but their presence in our world was an existence-proof for 3D self-assembly long before the notion was voiced. By a *non-covalent* interaction, we mean the formation of several non-covalent weak chemical bonds between molecules, including hydrogen bonds, ionic bonds and van der Waals interactions. These interactions (of the order of 1-5 kcal/mol) can be considered *reversible* at normal temperatures, while covalent interactions (typically $>$ 50 kcal/mol) are regarded as *irreversible.*

The self-association process leads the molecules to form stable hierarchical macroscopic structures. Even if the bonds themselves are rather weak, their collective interaction often results in very stable assemblies; think, for example, of an ice cube, held together by hydrogen bonds. Two important elements of molecular self-assembly are *complementarity* and *self-stability*, where both the size and the correct orientation of the molecules are crucial in order to have a complementary and compatible fitting.

N. Jonoska et al. (Eds.): Molecular Computing (Head Festschrift), LNCS 2950, pp. 61–83, 2004.

The key engineering principle for molecular self-assembly is to design molecular building blocks that are able to undergo spontaneous stepwise interactions so that they self-assemble via weak bonding. This design is a type of "chemical programming", where the instructions are incorporated into the covalent structural framework of each molecular component, and where the running of the algorithm is based on the specific interaction patterns taking place among the molecules, their environment, and the intermediate stages of the assembly. The aim of the game is to induce and direct a controlled process.

Molecular self-assembly design is an art and to select from the vast virtual combinatorial library of alternatives is far from being an automatic task [19]. There are principles though, that could be mathematically analyzed and one of the purposes of this paper is to lead the reader towards such possibilities. We shall talk mainly about self-assembly from branched DNA-molecules, which in the last few years have produced considerable advances in the suggestion of potential biological materials for a wide range of applications [39]. Other directions using peptides and phospholipids have been also pursued successfully [57,35,4].

We shall start with an abstract overview of some of the principles governing self-assembly which have been investigated by chemists (for an introduction see also [27]), with a special emphasis on DNA self-assembly. With the desire to formalise in an appropriate mathematical language such principles and develop a combinatorial theory of self-assembly, we try to suggest mathematical structures that arise naturally from physical examples. All through the paper, we support our formalistic choices with experimental observations. A number of combinatorial and algorithmic problems are proposed. The word "tile" is used throughout the paper in a broad sense, as a synonym of "molecule" or of "combinatorial building block" leading to some assembly.

2 Examples of Molecular Self-assembly and Scales

Self-assembled entities may be either discrete constructions, or extended assemblies, potentially infinite, and in practice may reach very large sizes. These assemblies include such species as 1-dimensional polymolecular chains and fibers, or 2-dimensional layers and membranes, or 3-dimensional solids. Due to the exceptionally complicated cellular environment, the interplay of the different ligand affinities and the inherent complexity of the building blocks, it is not easy to predict, control and re-program *cellular* components. Proteins can in principle be engineered but to predict protein conformation is far from our grasp nowadays. At the other extreme lie chemical assemblies, such as organic or inorganic crystals, which are constituted by much simpler structural components that are not easily programmed. Within this spectrum of assembly possibilities, DNA self-assembly has revealed itself as the most tractable example of programmable molecular assembly, due to the high specificity of intermolecular Watson-Crick base-pairing, combined with the known structure formed by the components when they associate [31]. This has been demonstrated in recent years both theoretically and experimentally as we shall discuss later.

3 Molecular Self-assembly Processes

There are three basic steps that define a process of molecular self-assembly:

1. *molecular recognition*: elementary molecules selectively bind to others;
2. *growth*: elementary molecules or intermediate assemblies are the building blocks that bind to each other following a sequential or hierarchical assembly; cooperativity and non-linear behavior often characterize this process;
3. *termination*: a built-in halting feature is required to specify the completion of the assembly. Without it, assemblies can potentially grow infinitely; in practice, their growth is interrupted by physical and/or environmental constraints.

Molecular self-assembly is a *time-dependent process* and because of this, temporal information and kinetic control may play a role in the process, before thermodynamic stability is reached. For example, in a recent algorithmic self-assembly simulating a circuit constituted by a sequence of XOR gates [30], a template describing the input for the circuit, assembled first from DNA tiles as the temperature was lowered, because these tiles were programmed to have stronger interactions; the individual tiles that performed the gating functions, i.e. the actual computation of each XOR gate, assembled on the template later (at a lower temperature), because they interacted more weakly. If, as in this example, the kinetic product is an intermediate located on the pathway towards the final product, such a process is *sequential.* If not, then the process is said to *bifurcate.*

Molecular self-assembly is also a *highly parallel process*, where many copies of different molecules bind simultaneously to form intermediate complexes. One might be seeking to construct many copies of the same complex at the same time, as in the assembly of periodic 1D or 2D arrays; alternatively, one might wish to assemble in parallel different molecules, as in DNA-based computation, where different assemblies are sought to test out the combinatorics of the problem [1,22]. A sequential (or *deterministic*) process is defined as a sequence of highly parallel instruction steps.

Programming a system that induces *strictly sequential assembly* might be achieved, depending on the sensitivity of the program to perturbations. In a robust system, the instructions (that is the coding of the molecular interactions) are strong enough to ensure the stability of the process against interfering interactions or against the modification of parameters. Sensitivity to perturbations limits the operational range, but on the other hand, it ensures control on the assembly.

An example of *strong instructions* is the "perfect" pairing of strands of different length in the assembly of DNA-tiles due to Watson-Crick interacting sequences. The drawback in sequential assembly of DNA-tiles is due to the complex combinatorics of sequences which are needed to construct objects with discrete asymmetric shapes, or aperiodic assemblies. The search for multiple sequences which pair in a controlled way and avoid unwanted interactions is far from being an obvious task. Alternative approaches concern besides tile design, self-assembly algorithms and protocols (Section 7).

A sequential process might either be *commutative*, if the order of the assembly steps can be interchanged along the pathway leading to the final assembly, or it might be non-commutative, if the intermediates need to interact in a fixed consecutive manner. DNA-based computations, such as the assembly of graphs [34] are commutative: a series of branched junctions can come together in any order to yield the final product (as discussed in Section 6 for 3-color graphs). An example of a non-commutative process is the construction of DNA tiles along the assembly of a periodic 2D array: single stranded DNA sequences are put in a pot at once, and since the tiles melt at a temperature higher than the intermolecular interactions, tiles are "prepared" first, before the 2D assembly takes place. Even if indirectly, these physical conditions imply non-commutativity. Later on, the 2D lattice can assemble with gaps that can later be filled in from the 3rd direction. Commutativity, in this latter step, may create irregularities when 3D arrays are considered instead, since gaps might get sealed in as a defect. Any hierarchical construction, such as solid-support-based DNA object synthesis [58] is non-commutative. Another example of a non-commutative assembly is a frame-based construction [32], wherein an assembly is templated by a "frame" that surrounds it: tiles assemble within the boundaries of the frame and they are guided by the code of the tiles forming the frame. It is non-commutative, in that the frame has to be available first.

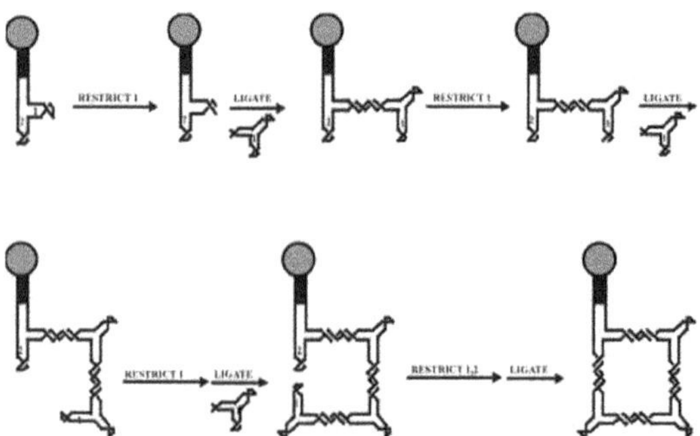

Fig. 1. *Protocol for the Synthesis of a Quadrilateral.* The intermolecular additions of corners is repetitive, but a different route leads to intramolecular closure.

Another characteristic of a molecular self-assembly is that the *hierarchical build-up* of complex assemblies, allows one to intervene at each step, either to suppress the following one, or to orient the system towards a different pathway. For example, the construction of a square from identical units using the solid-support method entailed the same procedures to produce an object with 2, 3, or 4 corners. Once the fourth corner was in place, a different pathway was taken to close the square [58], as shown in Figure 1. A pentagon, hexagon or higher

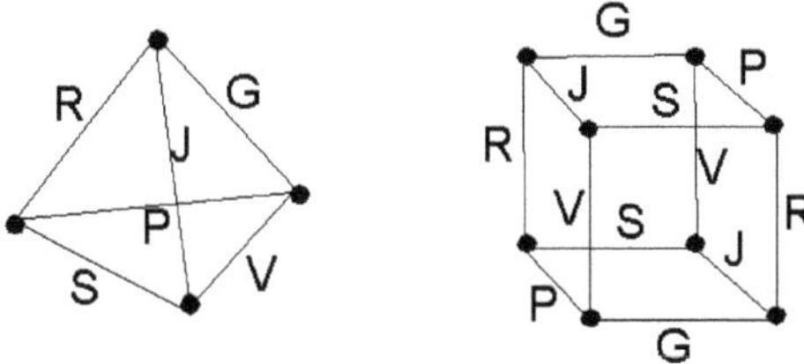

Fig. 2. Triplet junctions GPV, JRV, JGS and PSR can combine in different configurations. The two smallest ones are a tetrahedron and a cube.

polygon could have been made by the same procedure, just by choosing to close it after adding more units.

Instructions might be strong but still allow for different objects to appear. The same set of tiles might assemble into objects with different geometrical shapes and different sizes, that satisfy the same designed combinatorial coding. For instance, consider chemical "three-arm junction nodes" (name them GPV, JRV, JGS, PSR) accepting 6 kinds of "rods", called G, P, V, J, S and R. Several geometrical shapes can be generated from these junctions and rods, in such a way that all junctions in a node are occupied by some rod. Two such shapes are illustrated in Figure 2. In general, there is no way to prevent a given set of strands from forming dimers, trimers, etc. Dimers are bigger than monomers, trimers bigger than dimers, and so on, and size is an easy property for which to screen. However, as a practical matter, entropy will favor the species with the smallest number of components, such as the tetrahedral graph in Figure 2; it can be selected by decreasing the concentration of the solution. If, under convenient conditions, a variety of products results from a non-covalent self-assembly, it is possible to obtain only one of them by converting the non-covalent self-assembly to a covalent process (e.g., [16]). Selecting for specific shapes having the same number of monomers though, might be difficult. It is a combinatorial question to design a coding for a set of tiles of fixed shape, that gives rise to an easily screenable solution set.

4 Molecular Tiling: A Mathematical Formulation

Attempts to describe molecular assembly, and in particular DNA self-assembly, in mathematical terms have been made in [2,6]. Here, we discuss some algorithmic and combinatorial aspects of self-assembly keeping in mind the physics behind the process.

Tiles and self-assembly. Consider a connected subset T (tile) in $\mathbb{R}^3$, for example a convex polyhedron, with a distinguishable subset of mutually complementary (possibly overlapping) non-empty domains on the boundary, denoted $D_b, D'_b \subset \partial T$, where b runs over a (possibly infinite) set B. We are interested in assemblies generated by T, that are subsets A in the Euclidean space, decomposed into a

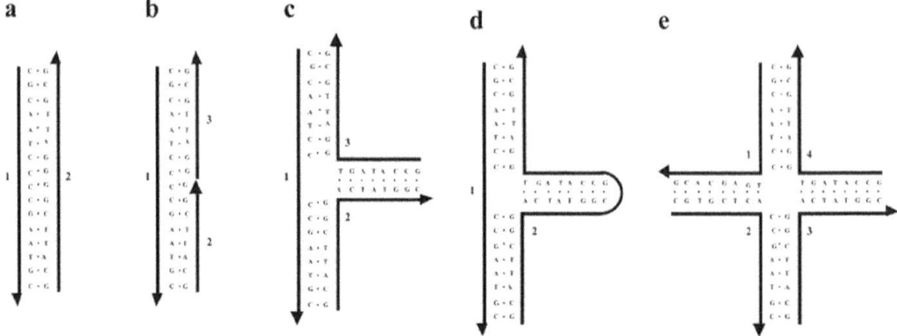

Fig. 3. *A Variety of Complements to a Single Strand.* Panel (a) illustrates a conventional Watson-Crick duplex, where strand 2 complements strand 1. Panels (b-e) illustrates a variety of complements to strand 1.

union of congruent copies of T, where two copies may intersect only at their boundaries and have a "tendency" to meet across complementary domains on the boundary. It is important to recognize that in the case of DNA, there are many forms of complementarity, as a function of motif structure [41]. Figure 3 illustrates a DNA strand (named 1) complementary to a variety of other DNA strands; more complex types of complementarity exist, such as complementarity in the PX sense [60,46] or in the anti-junction sense [12,60].

We want to consider a biological macromolecule T (e.g., a protein or a nucleic acid motif), with complementary binding sites D_b, D'_b such that different copies of T bind along complementary domains and self-assemble into complexes. In the geometric context we specify the binding properties by introducing (binding) isometries $b : \mathbb{R}^3 \to \mathbb{R}^3$ to each $b \in B$ such that T and $b(T)$ intersect only at the boundary, and $b(D_b) = D'_b$. From now on B is understood as a subset in the Euclidean isometry group $Iso(\mathbb{R}^3)$.

Accordingly, we define an *assembly A associated with* (T, B) by the following data:

1. a connected graph $G = G_A$ with the vertex set $1 \dots N$,
2. subsets T_i in $\mathbb{R}^3$, where $i = 1 \dots N$, which may mutually intersect only at their boundaries,
3. an isometry $b_{k,l} : \mathbb{R}^3 \to \mathbb{R}^3$ moving T_k onto T_l, for each edge (k, l) in G, such that there exists an isometry $a_{k,l}$ which moves T_k to T and conjugates $b_{k,l}$ to some binding isometry b in B. Notice that this b is uniquely determined by $b_{k,l}$ up to conjugation.

Given a graph G_A and a tile T, the assemblies described by G_A and T might not be unique. The assembly is unambiguously described by the isometries associated to the edges of G_A (i.e. condition (3) above). See Figure 4 for an example.

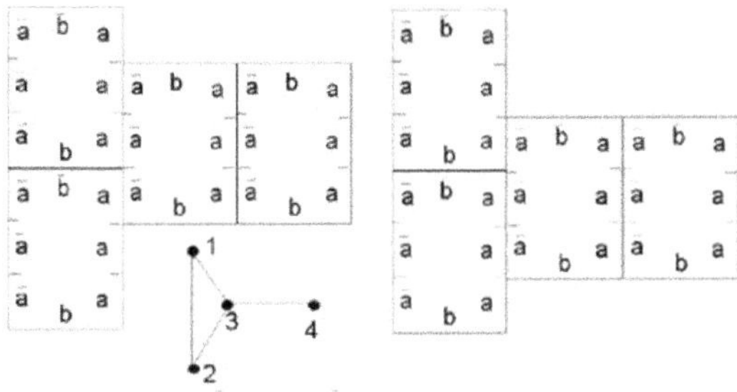

Fig. 4. Four copies of the same tile are arranged in two different assemblies that correspond to the same graph G_A. The labels $a, \bar{a}, b, \bar{b}$ correspond to codes for edges.

Several tiles. If we start with several different tiles $T^1, \ldots, T^n$ rather than with a single T, we consider the sets of pairs of binding isometries $B^{i,j} \subset Iso(\mathbb{R}^3) \times Iso(\mathbb{R}^3)$ such that $b_1^{i,j}(T^i)$ and $b_2^{i,j}(T^j)$ intersect only at their boundaries and their intersection is non-empty. The definition of an assembly associated to $(\{T^i\}, \{B^{i,j}\})$ goes as above with the following modifications: the graph G has vertices colored by the index set $1 \ldots n$, the corresponding subsets in $\mathbb{R}^3$ are denoted T_k^i where $i = 1 \ldots n$ and $k = 1 \ldots N_i$, and finally, we forfeit the isometries $b_{k,l}$ and for each edge (k^i, l^j) we emphasize an isometry of $\mathbb{R}^3$ which moves T_k^i to $b_1^{i,j}(T^i)$ and T_l^j to $b_2^{i,j}(T^j)$.

In what follows, we refer to the union of tiles defined above, as an assembly.

Qualities of an assembly. The *tightness* of the tiling is one quality that chemists appreciate. This can be measured by the number of cycles in the graph G, or equivalently by the negative Euler characteristic of the graph.

The *imperfection* of a tiling is measured by the "unused" areas of the boundaries of the tiles. First define the *active domain* $\partial_{act}(T) \subset \partial T$ as the union of the intersections of ∂T with $b(T)$ for all $b \in B$. Then define the "unused boundary" $\partial_{un}(A = \cup T_i)$ as the union $\cup_{i=1}^{N} \partial_{act}(T_i)$ minus the union of the pairwise intersections $\cup_{(k,l) \in G} T_k \cap T_l$. An assembly is called *perfect* if the area of the imperfection equals zero. We say that an assembly contained in a *given* subset $X \subset \mathbb{R}^3$ is *perfect* with respect to ∂X, if $\partial_{un}(A) \subset \partial X$.

The *uniqueness* refers to the uniqueness of an assembly subject to some additional constraints. For example, given an $X \subset \mathbb{R}^3$, one asks first if X can be tiled by (T, B) and then asks for the uniqueness of such a tiling. We say that (T, B) generates an *unconditionally unique* assembly if every imperfect assembly uniquely extends to a perfect assembly.

The essential problem of tiling engineering is designing a relatively simple tile or a few such tiles which assemble with high quality into large and complicated subsets in $\mathbb{R}^3$. Here is a specific example for the unit sphere S^2 rather than S^3, where one uses the obvious extension of the notion of tilings to homogeneous spaces. Given $\epsilon, \delta > 0$, consider triangulations of the sphere into triangles Δ with $Diam(\Delta) \leq \epsilon$ and $area(\Delta) \geq \delta Diam^2(\Delta)$. It is easy to see that the number of mutually non-congruent triangles in such a triangulation, call it $n(\epsilon, \delta)$, goes to ∞ for $\epsilon \to 0$ and every fixed $\delta > 0$. The problem is to evaluate the asymptotic behavior of $n(\epsilon, \delta)$ for $\epsilon \to 0$ and for either a fixed δ or $\delta \to 0$.

Complementarity of the domains. Two tiles T_1 and T_2 have complementary sites, D_b^1, D_b^2, when they can bind along their boundaries to each other forming a connected subset of $\mathbb{R}^3$. In physical terms, the two overlapping parts D_b^1, D_b^2 can have complementary geometrical shape (e.g. think of the concave surface of a protein and of the convex surface of a ligand binding to it, much as a classical 'lock and key'), but might also correspond to Watson-Crick complementary sequences (e.g. $5' - ATTCGA - 3'$ and $3' - TAAGCT - 5'$, where A is complementary to T and C to G as discussed before; see Figure 3).

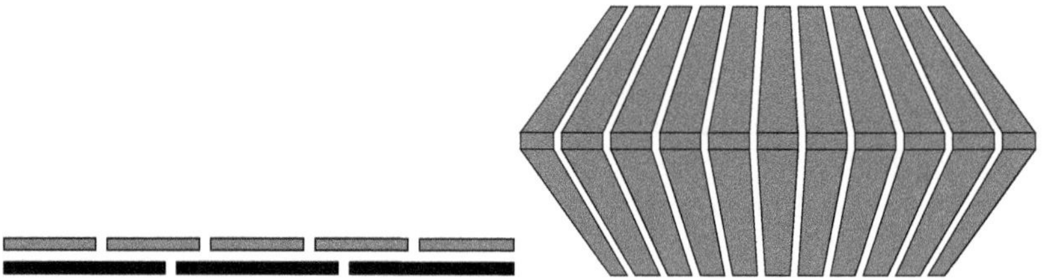

Fig. 5. Left: Rodlike tiles differing in length form an assembly that grows until the ends exactly match. Right: polymeric structure growing until the energy required to fit new subunits becomes too large.

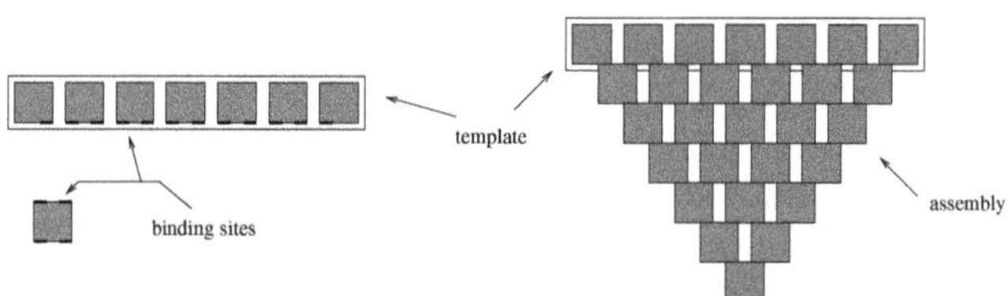

Fig. 6. A tile is stable in the assembly only if it binds at two adjacent binding sites. The stability of the whole assembly is insured by the enforced stability of the template. The formal description of this example is not completely captured by our model.

Real life examples. It remains unclear, in general, how cells control the size of (imperfect, with some unused boundary) assemblies, but certain mechanisms are understood. For example, out of two rod-like molecules of length three and five, one gets a double rod of length 15 as illustrated in Figure 5 (left). Another strategy is starting an assembly from a given template (see Figure 6 for a specific

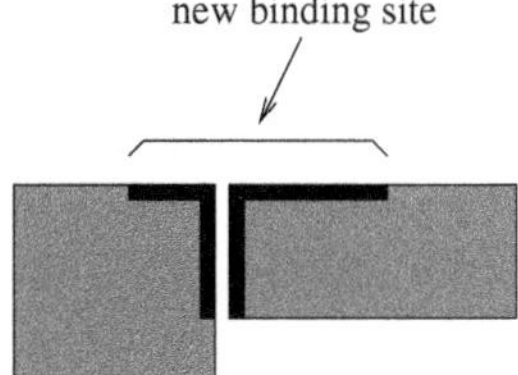

Fig. 7. Tiles which differ in shape and binding sites. Their binding generates a new contiguous binding site.

design). Sometimes, tiling is non-isometric: tiles distort slightly in order to fit, and the assembly terminates when the bending energy becomes too costly or when the accumulated distortion deforms and deactivates the binding sites (see Figure 5 (right)). Also, the binding of a ligand to an active site might change the shape of the molecule and thus influence the binding activity of other sites. Another possibility is the creation of a new binding site distributed over two or more tiles bound together at an earlier stage of the assembly (see Figure 7). These mechanisms may produce a non-trivial dynamics in the space of assemblies in the presence of free-energy. In particular, one may try to design a system which induces a periodic motion of a tile over a template, something in the spirit of RNA-polymerase cycling around a circle of DNA [11].

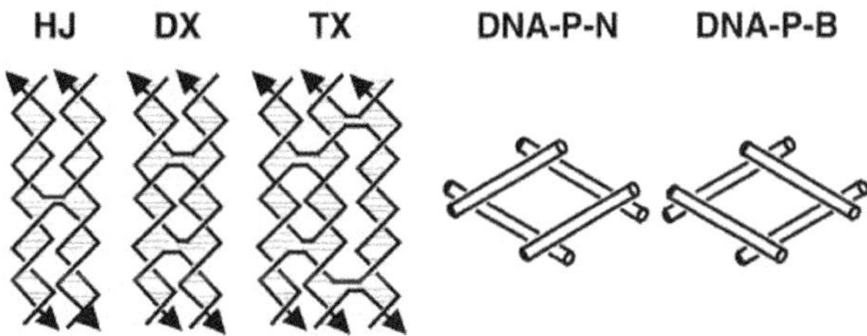

Fig. 8. Key Motifs in Structural DNA Nanotechnology. On the left is a Holliday junction (HJ), a 4-arm junction that results from a single reciprocal exchange between double helices. To its right is a double crossover (DX) molecule, resulting from a double exchange. To the right of the DX is a triple crossover (TX) molecule, that results from two successive double reciprocal exchanges. The HJ, the DX and the TX molecules all contain exchanges between strands of opposite polarity. To the right of the TX molecule are a pair of DNA parallelograms, *DNA-P-N* [29], constructed from normal DNA, and *DNA-P-B* [43], constructed from Bowtie junctions, containing 5', 5' and 3', 3' linkages in their crossover strands.

DNA tiles and tensegrity. Molecules of some nanometer size, made out of DNA strands, have been proposed in a variety of different shapes. See Figure 8 for a representative collection of shapes. Algorithms have been developed successfully to produce that self-assemble into these and other motifs [36]. *Branched molecules* are tiles constituted by several single strands which self-assemble along their coding sequences in a "star-like" configuration, where a tip of the star is a branch [36,51,38] (Figure 3a,c,e illustrate 2, 3 and 4-arm branched molecules). Theoretically, one might think to construct k-armed branched molecules, for any $k > 2$, where each strand is paired with two other strands to form a pair of double-helical arms; in practice, molecules with 6 arms have been reported, but larger ones are under construction. The angles between the arms are known to be flexible in most cases. If one adds *sticky ends* to a branched molecule, i.e. single stranded extensions of a double helix, a cluster is created that contains specifically addressable ends [36]. This idea is illustrated in Figure 9, where a 4-arm branched junction with two complementary pairs of sticky ends self-assembles to produce a quadrilateral.

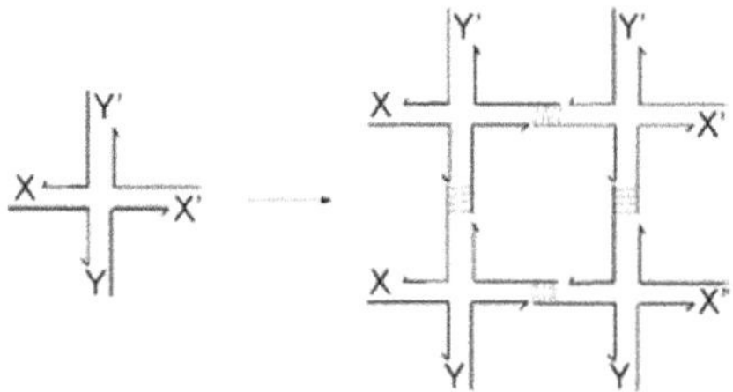

Fig. 9. Formation of a 2-dimensional lattice (right) from a 4-arm branched junction (left). X is a sticky end and X' is its complement. The same relationship holds for Y and Y'. X and Y are different from each other.

The search for motifs based on DNA branched junctions that behave as though they are "rigid" while in the test tube, led to the design of several DNA-molecules, and some are illustrated in Figure 8. Rigid shapes impose strong limitations on the design of suitable molecular tiles; roughly speaking, a *rigid*, or *tense*, object is a 3-dimensional solid that does not undergo deformations: we ask that if its 1-dimensional faces do not undergo deformation, then no deformation exists. For a tetrahedron or any convex deltahedron, it is easy to see that no change of the angles between edges (edges are 1-dimensional faces for the tetrahedron) can take place without the edges be deformed. On the other hand, a cube is an example of a non-tense object since we can fix the edges (1-faces) of the cube not to undergo any deformation and still be able to deform the angles between them.

Geometry of the boundaries: smooth deformations of tiles. It might be appropriate to consider assemblies which are affected by an ϵ-deformation in the shape of the tiles after binding. More precisely, a tile $T \subseteq \mathbb{R}^3$ is mapped in $\mathbb{R}^3$ by some *ϵ-deformation* as follows: there is an embedding $\epsilon \; : \; T \subseteq \mathbb{R}^3 \mapsto T' \subseteq \mathbb{R}^3$ such that for all points $x \in T$ there is a point $y \in T'$ such that the Euclidean distance

$d(x, y) < \epsilon$. The definitions of isometry and binding site given at the beginning of Section 4 need to be adjusted accordingly into new notions of ϵ-isometry and ϵ-binding site, which intuitively correspond to the original notions up to some ϵ-variation. One needs to establish whether an ϵ-deformation affects a binding site or not, and give thresholds on the amount of deformation which is accepted to affect non-empty domains of the boundary.

The growth of the assembly affected by ϵ-deformation asks for the estimation of bounds in the size of the construction. The instability of the system comes from a narrow range of conditions on which the assembly takes place. The formation of singularities and of bifurcation points between different assemblies, might lead to the disruption of the assembly, but might also lead to variety in the complexity.

Physical considerations on the shape of tiles. In addition to the need to observe appropriate solution conditions that encourage self-assembly, it is important to realize that there are physical constraints on the assembly of real tiles that do not affect virtual tiles. For example, the helicity of random-sequence DNA is ≈ 10.5 nucleotide pairs per turn in solution. This value makes it easy to make TX molecules (Figure 8) whose three helix axes form an angle of $120°$, but $90°$ is much harder, unless one is able (perhaps through sequence variation) to change the repeat to 10.4 nucleotide pairs per turn [45].

In a similar vein, a likely form of shaped 3D arrays will entail polyhedra whose edges contain DX molecules (Figure 8) [40]. It might appear that a tetrahedron would be a good polyhedron to use as the basis of such a 3D tile. However, although the edges of a tetrahedron obviously span 3-space, there is no group of three edges to which one can attach a single extra helix (i.e. to make those edges DX molecules instead of single DNA helices, with the extra helices outside the helices defining the tetrahedron) to produce the needed vectors: their diameters would cause them to clash stereochemically when extended beyond the boundaries of the tetrahedron. Notice that extra-hedral domains on adjacent edges inherently clash, and there is no group of three edges in a tetrahedron that does not include an adjacent pair.

5 An Abstract Model to Describe the Dynamics of Self-assembly

A formal description of the dynamics of a self-assembly on a space S, where S can be either $\mathbb{R}$, $\mathbb{R}^2$, $\mathbb{R}^3$ or any discrete approximation of those, can be formulated by a simple iterative process as follows. Consider n tiles $T_1, \ldots, T_n$, and take a finite number of copies of each T_i, for all $i = 1 \ldots n$. At stage 1, randomly assign to each physical tile a specific position in S in such a way that no two tiles overlap and that only tiles lying side-by-side and having complementary boundary stuck together. The set of complexes containing more than one tile with their position in S, define a *configuration* on S; single tiles are removed from S and used to re-iterate the random assignment on the next stage: the configuration of tiles lying in S which one reaches at stage i, is filled up further by new non-overlapping complementary adjacent tiles at stage $i + 1$. The process is repeated until all tiles are used or when a sufficiently large connected area in S is filled (e.g. area $> N$, for some large N).

Different outputs might result from this random process: they go from very tight assemblies, to assemblies with several unfilled regions, to disconnected surfaces, and so on. The resulting configurations and the time for reaching a configuration strongly depend on the coding hypothesis, e.g. whether new binding sites can appear or not by the combination of several tiles, whether "holes" can be filled or not, how many different competing boundary sites are in the system, how many tiles are in the system, whether connected regions can undergo translations in S while the process takes place, whether connected tiles might become disconnected, etc.

The process could start from a specific configuration of S instead of using the first iteration step to set a random configuration. Such an initial configuration, if connected, would play the role of a *template* for the random process described by the iteration steps. Figure 6 illustrates an example of templating, where a 1-dimensional array of n molecules disposed in a row is expected to play the role of a template and interact with single molecules following a schema coded by the boundary of the tiles. Another example is the "nano-frame" proposed in [32], a border template constraining the region of the tiling assemblies.

6 Closed Assemblies and Covering Graphs

Closed tiling systems are assemblies whose binding sites have been all used. More formally, this amounts to saying that the graph G_A underlying the assembly A is such that all tiles T_i corresponding to its vertices, where $i = 1 \dots N$, intersect on all their boundary sites T_i. This means also that the degree of connection of each node i of G_A corresponds to the number of available interaction sites of T_i, and that each edge (i, l) departing from i corresponds to an isometry $b_{i,l}$ moving T_i onto T_l which fixes some binding site. Many graphs might be locally embeddable in G_A, and we call them *covering graphs* of G_A: a graph G is a covering graph of G_A if there is a map $p : G \to G_A$ such that

1. p is surjective, i.e. for all nodes $y \in G_A$ there is a node $x \in G$ such that $p(x) = y$,
2. if $x \to y$ in G then $p(x) \to p(y)$ in G_A,
3. $degree(x) = degree(p(x))$, for all nodes x in G,
4. $b_{x,y} = b_{p(x),p(y)}$, for each edge $x \to y \in G$,
5. $\{b_{x,y} : x \to y \in G\} = \{b_{p(x),z} : p(x) \to z \in G_A\}$, for each node $x \in G$.

Condition (4), saying that the binding site between T_x and T_y is the same as the binding site between $T_{p(x)}$ and $T_{p(y)}$, and condition (5), saying that the binding sites of T_x are the same as the binding sites of $T_{p(x)}$, ensure that G and G_A underly tiling systems for the same set of tiles. The graph on the left hand side of Figure 10 does not satisfy (5) and provides a counterexample. If G_A represents a *closed* tiling system for a set of tiles $\mathcal{S}$, then each covering graph of G_A represents a closed tiling system for $\mathcal{S}$ also.

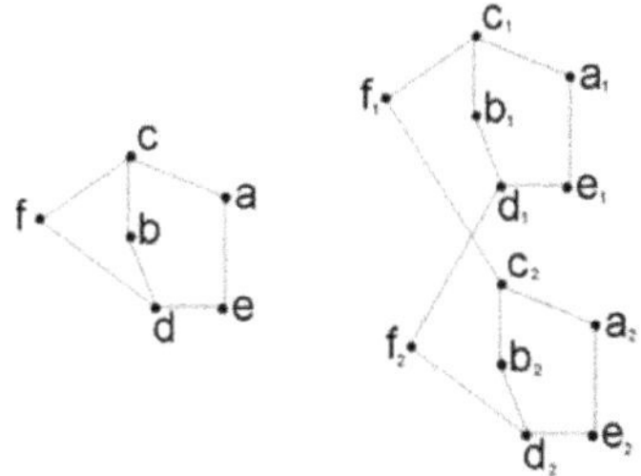

Fig. 10. Given a set of six distinct tiles whose binding sites are specific to each pair of tile interaction described by edges in the graph G_A (left), notice that G (right) is not a covering graph for G_A since it satisfies conditions (1) – (3) but it does not satisfy (5) (see text). To see this, consider the mapping p between nodes of G and G_A which is suggested by the labels of the nodes. We want to think of f in G_A as representing a tile T_f with two *distinct* binding sites, one interacting with T_c and the other with T_d. Node f_1 is linked to two copies of c and node f_2 is linked to two copies of d; this means that T_{f_1} (T_{f_2}), having the same binding sites as T_f, should bind to T_{c_1}, T_{c_2} (T_{d_1}, T_{d_2}). But this is impossible because the binding would require the existence of two *identical* sites in T_{f_1} (T_{f_2}).

Given a set of tiles one would like to characterize the family of closed assemblies, or equivalently, of covering graphs, if any. An important application is in the solution of combinatorial problems.

Example 1. [22]. A graph $G = (V, E)$ is said to be *3-colorable* if there is a surjective function $f : V \to \{a, b, c\}$ such that if $v \to w \in E$, then $f(v) \neq f(w)$. Imagine constructing the graph G with two kinds of molecules, one coding for the *nodes* and one for the *edges*. Node-molecules are branched molecules, where the number of branches is the degree of the node, and edge-molecules are two-branched molecules. Each branch of a node-molecule has a sticky end whose code contains information on the node of the graph that connects to it and on a color for the node. The n branches of a same node-molecule are assumed to have the same code. Edge-molecules have two sticky ends and their code contains information on the origin and target nodes as well as on the colors of such nodes. The two colors are supposed to be different.

To consider three colors in the physical realization of the graph G, one constructs a node-molecule for each one of the three colors, together with all possible combinations of pairs of different colors for edge-molecules.

By combining several identical copies of these molecules and ligating them, open and possibly closed assemblies will form. Open assemblies are discharged (this can be done with the help of exonuclease enzymes that digest molecules with free ends) and closed assemblies, if any, ensure that the graph is 3-colorable. The only closed assemblies that can be formed in the test tube are covering graphs.

7 A Random Model for Sequential Assembly

The random model introduced in Section 5 needs to be adjusted slightly to simulate *sequential assembly*. Sequentiality presupposes the formation of specific intermediates, i.e. complexes, at specific moments along the assembly process. This means that one can start from a random configuration in S, let the input tiles form complexes at random, remove from S isolated tiles and use them as input tiles to re-iterate the process until a sufficiently large number of specific intermediates is formed. This will provide one step of the sequential assembly, and in the *simplest* case, this step will be re-iterated to model the next steps of the sequential process, until all steps are realized. Different types of tiles might be used as input tiles to perform different steps of the sequential process.

In more *complicated* cases, the above model, might need to integrate new kinds of steps. It might be that some of the steps of the sequential process require the intervention of specific enzymes, cleaving or ligating DNA tiles. Such operations are random and their effect on tiles and complexes can be described rigorously. Also, one might need to consider that tiles forming a complex at step i, disassemble in step $i+1$ because of the interaction with new molecular tiles. This process is also random and can be formally described.

As mentioned in Section 3, the difficulty in inducing a sequential assembly comes from the complex combinatorics needed to realize objects of irregular but well-defined shape or aperiodic assemblies. A number of solutions have been proposed to overcome these combinatorial difficulties; they concern tile design (1)-(2), the algorithm for self-assembly (3)-(4) and the engineering protocol (5):

1. a variety of different forms of cohesion have been proposed, such as *sticky ended cohesion*, where single-stranded overhangs cohere between molecules [10]; *PX cohesion*, where topologically closed molecules cohere in a double-stranded interaction [60]; *edge-sharing*, where the interactions are characterized by lateral interactions [46]; *tecto-RNA*, where RNA domains pair laterally through loop osculations [21];
2. one can allow different forms of coding within the same molecule, which can involve the Watson-Crick sequences as well as the geometry of the molecule [9];
3. one can use "instructed gluing" elements together with DNA-tiles [7]; the idea is to add structural sub-units, as gluing elements between tiles, along the self-assembly process; in many cases, the use of such sub-units decreases the complexity of the protocol: the number of elementary molecules becomes smaller and the assembly algorithms becomes more specific;
4. the use of protecting groups, through which some of the potential interaction sites of the molecules are momentarily inhibited, is inherently a sequential protocol, applicable both to DNA objects [58] and to fractal assemblies [9];
5. the solid-support methodology in DNA nanotechnology [58] is an example of sequential assembly; it was used to construct the most complex DNA object to date, a truncated octahedron [59]; the step-wise synthesis of a square is illustrated in Figure 1 – here, enzymes intervene in some of the sequential steps.

8 Hierarchical Tiling

A set of tiles $\{T_1, \ldots, T_n\}$ is *self-consistent* if for each T_i with binding site D_i there is a tile T_j with binding site D_j such that $b(D_i) = D_j$, for some isometry b. Notice that i need not be different from j. In particular, a single tile T is *self-consistent with itself* if it has at least two binding sites which are complementary to each other.

Let $\{T_1, \ldots, T_n\}$ be basic elementary tiles which assemble in a variety of *tile complexes* $S_1, \ldots, S_l$, i.e. finite assemblies S_i with *unused* binding sites. A set of tile complexes is *self-consistent* if for each S_i with binding site D_i there is a tile complex S_j with binding site D_j such that $b(D_i) = D_j$, for some isometry b defined on tile complexes. New binding sites D_i generated from the assembly of a tile complex (as in Figure 7) are allowed.

A *hierarchical tiling* is an assembly X of tiles $\{T_1, \ldots, T_n\}$ that is obtained by successive steps of assembly generating intermediary sets of tile complexes $\mathcal{F}_0, \ldots, \mathcal{F}_m$ such that:

1. $\mathcal{F}_0 = \{T_1, \ldots, T_n\}$;
2. $\mathcal{F}_i = \{S_{i,1}, ..., S_{i,l_i}\}$, for $i = 1 \ldots m$, where each $S_{i,j}$ is a tile complex in $\mathcal{F}_{i-1}$;
3. $\mathcal{F}_i$ is a self-consistent set of tile complexes;
4. X is an assembly of $S_{m,1}, \ldots, S_{m,l_m}$.

The value of m is called *order* of the hierarchical tiling. A hierarchical tiling is *non-trivial* if for each family $\mathcal{F}_i$ there is at least one tile complex $S_{i,j}$ which is not in $\mathcal{F}_{i-1}$ already. Notice that not all assemblies can be defined as hierarchical assemblies of order m, for $m > 1$.

A dynamical model of hierarchical tiling. It can be defined by a repeated iteration of the random model for self-assembly presented in Section 5, where the tile complexes used as input tiles at step $i+1$ are the complexes formed in S at the end of step i. In general, a hierarchical assembly is *not* a sequential assembly. It might happen though, that certain assembly processes are defined by a combination of sequential steps during the hierarchical self-assembly.

Some concrete examples of hierarchical assembly. Suitable selection of structural units allows the design of molecular entities undergoing self-organisation into well-defined architectures, which subsequently may self-assemble into supramolecular *fibrils* and *networks*. The assembly of "infinite" *tubes* and *spheres* has been realized many times and in many laboratories with different kinds of molecules. A basic approach is to design a rod-like molecule with an hydrophobic end and a hydrophilic one. Then, one puts the molecules in different media and observes the formation of spheres, where the hydrophilic side of the molecules lies either inside or outside the sphere, depending on the properties of the medium. Alternatively, one might observe the formation of a long tube where, again, on the surface one finds sides with identical hydrophilic/hydrophobic properties. The formation of spheres and tubes leads us to ask how these shapes might assemble among themselves into supramolecular periodic or aperiodic structures. What other shapes do allow for the assembly of 1D, 2D and 3D arrays of such tile complexes?

Besides spheres, tubes and networks, chemists work on the design of synthetic molecules which lead to *helical* architectures of both molecular and supramolecular nature by hierarchical self-organization, or again to the formation of *mushroom-like* shapes and to a consequent assembly of such complexes into 3D arrays [18] (these arrangements are not *regular*, in the sense that they are not crystals). Mimicking nucleic-acid sequences, specific sequences of hydrogen bonding residues are led to act as structure-inducing codons, and such structural coding allows for the spontaneous but controlled generation of organized materials, e.g. [18].

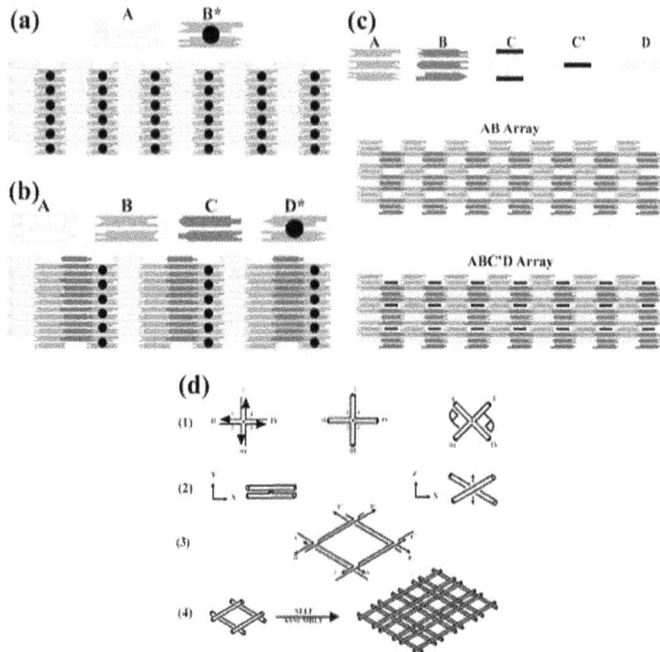

Fig. 11. *A variety of two-dimensional arrays that have been formed from DNA tiles.* Panels (a) and (b) illustrate 2D arrays composed of DX and $DX + J$ molecules. Panel (c) illustrates patterns obtained from TX molecules. Panel (d) illustrates an array made of DNA parallelograms.

At a different scale, for nanoscale molecules, like DNA, a broader range of possibilities can be explored since all of the contacts can be forced to be of a Watson-Crick form, although many other types of interaction are possible (e.g.,

[60]). The ifferent shapes of tiles introduced at the end of Section 4, enabled the assembly of several different kinds of periodic 1D and 2D arrays (see Figure 11). These hierarchical assemblies have order 2: single strands make the starting set of tiles, which assemble into specific intermediary molecular tiles (described in Section 4), and finally these molecular tiles self-assemble into a 2D array.

Periodic assemblies in 3 dimensions are still an open problem. Protocols for the assembly have been proposed, but highly ordered programmed 3D arrangements have not yet been realized to resolutions below 1 nm in the laboratory for DNA tiles. Aperiodic arrangements, typically harder to assemble and analyze than periodic assemblies, present an even greater challenge, because their characterization cannot rely on diffraction analysis in a simple fashion.

Example 2. Fractal assemblies. [8,9] Fractal constructions are a special case of aperiodic assemblies. The algorithm here is simple: from a starting molecular shape, which can look like a square or a triangle, and is designed to interact with copies of itself, one constructs a molecule with the same shape but a larger size, and re-iterates the process to get larger and larger assemblies of the given shape. The difficulty lies in the design of a set of basic shapes which can self-assemble into new self-similar shapes of larger sizes, and whose binding sites are coded by self-similar coding. An appropriate coding is important to ensure that tile complexes will self-assemble and that undesired binding is avoided. The order of this hierarchical tiling, corresponding to the number of iterations of the algorithm, is m, for potentially any value of m. In practice, a chemist would be happy with $m = 4, 5$.

These examples lead to some questions: within the set of feasible shapes and interactions, can we *classify* potential complexes? Once a complex is formed, can it be used as a building block to construct larger 1D, 2D or 3D arrays?

9 Size of the Assembly

How can the size of an assembly be controlled?

Rough termination is easy to induce. An obvious way is to limit the number of molecules in the solution. Another way is to use *protecting groups*, i.e. DNA molecules, which might be single strands for instance, whose binding sites are complementary to the binding sites of the tiles used in the assembly. The idea being that protecting groups might be added to the solution during the process of self-assembly to prevent new tiles from binding to available sites.

Exact termination is a consequence of the coding for the termination. If a synthesis or an assembly is templated, it is always possible to limit growth, by leaving out the constituent that is coded at the terminal position, for instance. The algorithmic synthesis of triangular circuits illustrated in Figure 6, provides another example where this is done [7]. In general, exact size control of a DNA self-assembly is hard to achieve. A few protocols have been presented so far.

In theory, DNA tiles can be used to "count" by creating boundaries with programmable sizes for 1D, 2D and possibly 3D periodic assemblies. The idea is to build periodic arrays of size $n \times m$ by generating repeatedly the Boolean truth table for n entries until m rows of the table have been filled [54,56]. If this schema can be physically implemented, then self-assembly of precisely-sized nanoscale arrays will be possible.

Fractal assemblies [8,9] can be thought of as a way to generate fixed geometrical shapes of controlled size. Besides the rectangular shapes of $n \times m$ arrays, one would like to have a way to grow assemblies with other shapes such as triangles, hexagons, etc. Fractal assembly allows us to do so by constructing objects with fixed sizes that are powers of some value: for instance, for the Sierpinski fractal, the size of the squares is 3^k, where k is the dimension.

10 Algorithmic Assembly

The combination of different instructions in a "molecular program" has been used to design self-assembly systems which follow specific assembly pathways. This idea has its mathematical analogue in the work of Wang [48,49,50], who proposed a finite set of tiles mimicking the behavior of any Turing Machine.

Wang tiles are squared tiles in $\mathbb{R}^2$ whose binding sites are the four sides of the square, and whose interaction is possible on binding sites labelled by the same color. If $T_1, T_2, \ldots, T_n$ are Wang tiles, then one asks that $\{T_1, T_2, \ldots, T_n\}$ be a self-consistent set. Once a set of Wang tiles is given, one asks whether the plane can be tiled with it, and what are the properties of the tiling, namely if the set generates periodic tiling only, or both periodic and non-periodic tiling, or aperiodic tiling only.

The molecular realization of Wang tiles (where a square becomes a 4-arm branched molecule with Watson-Crick complementary sticky ends as binding sites) can, theoretically, be used to make computations [54]. This notion has not yet been realized experimentally in more than one dimension [30]. A three-dimensional framework for computing 2D circuits and constructing DNA-objects with given shapes, has been suggested [7], where again, DNA tiles mimic Wang tiles. It is important to stress that molecular tiles are not conceived to generate only uniform tiling of the plane, but on the contrary, they can be used to induce the assembly of objects of arbitrary shapes.

Combinatorial optimisation problems: fixing a single shape. Two combinatorial problems have been stated in [3]. The first concerns minimum tile sets, i.e. given a shape, find the tile system with the minimum number of tile types that can uniquely self-assemble into this shape. The second concerns tile concentration, i.e. given a shape and a tile system that uniquely produces the given shape, assign concentrations to each tile-type so that the expected assembly time for the shape is minimized. The first combinatorial problem is NP-complete and the second is conjectured to be #P [3]. These problems have been formulated for any given shape even though only square tiles, i.e. Wang tiles, have been studied until now.

Templates and fixed shapes. Can one find a small set of relatively simple tiles such that, starting from a template supporting a linear code (that may be a DNA or RNA molecule incorporated into a macromolecular complex), the assembly process will create a given three dimensional shape in the space? We think here of interacting tiles performing a transformation from labeled templates into three dimensional structures and we ask what kind of transformations can be realized in this way [7]. Also, one wants to understand how much the complexity of the construction depends on the complexity of the tiles, where the latter can be measured by the number of the binding sites of the tiles, the size of the sets $B_{i,j}$, etc.

Combinatorial optimisation problems: fixing a "family" of shapes. Fractal assembly provides an example of an iterative algorithm for self-assembly which generates fractals of arbitrary dimension and not just a single shape with a given size. For each dimension, the building blocks necessary to build the corresponding fractal shape need to satisfy the same self-similar properties, and the design of a tile set which satisfies these properties is not obvious. For instance, given a Sierpinski square fractal and an iterative algorithm that produces arbitrarily large instances of this shape, is there a set of Wang tiles that can uniquely assemble into any fractal size? It is not at all clear that a set of Wang tiles with self-similar coding exists. In [9] a set of tiles, whose boundaries are characterized by both a coding sequence and a *geometrical* shape, is proposed. Does geometry have to be included in the coding of the tile boundaries to impose extra control on the assembly? What is the minimum number of tiles necessary to generate a *family* of shapes?

In general, let an algorithm for self-assembly be fixed. What are the properties of the tiles which are necessary to realize the algorithm?

Dynamic tiling. A molecular feature that has been used in algorithmic self-assembly is the possibility to program and change the status of a molecule. This means that the molecule passes in time through several possible physical conformations, i.e. geometrical shapes. In DNA nanotechnology, this has been done by using "template" molecules (programmable tiles) that interact with DNA single strands [47,46]: the pairing of the single stranded DNA present in the solution to a single strand subsequence of the tile induces this latter to change its conformation. Because of these conformational changes, tiles get a different status during the assembly, with the effect that one is able to control the dynamics of the algorithm and the direction of the assembly. As a result, one can generate different architectures out of the same set of tiles by varying their conformations.

Example 3. One can imagine a basic molecular system that is fundamentally a layer of *programmable* tiles which can guide the assembly of multiple layers of tiles above it [7]. In the 2-dimensional case this device can compute tree-like boolean circuits, and in 3D, it can induce finite regular and less regular shapes. *Multiple regular layers* are obtained by completely filling up the template board: a new layer of tiles will cover-up the old one and will play the role of a new board in allowing the creation of a third layer, and so on. *"Walls"* with specified "height", or *discrete irregular shapes* are obtained by *partially* filling-up the board, and this can be achieved by inserting appropriate coding in the programmable tiles that form the template board. The coding will discriminate what are the tiles that will interact with new ones and what are those that will avoid interaction.

In the example, a change in the programming of the board induces the formation of different shapes out of the same input set. This suggests that a formal notion of *complexity* describing self-assembly of molecular systems cannot be based merely on the variety of shapes that potentially can be assembled, but rather on the much larger variety of algorithms that allow their assembly.

DNA computing. Last, we want to mention the effort in designing algorithms for DNA-computation. The landmark step is in [1], where DNA is used to solve an instance of the Hamiltonian Path problem, asking to establish whether there is a path between two cities, given an incomplete set of available roads. A set of strands of DNA is used to represent cities and roads (similar to the description of the 3-coloring problem in Section 6), and the coding is such that a strand representing a road would connect (according to the rules of base-pairing) to any two strands representing a city. By mixing together strands, joining the cities connected by roads, and weeding out any wrong answers, it has been shown that the strands could self-assemble to solve the problem.

The first link between DNA-nanotechnology and DNA-computation was established in [54] with the suggestion that short branched DNA-molecules could be "programmed" to undergo algorithmic self-assembly and thus serve as the basis of computation. Other work has followed as [30,34,25].

11 Discussion

Most examples in this paper were based on Watson-Crick interactions of DNA molecules. Other kinds of interaction, usually referred to as *tertiary interactions*, can be used to lead a controlled behavior in the assembly of DNA molecules, for example, DNA triplexes [15], tecto-RNA [21] and G-quartet formation [42]. In the combinatorial DNA constructions that we presented, tertiary interactions were carefully avoided with the goal of maximizing control on the dynamics of the assembly. Tertiary interactions are not as readily controlled as Watson-Crick interactions. The next generation of structural DNA nanotechnologists will be likely to exploit this wider range of structural possibilities and it appears possible that new combinatorics might arise from these options.

Acknowledgement

The authors would like to thank Roberto Incitti and Natasha Jonoska for their comments on preliminary versions of this manuscript.

This research has been supported by grants GM-29554 from the National Institute of General Medical Sciences, N00014-98-1-0093 from the Office of Naval Research, grants DMI-0210844, EIA-0086015, DMR-01138790, and CTS-0103002 from the National Science Foundation, and F30602-01-2-0561 from DARPA/AFSOR.

References

1. L.M. Adleman. Molecular computation of solutions to combinatorial problems. *Science*, 266:1021-4, 1994.
2. L.M. Adleman. Toward a mathematical theory of self-assembly. Technical Report 00-722, Department of Computer Science, University of Southern California, 2000.
3. L.M. Adleman, Q. Cheng, A. Goel, M-D. Huang, D. Kempe, P. Moisset de Espanés, P.W.K. Rothemund. Combinatorial optimisation problems in self-assembly. *STOC'02 Proceedings*, Montreal Quebec, Canada, 2002.
4. A. Aggeli, M. Bell, N. Boden, J.N. Keen, P.F. Knowles, T.C.B. McLeish, M. Pitkeathly, S.E. Radford. Responsive gels formed by the spontaneous self-assembly of peptides into polymeric beta-sheet tapes, *Nature* 386:259-62, 1997.
5. P. Ball Materials Science: Polymers made to measure, *Nature*, 367:323-4, 1994.
6. A. Carbone, M. Gromov. Mathematical slices of molecular biology, *La Gazette des Mathématiciens*, Numéro Spéciale, 88:11–80, 2001.
7. A. Carbone, N.C. Seeman. Circuits and programmable self-assembling DNA structures. *Proceedings of the National Academy of Sciences USA*, 99:12577-12582, 2002.
8. A. Carbone, N.C. Seeman. A route to fractal DNA-assembly, *Natural Computing*, 1:469-480, 2002.
9. A. Carbone, N.C. Seeman. Coding and geometrical shapes in nanostructures: a fractal DNA-assembly, *Natural Computing*, 2003. In press.
10. S.N. Cohen, A.C.Y. Chang, H.W. Boyer, R.B. Helling. Construction of biologically functional bacterial plasmids in vitro, *Proceedings of the National Academy of Science USA*, 70:3240-3244, 1973.
11. A.M. Diegelman, E.T. Kool. Generation of circular RNAs and trans-cleaving catalytic RNAs by rolling transcription of circular DNA oligonucleotides encoding hairpin ribozymes, *Nucleic Acids Research*, 26, 3235-3241, 1998.
12. S.M. Du, S. Zhang, N.C. Seeman. DNA Junctions, Antijunctions and Mesojunctions, *Biochemistry*, 31:10955-10963, 1992.
13. I. Duhnam, N. Shimizu, B.A. Roe et al. The DNA sequence of human chromosome 22, *Nature*, 402:489–95, 1999.
14. B.F. Eichman, J.M. Vargason, B.H.M. Mooers, P.S. Ho. The Holliday junction in an inverted repeat DNA sequence: Sequence effects on the structure of four-way junctions, *Proceedings of the National Academy of Science USA*, 97:3971-3976, 2000.
15. G. Felsenfeld, D.R. Davies, A. Rich. Formation of a three-stranded polynucleotide molecule, *J. Am. Chem. Soc.*, 79:2023-2024, 1957.
16. T.-J. Fu, B. Kemper and N.C. Seeman. Endonuclease VII cleavage of DNA double crossover molecules, *Biochemistry* 33:3896-3905, 1994.
17. B. Grünbaum, G.C. Shephard. *Tilings and Patterns*, W.H. Freeman and Company, 1986.
18. J.D. Hartgerink, E. Beniash, and S.I. Stupp. Self-assembly and mineralization of peptide-amphiphile nanofibers, *Science*, 294:1684, 2001.
19. I. Huck, J.M. Lehn. Virtual combinatorial libraries: dynamic generation of molecular and supramolecular diversity by self-assembly, *Proceedings of the National Academy of Sciences USA*, 94:2106-10, 1997.
20. S. Hussini, L. Kari, S. Konstantinidis. Coding properties of DNA languages, *Theoretical Computer Science*, 290:1557-1579, 2003.
21. L. Jaeger, E. Westhof, N.B. Leontis. TectoRNA: modular assembly units for the construction of RNA nano-objects, *Nucleic Acids Research*, 29:455-463, 2001.

22. N. Jonoska. 3D DNA patterns and Computing, *Pattern formation in Biology, Vision and Dynamics*, edited by A. Carbone, M. Gromov, P. Prusinkiewicz, World Scientific Publishing Company, 310-324, 2000.
23. G. von Kiedrowski. Personal communication, February 2003.
24. T.H. LaBean, H. Yan, J. Kopatsch, F. Liu, E. Winfree, J.H. Reif, N.C. Seeman. The construction, analysis, ligation and self-assembly of DNA triple crossover complexes, *J. Am. Chem. Soc.* 122:1848-1860, 2000.
25. T.H. LaBean, E. Winfree, J.H. Reif. Experimental progress in computation by self-assembly of DNA tilings, *Proc. DNA Based Computers V*, DIMACS Series in Discrete Mathematics and Theoretical Computer Science, E. Win- free and D.K. Gifford (eds), American Mathematical Society, Providence, RI, vol.54:123-140, 2000.
26. J.M. Lehn. Sopramolecular Chemistry, *Science*, 260:1762-3, 1993
27. J.M. Lehn. Toward complex matter: Supramolecular chemistry and self-organisation. *Proceedings of the National Academy of Science USA*, 99(8):4763-4768, 2002.
28. F. Liu, R. Sha and N.C. Seeman. Modifying the surface features of two-dimensional DNA crystals, *Journal of the American Chemical Society* 121:917-922, 1999.
29. C. Mao, W. Sun and N.C. Seeman. Designed two-dimensional DNA Holliday junction arrays visualized by atomic force microscopy, *Journal of the American Chemical Society*, 121:5437-5443, 1999.
30. C. Mao, T. LaBean, J.H. Reif, N.C. Seeman. Logical computation using algorithmic self-assembly of DNA triple-crossover molecules. *Nature* 407:493-496, 2000; Erratum: *Nature* 408:750-750, 2000.
31. H. Qiu, J.C. Dewan and N.C. Seeman. A DNA decamer with a sticky end: The Ccystal structure of d-CGACGATCGT, *Journal of Molecular Biology*, 267:881-898, 1997.
32. J.H. Reif. Local parallel biomolecular computation, *DNA Based Computers, III*, DIMACS Series in Discrete Mathematics and Theoretical Computer Science, Vol 48 (ed. H. Rubin), American Mathematical Society, 217-254, 1999.
33. J.H. Reif. Molecular assembly and computation: from theory to experimental demonstrations, *29-th International Colloquium on Automata, Languages, and Programming(ICALP)*, Málaga, Spain (July 8, 2002). Volume 2380 of Lecture Notes in Computer Science, New York, pages 1-21, 2002.
34. P. Sa-Ardyen, N. Jonoska, N.C. Seeman. Self-assembling DNA graphs, *DNA Computing, 8th International Workshop on DNA-based Compters*, LNCS 2568:1-9, 2003.
35. J.M. Schnur. Lipid tubules: a paradigm for molecularly engineered structures, *Science*, 262:1669-76, 1993.
36. N.C. Seeman. Nucleic acid junctions and lattices, *J. Theor. Biol.*, 99:237-247, 1982.
37. N.C. Seeman, N.R. Kallenbach. Design of immobile nucleic acid junctions, *Biophysical Journal*, 44:201-209, 1983.
38. N.C. Seeman, N.R.Kallenbach. Nucleic-acids junctions: a successfull experiment in macromolecular design, *Molecular Structure: Chemical Reactivity and Biological Activity*, J.J. Stezowski, J.L. Huang, M.C. Shao (eds.), Oxford University Press, Oxford, 189-194, 1988.
39. N.C. Seeman. DNA engineering and its application to nanotechnology, *Trends in Biotech.*, 17:437-443, 1999.

40. N.C. Seeman. DNA nanotechnology: from topological control to structural control, *Pattern formation in Biology, Vision and Dynamics*, edited by A. Carbone, M. Gromov, P. Prusinkiewicz, World Scientific Publishing Company, 310-324, 2000.
41. N.C. Seeman. In the nick of space: Generalized nucleic acid complementarity and the development of DNA nanotechnology, *Synlett*, 1536-1548, 2000.
42. D. Sen, W. Gilbert, Formation of parallel four-stranded complexes by guanine-rich motifs in DNA and applications to meiosis. *Nature*, 334:364-366, 1988.
43. R. Sha, F. Liu, D.P. Millar and N.C. Seeman. Atomic force microscopy of parallel DNA branched junction arrays, *Chemistry & Biology* 7:743-751, 2000.
44. Z. Shen. DNA Polycrossover Molecules and their Applications in Homology Recognition. Ph.D. Thesis, New York University, 1999.
45. W.B. Sherman, N.C. Seeman. (in preparation)
Abstract "The design of nucleic acid nanotubes" appeared in *Journal of Biomolecular Structure & Dynamics*, online at `http://www.jbsdonline.com/index.cfm?search=seeman&d=3012&c=4096&p11491&do=detail`, 2003.
46. H. Yan, X. Zhang, Z. Shen, N.C. Seeman. A robust DNA mechanical device controlled by hybridization topology. *Nature*, 415:62-5, 2002.
47. B. Yurke, A.J. Turberfield, A.P.Jr. Mills, F.C. Simmel, J.L. Neumann. A DNA-fuelled molecular machine made of DNA. *Nature*, 406:605-608, 2000.
48. H. Wang. Proving theorems by pattern recognition, *Bell System Tech. J.*, 40:1-42, 1961.
49. H. Wang. Dominos and the AEA case of the decision problem. In *Proceedings of the Symposium on the Mathematical Theory of Automata*, 23-56, Polytechnic, New York, 1963.
50. H. Wang. Games, logic and computers, *Scientific American*, November, 98–106, 1965.
51. Y. Wang, J.E. Mueller, B. Kemper, and N.C. Seeman. The assembly and characterization of 5-arm and 6-arm DNA junctions, *Biochemistry* 30:5667-5674, 1991.
52. G.M. Whitesides, J.P. Mathias, C.T. Seto. Molecular self-assembly and nanochemistry: a chemical strategy for the synthesis of nanostructures, *Science*, 254:1312-9, 1991.
53. W.R. Wikoff, L. Liljas, R.L. Duda, H. Tsuruta, R.W. Hendrix, J.E. Johnson. Topologically linked protein rings in the bacteriophage HK97 caspid. *Science*, 289:2129-2133, 2000.
54. E. Winfree. On the computational power of DNA annealing and ligation, *DNA based computers, Proceedings of a DIMACS workshop, April ???4, 1995, Princeton University*, eds. R.J. Lipton and E.B. Baum, AMS Providence, 199-219, 1996.
55. E. Winfree, F. Liu, L.A. Wenzler, N.C. Seeman. Design and self-assembly of two-dimensional DNA crystals, *Nature*, 394:539–44, 1998.
56. E. Winfree. Algorithmic self-assembly of DNA: theoretical motivations and 2D assembly experiments. *J. Biol. Mol. Struct. Dynamics Conversat.*, 2:263-270, 2000.
57. S. Zhang, T. Holmes, C. Lockshin, A. Rich. Spontaneous assembly of a self-complementary oligopeptide to form a stable macroscopic membrane. *Proceeding of the National Academy of Sciences USA*, 90:3334-8, 1993.
58. Y. Zhang, N.C. Seeman. A solid-support methodology for the construction of geometrical objects from DNA, *J. Am. Chem. Soc.*, 114:2656-2663, 1992.
59. Y. Zhang, N.C. Seeman. The construction of a DNA truncated octahedron, *J. Am. Chem. Soc.*, 116:1661-1669, 1994.
60. X. Zhang, H. Yan, Z. Shen and N.C. Seeman. Paranemic cohesion of topologically-closed DNA molecules, *J Am. Chem. Soc.*, 124:12940-12941, 2002.

On Some Classes of Splicing Languages*

Rodica Ceterchi[1], Carlos Martín-Vide[2], and K.G. Subramanian[3]

[1] Faculty of Mathematics, University of Bucharest
14, Academiei st., 70109 Bucharest, Romania
rc@funinf.math.unibuc.ro, rc@fll.urv.es
[2] Research Group in Mathematical Linguistics
Rovira i Virgili University
Pl. Imperial Tàrraco 1, 43005 Tarragona, Spain
cmv@astor.urv.es
[3] Department of Mathematics
Madras Christian College
Tambaram, Chennai 600059, India
kgsmani@vsnl.net

Abstract. We introduce some classes of splicing languages generated with simple and semi-simple splicing rules, in both, the linear and circular cases. We investigate some of their properties.

1 Introduction

The operation of *splicing* was introduced as a generative mechanism in formal language theory by Tom Head in [9], and lays at the foundation of the field of DNA based computing (see [11] and [14].)

We are concerned in this paper with some classes of languages which arise when considering very simple types of splicing rules. Our main inspiration is the study of [12], where the notion of *simple H* system and the associated notion of *simple splicing languages* have been introduced.

A *simple* splicing rule is a rule of the form $(a, 1; a, 1)$ with $a \in A$ a symbol called *marker*. More precisely, we will call such a rule a $(1,3)$-simple splicing rule, since the marker a appears on the 1 and 3 positions of the splicing rule, and since one can conceive of (i, j)-simple splicing rules for all pairs (i, j) with $i = 1, 2$, $j = 3, 4$. We denote by $SH(i, j)$ the class of languages generated by simple splicing rules of type (i, j). The paper [12] focuses basically on the study of the $SH(1, 3)$ class (which is equal to the $SH(2, 4)$ class and is denoted by SH). Only towards the end of the paper the authors of [12] show that there are three such distinct and incomparable classes, $SH = SH(1, 3) = SH(2, 4)$, $SH(2, 3)$ and $SH(1, 4)$, of (linear) simple splicing languages. They infer that most of the results proven for the SH class will hold also for the other classes, and point out towards studying one-sided splicing rules of radius at most k: rules $(u_1, 1; u_3, 1)$

* This work was possible thanks to the grants SAB2000-0145, and SAB2001-0007, from the Secretaría de Estado de Educación y Universidades, Spanish Ministry for Education, Culture and Sport.

N. Jonoska et al. (Eds.): Molecular Computing (Head Festschrift), LNCS 2950, pp. 84–105, 2004.

with $|u_1| \leq k$, $|u_3| \leq k$. Note that the "one-sidedness" they are considering is of the $(1,3)$ type, and that other (i,j) types can be considered, provided they are distinct from $(1,3)$.

The notion of *semi-simple* splicing rules introduced by Goode and Pixton in their paper [8] seems to follow the research topic proposed by [12] considering the simplest case, with $k = 1$. More precisely, semi-simple splicing rules are rules of the form $(a, 1; b, 1)$ with $a, b \in A$, symbols which again we will call markers. Again the one-sidedness is of the $(1,3)$ type, and we denote by $SSH(1,3)$ the class of languages generated by such rules.

Goode and Pixton state explicitly that their interest lies in characterizing regular languages which are splicing languages of this particular type (we will be back to this problem later), and not in continuing the research along the lines of [12].

Instead, we are interested here in analyzing different classes, for their similarities or dissimilarities of behavior. We show that the classes $SSH(1,3)$ and $SSH(2,4)$ are incomparable, and we provide more examples of languages of these types.

The connections between (linear) splicing languages and regular languages are well explored in one direction: it has been shown that iterated splicing which uses a finite set of axioms and a finite set of rules generates only regular languages (see [6] and [15]). On the other hand, the problem of *characterizing* the regular languages which are splicing languages is still open. Very simple regular languages, like $(aa)^*$, are *not* splicing languages. The problem is: starting from a regular language $L \subset V^*$, can we find a set of rules R, and a (finite) set of axioms $A \subset V^*$, such that L is the splicing language of the H system (V, A, R)? This problem is solved for simple rules in [12] with algebraic tools. The problem is also addressed by Tom Head in [10], and solved for a family of classes, a family to which we will refer in Section 5. In [8] the problem is solved for $(1,3)$-semi-simple rules. The characterization is given in terms of a certain directed graph, the $(1,3)$ arrow graph canonically associated to a regular language respected by a set of $(1,3)$-semi-simple rules. We show in Section 4 that their construction can be dualized: we construct the $(2,4)$ arrow graph for a language respected by $(2,4)$-semi-simple splicing rules. We obtain thus an extension of the characterization from [8]. Among other things, this enables us to prove for languages in $SSH(2,4)$ properties valid for languages in $SSH(1,3)$. The construction also makes obvious the fact that the other two semi-simple types have to be treated differently.

In Section 5 we try to find more connections to the study in [10] and point out certain open problems in this direction.

The second part of this paper is concerned with *circular splicing*. Splicing of circular strings has been considered as early as the pioneering paper [9], but it does not occupy in the literature the same volume as that occupied by linear splicing. With respect to the computational power, it is still unknown precisely what class can be obtained by circular splicing, starting from a finite set of rules and a finite set of axioms (see [1], [15]). It is known that finite circular splicing systems can generate non-regular circular languages (see [17],[15]). On the other hand, there are regular circular languages (for instance the language $\hat{}(aa)^*b$) which cannot be generated by any finite circular splicing system (with Păun type of rules) (see [2], [3]). The fact that the behavior in the circular case is by no means predictable from the behavior in the linear case is made apparent by the example of the $\hat{}(aa)^*$ circular language, which is both a regular and a splicing language (see [2], [3]).

This has motivated the beginning of the study of simple circular H systems in the paper [5]. We recall briefly in Section 6 some results of [5], which are completed here by some closure properties.

In Section 7 we introduce the semi-simple circular splicing systems and first steps in the study of such systems are made by proving the incomparability of classes $SSH^\circ(1,3)$ and $SSH^\circ(2,4)$, a situation which is totally dissimilar to the simple circular case, where more classes colapse than in the simple linear case.

The following table summarizes the notations used for the different types of H systems and the respective classes of languages considered throughout the paper, and the main references for the types already studied in the literature.

Table 1. Types of splicing systems and languages

	linear case, V^*	circular case, V°
simple	SH [12]	SH° [5]
semi-simple	SSH [8]	SSH°
k-simple	S_kH [10]	
k-semi-simple	SS_kH	

2 Preliminaries and Notations

For a finite alphabet V we denote by V^* the free monoid over V, by 1 the empty string, and we let $V^+ = V^* \setminus \{1\}$. For a string $w \in V^*$ we denote by $|w|$ its length, and if $a \in V$ is a letter, we denote by $|w|_a$ the number of occurrences of a in w. We let $mi : V^* \longrightarrow V^*$ denote the mirror image function, and we will denote still by mi its restriction to subsets of V^*. For a word $w \in V^*$, we call $u \in V^*$ a *factor of* w, if there exist $x, y \in V^*$ such that $w = xuy$. We call $u \in V^*$ a *factor of a language* L, if u is a factor of some word $w \in L$.

On V^* we consider the equivalence relation given by $xy \sim yx$ for $x, y \in V^*$ (the equivalence given by the circular permutations of the letters of a word).

A *circular* string over V will be an equivalence class w.r.t. the relation above. We denote by $\hat{}w$ the class of the string $w \in V^*$. We denote by V° the set of all circular strings over V, i.e., $V^\circ = V^*/\sim$. The empty circular string will be denoted $\hat{}1$, and we let $V^\oplus = V^\circ \setminus \{\hat{}1\}$. Any subset of V° will be called a *circular language*.

For circular words, the notions of length and number of occurrences of a letter, can be immediately defined using representatives: $|\hat{}w| = |w|$, and $|\hat{}w|_a =$

$|w|_a$. Also the mirror image function $mi : V^* \longrightarrow V^*$ can be readily extended to $mi : V^\circ \longrightarrow V^\circ$; if $\hat{}w$ is read in a clockwise manner, $mi(\hat{}w) = \hat{}mi(w)$ consists of the letters of w read in a counter-clockwise manner. We will denote composition of functions algebraically.

To a language $L \subseteq V^*$ we can associate the circular language $Cir(L) = \{\hat{}w \mid w \in L\}$ which will be called *the circularization of* L. The notation L° is also used in the literature.

To a circular language $C \subseteq V^\circ$ we can associate several *linearizations*, i.e., several languages $L \subseteq V^*$ such that $Cir(L) = C$. The *full linearization* of the circular language C will be $Lin(C) = \{w \mid \hat{}w \in C\}$.

Having a family FL of languages we can associate to it its circular counterpart $FL^\circ = \{Cir(L) \mid L \in FL\}$. We can thus speak of FIN°, REG°, etc. A circular language is in REG° if some linearization of it is in REG.

Next we recall some notions on splicing. A *splicing rule* is a quadruple $r = (u_1, u_2; u_3, u_4)$ with $u_i \in V^*$ for $i = 1, 2, 3, 4$. The action of the splicing rule r on linear words is given by

$$(x_1u_1u_2x_2, y_1u_3u_4y_2) \vdash_r x_1u_1u_4y_2.$$

In other words, the string $x = x_1u_1u_2x_2$ is cut between u_1 and u_2, the string $y = y_1u_3u_4y_2$ between u_3 and u_4, and two of the resulting substrings, namely x_1u_1 and u_4y_2, are pasted together producing $z = x_1u_1u_4y_2$.

For a language $L \subseteq V^*$ and a splicing rule r, we denote

$$r(L) = \{z \mid (x, y) \vdash_r z \text{ for some } x, y \in L\}.$$

We say that a splicing rule r *respects* a language L if $r(L) \subseteq L$. In other words, L is closed w.r.t. the rule r. A set of rules R respects the language L if $R(L) \subseteq L$.

For an alphabet V, a (finite) $A \subset V^*$ and a set R of splicing rules, a triple $S = (V, A, R)$ is called a *splicing system*, or *H system*. The pair $\sigma = (V, R)$ is called a splicing (or H) *scheme*.

For an arbitrary language $L \subseteq V^*$, we denote $\sigma^0(L) = L$, $\sigma(L) = L \cup \{z \mid \text{for some } r \in R, x, y \in L, (x, y) \vdash_r z\}$, and for any $n \geq 1$ we define $\sigma^{n+1}(L) = \sigma^n(L) \cup \sigma(\sigma^n(L))$. The *language* generated by the H scheme σ starting from $L \subseteq V^*$ is defined as

$$\sigma^*(L) = \bigcup_{n \geq 0} \sigma^n(L).$$

The *language* generated by the H system $S = (V, A, R)$, denoted $L(S)$, is $L(S) = \sigma^*(A)$, i.e., what we obtain by iterated splicing starting from the strings in the axiom set A. We recall that $L(S)$ can be also characterized as being the smallest language $L \subseteq V^*$ such that:

(i) $A \subseteq L$;

(ii) $r(L) \subseteq L$ for any $r \in R$.

In a similar manner we define splicing on circular languages. The only difference is that we start from circular strings as axioms, $A \subset V^\circ$, and the application of a splicing rule produces a circular string from two circular strings (as we will formally define in section 6).

Other notations and definitions will be introduced, whenever necessary, throughout the paper.

3 Simple Linear *H* Systems and Languages

We recall in this section some of the results of [12] on simple linear H systems.

A *simple* (linear) splicing rule is a splicing rule of one of the four following forms: $(a,1;a,1)$, $(1,a;1,a)$, $(1,a;a,1)$, $(a,1;1,a)$, where a is a symbol in V, called *marker*. The action of the four types of simple splicing rules on strings is illustrated below:

$$\begin{aligned}
type(1,3):&\ (xax', yay') \vdash_{(a,1;a,1)} xay',\\
type(2,4):&\ (xax', yay') \vdash_{(1,a;1,a)} xay',\\
type(2,3):&\ (xax', yay') \vdash_{(1,a;a,1)} xy',\\
type(1,4):&\ (xax', yay') \vdash_{(a,1;1,a)} xaay'.
\end{aligned}$$

We note that the action of rules of type $(1,3)$ and $(2,4)$ coincide.

A *simple H system* is a triple $S = (V, A, R)$ consisting of an alphabet V, a finite set $A \subset V^*$ called the set of *axioms* (or *initial strings*), and a finite set R of simple splicing rules of one of the types (i,j), with $i = 1,2$, $j = 3,4$.

The *language* generated by a simple H system S as above is defined as usual for splicing schemes

$$L = \sigma^*(A).$$

where $\sigma = (V, R)$. If the rules in R are of type (i,j), this language will be called an *(i,j)-simple splicing language*. The class of all (i,j)-simple splicing languages will be denoted by $SH(i,j)$. We have $SH(1,3) = SH(2,4)$ and let SH denote this class.

Theorem 1. *(Theorem 11 of [12]) Each two of the families SH, $SH(1,4)$ and $SH(2,3)$ are incomparable.*

We have, from the more general results of [15] and [6], that all three classes contain only regular languages, i.e., $SH, SH(1,4), SH(2,3) \subset REG$. In [12] the inclusion $SH \subset REG$ is also proved directly, using a representation result.

The inverse problem, of characterizing those regular languages which are simple splicing languages is solved via an algebraic characterization.

4 Semi-simple Linear *H* Systems and Languages

Semi-simple splicing rules were first considered by Goode and Pixton in [8]. In [12] it is mentioned that "it is also of interest to consider one-sided splicing rules of radius at most k: $(u_1,1;u_2,1)$ with $|u_1| \leq k$, $|u_2| \leq k$". The semi-simple splicing rules are one-sided splicing rules of radius $k = 1$. Moreover, the "one-sidedness" of the rules considered by Goode and Pixton is of the $(1,3)$ type. We present the basic definitions for the four possible types.

For two symbols $a, b \in V$ a *semi-simple* (linear) splicing rule with markers a and b is a splicing rule of one of the following four types: $(a,1;b,1)$, $(1,a;1,b)$, $(1,a;b,1)$, $(a,1;1,b)$.
The action of the four rules on strings is obvious.

A *semi-simple H system* is an H system $S = (V, A, R)$ with all rules in R semi-simple splicing rules of one of the types (i,j), with $i = 1,2$, $j = 3,4$.

The *language* generated by a semi-simple H system S as above is defined as usual for splicing schemes, $L = \sigma^*(A)$, where $\sigma = (V, R)$. If the rules in R are semi-simple rules of type (i,j), this language will be called an (i,j)*-semi-simple splicing language*. The class of all (i,j)-semi-simple splicing languages will be denoted by $SSH(i,j)$.

In [8] only the class $SSH(1,3)$ is considered. Since in [8] only one example of a semi-simple language of the $(1,3)$ type is given, we find it useful to provide some more.

Example 1 *Consider* $S_3 = (\{a,b\}, \{aba\}, \{(a,1;b,1)\})$. *The language generated by* S_3 *is* $L_3 = L(S_3) = aa^+ \cup aba^+ \in SSH(1,3)$.

Example 2 *Consider* $S_4 = (\{a,b\}, \{aba\}, \{(1,a;1,b)\})$. *The language generated by* S_4 *is* $L_4 = L(S_4) = b^+a \cup ab^+a \in SSH(2,4)$.

The following result shows that, even for the types $(1,3)$ and $(2,4)$, which share some similarity of behavior, unlike in the case of simple rules, the respective classes of languages are incomparable.

Theorem 2. *The classes* $SSH(1,3)$ *and* $SSH(2,4)$ *are incomparable.*

Proof: Note first that simple splicing languages in the class SH are in the intersection $SSH(1,3) \cap SSH(2,4)$.

The language $L_1 = a^+ \cup a^+ab \cup aba^+ \cup aba^+b$ belongs to $SSH(1,3)$, but not to $SSH(2,4)$.

For the first assertion, L_1 is generated by the $(1,3)$-semi-simple H system $S_1 = (\{a,b\}, \{abab\}, \{(a,1;b,1)\})$. We sketch the proof of $L_1 = L(S_1)$. Denote by r the unique splicing rule, $r = (a,1;b,1)$. For the inclusion $L_1 \subseteq L(S_1)$, note first that

$$(a|bab, abab|) \vdash_r a,$$
$$(a|bab, ab|ab) \vdash_r a^2b.$$

Next, if $a^nb \in L(S_1)$, then $a^n, a^{n+1}b \in L(S_1)$, for any natural $n \geq 2$, since

$$(a^n|b, abab|) \vdash_r a^n,$$
$$(a^n|b, ab|ab) \vdash_r a^{n+1}b.$$

Thus, by an induction argument, it follows that $a^+ \subseteq L(S_1)$ and $a^+ab \subseteq L(S_1)$.

Similarly, using induction, from

$$(aba|b, abab|) \vdash_r aba,$$
$$(aba|b, ab|a^n) \vdash_r aba^{n+1}, \text{ for any } n \geq 1,$$

it follows that $aba^+ \subseteq L(S_1)$, and from

$$(aba|b, ab|ab) \vdash_r aba^2b,$$
$$(aba^2|b, ab|ab) \vdash_r aba^{n+1}b, \text{ for any } n \geq 2,$$

it follows that $aba^+b \subseteq L(S_1)$.

For the other inclusion, $L(S_1) \subseteq L_1$, we use the characterization of $L(S_1)$ as the smallest language which contains the axioms, and is respected by the splicing rules. It is a straightforward exercise to prove that L_1 contains the axiom of S_1, and that it is respected by the splicing rule of S_1.

To show that $L_1 \notin SSH(2,4)$, note that each word in L_1 has at most two occurrences of b. Suppose L_1 were in $SSH(2,4)$. Then it would have been generated by (and thus closed to) rules of one of the forms: $(1,a;1,b)$, $(1,b;1,a)$, $(1,a;1,a)$, $(1,b;1,b)$. But we have:

$$(ab|ab, a|bab) \vdash_{(1,a;1,b)} abbab \notin L_1,$$
$$(aba|b, |abab) \vdash_{(1,b;1,a)} abaabab \notin L_1,$$
$$(ab|ab, |abab) \vdash_{(1,a;1,a)} ababab \notin L_1,$$
$$(aba|b, a|bab) \vdash_{(1,b;1,b)} ababab \notin L_1,$$

where we have marked the places where the cutting occurs, and all the words obtained are not in L_1 because they contain three occurrences of b.

The language $L_2 = b^+ \cup abb^+ \cup b^+ab \cup ab^+ab$ belongs to $SSH(2,4)$, but not to $SSH(1,3)$.

For the first assertion, L_2 is generated by the $(2,4)$-semi-simple H system $S_2 = (\{a,b\}, \{abab\}, \{(1,a;1,b)\})$. The proof is along the same lines as the proof of $L_1 \in SSH(1,3)$.

For the second assertion, note that each word in L_2 has at most two occurrences of a. If L_2 were in $SSH(1,3)$ it would be closed to rules of one of the forms $(a,1;b,1)$, $(b,1;a,1)$, $(a,1;a,1)$, $(b,1;b,1)$. But, even starting from the axiom, we can choose the cutting places in such a way as to obtain words with three occurrences of a, and thus not in L_2:

$$(aba|b, ab|ab) \vdash_{(a,1;b,1)} abaab \notin L_2,$$
$$(abab|, a|bab) \vdash_{(b,1;a,1)} abab^2ab \notin L_2,$$
$$(aba|b, a|bab) \vdash_{(a,1;a,1)} ababab \notin L_2,$$
$$(abab|, ab|ab) \vdash_{(b,1;b,1)} ababab \notin L_2.$$

□

The languages in the proof of Theorem 2 also provide examples of semi-simple splicing languages which are not simple.

For the $(1,3)$ type, we have $L_1 \in SSH(1,3) \setminus SH(1,3)$. L_1 is not respected by the simple rules $(a,1;a,1)$ and $(b,1;b,1)$; for instance,

$$(aba^m b|, ab|a^n) \vdash_{(b,1;b,1)} aba^m ba^n \notin L_1,$$
$$(aba^k|a^s b, a|ba^n b) \vdash_{(a,1;a,1)} aba^k ba^n b \notin L_1.$$

In a similar manner one can show that $L_2 \in SSH(2,4) \setminus SH(2,4)$.

Splicing rules can be seen as functions from $V^* \times V^*$ to 2^{V^*}: if $r = (u_1, u_2; u_3, u_4)$ and $x, y \in V^*$, then $r(x,y) = \{z \in V^* \mid (x,y) \vdash_r z\}$. Of course, if there is no z such that $(x,y) \vdash_r z$ (that is, x and y cannot be spliced by using rule r), then $r(x,y) = \emptyset$. We denote by inv the function $inv : A \times B \longrightarrow B \times A$, defined by $inv(x,y) = (y,x)$, for $x \in A$, $y \in B$, and arbitrary sets A and B.

We have a strong relationship between semi-simple splicing rules of types $(1,3)$ and $(2,4)$.

Proposition 1. *Let $a, b \in V$ be two symbols. The following equality of functions from $V^* \times V^*$ to 2^{V^*} holds:*

$$(a,1;b,1)mi = inv(mi \times mi)(1,b;1,a).$$

Proof. Let $x, y \in V^*$. If $|x|_a = 0$ or $|y|_b = 0$, or both, then, clearly, both $(a,1;b,1)mi$ and $inv(mi \times mi)(1,b;1,a)$ return $\emptyset$. For any $x_1 a x_2 \in V^* a V^*$ and $y_1 b y_2 \in V^* b V^*$, we have, on one hand:

$$(x_1 a|x_2, y_1 b|y_2) \vdash_{(a,1;b,1)} x_1 a y_2 \longrightarrow_{mi} mi(y_2) a\, mi(x_1).$$

On the other hand, $inv(mi \times mi)(x_1 a x_2, y_1 b y_2) = (mi(y_2) b\, mi(y_1), mi(x_2) a\, mi(x_1))$, and

$$(mi(y_2)|b\, mi(y_1), mi(x_2)|a\, mi(x_1)) \vdash_{(1,b;1,a)} mi(y_2) a\, mi(x_1).$$ □

Corollary 1. *There exists a bijection between the classes of languages $SSH(1,3)$ and $SSH(2,4)$.*

Proof. We construct $\varphi : SSH(1,3) \longrightarrow SSH(2,4)$ and $\psi : SSH(2,4) \longrightarrow SSH(1,3)$. First, making an abuse of notation, we define φ and ψ on types of semi-simple splicing rules, by:

$$\varphi(a,1;b,1) = (1,b;1,a),$$
$$\psi(1,b;1,a) = (a,1;b,1), \text{ for all } a,b \in V^*.$$

φ transforms a $(1,3)$ rule into a $(2,4)$ one, and ψ makes the reverse transformation. Obviously, $\psi(\varphi(r)) = r$, and $\varphi(\psi(r')) = r'$, for any $(1,3)$ rule r, and any $(2,4)$ rule r'. For a set of $(1,3)$ rules, R, let $\varphi(R)$ denote the corresponding set of $(2,4)$ rules, and for R', set of $(2,4)$ rules, let $\psi(R')$ denote the corresponding set of $(1,3)$ rules.

Let $S = (V, A, R)$ be a $(1,3)$ semi-simple splicing system. Again, making an abuse of notation, we let $\varphi(S)$ denote the $(2,4)$ semi-simple splicing system $\varphi(S) = (V, mi(A), \varphi(R))$. For $S' = (V, A', R')$ a $(2,4)$ semi-simple splicing system, let $\psi(S')$ denote the $(1,3)$ semi-simple splicing system $\psi(S') = (V, mi(A'), \psi(R'))$.

Now we define φ and ψ on the respective classes of languages by: $\varphi(L(S)) = L(\varphi(S))$, for $L(S) \in SSH(1,3)$, and $\psi(L(S')) = L(\psi(S'))$, for $L(S') \in SSH(2,4)$. Note that $L(\varphi(S)) = mi(L(S))$, and also $L(\psi(S')) = mi(L(S'))$, thus φ and ψ are given by the mirror image function, and are obviously inverse to each other. □

The above results emphasize the symmetry between the $(1,3)$ and the $(2,4)$ cases, symmetry which makes possible the construction which follows next.

Let V be an alphabet, $L \subset V^*$ a language, and $R(L)$ the set of splicing rules (of a certain predefined type) which respects L. For a subset of rules $R \subseteq R(L)$ we denote by $\sigma = (V, R)$ a splicing scheme. We are looking for necessary and sufficient conditions for the existence of a finite set $A \subset V^*$ such that $L = \sigma^*(A)$. For $R(L)$ the set of $(1,3)$-semi-simple splicing rules which respect L, Goode and Pixton have given such a characterization, in terms of a certain directed graph, the $(1,3)$-*arrow graph* canonically associated to a pair (σ, L). Their construction can be modified to accomodate the $(2,4)$-semi-simple splicing rules.

We present the main ingredients of this second construction, which we will call the $(2,4)$-*arrow graph* associated to a pair (σ, L), where $\sigma = (V, R)$ is a $(2,4)$-semi-simple H scheme (all rules in R are of the $(2,4)$-semi-simple type).

Enrich the alphabet with two new symbols, $\bar{V} = V \cup \{S, T\}$, enrich the set of rules $\bar{R} = R \cup \{(1, S; 1, S)\} \cup \{(1, T; 1, T)\}$, take $\bar{\sigma} = (\bar{V}, \bar{R})$, and $\bar{L} = SLT$. The $(2,4)$-arrow graph of (σ, L) will be a directed graph G, with the set of vertices $V(G) = \bar{V}$, and a set of $(2,4)$-*edges* $E(G) \subset \bar{V} \times V^* \times \bar{V}$ defined in the following way: a triple $e = (b', w, b)$ is an edge (from b' to b) if there exists a $(2,4)$ rule $r = (1, a; 1, b) \in \bar{R}$ such that $b'wa$ is a factor of $\bar{L}$.

In order to stress the "duality" of the two constructions, let us recall from [8] the definition of a $(1,3)$-edge: a triple $e = (a, w, a')$ is an edge (from a to a') if there exists a $(1,3)$ rule $r = (a, 1; b, 1) \in \bar{R}$ such that bwa' is a factor of $\bar{L}$.

We list below some results and notions, which are analogous to the results obtained for the $(1,3)$ case in [8].

Lemma 1. *We have $\bar{\sigma}(\bar{L}) \subset \bar{L}$. If $A \subset V^*$ then $\bar{\sigma}^*(SAT) = S\sigma^*(A)T$.*

Proof. For $x = Sw_1T$ and $y = Sw_2T$ two words in $\bar{L}$ we have

$$(x, y) \vdash_{(1,S;1,S)} y \in \bar{L},$$
$$(x, y) \vdash_{(1,T;1,T)} x \in \bar{L},$$
$$(x, y) \vdash_r Sr(w_1, w_2)T \in \bar{L}, \text{ for any other } r \in R.$$

□

Lemma 2. *(S, w, T) is a $(2,4)$ edge iff $SwT \in \bar{L}$.*

Proof. By definition, if (S, w, T) is a $(2,4)$ edge, then there exists a $(2,4)$ rule $r = (1, a; 1, T) \in \bar{R}$ such that Swa is a factor of $\bar{L}$. The only rule in $\bar{R}$ involving T is $(1, T; 1, T)$, thus $a = T$, and SwT is a factor of $\bar{L}$. But $\bar{L} \subset SV^*T$, thus $SwT \in \bar{L}$. Conversely, if $SwT \in \bar{L}$, then SwT is a factor of $\bar{L}$, and since rule $(1, T; 1, T) \in \bar{R}$, we have that (S, w, T) is a $(2,4)$ edge. □

The *product* of two adjacent edges in G, $e_1 = (b_0, w_1, b_1)$ and $e_2 = (b_1, w_2, b_2)$, is defined as the triple $e_1 e_2 = (b_0, w_1 b_1 w_2, b_2)$.

Lemma 3. *(The closure property) Whenever e_1 and e_2 are adjacent edges in G, $e_1 e_2$ is an edge of G.*

Proof. If $e_1 = (b_0, w_1, b_1)$ and $e_2 = (b_1, w_2, b_2)$ are adjacent $(2,4)$ edges in G, then there exist the $(2,4)$ rules $r_1 = (1, a_1; 1, b_1)$ and $r_2 = (1, a_2; 1, b_2)$ in $\bar{R}$, and there exist the strings $x_0, y_1, x_1, y_2 \in \bar{V}$, such that the words $x_0 b_0 w_1 a_1 y_1$ and $x_1 b_1 w_2 a_2 y_2$ are in $\bar{L}$. Using r_1 to splice these words we obtain

$$(x_0 b_0 w_1 | a_1 y_1, x_1 | b_1 w_2 a_2 y_2) \vdash_{r_1} x_0 b_0 w_1 b_1 w_2 a_2 y_2 \in \bar{L},$$

so $b_0 w_1 b_1 w_2 a_2$ is a factor of $\bar{L}$. This fact, together with having rule r_2 in $\bar{R}$, makes $e_1 e_2 = (b_0, w_1 b_1 w_2, b_2)$ an edge of G. □

A *path* in G is a sequence $\pi =< e_1, \cdots, e_n >$ of edges $e_k = (b_{k-1}, w_k, b_k)$, $1 \leq k \leq n$, every two consecutive ones being adjacent. The label of a path as above is $\lambda(\pi) = b_0 w_1 b_1 \cdots w_n b_n$. A single edge e is also a path, $< e >$, thus its label $\lambda(e)$ is also defined.

Lemma 4. *For $\pi =< e_1, \cdots, e_n >$ a path in G, the product $e = e_1 \cdots e_n$ exists and is an edge of G, whose label equals the label of the path, i.e., $\lambda(e) = \lambda(\pi)$.*

Proof. Using Lemma 3 and the definitions, one can prove by straightforward calculations that the product of adjacent edges is associative, hence the product of n edges of a path can be unambiguously defined, and is an edge. □

The *language of the* $(2,4)$ *graph* G, denoted $L(G)$, is the set of all labels of paths in G, from S to T.

Lemma 5. $L(G) = \bar{L}$.

Proof. From Lemma 2 we have $\bar{L} \subseteq L(G)$, since $SwT \in \bar{L}$ implies (S, w, T) is an edge, and $\lambda(S, w, T) = SwT \in L(G)$ by definition. For the other implication, if π is a path in G from S to T, then, according to Lemma 4 there exists the product edge $e = (S, w, T)$ of all edges in π and $\lambda(\pi) = \lambda(e) = SwT \in \bar{L}$. □

The following is the analogue of the prefix edge property of $(1,3)$ arrow graphs from [8].

Lemma 6. *(Suffix edge) If $(b'', ub'v, b)$ is an edge of G, with $b' \in V$, then (b', v, b) is an edge of G.*

Proof. From $(b'', ub'v, b)$ being an edge, there exists a rule $r = (1, a; 1, b) \in \bar{R}$ such that $b''ub'va$ is a factor of $\bar{L}$. But this makes $b'va$ a factor of $\bar{L}$, which, together with the existence of rule r, establishes that (b', v, b) is an edge. □

By a *prime* edge we mean an edge which is not decomposable into a product of edges. Similar to the $(1,3)$ case, a (unique) decomposition into prime edges is possible.

Lemma 7. *For every edge e there exists a sequence of prime adjacent edges $e_1, \cdots, e_n$, such that $e = e_1 \cdots e_n$.*

The *prime* $(2,4)$*-arrow graph* is the subgraph G_0 of G with all edges prime.

Lemma 8. $L(G_0) = \bar{L}$.

Proof. From Lemma 7, $L(G_0) = L(G)$, and from Lemma 5, $L(G) = \bar{L}$. □

Lemma 9. *If there exists a set $A \subset V^*$ such that $L = \sigma^*(A)$, then , if (b_0, w, b_1) is a prime edge in G, we have that w is a factor of A.*

Proof. For any edge $e = (b_0, w, b_1)$ there is a rule $r = (1, a_1; 1, b_1) \in \bar{R}$ such that b_0wa_1 is a factor of $\bar{L} = \bar{\sigma}^*(\bar{A})$. We can thus define

$$N(e) = min\{n \mid \text{there exists } r = (1, a_1; 1, b_1) \in \bar{R} \text{ such that } b_0wa_1 \text{ is a factor of } \bar{\sigma}^n(\bar{A})\}.$$

For e prime with $N(e) = 0$, b_0wa_1 is a factor of A, thus w is a factor of A, so the assertion is true. Let $e = (b_0, w, b_1)$ be a prime edge with $N(e) = n > 0$, and suppose the assertion is true for all prime edges e' with $N(e') < n$. Then, there exist a rule $r_1 = (1, a_1; 1, b_1) \in \bar{R}$, and a string $z_0 = x_0b_0wa_1y_1$ in $\bar{\sigma}^n(\bar{A})$. Because $n > 0$, there exist a rule $r = (1, a; 1, b) \in \bar{R}$, and strings $z = xay'$, $z' = x'by \in \bar{\sigma}^{n-1}(\bar{A})$, such that $(z, z') \vdash_r z_0$. We analyze now the equality of the two decompositions of z_0,

$$z_0 = xby = x_0b_0wa_1y_1,$$

by comparing the suffixes:

Case 1: $|by| \geq |b_0wa_1y_1|$. Then b_0wa_1 is a factor of by, which makes it a factor of $z' = x'by \in \bar{\sigma}^{n-1}(\bar{A})$, which contradicts $N(e) = n$.

Case 2: $|by| < |a_1y_1|$. Then b_0wa_1 is a factor of x, which makes it a factor of $z = xay' \in \bar{\sigma}^{n-1}(\bar{A})$, which again contradicts $N(e) = n$.

Case 3: $|a_1y_1| < |by| < |b_0wa_1y_1|$. Then (we have an occurrence of b inside w) $w = ubv$, from which follows the existence of the suffix edge of e, $e_2 = (b, v, b_1) \in E(G)$. Formally, $e = (b_0, u, b)e_2$, with e and e_2 edges. If $e_1 = (b_0, u, b)$ is also an edge, this would contradict the primality of e. But, from $xby = x_0b_0ubva_1y_1$, in this case, it follows that $x = x_0b_0u$, thus $z = x_0b_0uay'$. This makes b_0ua a factor of z, which, together with the existence of rule $r \in \bar{R}$, makes e_1 an edge, and thus leads to contradiction. The only remaining case is:

Case 4: $|by| = |a_1y_1|$. Then $x = x_0b_0w$. Thus $z = xay' = x_0b_0way$, making b_0wa a factor of z, which, together with the existence of rule r in $\bar{R}$, make $e' = (b_0, w, b)$ an edge. Moreover, e' is a prime edge. (If it were not, there would exist edges e_1, e_2 such that $e' = e_1e_2$. If $e_1 = (b_0, u, b')$ and $e_2 = (b', v, b)$, then $w = ub'v$. It would follow that $e'_2 = (b', v, b_1)$ is a suffix edge of e, and thus e would admit the decomposition $e = e_1e'_2$, contradicting its primality.) Since $N(e') \leq n - 1$, the assertion is true for e', so w is a factor of A. □

The following characterization of $(2,4)$-semi-simple splicing languages in terms of $(2,4)$-arrow graphs holds.

Theorem 3. *(Analogue of Theorem 30.1 of [8]) For a language $L \subset V^*$, $R \subset R(L)$, where $R(L)$ is the set of $(2,4)$-semi-simple rules which respect L, take $\sigma = (V, R)$ and consider G_0 the prime $(2,4)$-arrow graph for (σ, L). Then there exists a finite $A \subset V^*$ such that $L = \sigma^*(A)$ if and only if G_0 is finite.*

Proof. From Lemma 9, if A is finite, then so is G_0.

Conversely, suppose G_0 is finite. For each prime edge $e = (b_1, w, b_0)$ and each rule $(1, a_0; 1, b_0)$ in $\bar{R}$ select one word in $\bar{L}$ which has b_1wb_0 as a factor. Let $\bar{A} = SAT$ be the set of these strings. We prove next that $\bar{L} = \bar{\sigma}^*(\bar{A})$, thus $L = \sigma^*(A)$.

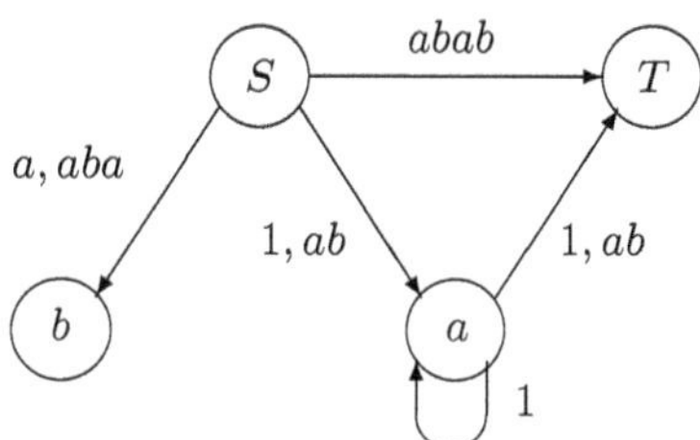

Fig. 1. The $(1,3)$ prime arrow graph of $L_1 = a^+ \cup a^+ab \cup aba^+ \cup aba^+b$

Take any edge in G from S to T, and consider $e = e_ne_{n-1}\cdots e_1$ its prime factorization, with $e_k = (b_k, w_k, b_{k-1})$, $1 \leq k \leq n$. For every k, select a rule $r_k = (1, a_{k-1}; 1, b_{k-1})$ and a word z_k in $\bar{A}$ which has $b_kw_ka_{k-1}$ as a factor. For $k = 1$, $b_0 = T$, thus $a_0 = T$, and b_1w_1T is a factor (actually a suffix) of z_1. For $k = n$, $b_n = S$, thus Sw_na_{n-1} is a factor (actually a prefix) of z_n. We splice z_2 and z_1 using rule r_1, next z_3 and the previously obtained string using r_2, and so on, until splicing of z_n and the previously obtained string, using r_{n-1}, finally gives us $\lambda(e)$. □

Corollary 2. *A language $L \subseteq V^*$ belongs to the class $SH(2,4)$ if and only if the prime $(2,4)$ arrow graph constructed from $(\hat{\sigma}, L)$, with $(\hat{\sigma} = (V, R(L))$, is finite.*

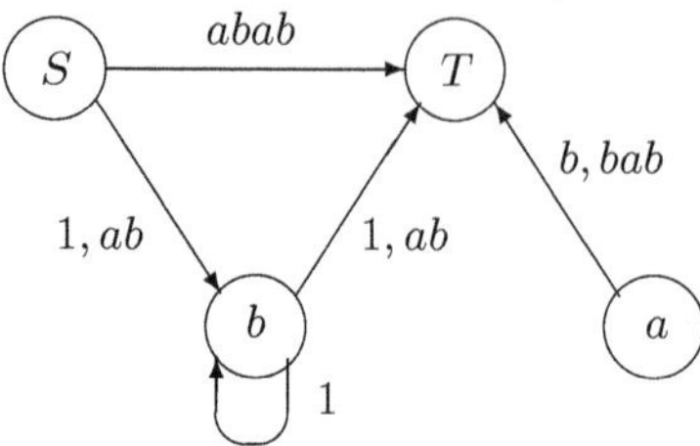

Fig. 2. The $(2,4)$ prime arrow graph of $L_2 = b^+ \cup abb^+ \cup ab^+ab \cup b^+ab$

The languages from Theorem 2 have their prime arrow graphs depicted in Figures 1 and 2.

The languages from Examples 1 and 2 have their prime arrow graphs depicted in Figures 3 and 4.

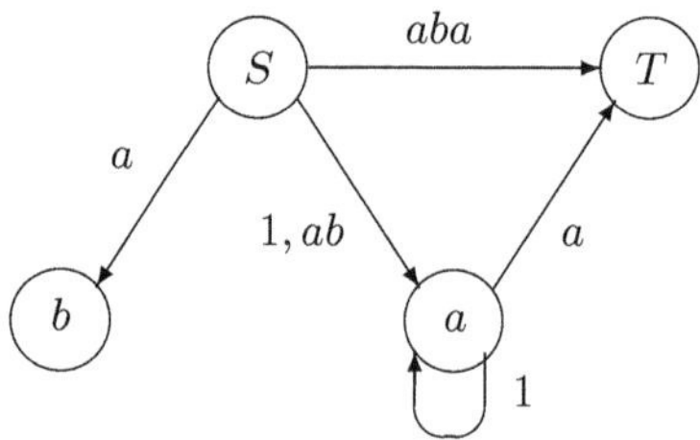

Fig. 3. The $(1,3)$ prime arrow graph of $L_3 = aa^+ \cup aba^+$

For *simple* splicing languages in the class $SH{=}SH(1,3) = SH(2,4)$, the $(1,3)$ arrow graph and the $(2,4)$ arrow graph coincide, because:

- (a,w,b) is a $(1,3)$-edge iff there exists a $(1,3)$ rule $(a,1;a,1)$ such that awb is a factor of L;
- (a,w,b) is a $(2,4)$-edge iff there exists a $(2,4)$ rule $(1,b;1,b)$ such that awb is a factor of L.

Example 3 *([12]) Let $S_1 = (\{a,b,c\},\{abaca, acaba\},\{(b,1;b,1),(c,1;c,1)\})$ be a $(1,3)$ simple splicing system. The generated language is:*

$$L = L(S_1) = (abac)^+a \cup (abac)^*aba \cup (acab)^+a \cup (acab)^*aca.$$

The same language can be generated by the $(2,4)$ simple splicing system $S_2 = (\{a,b,c\},\{abaca, acaba\},\{(1,b;1,b),(1,c;1,c)\})$, i.e., we have $L = L(S_2)$.

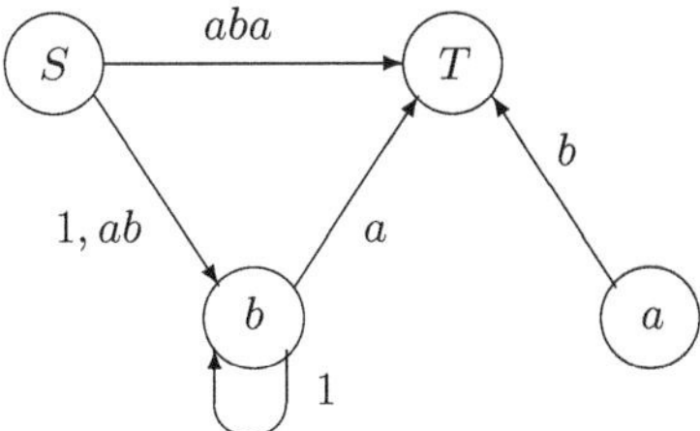

Fig. 4. The $(2,4)$ prime arrow graph of $L_4 = b^+a \cup ab^+a$

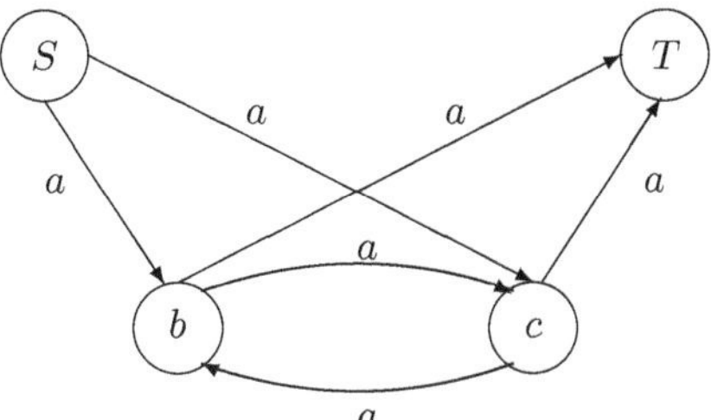

Fig. 5. The $(1,3)$ and also the $(1,4)$ prime arrow graph of $L = (abac)^+a \cup (abac)^*aba \cup (acab)^+a \cup (acab)^*aca$

Note that the axiom set of S_1 is closed to mirror image, and equals the axiom set of S_2. The $(1,3)$ arrow graph of L coincides with its $(2,4)$ arrow graph, and is depicted in Figure 5.

Due to the construction of the $(2,4)$-arrow graph, other results of [8] can be extended from languages in $SSH(1,3)$ to languages in the class $SSH(2,4)$. We give the following, without proofs, since the proofs in the $(1,3)$ case are all based on the prime $(1,3)$ arrow graph construction, and can be replaced by similar proofs, based on the prime $(2,4)$ arrow graph.

First, some relations with the notion of *constant* introduced by Schützenberger [16]. A *constant of a language* $L \subseteq V^*$ is a string $c \in V^*$ such that for all $x, y, x', y' \in V^*$, if xcy and $x'cy'$ are in L, then xcy' is in L. The relation with the splicing operation is obvious, since $xcy' = (c,1;c,1)(xcy, x'cy') = (1,c;1,c)(xcy, x'cy')$. Thus c is a constant of L iff $r(L) \subseteq L$ for $r = (c,1;c,1)$, or iff $r'(L) \subseteq L$ for $r' = (1,c;1,c)$.

Theorem 4. *A language $L \subseteq V^*$ is a simple splicing language (in the class SH=$SH(1,3) = SH(2,4)$) iff there exists an integer K such that every factor of L of length at least K contains a symbol which is a constant of L.*

There are many constants in the $(1,3)$ and the $(2,4)$ semi-simple cases.

Theorem 5. *If $L \in SSH(1,3) \cup SSH(2,4)$ then there exists a positive integer K such that every string in V^* of length at least K is a constant of L.*

For $L \in SSH(1,3)$ the result is proved in [8]. For $L \in SSH(2,4)$ the same proof holds, replacing prefix edge with suffix edge.

According to [7], a language L is *strictly locally testable* if and only if there exists a positive integer K such that every string in V^* of length at least K is a constant of L. We have thus:

Corollary 3. *(extension of Corollary 30.2 of [8]) If L is a* $(1,3)$ *or a* $(2,4)$*-semi-simple splicing language, then L is strictly locally testable.*

5 A Hierarchy of Classes of Semi-simple Languages

To illustrate the number and variety of open problems in this area, let us note that there is no connection formulated yet between the semi-simple languages as introduced by Goode and Pixton in [8], and the classes introduced by Tom Head in [10].

Recall that, inspired by the study of simple splicing languages introduced in [12], Head introduced the families of splicing languages $\{S_kH \mid k \geq -1\}$ using the *null context splicing rules* defined in [9]. A *splicing rule*, according to the original definition of [9], is a sixtuple $(u_1, x, u_2; u_3, x, u_4)$ (x is called the cross-over of the rule) which acts on strings precisely as the (Păun type) rule $(u_1x, u_2; u_3x, u_4)$. (For a detailed discussion of the three types of splicing – definitions given by Head, Păun, and Pixton – and a comparison of their generative power, we send the reader to [1], [2], [3], [4].)

A *null context splicing rule* is a rule of the particular form $(1, x, 1; 1, x, 1)$ (all contexts u_i are the empty string 1) and thus will be precisely a rule of type $(x, 1; x, 1)$ with $x \in V^*$. Such a rule is identified by Tom Head with the string $x \in V^*$. For a fixed $k \geq -1$, a splicing system (V, A, R) with rules $(x, 1; x, 1) \in R$ such that $|x| \leq k$ is called an S_kH system, and languages generated by an S_kH system are the S_kH *languages* (more appropriately, $S_kH(1,3)$ languages).

Note that the class $S_1H(1,3)$ is precisely the class SH of simple H systems of [12].

For $k = -1$, the $S_{-1}H(1,3)$ systems are precisely those for which the set of rules R is empty. Thus the languages in the class $S_{-1}H(1,3)$ are precisely the finite languages, i.e., $S_{-1}H(1,3) = FIN$.

For $k = 0$, the $S_0H(1,3)$ systems have either R empty, or $R = \{(1,1;1,1)\}$. Thus infinite languages in the class $S_0H(1,3)$ are generated by H systems of the form $S = (V, A, \{(1,1;1,1)\})$, with $A \neq \emptyset$, and it can be easily shown that $L(S) = alph(A)^*$, where $alph(A)$ denotes the set of symbols appearing in strings of A.

The union of all classes S_kH is precisely the set of null context splicing languages. It is shown in [10] that this union coincides with the family of strictly locally testable languages (in the sense of [7]). Also from [10] we have the following result:

Lemma 10. *The sequence* $\{S_kH(1,3) \mid k \geq -1\}$ *is a strictly ascending infinite hierarchy of language families for alphabets of two or more symbols.*

Proof. The inclusions are obvious, their strictness is established via an example:

Example 4 *[10] Let $V = \{a,b\}$. For each $k \geq -1$ let $L_k = a^k(b^k a^{(k-1)} b^k a^k)^*$. Note that $L_k = L(V, A_k, R_k)$ where $A_k = \{a^k b^k a^{(k-1)} b^k a^k\}$ and $R_k = (a^k, 1; a^k, 1)$. Thus $L_k \in S_k H(1,3)$ and is not in $S_j H(1,3)$ for any $j < k$.* □

A "semi-simple" extension of the above concepts would be to consider rules of type $(u_1, 1; u_3, 1)$ with "radius" k, i.e., $|u_1|, |u_3| \leq k$ (as suggested also in [12]). Let us call such a rule a *k-fat semi-simple rule of type* $(1,3)$. For a fixed $k \geq -1$, a splicing system (V, A, R) with rules $r \in R$ of the above form is called a *k-fat semi-simple* $(1,3)$ *H system*, or briefly an $SS_k H(1,3)$ system, and languages generated by such systems are the $SS_k H(1,3)$ *languages.*

Note that the $SS_1 H(1,3)$ class is precisely the class $SSH(1,3)$ considered in [8].

For $k = -1$, similarly to the simple case, the $SS_{-1} H(1,3)$ systems are precisely those for which the set of rules R is empty, and thus the languages in the class $SS_{-1} H(1,3)$ are precisely the finite languages, $SS_{-1} H(1,3) = S_{-1} H(1,3) = FIN$.

For $k = 0$, again reasoning as in the simple case, we have that $SS_0 H(1,3)$ systems have either R empty, or $R = \{(1,1;1,1)\}$. Thus we have the equality of classes, $SS_0 H(1,3) = S_0 H(1,3)$, and thus $SS_0 H(1,3)$ contains only finite languages and languages of the form W^* for some subalphabet $W \subseteq V$.

We have the following analogue of Lemma 10:

Lemma 11. *The sequence $\{SS_k H(1,3) \mid k \geq -1\}$ is a strictly ascending infinite hierarchy of language families for alphabets of three or more symbols.*

Proof. The inclusions are obvious, their strictness follows from the example below.

Example 5 *Let $V = \{a,b,c\}$. For each $k \geq -1$ let*

$$L_k = a^k(b^k a^{(k-1)} b^k a^k)^+ \cup c^k(b^k a^{(k-1)} b^k a^k)^+.$$

Note that $L_k = L(V, A_k, R_k)$ where $A_k = \{a^k b^k a^{(k-1)} b^k a^k, c^k b^k a^{(k-1)} b^k a^k\}$ and $R_k = (a^k, 1; c^k, 1)$. Thus $L_k \in SS_k H(1,3)$ and is not in $SS_j H(1,3)$ for any $j < k$. □

It is unknown whether the result above holds for an alphabet of two symbols.

It remains to be investigated whether other results obtained in [10] for $S_k H$ languages can be extended to $SS_k H$ languages. The relationship with the hierarchy of regular languages in [13] also remains to be studied.

Considering other types (i,j), different from $(1,3)$, is also worth exploring further.

For two words $u, v \in V^*$, such that $|u|, |v| \leq k$, a *semi-simple k-fat (linear) splicing rule*, with markers u and v, is a splicing rule of one of the following four types: $(u,1;v,1)$, $(1,u;1,v)$, $(1,u;v,1)$, $(u,1;1,v)$. The corresponding $SS_k H(i,j)$

systems and languages can be defined as usual. For instance, a *k-fat semi-simple* (2,4) *H system*, or briefly an $SS_kH(2,4)$ system, is an H system with rules of the form $(1,x;1,y)$, with $x,y \in V^*$, and $|x|,|y| \leq k$. For every $k \geq -1$ we will have the corresponding class of languages, $SS_kH(2,4)$. One can readily prove analogues of Proposition 1 and Corollary 1 for arbitrary $k \geq -1$.

Proposition 2. *Let $u,v \in V^*$ be two strings. The following equality of functions from $V^* \times V^*$ to 2^{V^*} holds:*

$$(u,1;v,1)mi = inv(mi \times mi)(1,mi(v);1,mi(u)).$$

Corollary 4. *For every $k \geq -1$, there exists a bijection between the classes of languages $SSH_k(1,3)$ and $SSH_k(2,4)$.*

The bijection between H systems will be given by standard passing from a k-fat (1,3) rule $(u,1;v,1)$, to the k-fat (2,4) rule $(1,mi(v);1,mi(u)$, and by mirroring the axiom set. On languages it will again be the mirror image.

Using this fact, an analogue of Lemma 11 can be readily proved for the sequence $\{SS_kH(2,4) \mid k \geq -1\}$.

6 Simple Circular *H* Systems

Splicing for circular strings was considered by Tom Head in [9]. In [11] the circular splicing operation which uses a rule of the general type $r = (u_1,u_2;u_3,u_4)$ is defined by:

$$(\hat{}xu_1u_2, \hat{}yu_3u_4) \vdash_r \hat{}xu_1u_4yu_3u_2$$

and is depicted in Figure 6.

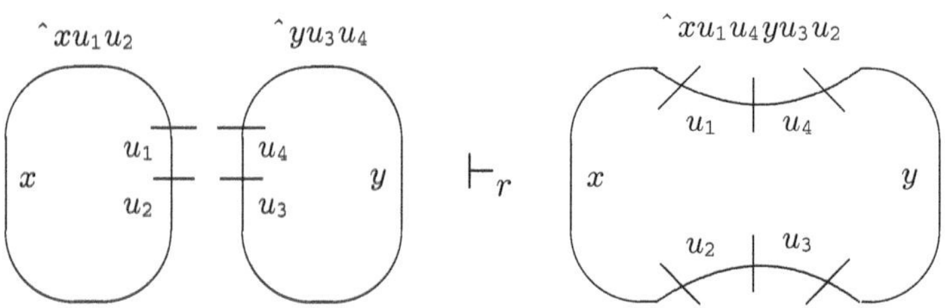

Fig. 6. Circular splicing

We mention the following easy to prove properties of circular splicing, which are not shared with the linear splicing.

Lemma 12. *For every splicing rule $(u_1,u_2;u_3,u_4)$, and every $\hat{}x, \hat{}y, \hat{}z \in V^\circ$ such that $(\hat{}x,\hat{}y) \vdash_{(u_1,u_2;u_3,u_4)} \hat{}z$ we have:*
the length preserving property: $|\hat{}x| + |\hat{}y| = |\hat{}z|$,
the symbol preserving property: $|\hat{}x|_b + |\hat{}y|_b = |\hat{}z|_b$ for every $b \in V$.

For a fixed symbol $a \in V$ (a marker) we can consider four types of *simple circular splicing rules*: $(a, 1; a, 1)$, $(1, a; 1, a)$, $(1, a; a, 1)$, $(a, 1; 1, a)$. Their action on circular strings is illustrated below:

$$\begin{aligned} type(1,3) &: (\hat{\ }xa, \hat{\ }ya) \vdash_{(a,1;a,1)} \hat{\ }xaya, \\ type(2,4) &: (\hat{\ }ax, \hat{\ }ay) \vdash_{(1,a;1,a)} \hat{\ }axay, \\ type(2,3) &: (\hat{\ }ax, \hat{\ }ya) \vdash_{(1,a;a,1)} \hat{\ }axya, \\ type(1,4) &: (\hat{\ }xa, \hat{\ }ay) \vdash_{(a,1;1,a)} \hat{\ }xaay. \end{aligned}$$

For $i = 1, 2$ and $j = 3, 4$ an *(i, j)-simple circular H system* is a circular H system $S = (V, A, R)$, with the set of rules R consisting only of simple rules of type (i, j) for some $a \in V$.

We denote by $SH^\circ(i, j)$ the class of all (i, j)-simple circular languages. Only two out of these four classes are distinct.

Theorem 6. *(See Theorems 4.1 and 4.2 of [5]) We have:*
(i) $SH^\circ(1,3) = SH^\circ(2,4)$.
(ii) $SH^\circ(2,3) = SH^\circ(1,4)$.
(iii) The classes $SH^\circ(1,3)$ *and* $SH^\circ(2,3)$ *are incomparable.*

Several other properties which emphasize the difference between the case of *linear* simple splicing and that of *circular* simple splicing (for instance the behavior over the one-letter alphabet) have been presented in [5].

We have $SH^\circ(1,3) \subset REG^\circ$ as proved in [5], and in [17] in a different context. The relationship between $SH^\circ(1,4)$ and REG° is still an open problem.

We give next some closure properties of $SH^\circ(1,3)$.

Theorem 7. *We have the following:*
(i) $SH^\circ(1,3)$ *is* not *closed under union.*
(ii) $SH^\circ(1,3)$ *is* not *closed under intersection with* REG°.
(iii) $SH^\circ(1,3)$ *is* not *closed under morphisms.*
(iv) $SH^\circ(1,3)$ *is* not *closed under inverse morphisms.*

Proof. For assertion (i), take $L_1 = \{\hat{\ }a^n \mid n \geq 1\} \in SH^\circ(1,3)$, and $L_2 = \{\hat{\ }(ab)^n \mid n \geq 1\} \in SH^\circ(1,3)$. Suppose $L_1 \cup L_2 \in SH^\circ(1,3)$, then it would be closed under some $(1,3)$ rules. But with rule $(a, 1; a, 1)$ we get:

$(\hat{\ }(a)^n, \hat{\ }(ab)^n) \vdash_{(a,1;a,1)} \hat{\ }a^n(ab)^n \notin L_1 \cup L_2$,

and then, the only simple splicing rule which could be used to generate $L_1 \cup L_2$, which is infinite, would be $(b, 1; b, 1)$. But the words $\hat{\ }a^n$ cannot be generated using this rule.

For (ii), take $V^\circ \in SH^\circ(1,3)$ and $L = \{\hat{\ }a^n b \mid n \geq 1\}$ which is in REG° but not in $SH^\circ(1,3)$. We have $V^\circ \cap L = L \notin SH^\circ(1,3)$.

For (iii), take $L = \{\hat{\ }a^n \mid n \geq 1\} \cup \{\hat{\ }b^n \mid n \geq 1\} \in SH^\circ(1,3)$. Take the morphism h with $h(a) = a$ and $h(b) = ab$. Then $h(L) = \{\hat{\ }a^n \mid n \geq 1\} \cup \{\hat{\ }(ab)^n \mid n \geq 1\} \notin SH^\circ(1,3)$ (as proved in (i) above).

Fot (iv), take $V = \{a, b\}$ and $L = \{\hat{\ }a\} \in FIN^\circ \subset SH^\circ(1,3)$. Take morphism h defined by $h(a) = a$, $h(b) = 1$. Then $h^{-1}(L) = \{\hat{\ }b^n a \mid n \geq 0\} \notin SH^\circ(1,3)$. □

7 Semi-simple Circular Splicing Systems

For fixed letters $a, b \in V$, we will consider the following four types of *semi-simple circular splicing rules*: $(a,1;b,1)$, $(1,a;1,b)$, $(1,a;b,1)$, $(a,1;1,b)$.

The action of the four rules on circular strings is depicted in Figures 7, 8, 9 and 10.

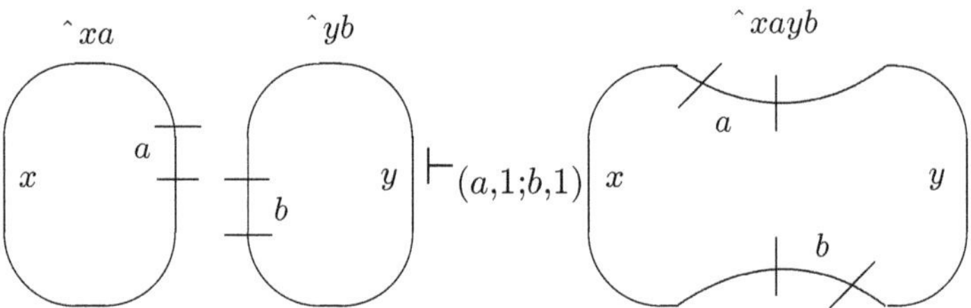

Fig. 7. Circular splicing using the semi-simple rule (a,1;b,1)

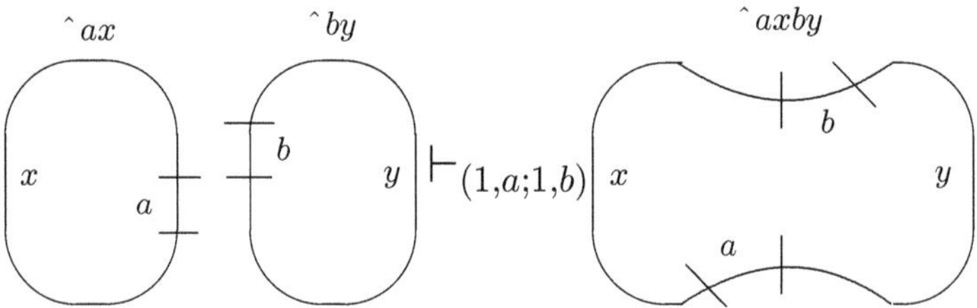

Fig. 8. Circular splicing using the semi-simple rule (1,a;1,b)

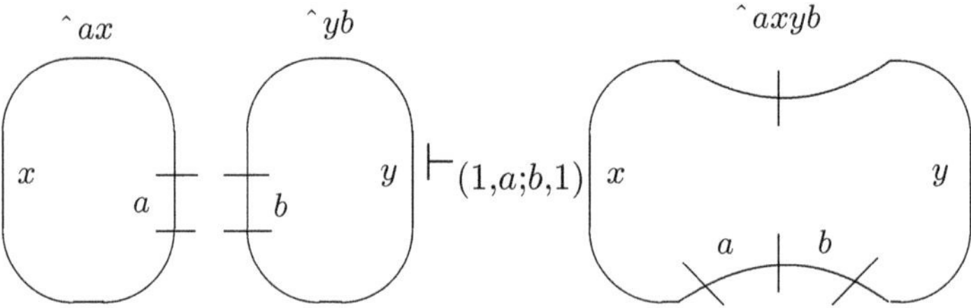

Fig. 9. Circular splicing using the semi-simple rule (1,a;b,1)

We will consider the semi-simple splicing rules as functions from $V^\circ \times V^\circ$ to 2^{V°. Using the mirror image function we can express some relationships between the four types of rules seen as functions.

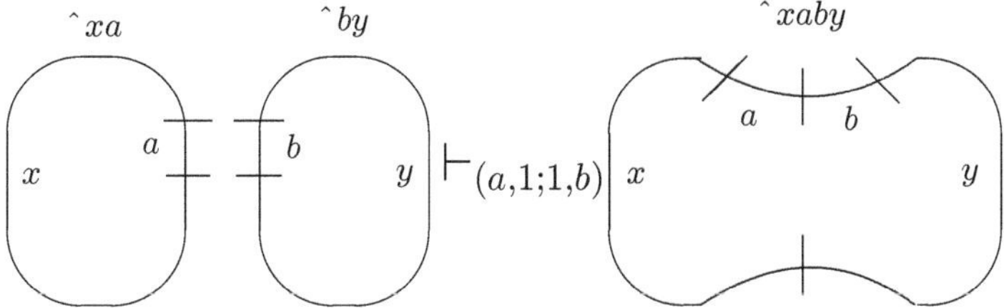

Fig. 10. Circular splicing using the semi-simple rule (a,1;1,b)

Lemma 13. *The following equalities of functions from $V_a^\circ \times V_b^\circ$ to 2^{V° hold:*

$$(a,1;b,1)mi = (mi \times mi)(1,a;1,b),$$
$$(a,1;1,b)mi = (mi \times mi)(1,a;b,1).$$

Theorem 8. *The classes $SSH^\circ(1,3)$ and $SSH^\circ(2,4)$ are incomparable.*

Proof. Note first that simple circular splicing languages (in $SH^\circ(1,3) = SH^\circ(2,4)$) are in the intersection $SSH^\circ(1,3) \cap SSH^\circ(2,4)$.

The semi-simple H system $S_1 = (\{a,b,c\}, \{\hat{}aac, \hat{}b\}, \{(c,1;b,1)\})$ generates a language $L_1 = L(S_1) \in SSH^\circ(1,3) \setminus SSH^\circ(2,4)$.

Note that the words of L_1 have the following two properties:

(I) All words in L_1, with the exception of the two axioms, have occurrences of all three letters a, b, and c.

(II) All occurrences of a in the words of L_1 are in subwords of the form aac.

Now, suppose L_1 is $(2,4)$ generated. Then it would be respected by one or several of the rules of the forms:

(i) $(1,c;1,b)$, (ii) $(1,b;1,c)$, (iii) $(1,a;1,b)$, (iv) $(1,b;1,a)$,
(v) $(1,c;1,a)$, (vi) $(1,a;1,c)$, (vii) $(1,a;1,a)$, (viii) $(1,b;1,b)$, (ix) $(1,c;1,c)$.

But we have:
(v) $(\hat{}aac, \hat{}aac) \vdash_{(1,c;1,a)} \hat{}aacaca \notin L_1$ (no b's),
(vi) $(\hat{}aac, \hat{}aac) \vdash_{(1,a;1,c)} \hat{}aacaca \notin L_1$ (no b's),
(vii) $(\hat{}aac, \hat{}aac) \vdash_{(1,a;1,a)} \hat{}caacaa \notin L_1$ (no b's),
(ix) $(\hat{}aac, \hat{}aac) \vdash_{(1,c;1,c)} \hat{}caacaa \notin L_1$ (no b's),
(viii) $(\hat{}b, \hat{}b) \vdash_{(1,b;1,b)} \hat{}bb \notin L_1$ (no a's, no c's),
(i) $(\hat{}aac, \hat{}b) \vdash_{(1,c;1,b)} \hat{}caab \notin L_1$ (contradicts (II)),
(ii) $(\hat{}b, \hat{}aac) \vdash_{(1,b;1,c)} \hat{}bcaa \notin L_1$ (contradicts (II)),
(iii) $(\hat{}aac, \hat{}b) \vdash_{(1,a;1,b)} \hat{}acab \notin L_1$ (contradicts (II)),
(iv) $(\hat{}b, \hat{}aac) \vdash_{(1,b;1,a)} \hat{}baca \notin L_1$ (contradicts (II)).

The semi-simple H system $S_2 = (\{a,b,c\}, \{\hat{}aac, \hat{}b\}, \{(1,c;1,b)\})$ generates a language $L_2 = L(S_2) \in SSH^\circ(2,4) \setminus SSH^\circ(1,3)$. The proof that L_2 is not in $SSH^\circ(1,3)$ is similar to the above one. We note that the words of L_2 have the following properties:

(I) All words in L_2, with the exception of the two axioms, have occurrences of all three letters a, b, and c.

(II) All occurrences of a in the words of L_2 are in subwords of the form caa.

If we suppose L_2 is $(1,3)$ generated, then it would be respected by one or several of the rules of the forms:

(i) $(c,1;b,1)$, (ii) $(b,1;c,1)$, (iii) $(a,1;b,1)$, (iv) $(b,1;a,1)$,

(v) $(c,1;a,1,)$ (vi) $(a,1;c,1)$, (vii) $(a,1;a,1)$, (viii) $(b,1;b,1)$, (ix) $(c,1;c,1)$.

But we have:

(v) $(\hat{}aac,\hat{}aac) \vdash_{(c,1;a,1)} \hat{}aacaca \notin L_2$ (no b's, contradicting (I)),
(vi) $(\hat{}aac,\hat{}aac) \vdash_{(a,1;c,1)} \hat{}acaaac \notin L_2$ (no b's),
(vii) $(\hat{}aac,\hat{}aac) \vdash_{(a,1;a,1)} \hat{}caacaa \notin L_2$ (no b's),
(ix) $(\hat{}aac,\hat{}aac) \vdash_{(c,1;c,1)} \hat{}caacaa \notin L_2$ (no b's),
(viii) $(\hat{}b,\hat{}b) \vdash_{(b,1;b,1)} \hat{}bb \notin L_2$ (no a's, no c's),
(i) $(\hat{}aac,\hat{}b) \vdash_{(c,1;b,1)} \hat{}aacb \notin L_2$ (contradicts (II)),
(ii) $(\hat{}b,\hat{}aac) \vdash_{(b,1;c,1)} \hat{}baac \notin L_2$ (contradicts (II)),
(iii) $(\hat{}aac,\hat{}b) \vdash_{(a,1;b,1)} \hat{}acab \notin L_2$ (contradicts (II)),
(iv) $(\hat{}b,\hat{}aac) \vdash_{(b,1;a,1)} \hat{}baca \notin L_2$ (contradicts (II)). □

8 Conclusions and Further Research

We have introduced some particular classes of splicing languages, continuing the studies of [12], [8], and [10]. Our emphasis was on *classes* associated to *types*, different from the $(1,3)$ type, which has been more extensively studied. We have also proposed extensions of these concepts to the circular case, continuing the research started in [5]. We have pointed out several open problems which remain to be investigated.

Following the lines of the extensive study in [12], many other problems can be formulated and investigated for the other classes as well: closure properties, decidability problems, descriptional complexity questions, characterization problems. The interesting notion of arrow graph is also worth exploring further. In particular, the behavior of the classes which arise from considering splicing rules of types $(2,3)$ and $(1,4)$ is expected to deviate from the "standard" of the $(1,3)$ and $(2,4)$ classes.

Acknowledgement. Part of the material in this paper was presented at the First Joint Meeting RSME-AMS, Sevilla, June 2003, as the conference talk entitled "Another Class of Semi-simple Splicing Languages".

References

1. P. Bonizzoni, C. De Felice, G. Mauri, R. Zizza, DNA and Circular Splicing, in *DNA Computing, LNCS vol.2054* (A. Condon, G. Rozenberg eds.), 2001, 117–129.
2. P. Bonizzoni, C. De Felice, G. Mauri, R. Zizza, Circular Splicing and Regularity, 2002, submitted.

3. P. Bonizzoni, C. De Felice, G. Mauri, R. Zizza, Decision Problems for Linear and Circular Splicing Systems, *Proceedings DLT02*, to appear.
4. P. Bonizzoni, C. Ferreti, G. Mauri, R. Zizza, Separating Some Splicing Models, *Information Processing Letters*, 79/6 (2001), 255–259.
5. R. Ceterchi, K. G. Subramanian, Simple Circular Splicing Systems (to appear *ROMJIST* 2003).
6. K. Culik, T. Harju, Splicing Semigroups of Dominoes and DNA, *Discrete Applied Mathematics*, 31(1991), 261–277.
7. A. DeLuca, A. Restivo, A Characterization of Strictly Locally Testable Languages and Its Application to Subsemigroups of a Free Semigroup, *Information and Control*, 44(1980), 300–319.
8. E. Goode, D. Pixton, Semi-simple Splicing Systems, in *Where Mathematics, Computer Science, Linguistics and Biology Meet* (C. Martín-Vide, V. Mitrana eds.) Kluwer Academic Publ., Dordrecht, 2001, 343–352.
9. T. Head, Formal Language Theory and DNA: An Analysis of the Generative Capacity of Specific Recombinant Behavior, *Bull. Math. Biol.*, 49 (1987), 737-759.
10. T. Head, Splicing Representations of Strictly Locally Testable Languages, *Discrete Applied Math.*, 87 (1998), 139–147.
11. T. Head, G. Păun, D. Pixton, Language Theory and Molecular Genetics: Generative Mechanisms Suggested by DNA Recombination, in *Handbook of Formal Languages* (G. Rozenberg, A Salomaa, eds.), Springer-Verlag, Heidelberg, Vol. 2, 1997, 295–360.
12. A. Mateescu, G. Păun, G. Rozenberg, A. Salomaa, Simple Splicing Systems, *Discrete Applied Math.*, 84 (1998), 145–163.
13. G. Păun, On the Splicing Operation, *Discrete Applied Math.*, 70(1996), 57–79.
14. G. Păun, G. Rozenberg, A. Salomaa, *DNA Computing. New Computing Paradigms*, Springer-Verlag, Berlin, 1998.
15. D. Pixton, Regularity of Splicing Languages, *Discrete Applied Math.*, 69 (1996), 101–124.
16. M. P. Schützenberger, Sur certaines operations de fermeture dans les langages rationnels, *Symposia Math.*, 15 (1975), 245–253.
17. R. Siromoney, K. G. Subramanian, V. R. Dare, Circular DNA and Splicing Systems, in *Proc. of ICPIA, LNCS vol. 654*, Springer-Verlag, Berlin, 1992, 260–273.

The Power of Networks of Watson-Crick D0L Systems

Erzsébet Csuhaj-Varjú[1] and Arto Salomaa[2]

[1] Computer and Automation Research Institute
Hungarian Academy of Sciences
Kende u. 13-17
H-1111 Budapest, Hungary
csuhaj@sztaki.hu

[2] Turku Centre for Computer Science
Lemminkäisenkatu 14 A
FIN -20520 Turku, Finland
asalomaa@it.utu.fi

Abstract. The notion of a network of Watson-Crick D0L systems was recently introduced, [7]. It is a distributed system of deterministic language defining devices making use of Watson-Crick complementarity. The research is continued in this paper, where we establish three results about the power of such networks. Two of them show how it is possible to solve in linear time two well-known NP-complete problems, the Hamiltonian Path Problem and the Satisfiability Problem. Here the characteristic feature of DNA computing, the massive parallelism, is used very strongly. As an illustration we use the propositional formula from the celebrated recent paper, [3]. The third one shows how in the very simple case of four-letter DNA alphabets we can obtain weird (not even Z-rational) patterns of population growth.

1 Introduction

Many mathematical models of DNA computing have been investigated, some of them already before the fundamental paper of Adleman, [1]. The reader is referred to [11,2] for details. Networks of language generating devices, in the sense investigated in this paper, were introduced in [5,6,7]. This paper continues the work begun in [7]. The underlying notion of a Watson-Crick D0L system was introduced and studied further in [10,8,16,17,18,20,4].

The reader can find background material and motivation in the cited references. Technically this paper is largely self-contained. Whenever need arises, [14,13] can be consulted in general matters about formal languages, [12] in matters dealing with Lindenmayer systems, and [19,9] in matters dealing with formal power series.

Watson-Crick complementarity is a fundamental concept in DNA computing. A notion, called Watson-Crick D0L system, where the paradigm of complementarity is considered in the operational sense, was introduced in [10].

N. Jonoska et al. (Eds.): Molecular Computing (Head Festschrift), LNCS 2950, pp. 106–118, 2004.

A Watson-Crick D0L system (a WD0L system, for short) is a D0L system over a so-called DNA-like alphabet Σ and a mapping ϕ called the mapping defining the trigger for complementarity transition. In a DNA-like alphabet each letter has a complementary letter and this relation is symmetric. ϕ is a mapping from the set of strings (words) over the DNA-like alphabet Σ to $\{0,1\}$ with the following property: the ϕ-value of the axiom is 0 and whenever the ϕ-value of a string is 1, then the ϕ-value of its complementary string must be 0. (The complementary string of a string is obtained by replacing each letter in the string with its complementary letter.) The derivation in a Watson-Crick D0L system proceeds as follows: when the new string has been computed by applying the homomorphism of the D0L system, then it is checked according to the trigger. If the ϕ-value of the obtained string is 0 (the string is a correct word), then the derivation continues in the usual manner. If the obtained string is an incorrect one, that is, its ϕ-value is equal to 1, then the string is changed for its complementary string and the derivation continues from this string.

The idea behind the concept is the following: in the course of the computation or development things can go wrong to such extent that it is of worth to continue with the complementary string, which is always available. This argument is general and does not necessarily refer to biology. Watson-Crick complementarity is viewed as an *operation*: together with or instead of a word w we consider its complementary word.

A step further was made in [7]: *networks* of Watson-Crick D0L systems were introduced. The notion was a particular variant of a general paradigm, called networks of language processors, introduced in [6] and discussed in details in [5].

A network of Watson-Crick D0L systems is a finite collection of Watson-Crick D0L systems over the same DNA-like alphabet and with the same trigger. These WD0L systems act on their own strings in a synchronized manner and after each derivation step communicate some of the obtained words to each other. The condition for communication is determined by the trigger for complementarity transition. Two variants of communication protocols were discussed in [7]. In the case of protocol (a), after performing a derivation step, the node keeps every obtained correct word and the complementary word of each obtained incorrect word (each corrected word) and sends a copy of each *corrected* word to every other node. In the case of protocol (b), as in the previous case, the node keeps all the correct words and the corrected ones (the complementary strings of the incorrect strings) but communicates a copy of each *correct* string to each other node. The two protocols realize different strategies: in the first case, if some error is detected, it is corrected but a note is sent about this fact to the others. In the second case, the nodes inform the other nodes about their correct strings and keep for themselves all information which refers to the correction of some error.

The purpose of this paper is to establish three results about the power of such networks, where the trigger ϕ is defined in a very natural manner. The results are rather surprising because the underlying D0L systems are very simple. Two of them show how it is possible to solve in linear time two well-known NP-complete problems, the Hamiltonian Path Problem and the Satisfiability Problem. Here the characteristic feature of DNA computing, the massive parallelism, is used very strongly. As an illustration we use the formula from the celebrated recent paper, [3]. The third result shows how in the very simple case of four-letter DNA alphabets we can obtain weird (not even Z-rational) patterns of population growth.

2 Definitions

By a D0L system we mean a triple $H = (\Sigma, g, w_0)$, where Σ is an alphabet, g is an endomorphism defined on Σ^* and $w_0 \in \Sigma^*$ is the axiom. The word sequence $S(H)$ of H is defined as the sequence of words $w_0, w_1, w_2, \ldots$ where $w_{i+1} = g(w_i)$ for $i \geq 0$.

In the following we recall the basic notions concerning Watson-Crick D0L systems, introduced in [10,16].

By a DNA-like alphabet Σ we mean an alphabet with $2n$ letters, $n \geq 1$, of the form $\Sigma = \{a_1, \ldots, a_n, \bar{a}_1, \ldots, \bar{a}_n\}$. Letters a_i and $\bar{a}_i$, $1 \leq i \leq n$, are said to be complementary letters; we also call the non-barred symbols purines and the barred symbols pyrimidines. The terminology originates from the basic DNA alphabet $\{A, G, C, T\}$, where the letters A and G are for purines and their complementary letters T and C for pyrimidines.

We denote by h_w the letter-to-letter endomorphism of a DNA-like alphabet Σ mapping each letter to its complementary letter. h_w is also called the Watson-Crick morphism.

A Watson-Crick D0L system (a WD0L system, for short) is a pair $W = (H, \phi)$, where $H = (\Sigma, g, w_0)$ is a D0L system with a DNA-like alphabet Σ, morphism g and axiom $w_0 \in \Sigma^+$, and $\phi : \Sigma^* \to \{0, 1\}$ is a recursive function such that $\phi(w_0) = \phi(\lambda) = 0$ and for every word $u \in \Sigma^*$ with $\phi(u) = 1$ it holds that $\phi(h_w(u)) = 0$.

The word sequence $S(W)$ of a Watson-Crick D0L system W consists of words $w_0, w_1, w_2, \ldots$, where for each $i \geq 0$

$$w_{i+1} = \begin{cases} g(w_i) & \text{if } \phi(g(w_i)) = 0 \\ h_w(g(w_i)) & \text{if } \phi(g(w_i)) = 1. \end{cases}$$

The condition $\phi(u) = 1$ is said to be the trigger for complementarity transition. In the following we shall also use this term: a word $w \in \Sigma^*$ is called correct according to ϕ if $\phi(w) = 0$, and it is called incorrect otherwise. If it is clear from the context, then we can omit the reference to ϕ.

An important notion concerning Watson-Crick D0L systems is the *Watson-Crick road.* Let $W = (H, \phi)$ be a Watson-Crick D0L system, where $H = (\Sigma, g, w_0)$. The Watson-Crick road of W is an infinite binary word α over $\{0, 1\}$ such that the ith bit of α is equal to 1 if and only if at the ith step of the computation in W a transition to the complementary takes place, that is, $\phi(g(w_{i-1})) = 1$, $i \geq 1$, where $w_0, w_1, w_2, \ldots$ is the word sequence of W.

Obviously, various mappings ϕ can satisfy the conditions of defining a trigger for complementarity transition. In the following we shall use a particular variant and we call the corresponding Watson-Crick D0L system *standard.*

In this case a word w satisfies the trigger for turning to the complementary word (it is incorrect) if it has more occurrences of pyrimidines (barred letters) than purines (non-barred letters). Formally, consider a DNA-like alphabet $\Sigma = \{a_1, \ldots, a_n, \bar{a}_1, \ldots, \bar{a}_n\}$, $n \geq 1$. Let $\Sigma_{PUR} = \{a_1, \ldots, a_n\}$ and $\Sigma_{PYR} = \{\bar{a}_1, \ldots, \bar{a}_n\}$. Then, we define $\phi : \Sigma^* \to \{0, 1\}$ as follows: for $w \in \Sigma^*$

$$\phi(w) = \begin{cases} 0 & \text{if } |w|_{\Sigma_{PUR}} \geq |w|_{\Sigma_{PYR}} \text{ and} \\ 1 & \text{if } |w|_{\Sigma_{PUR}} < |w|_{\Sigma_{PYR}}. \end{cases}$$

Following [7], we now define the basic notion in this paper, a *network* of Watson-Crick D0L systems (an NWD0L system). It is a finite collection of WD0L systems over the same DNA-like alphabet and with the same trigger, where the component WD0L systems act in a synchronized manner by rewriting their own sets of strings in the WD0L manner and after each derivation step communicate some of the obtained words to each other. The condition for communication is determined by the trigger for complementarity transition.

Definition 1 *By an N_rWD0L system (a network of Watson-Crick D0L systems) with r components or nodes, where $r \geq 1$, we mean an $r+2$-tuple*

$$\Gamma = (\Sigma, \phi, (g_1, \{A_1\}), \ldots, (g_r, \{A_r\})),$$

where

- $\Sigma = \{a_1, \ldots, a_n, \bar{a}_1, \ldots, \bar{a}_n\}$, $n \geq 1$, *is a DNA-like alphabet, the alphabet of the system,*
- $\phi : \Sigma^* \to \{0, 1\}$ *is a mapping defining a trigger for complementarity transition, and*
- $(g_i, \{A_i\})$, $1 \leq i \leq r$, *called the ith component or the ith node of Γ, is a pair where g_i is a D0L morphism over Σ and A_i is a correct nonempty word over Σ according to ϕ, called the axiom of the ith component.*

If the number of the components in the network is irrelevant, then we speak of an NWD0L system. An NWD0L system is called *standard* if ϕ is defined in the same way as in the case of standard WD0L systems.

Definition 2 *For an N_rWD0L system $\Gamma = (\Sigma, \phi, (g_1, \{A_1\}), \ldots, (g_r, \{A_r\}))$, $r \geq 1$, the r-tuple $(L_1, \ldots, L_r)$, where L_i, $1 \leq i \leq r$, is a finite set of correct strings over Σ according to ϕ, is called a state of Γ. L_i, $1 \leq i \leq r$, is called the state or contents of the ith component. $(\{A_1\}, \ldots, \{A_r\})$ is said to be the initial state of Γ.*

NWD0L systems change their states by direct derivation steps. A direct change of a state to another one means a rewriting step followed by communication according to the given protocol of the system. In the following we define two variants of communication protocols, representing different communication philosophies.

Definition 3 *Let* $s_1 = (L_1, \ldots, L_r)$ *and* $s_2 = (L'_1, \ldots, L'_r)$ *be two states of an* N_r *WD0L system* $\Gamma = (\Sigma, \phi, (g_1, \{A_1\}), \ldots, (g_r, \{A_r\}))$, $r \geq 1$.

- *We say that* s_1 *directly derives* s_2 *by protocol* (a), *written as* $s_1 \Longrightarrow_a s_2$, *if*

$$L'_i = C'_i \cup_{j=1}^{r} h_w(B'_j),$$

where
$C'_i = \{g_i(v) \mid v \in L_i, \phi(g_i(v)) = 0\}$ *and*
$B'_j = \{g_j(u) \mid u \in L_j, \phi(g_j(u)) = 1\}$.
- *We say that* s_1 *directly derives* s_2 *by protocol* (b), *written as* $s_1 \Longrightarrow_b s_2$, *if*

$$L'_i = h_w(B'_i) \cup_{j=1}^{r} C'_j,$$

where
$B'_i = \{g_i(u) \mid u \in L_i, \phi(g_i(u)) = 1\}$ *and*
$C'_j = \{g_j(v) \mid v \in L_j, \phi(g_j(v)) = 0\}$ *holds.*

Thus, in the case of both protocols, after applying a derivation step in the WD0L manner the node keeps the correct words and the corrected words (the complementary words of the incorrect ones), and in the case of protocol (a) it sends a copy of every *corrected* word to each other node, while in the case of protocol (b) it communicates a copy of every *correct* word to each other node. The two protocols realize different communication strategies: In the first case the nodes inform each other about the correction of the detected errors, while in the second case the nodes inform each other about the obtained correct words.

Definition 4 *Let* $\Gamma = (\Sigma, \phi, (g_1, \{A_1\}), \ldots, (g_r, \{A_r\}))$, *for* $r \geq 1$, *be an* N_r *WD0L system with protocol* (x), $x \in \{a, b\}$.

The state sequence $S(\Gamma)$ *of* Γ, $S(\Gamma) = s(0), s(1), \ldots$, *is defined as follows:* $s(0) = (\{A_1\}, \ldots, \{A_r\})$ *and* $s(t) \Longrightarrow_x s(t+1)$ *for* $t \geq 0$.

The notion of a *road* is extended to concern NWD0L systems and their components in the natural fashion. For formal details we refer to [7].

3 Satisfiability Problem

It is well known that the satisfiability problem SAT of propositional formulas is NP-complete, [13,15]. We will consider propositional formulas α in conjunctive (not necessarily 3-conjunctive) normal form. Thus α is a conjunction of disjunctions whose terms are *literals*, that is, variables or their negations. The disjunctions are referred to as *clauses* of α. We assume that α contains v variables $x_i, 1 \leq i \leq v$, and c clauses. The formula α is *satisfiable* if there is a truth-value assignment for the variables (that is, an assignment of T or F) giving α the value T. When we speak of computions (dealing with α) in *linear time*, we refer to functions linear either in v or c.

To illustrate our arguments, we use the propositional formula β from [3]. A DNA computer was constructed in [3] to solve the satisfiability problem of β. The formula β, given below, is in 3-conjunctive normal form, and involves 20 variables and 24 clauses.

$$\begin{aligned}
&(\sim x_3 \vee \sim x_{16} \vee x_{18}) \wedge (x_5 \vee x_{12} \vee \sim x_9)\\
&\wedge(\sim x_{13} \vee \sim x_2 \vee x_{20}) \wedge (x_{12} \vee \sim x_9 \vee \sim x_5)\\
&\wedge(x_{19} \vee \sim x_4 \vee x_6) \wedge (x_9 \vee x_{12} \vee \sim x_5)\\
&\wedge(\sim x_1 \vee x_4 \vee \sim x_{11}) \wedge (x_{13} \vee \sim x_2 \vee \sim x_{19})\\
&\wedge(x_5 \vee x_{17} \vee x_9) \wedge (x_{15} \vee x_9 \vee \sim x_{17})\\
&\wedge(\sim x_5 \vee \sim x_9 \vee \sim x_{12}) \wedge (x_6 \vee x_{11} \vee x_4)\\
&\wedge(\sim x_{15} \vee \sim x_{17} \vee x_7) \wedge (\sim x_6 \vee x_{19} \vee x_{13})\\
&\wedge(\sim x_{12} \vee \sim x_9 \vee x_5) \wedge (x_{12} \vee x_1 \vee x_{14})\\
&\wedge(x_{20} \vee x_3 \vee x_2) \wedge (x_{10} \vee \sim x_7 \vee \sim x_8)\\
&\wedge(\sim x_5 \vee x_9 \vee \sim x_{12}) \wedge (x_{18} \vee \sim x_{20} \vee x_3)\\
&\wedge(\sim x_{10} \vee \sim x_{18} \vee \sim x_{16}) \wedge (x_1 \vee \sim x_{11} \vee \sim x_{14})\\
&\wedge(x_8 \vee \sim x_7 \vee \sim x_{15}) \wedge (\sim x_8 \vee x_{16} \vee \sim x_{10}).
\end{aligned}$$

For each variable x_i, we introduce two auxiliary letters t_i and f_i. Intuitively, t_i (resp. f_i) indicates that the value T (resp. F) is assigned to x_i. The letter t_i (resp. f_i) is *characteristic* for the clause C if the variable x_i appears unnegated (resp. negated) in C. Thus, the letters f_3, f_{16}, t_{18} are characteristic for the first clause in the formula β above, whereas the letters f_8, f_{10}, t_{16} are characteristic for the last clause.

Theorem 1 *The satisfiability problem can be solved in linear time by standard NWD0L systems.*

Proof. Consider a propositional formula α in conjunctive normal form, with v variables $x_i, 1 \leq i \leq v$, and c clauses $C_i, 1 \leq i \leq c$. We construct a standard N_{2v+c+1}WD0L system Γ as follows. The nodes of Γ are denoted by

$$M_i, M_i', \ 1 \leq i \leq v, \ N_i, \ 1 \leq i \leq c, \ P.$$

The alphabet consists of the letters in the two alphabets

$$V_1 = \{S_i \mid 0 \leq i \leq v-1\} \cup \{R_i \mid 1 \leq i \leq c\} \cup \{E, G, G^v\}$$

and

$$V_2 = \{t_i, f_i \mid 1 \leq i \leq v\},$$

as well as of their barred versions.

The communication protocol (*b*) will be followed, that is, correct words will be communicated. Intuitively, the letters S_i are associated to the variables, the letters R_i to the clauses, and the letters t_i, f_i to the truth-values. The letter E is a special ending letter, and the letter G a special garbage letter.

We now define the D0L productions for each node. We begin with the letters of V_2. In all nodes M_i, M'_i, $1 \leq i \leq v$, as well as in the node P, we have the productions

$$t_j \to t_j,\ f_j \to f_j,\ 1 \leq j \leq v.$$

In all nodes N_i, $1 \leq i \leq c$, we have the productions

$$t_j \to t_j\overline{t_j},\ f_j \to f_j\overline{f_j},\ \overline{t_j} \to \lambda,\ \overline{f_j} \to \lambda,\ 1 \leq j \leq v,$$

except in the case that t_j or f_j is characteristic for the clause C_i the barred letters are removed from the former productions, giving rise to the production $t_j \to t_j$ or $f_j \to f_j$.

We then define the productions for the remaining letters. Each node M_i (resp. M'_i), $1 \leq i \leq v-1$, has the production $S_{i-1} \to t_i S_i$ (resp. $S_{i-1} \to f_i S_i$). The node M_v (resp. M'_v) has the production $S_{v-1} \to t_v\overline{R_1}$ (resp. $S_{v-1} \to f_v\overline{R_1}$). Each node N_i, $1 \leq i \leq c-1$, has the production $\overline{R_i} \to \overline{R_{i+1}}$. The node N_c has the production $\overline{R_c} \to \overline{E}$. The node P has the production $\overline{E} \to \lambda$. Each letter x, barred or nonbarred, whose production has not yet been defined in some node, has in this node the production $x \to \overline{G^v}$. This completes the definition of the standard network Γ.

The formula α is satisfiable exactly in case, after $v+c+1$ computation steps, a word w over the alphabet V_2 appears in the node P. Each such word w indicates a truth-value assignment satisfying α. Moreover, because of the communication, each such word w appears in all nodes in the next steps.

The verification of this fact is rather straightforward. There is only one "proper" path of computation. In the first v steps a truth-value assignment is created. (Actually all possible assignments are created!) In the next c steps it is checked that each of the clauses satisfies the assignment. In the final step the auxiliary letter is then eliminated. Any deviation from the proper path causes the letter G to be introduced. This letter can never be eliminated. We use the production $G \to \overline{G^v}$ instead of the simple $G \to G$ to avoid the unnecessary communication of words leading to nothing. Thus, we have completed the proof of our theorem. □

Coming back to our example, the required network possesses 65 nodes. The alphabet V_1 has 46 and the alphabet V_2 40 letters. The productions for the node M_1, for instance, are

$$S_0 \to t_1 S_1,\ t_j \to t_j,\ f_j \to f_j,\ 1 \leq j \leq 20,$$

and $x \to \overline{G^{20}}$ for all other letters x. For the node N_1 the productions are

$$\overline{R_1} \to \overline{R_2},\ f_3 \to f_3,\ f_{16} \to f_{16},\ t_{18} \to t_{18},\ x \to x\overline{x},$$

for other letters x in V_2. Furthermore, N_1 has the production $\overline{x} \to \lambda$, for all letters x in V_2, and the production $x \to \overline{G^{20}}$ for all of the remaining letters x.

After 46 computation steps the word

$$w = f_1t_2f_3f_4f_5f_6t_7t_8f_9t_{10}t_{11}t_{12}f_{13}f_{14}t_{15}t_{16}t_{17}f_{18}f_{19}f_{20}$$

appears in all nodes. This word indicates the only truth-value assignment satisfying the propositional formula β.

The assignment and the word w can be found by the following direct argument. (The argument uses certain special properties of β. The above proof or the construction in [3] are completely independent of such properties.) The conjunction of the clauses involving $\sim x_9$ is logically equivalent to $\sim x_9$. This implies that x_9 must assume the value F. Using this fact and the conjunction of the clauses involving $\sim x_5$, we infer that x_5 must assume the value F. (x_9 and x_5 are detected simply by counting the number of occurrences of each variable in β.) From the 9th clause we now infer that x_{17} must have the value T. After this the value of each remaining variable can be uniquely determined from a particular clause. The values of the variables can be obtained in the following order (we indicate only the index of the variable):

$$9, 5, 17, 15, 7, 8, 10, 16, 18, 3, 20, 2, 13, 19, 6, 4, 11, 1, 14, 12.$$

4 Hamiltonian Path Problem

In this section we show how another well-known NP-complete problem, namely the Hamiltonian Path Problem (HPP) can be solved in linear time by standard NWD0L systems. In this problem one asks whether or not a given directed graph $\gamma = (V, E)$, where V is the set of vertices or nodes of γ, and E denotes the set of its edges, contains a Hamiltonian path, that is, a path which starting from a node V_{in} and ending at a node V_{out} visits each node of the graph exactly once. Nodes V_{in} and V_{out} can be chosen arbitrarily. This problem has a distinguished role in DNA computing, since the famous experiment of Adleman in 1994 demonstrated the solution of an instance of a Hamiltonian path problem in linear time.

Theorem 2 *The Hamiltonian Path Problem can be solved in linear time by standard NWD0L systems.*

Proof. Let $\gamma = (V, E)$ be a directed graph with n nodes $V_1, \ldots, V_n$, where $n \geq 1$, and let us suppose that $V_{in} = V_1$ and $V_{out} = V_n$. We construct a standard $NWD0L_{2n+1}$ system Γ such that any word in the language of Γ identifies a Hamiltonian path in γ in a unique manner. If the language is the empty set, then there is no Hamiltonian path in γ. Moreover, the computation process of any word in the language $L(\Gamma)$ of Γ ends in $2n + 1$ steps.

Let us denote the nodes of Γ by $M_1, \ldots, M_{2n+1}$, and let the alphabet Σ of Γ consist of letters $a_i, \$_i$, for $1 \leq i \leq n$, Z_i, with $1 \leq i \leq n$, X_k, with $n + 1 \leq k \leq 2n$, $Z, F, \$$, as well as of their barred versions.

We follow the communication protocol (b), that is, the copies of the correct words are communicated.

Now we define the productions for the nodes of Γ.

For i with $1 \leq i \leq n$, the node M_i has the following D0L productions: $a_j \to a_j$, for $1 \leq j \leq n$, $Z_j \to Z_{j+1}$ for $1 \leq j \leq n-1$, $Z \to Z_1$; $\$_k \to a_i\$_i$, if γ has a directed edge from V_k to V_i, for $1 \leq k \leq n, k \neq i$; $\$ \to a_1\$_1$.

Each node M_i, where $n+1 \leq i \leq 2n$, is given with the following set of D0L productions: $\$_n \to \bar{X}_i$, and $Z_n \to \lambda$, for $i = n+1$; $\bar{X}_{i-1} \to \bar{X}_i$, for $n+2 \leq i \leq 2n$, $a_j \to a_j\bar{a}_j$ for $j \neq i-n$, where $1 \leq j \leq n$, $\bar{a}_j \to \lambda$, for $1 \leq j \leq n$, $a_j \to a_j$, for $j = i-n$.

Finally, the node M_{2n+1} has the following productions:

$\bar{X}_{2n} \to \lambda$, $a_i \to a_i$, for $1 \leq i \leq n$, $\bar{a}_i \to \lambda$, for $1 \leq i \leq n$.

Furthermore, the above definition is completed by adding the production $B \to F$, for all such letters B of Σ whose production was not specified above.

Let the axiom of the node M_1 be $\$Z$, and let the axiom of the other nodes be F.

We now explain how Γ simulates the procedure of looking for the Hamiltonian paths in γ. Nodes M_i, $1 \leq i \leq n$, are responsible for simulating the generation of paths of length n in γ, nodes M_j, $n+1 \leq j \leq 2n$, are for deciding whether each node is visited by the path exactly once, while node M_{2n+1} is for storing the result, namely the descriptions of the possible Hamiltonian paths. The computation starts at the node M_1, by rewriting the axiom $\$Z$ to $a_1\$_1Z_1$. The strings are sent to the other nodes in Γ. String $a_1\$_1Z_1$ refers to the fact that the node V_1 has been visited, the first computation step has been executed - Z_1 - and the string was forwarded to the node M_k. This string, after arriving at the node M_k, will be rewritten to a_1FZ_2 if there is no direct edge in γ from V_1 to V_k, otherwise the string will be rewritten to $a_1a_k\$_kZ_2$, and the number of visited nodes is increased by one (Z_2). Meantime, the strings arriving at nodes M_l, with $n+1 \leq l \leq 2n+1$, will be rewritten to a string with an occurrence of F, and thus, will never represent a Hamiltonian path in γ. Repeating the previous procedure, at the nodes M_i, for $1 \leq i \leq n$, after performing the first k steps, where $k \leq n$, we have strings of the form $ua_ia_l\$_lZ_k$, where u is a string consisting of letters from $\{a_1, \ldots, a_n\}$, representing a path in γ of length $k-2$, and the string will be forwarded to the node M_l. After the nth step, strings of the form $v\$_nZ_n$, where v is a string of length n over $\{a_1, \ldots, a_n\}$ representing a path of length n in γ, will be checked at nodes of M_j, $n+1 \leq j \leq 2n$, to find out whether or not the paths satisfy the conditions of being Hamiltonian paths. At the step $n+1$, the string $v\$_nZ_n$ at the node M_{n+1} will be rewritten to $v'\bar{X}_{n+1}$, where $\bar{X}_{n+1}$ refers to that the $(n+1)$st step is performed, and v' is obtained from v by replacing all letters a_j, $1 \leq j \leq n, j \neq 1$, with $a_j\bar{a}_j$, whereas the possible occurrences of a_1 are replaced by themselves. If the string does not contain a_1, then the new string will not be correct, and thus its complement contains the letter X_{n+1}. Since letters X_j are rewritten to the trap symbol F, these strings will never be rewritten to a string representing a Hamiltonian path in γ. By the form of their productions, the other nodes will generate strings with an occurrence of F at the $(n+1)$st step. Continuing this procedure, during the next $n-1$ steps, in the same way as above, the occurrence of the letter a_i,

$2 \leq i \leq n$, is checked at the node M_{n+i}. Notice that after checking the occurrence of the letter a_i, the new strings will be forwarded to all nodes of Γ, so the strings representing all possible paths in γ are checked according to the criteria of being Hamiltonian. After the $(2n)$th step, if the node M_{2n} contains a string $v\bar{X}_{2n}$, where $v = a_{i_1} \dots a_{i_n}$, then this string is forwarded to the node M_{2n+1} and is rewritten to v. It will represent a Hamiltonian path in γ, where the string of indices $i_1, \dots, i_n$ corresponds to the nodes $V_{i_1}, \dots, V_{i_n}$, visited in this order. □

5 Population Growth in DNA Systems

The alphabet in the systems considered above is generally bigger than the four-letter DNA alphabet. We now investigate the special case of DNA alphabets. By a *DNA system*, [18], we mean a Watson-Crick D0L system whose alphabet equals the four-letter DNA alphabet. In this section we consider networks based on DNA systems only. It turns out that quite unexpected phenomena occur in such simple networks. For instance, the population growth can be very weird, a function that is not even Z- rational, although the simplicity of the four-letter nodes might suggest otherwise.

We will show in this section that it is possible to construct standard networks of DNA systems, where the population growth is not Z-rational. (By the *population growth function* $f(n)$ we mean the total number of words in all nodes at the nth time instant, $n \geq 0$. We refer to [7] for formal details of the definition.)

The construction given below can be carried out for any pair (p, q) of different primes. We give it explicitly for the pair $(2, 5)$. The construction resembles the one given in [18].

Theorem 3 *There is a network of standard DNA systems, consisting of two components, whose population growth is not Z-rational. This holds true independently of the protocol.*

Proof. Consider the network Γ of standard DNA systems, consisting of two components defined as follows. The first component has the axiom TG and rules

$$A \to A,\ G \to G,\ T \to T^2,\ C \to C^5.$$

The second component has the axiom A and rules

$$A \to A,\ G \to G,\ T \to T,\ C \to C.$$

Clearly, the second component does not process its words in any way.

The beginning of the sequence of the first component is

$$TG,\ A^2C,\ T^2G^5,\ \mathbf{T^4G^5},\ A^8C^5,\ T^8C^{25},$$

$$\mathbf{T^{16}G^{25}},\ A^{32}C^{25},\ T^{32}G^{125}, \mathbf{T^{64}G^{125}},\ A^{128}C^{125},$$

$$T^{128}G^{625},\ \mathbf{T^{256}G^{625}},\ \mathbf{T^{512}G^{625}},\ A^{1024}C^{625} \dots$$

We have indicated by boldface the words *not* obtained by a complementarity transition.

The beginning of the road of the first component is

$$110110110110011\ldots$$

Denote by $r_j, j \geq 1$, the jth bit in the road.

Consider the increasing sequence of numbers, consisting of the positive powers of 2 and 5:

$$2, 4, 5, 8, 16, 25, 32, 64, 125, 128, 256, 512, 625, 1024, \ldots$$

For $j \geq 1$, denote by m_j the jth number in this sequence.

The next two lemmas are rather immediate consequences of the definitions of the sequences r_j and m_j.

Lemma 1 *The road of the first component begins with 11 and consists of occurrences of 11 separated by an occurrence of 0 or an occurrence of 00.*

The distribution of the occurrences of 0 and 00 depends on the sequence m_j:

Lemma 2 *For each* $j \geq 1$, $r_j = r_{j+1} = 0$ *exactly in case* $m_j = 2m_{j-1} = 4m_{j-2}$.

The next lemma is our most important technical tool.

Lemma 3 *The bits r_j in the road of the first component do not constitute an ultimately periodic sequence.*

Proof of Lemma 3. Assume the contrary. By Lemma 2 we infer the existence of two integers a (*initial mess*) and p (*period*) such that, for any $i \geq 0$, the number of powers of 2 between 5^{a+ip} and $5^{a+(i+1)p}$ is constant, say q. Let 2^b be the smallest power of 2 greater than 5^a. Thus,

$$5^a < 2^b < 2^{b+1} < \ldots < 2^{b+q-1} < 5^{a+p} < 2^{b+q}$$

and, for all $i \geq 1$,

$$2^{b+iq-1} < 5^{a+ip} < 2^{b+iq}.$$

Denoting $\alpha = \frac{\log 5}{\log 2}$, we obtain

$$b + iq - 1 < \alpha(a + ip) < b + iq$$

and further

$$q + \frac{b-1}{i} < p\alpha + \frac{a\alpha}{i} < q + \frac{b}{i}.$$

Considering large enough i, we infer $q = p\alpha$, which is impossible. This contradiction proves Lemma 3.

We now return to the main proof of Theorem 3. We assume that the communication protocol is (b), where corrected words are communicated. The proof in the other case is a similar application of Lemma 3 and is left to the reader.

Denote the population growth function by $f(i)$, $i \geq 0$. Thus, the first few numbers in the population growth sequence are:

$$2, 3, 4, 4, 5, 6, 6, 7, 8, 8, 9, 10, 10, 10, 11, \ldots$$

Assume now that $f(i)$ is Z-rational. Then also the function

$$g(i) = f(i) - f(i-1), \ i \geq 1,$$

is Z-rational. But clearly $g(i) = 0$ exactly in case $r_i = 0$. Consequently, by the Skolem-Mahler-Lech Theorem (see [19], Lemma 9.10) 0's occur in the sequence r_i in an ultimately periodic fashion. But this contradicts Lemma 3 and, thus, Theorem 3 follows. □

References

1. L.M. Adleman, Molecular computation of solutions to combinatorial problems. *Science*, 266 (1994), 1021–1024.
2. M. Amos, Gh. Păun, G. Rozenberg and A. Salomaa, DNA-based computing: a survey. *Theoretical Computer Science*, 287 (2002), 3–38.
3. S. Braich, N. Chelyapov, C. Johnson, P.W.K. Rothemund and L. Adleman, Solution of a 20-variable 3-SAT problem on a DNA computer. *Sciencexpress*, 14 March 2002; 10.1126/science.1069528.
4. J. Csima, E. Csuhaj-Varjú and A. Salomaa, Power and size of extended Watson-Crick L systems. *Theoretical Computer Science*, 290 (2003), 1665-1678.
5. E. Csuhaj-Varjú, Networks of Language Processors. *EATCS Bulletin*, 63 (1997), 120–134. Appears also in Gh. Păun, G. Rozenberg, A. Salomaa (eds.), *Current Trends in Theoretical Computer Science*, World Scientific, Singapore, 2001, 771–790.
6. E. Csuhaj-Varjú and A. Salomaa, Networks of parallel language processors. In: *New Trends in Computer Science, Cooperation, Control, Combinatorics* (Gh. Păun and A. Salomaa, eds.), LNCS 1218, Springer-Verlag, Berlin, Heidelberg, New York, 1997, 299–318.
7. E. Csuhaj-Varjú and A. Salomaa, Networks of Watson-Crick D0L systems. *TUCS Report* 419, Turku Centre for Computer Science, Turku, 2001. To appear in M. Ito and T. Imaoka (eds.), *Words, Languages, Combinatorics III. Proceedings of the Third International Colloquium in Kyoto, Japan.* World Scientific Publishing Company, 2003.
8. J. Honkala and A. Salomaa, Watson-Crick D0L systems with regular triggers. *Theoretical Computer Science*, 259 (2001), 689–698.
9. W. Kuich and A. Salomaa, *Semirings, Automata, Languages.* EATCS Monographs on Theoretical Computer Science, Springer Verlag, Berlin, Heidelberg, New York, Tokyo, 1986.
10. V. Mihalache and A. Salomaa, Language-theoretic aspects of DNA complementarity. *Theoretical Computer Science*, 250 (2001), 163–178.
11. G. Păun, G. Rozenberg and A. Salomaa, *DNA Computing. New Computing Paradigms.* Springer-Verlag, Berlin, Heidelberg, New York, 1998.
12. G. Rozenberg and A. Salomaa, *The Mathematical Theory of L Systems.* Academic Press, New York, London, 1980.

13. G. Rozenberg and A. Salomaa, eds., *Handbook of Formal Languages, Vol. I-III.* Springer-Verlag, Berlin, Heidelberg, New York, 1997.
14. A. Salomaa, *Formal Languages.* Academic Press, New York, 1973.
15. A. Salomaa, *Computation and Automata.* Cambridge University Press, Cambridge, London, New York, 1985.
16. A. Salomaa, Watson-Crick walks and roads in D0L graphs. *Acta Cybernetica*, 14 (1999), 179–192.
17. A. Salomaa, Uni-transitional Watson-Crick D0L systems. *Theoretical Computer Science*, 281 (2002), 537–553.
18. A. Salomaa, Iterated morphisms with complementarity on the DNA alphabet. In M. Ito, Gh. Păun and S. Yu (eds.), *Words, Semigroups, Transductions*, World Scientific, 2001, 405–420.
19. A. Salomaa and M. Soittola, *Automata-Theoretic Aspects of Formal Power Series.* Text and Monographs in Computer Science. Springer-Verlag, Berlin, Heidelberg, New York, 1978.
20. A. Salomaa and P. Sosík, Watson-Crick D0L systems: the power of one transition. *Theoretical Computer Science*, 301 (2003), 187–200.

Fixed Point Approach to Commutation of Languages*

Karel Culik II[1], Juhani Karhumäki[2], and Petri Salmela[2]

[1] Department of Computer Science
University of South Carolina
Columbia, 29008 S.C., USA
kculik@sc.rr.com

[2] Department of Mathematics and
Turku Centre for Computer Science
University of Turku
20014 Turku, FINLAND
karhumak@cs.utu.fi, pesasa@utu.fi

Abstract. We show that the maximal set commuting with a given regular set – its centralizer – can be defined as the maximal fixed point of a certain language operator. Unfortunately, however, an infinite number of iterations might be needed even in the case of finite languages.

1 Introduction

The commutation of two elements in an algebra is among the most natural operations. In the case of free semigroups, i.e., words, it is easy and completely understood: two words commute if and only if they are powers of a common word, see, e.g., [11]. For the monoid of languages, even for finite languages the situation changes drastically. Many natural problems are poorly understood and likely to be very difficult. For further details we refer in general to [3], [9] or [6] and in connection to complexity issues to [10] and [4].

Commutation of languages X and Y means that the equality $XY = YX$ holds. It is an equality on sets, however to verify it one typically has to go to the level of words. More precisely, for each $x \in X$ and $y \in Y$ one has to find $x' \in X$ and $y' \in Y$ such that $xy = y'x'$. In a very simple setting this can lead to nontrivial considerations. An illustrative (and simple) example is a proof that for a two-element set $X = \{x, y\}$ with $xy \neq yx$, the maximal set commuting with X is X^+, see [1].

One can also use the above setting to define a computation. Given languages X and Y, for a word $x \in X$ define the rewriting rule

$$x \Rightarrow_C x' \text{ if there exists } x' \in X, y, y' \in Y \text{ such that } xy = y'x'.$$

Let $\Rightarrow_C^*$ be the transitive and reflexive closure of $\Rightarrow_C$. What can we say about this relation? Very little seems to be known. Natural unanswered (according to

* Supported by the Academy of Finland under grant 44087

N. Jonoska et al. (Eds.): Molecular Computing (Head Festschrift), LNCS 2950, pp. 119–131, 2004.

our knowledge) questions are: when is the closure of a word $x \in X$, i.e., its *orbit*, finite or is it always recursive for given regular languages X and Y?

We do not claim that the above operation is biologically motived. However, it looks to us that it resembles some of the natural operation on DNA-sequences, see [13]: the result is obtained by matching two words and then factorizing the result differently. Consequently, it provides a further illustration how computationally complicated are the operations based on matching of words.

Our goal is to consider a particular question on commutation of languages without any biological or other motivation. More precisely, we want to introduce an algebraic approach, so-called *fixed point approach*, to study *Conway's Problem*. The problem asks whether or not the maximal language commuting with a given regular language X is regular as well. The maximal set is called the *centralizer of* X. An affirmative answer is known only in very special cases, see, e.g., [15], [3], [14], [7] and [8]. In general, the problem seems to be very poorly understood – it is not even known whether the centralizer of a finite language X is recursive!

We show that the centralizer of any language is the largest fixed point of a very natural language operator. Consequently, it is obtained as the limit of a simple recursion. When started from a regular X all the intermediate approximations are regular, as well. However, as we show by an example, infinite number of iterations might be needed and hence the Conway's Problem remains unanswered.

One consequence of our results is that if Conway's Problem has an affirmative answer, even nonconstructively, then actually the membership problem for the centralizer of a regular language is decidable, i.e., it is recursive.

2 Preliminaries

We shall need only very basic notations of words and languages; for words see [12] or [2] and for languages [17] or [5].

Mainly to fix the terminology we specify the following. The *free semigroup* generated by a finite *alphabet* A is denoted by A^+. Elements of A^+ are called *words* and subsets of A^+ are called *languages*. These are denoted by lower case letters $x, y, \ldots$ and capital letters $X, Y, \ldots$, respectively. Besides standard operations on words and languages we especially need the operations of the quotients. We say that a word v is a *left quotient* of a word w if there exists a word u such that $w = uv$, and we write $v = u^{-1}w$. Consequently, the operation $(u, v) \to u^{-1}v$ is a partial mapping. Similarly we define *right quotients*, and extend both of these to languages in a standard way: $X^{-1}Y = \{x^{-1}y \mid x \in X, y \in Y\}$.

We say that two languages X and Y *commute* if they satisfy the equality $XY = YX$. Given an $X \subseteq A^+$ it is straightforward to see that there exists the unique maximal set $\mathcal{C}(X)$ commuting with X. Indeed, $\mathcal{C}(X)$ is the union of all sets commuting with X. It is also easy to see that $\mathcal{C}(X)$ is a subsemigroup of A^+. Moreover, we have simple approximations, see [3]:

Lemma 1. *For any* $X \subseteq A^+$ *we have* $X^+ \subseteq \mathcal{C}(X) \subseteq \mathrm{Pref}(X^+) \cap \mathrm{Suf}(X^+)$.

Here $\mathrm{Pref}(X^+)$ (resp. $\mathrm{Suf}(X^+)$) stands for all nonempty prefixes (resp. suffixes) of X^+.

Now we can state:

Conway's Problem. Is the centralizer of a regular X regular as well?

Although the answer is believed to be affirmative, it is known only in the very special cases, namely when X is a prefix set, binary or ternary, see [15], [3] or [7], respectively. This together with the fact that we do not know whether the centralizer of a finite set is even recursive, can be viewed as an evidence of amazingly intriguing nature of the problem of commutation of languages.

Example 1. (from [3]) Consider $X = \{a, ab, ba, bb\}$. Then, as can be readily seen, the centralizer $\mathcal{C}(X)$ equals to $X^+ \setminus \{b\} = (X \cup \{bab, bbb\})^+$. Hence, the centralizer is finitely generated but doesn't equal either to X^+ or $\{a, b\}^+$.

Finally, we note that in the above the centralizers were defined with respect to the semigroup A^+. Similar theory can be developed over the free monoid A^*.

3 Fixed Point Approach

As discussed extensively in [14] and [8], there has been a number of different approaches to solve the Conway's Problem. Here we introduce one more, namely so-called fixed point approach. It is mathematically quite elegant, although at the moment it does not yield into breakthrough results. However, it can be seen as another evidence of the challenging nature of the problem.

Let $X \subseteq A^+$ be an arbitrary language. We define recursively

$$\begin{aligned} X_0 &= \mathrm{Pref}(X^+) \cap Suf(X^+), \text{ and} \\ X_{i+1} &= X_i \setminus [X^{-1}(XX_i \Delta X_i X) \cup (XX_i \Delta X_i X)X^{-1}], \text{ for } i \geq 0, \end{aligned} \tag{1}$$

where Δ denotes the symmetric difference of languages. Finally we set

$$Z_0 = \bigcap_{i \geq 0} X_i. \tag{2}$$

We shall prove

Theorem 1. *Z_0 is the centralizer of X, i.e., $Z_0 = \mathcal{C}(X)$.*

Proof. The result follows directly from the following three facts:

(i) $X_{i+1} \subseteq X_i$ for all $i \geq 0$,
(ii) $\mathcal{C}(X) \subseteq X_i$ for all $i \geq 0$, and
(iii) $Z_0 X = X Z_0$.

Indeed, (iii) implies that $Z_0 \subseteq \mathcal{C}(X)$, while (ii) together with (2) implies that $\mathcal{C}(X) \subseteq Z_0$.

Claims (i)–(iii) are proved as follows. Claim (i) is obvious. Claim (ii) is proved by induction on i. The case $i = 0$ is clear, by Lemma 1. Let $z \in \mathcal{C}(X)$ and assume that $\mathcal{C}(X) \subseteq X_i$. Assume that $z \notin X_{i+1}$. Then

$$z \in X^{-1}(XX_i \Delta X_i X) \cup (XX_i \Delta X_i X)X^{-1}.$$

Consequently, there exists an $x \in X$ such that

$$xz \text{ or } zx \in (XX_i \Delta X_i X).$$

This, however, is impossible since $z \in \mathcal{C}(X) \subseteq X_i$ and $\mathcal{C}(X)X = X\mathcal{C}(X)$. For example, xz is clearly in XX_i, but also in X_iX due to the identity $xz = z'x'$ with $z' \in \mathcal{C}(X)$, $x' \in X$. So z must be in X_{i+1}, and hence (ii) is proved.

It remains to prove the condition (iii). If Z_0X and XZ_0 were inequal, then there would exist a word $w \in Z_0$, such that either $wX \not\subseteq XZ_0$ or $Xw \not\subseteq Z_0X$. By symmetry, we may assume the previous case. By the definition of Z_0 and (i) this would mean that begining from some index k we would have $wX \not\subseteq XX_i$, when $i \geq k$. However, $w \in Z_0 \subseteq X_i$ for every $i \geq 0$, especially for k, and hence $X_kX \neq XX_k$. This would mean that $w \in (XX_k \Delta X_k X)X^{-1}$ and hence $w \notin X_{k+1}$, and consequently $w \notin Z_0$, a contradiction.

Theorem 1 deserves a few remarks.

First we define the language operator φ by the formula

$$\varphi : Y \mapsto Y \setminus [X^{-1}(XY \Delta YX) \cup (XY \Delta YX)X^{-1}],$$

where X is a fixed language. Then obviously all languages commuting with X are fixed points of φ, and the centralizer is the maximal one. Second, in the construction of Theorem 1 it is not important to start from the chosen X_0. Any superset of $\mathcal{C}(X)$ would work, in particular A^+. Third, as can be seen by analyzing the proof, in formula (1) we could drop one of the members of the union. However, the presented symmetric variant looks more natural.

In the next section we give an example showing that for some languages X an infinite number of iterations are needed in order to get the centralizer. In the final concluding section we draw some consequenses of this result.

4 An Example

As an example of the case in which the fixed point approach leads to an infinite iteration we discuss the language $X = \{a, bb, aba, bab, bbb\}$. First we prove that the centralizer of this language is X^+. To do this we start by proving the following two lemmata. We consider the *prefix order* of A^*, and say that two words are *incomparable* if they are so with respect to this order.

Lemma 2. *Let X be a rational language including a word v incomparable with other words in X. If $w \in \mathcal{C}(X)$, then for some integer $n \in \{0, 1, 2, \ldots\}$ there exist words $t \in X^n$ and $u \in \mathrm{Suf}(X)$ such that $w = ut$ and $uX^nX^* \subseteq \mathcal{C}(X)$.*

Proof. If $w \in \mathcal{C}(X)$ and v is an incomparable element in X, then equation $X\mathcal{C}(X) = \mathcal{C}(X)X$ implies that $vw \in \mathcal{C}(X)X$ and therefore $vwv_1^{-1} \in \mathcal{C}(X)$ for some element $v_1 \in X$. Repeating the argument n times we obtain

$$v^n w(v_n \cdots v_2 v_1)^{-1} \in \mathcal{C}(X), \qquad v_i \in X,$$

where $t = v_n \cdots v_2 v_1$ and $w = ut$. Then $v^n u \in \mathcal{C}(X)$ for some integer $n \in \{0, 1, 2, \ldots\}$ and word $u \in \mathrm{Suf}(X) \cap \mathrm{Pref}(w)$. Since v is incomparable, we conclude that for every $s \in X^n$

$$v^n us \in \mathcal{C}(X)X^n = X^n\mathcal{C}(X),$$

and hence

$$us \in \mathcal{C}(X).$$

In other words, $uX^n \subseteq \mathcal{C}(X)$. Since $\mathcal{C}(X)$ is a semigroup, we have also the inclusion $uX^nX^* \subseteq \mathcal{C}(X)$.

For every proper suffix $u_i \in \mathrm{Suf}(X)$, including the empty word 1, there either exists a minimal integer n_i, for which $u_iX^{n_i} \subseteq \mathcal{C}(X)$, or $u_iX^n \nsubseteq \mathcal{C}(X)$ for every integer $n \geq 0$. Since Lemma 2 excludes the latter case, we can associate with every word $w \in \mathcal{C}(X)$ a word $u_i \in \mathrm{Suf}(X)$ and the minimal n_i such that $w \in u_iX^{n_i}X^*$.

Lemma 3. *If the finite language X contains an incomparable word, it has a rational centralizer. Moreover, the centralizer is finitely generated.*

Proof. If the language X is finite, then the set of proper suffixes of X is also finite. With the above terminology we can write

$$\mathcal{C}(X) = \bigcup_{i \in I} u_iX^{n_i}X^* = \underbrace{(\bigcup_{i \in I} u_iX^{n_i})}_{=G} X^* = GX^*,$$

where I is an index set defining suffixes u_i above. Here the language G is finite and $X \subseteq G$. Indeed if $u_0 = 1$, then $n_0 = 1$, and hence $u_0X^{n_0} = 1 \cdot X = X \subseteq G$.

Since $\mathcal{C}(X)$ is semigroup and X is included in G, we obtain

$$\mathcal{C}(X) = \mathcal{C}(X)^+ = (GX^*)^+ = (X + G)^+ = G^+.$$

Now we can prove that the centralizer of our language $X = \{a, bb, aba, bab, bbb\}$ is X^+. The word bab is incomparable. The set of proper suffixes of X is $\{1, a, b, ab, ba, bb\}$. We will consider all of these words separately:

$u_0 = 1 : 1 \cdot X \subseteq \mathcal{C}(X)$ so that $n_0 = 1$.
$u_1 = a : a \in X \subseteq \mathcal{C}(X)$ so that $n_1 = 0$.
$u_2 = b : b \cdot a^n \cdot a \notin X\mathcal{C}(X) = \mathcal{C}(X)X$ and therefore $b \cdot a^n \notin \mathcal{C}(X)$ for all $n \in \mathbb{N}$. This means that the number n_2 does not exist.

$u_3 = ab$: $a \cdot ab \cdot (bab)^n \notin \mathrm{Suf}(X^+)$ implies $aab(bab)^n \notin \mathcal{C}(X)$ so that $aab(bab)^n \notin X\mathcal{C}(X)$ and therefore $ab(bab)^n \notin \mathcal{C}(X)$ for all $n \in \mathbb{N}$. Hence the number n_3 does not exist.

$u_4 = ba$: $ba \cdot a^n \cdot a \notin X\mathcal{C}(X)$ and therefore $ba \cdot a^n \notin \mathcal{C}(X)$ for all $n \in \mathbb{N}$, and hence the number n_4 does not exist.

$u_5 = bb$: $bb \in X \subseteq \mathcal{C}(X)$ so that $n_5 = 0$.

As a conclusion $I = \{0, 1, 5\}$, and $G = \bigcup_{i \in I} u_i X^{n_i} = 1 \cdot X + a + bb = X$. This gives us the centralizer

$$\mathcal{C}(X) = GX^* = XX^* = X^+,$$

in other words we have established:

Fact 1 $\mathcal{C}(\{a, bb, aba, bab, bbb\}) = \{a, bb, aba, bab, bbb\}^+$.

Next we prove that the fixed point approach applied to the language X leads to an infinite loop of iterations. We prove this by showing that there exist words in $\mathcal{C}(X) \setminus X_i$ for every X_i of the iteration (1). To do this we take a closer look on the language $L = (bab)^* ab (bab)^*$. Clearly $L \subseteq X_0 = \mathrm{Pref}(X^+) \cap \mathrm{Suf}(X^+)$ and $L \cap X^+ = \emptyset$.

By the definition of the fixed point approach, word $w \in X_i$ is in X_{i+1} if and only if $Xw \subseteq X_i X$ and $wX \subseteq XX_i$. We will check this condition for an arbitrary word $(bab)^k ab (bab)^n \in L$ with $k, n \geq 1$. The first condition $Xw \subseteq X_i X$ leads to the cases:

$$\left.\begin{aligned} a \cdot (bab)^k ab(bab)^n &= (aba)(bb \cdot a)^{k-1}(bab)^{n+1} \in X^+X \subseteq X_iX, \\ bb \cdot (bab)^k ab(bab)^n &= (bbb)a(bb \cdot a)^{k-1}(bab)^{n+1} \in X^+X \subseteq X_iX, \\ aba \cdot (bab)^k ab(bab)^n &= a(bab)a(bb \cdot a)^{k-1}(bab)^{n+1} \in X^+X \subseteq X_iX, \\ bbb \cdot (bab)^k ab(bab)^n &= (bb)^2 a(bb \cdot a)^{k-1}(bab)^{n+1} \in X^+X \subseteq X_iX \end{aligned}\right\} \quad (3)$$

and

$$bab \cdot (bab)^k ab(bab)^n = (bab)^{k+1} ab(bab)^{n-1} \cdot bab \in X_iX.$$

However, the last one holds if and only if

$$(bab)^{k+1} ab(bab)^{n-1} \in X_i. \quad (4)$$

Similarly, the second condition $wX \subseteq XX_i$ yields us:

$$\left.\begin{aligned} (bab)^k ab(bab)^n \cdot a &= bab(bab)^{k-1}(a \cdot bb)^n aba \in XX^+ \subseteq XX_i, \\ (bab)^k ab(bab)^n \cdot bb &= bab(bab)^{k-1}(a \cdot bb)^n a \cdot bbb \in XX^+ \subseteq XX_i, \\ (bab)^k ab(bab)^n \cdot aba &= bab(bab)^{k-1}(a \cdot bb)^n a \cdot a \cdot bab \cdot a \in XX^+ \subseteq XX_i, \\ (bab)^k ab(bab)^n \cdot bbb &= bab(bab)^{k-1}(a \cdot bb)^n a \cdot a \cdot bb \cdot bb \in XX^+ \subseteq XX_i \end{aligned}\right\} \quad (5)$$

and

$$(bab)^k ab(bab)^n \cdot bab = bab \cdot (bab)^{k-1} ab(bab)^{n+1} \in XX_i.$$

Here the last one holds if and only if

$$(bab)^{k-1}ab(bab)^{n+1} \in X_i. \tag{6}$$

From (4) and (6) we obtain the equivalence

$$(bab)^k ab(bab)^n \in X_{i+1} \iff (bab)^{k+1}ab(bab)^{n-1}, (bab)^{k-1}ab(bab)^{n+1} \in X_i \tag{7}$$

Now, the result follows by induction, when we cover the cases $k = 0$ or $n = 0$.

In the case $k = 0$ and $n \geq 0$ we have that $ab(bab)^n \in X_0$, but $ab(bab)^n \notin X_1$, since $a \cdot ab(bab)^n \notin X_0X$. The same applies also for $(bab)^n ba$ by symmetry.

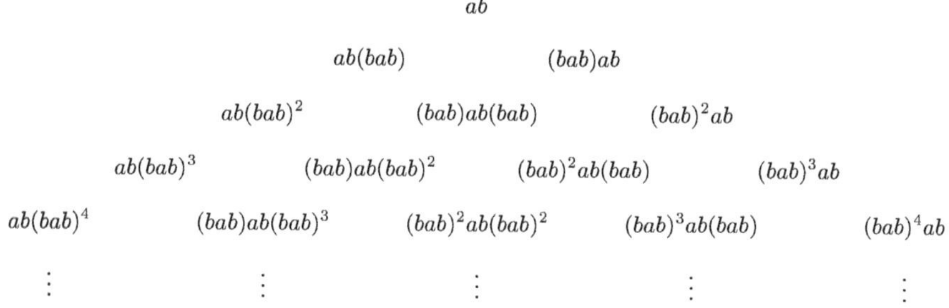

Fig. 1. Language $(bab)^*ab(bab)^*$ written as a pyramid.

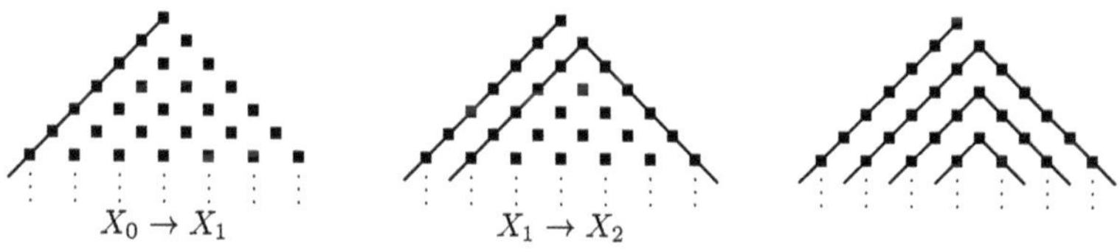

Fig. 2. Deleting words in $(bab)^*ab(bab)^*$ from languages X_i during the iteration.

In the case $n = 0$ and $k \geq 1$ we first note that $X(bab)^k ab \subseteq X_0X$, due to the equations in (3) and the fact that $bab \cdot (bab)^k ab = (bab)^k ba \cdot bab \in X_0X$. Similarly, $(bab)^k abX \subseteq XX_0$, due to the equations in (5) and the fact $(bab)^k ab \cdot bab = bab \cdot (bab)^{k-1}ab(bab) \in XX_0$. These together imply that $(bab)^k ab \in X_1$. On the other hand, since $(bab)^k ba \notin X_1$, then $bab \cdot (bab)^k ab \notin X_1X$, and hence $(bab)^k ab \notin X_2$.

Over all we obtain the following result: if $i = \min\{k, n+1\}$, then

$$(bab)^k ab(bab)^n \in X_i \quad \text{but} \quad (bab)^k ab(bab)^n \notin X_{i+1}.$$

The above can be illustrated as follows. If the language $(bab)^*ab(bab)^*$ is written in the form of a downwards infinite pyramid as shown in Figure 1, then Figure 2 shows how the fixed point approach deletes parts of this language during the iterations. In the first step $X_0 \to X_1$ only the words in $ab(bab)^*$ are deleted as drawn on the leftmost figure. The step $X_1 \to X_2$ deletes words in $(bab)ab(bab)^*$ and $(bab)^*ab$, and so on. On the step $X_i \to X_{i+1}$ the operation always deletes the remaining words in$(bab)^i ab(bab)^*$ and $(bab)^*ab(bab)^{i-1}$, but it never manages to delete the whole language $(bab)^*ab(bab)^*$. This leads to an infinite chain of steps as shown in the following

Fact 2 $X_0 \supset X_1 \supset \cdots X_i \supset \cdots \mathcal{C}(X)$.

When computing the approximations X_i the computer and available software packages are essential. We used *Grail+*, see [18]. For languages X and X^+ their minimal automata are shown in Figure 3.

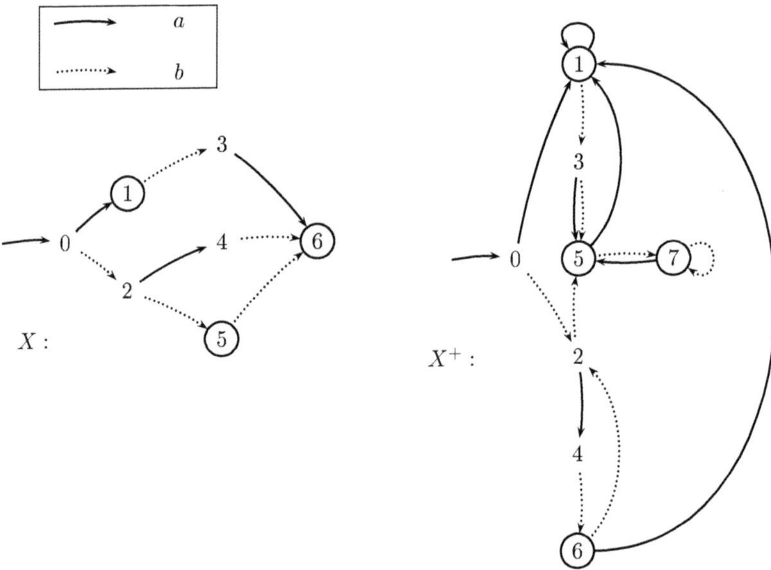

Fig. 3. Finite automata recognizing languages X and X^+

Let us consider the minimal automata we obtain in the iteration steps of the procedure, and try to find some common patterns in those. The automaton recognizing the starting language X_0 is given in Figure 4.

The numbers of states, finals states and transitions for a few first steps of the iteration are given in Table 1. From this table we can see that after a few steps the growth becomes constant. Every step adds six more states, three of those being final, and eleven transitions.

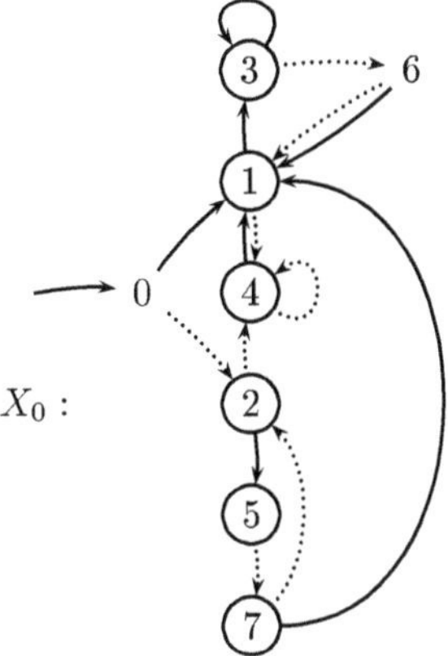

Fig. 4. Finite automaton recognizing the language X_0

When we draw the automata corresponding to subsequent steps from X_i to X_{i+1}, as in Figures 5 and 6, we see a clear pattern in the growth. See that automata representing languages X_5 and X_6 are built from two sequence of sets of three states. In the automaton of X_6 both sequences have got an additional set of three states. So there are totally six new states, including three new final ones, and eleven new transitions. In every iteration step the automata seem to use the same pattern to grow. When the number of iteration steps goes to infinity, the lengths of both sequences go also to infinity. Then the corresponding states can be merged together and the result will be the automaton recognizing the language X^+. This seems to be a general phenomena in the cases where an infinite number of iterations are needed. Intuitively that would solve the Conway's Problem. However, we do not know how to prove it.

5 Conclusions

Our main theorem has a few consequences. As theoretical ones we state the following two. These are based on the fact that formula (1) in the recursion is very simple. Indeed, if X is regular so are all the approximations X_i. Similarly, if X is recursive so are all these approximations. Of course, this does not imply that also the limit, that is the centralizer, should be regular or recursive.

	final states	states	transitions
X_0	6	8	15
X_1	5	9	17
X_2	6	13	24
X_3	9	19	35
X_4	12	25	46
X_5	15	31	57
X_6	18	37	68
X_7	21	43	79
X_8	24	49	90
X_9	27	55	101
X_{10}	30	61	112
X_{11}	33	67	123
X_{12}	36	73	134
X_{13}	39	79	145
X_{14}	42	85	156
X_{15}	45	91	167
X_{16}	48	97	178
X_{17}	51	103	189
X_{18}	54	109	200
X_{19}	57	115	211
X_{20}	60	121	222
X_{21}	63	127	233
X_{22}	66	133	244
X_{23}	69	139	255
X_{24}	72	145	266
X_{25}	75	151	277
X_{26}	78	157	288
X_{27}	81	163	299
X_{28}	84	169	310
X_{29}	87	175	321
X_{30}	90	181	332
X_{31}	93	187	343
X_{32}	96	193	354
X_{33}	99	199	365
X_{34}	102	205	376
X_{35}	105	211	387
X_{36}	108	217	398
X_{37}	111	223	409
X_{38}	114	229	420
X_{39}	117	235	431

Table 1. The numbers of states, final states and transitions of automata corresponding to the iteration steps for the language X.

What we can conclude is the following much weaker result, first noticed in [8]:

Theorem 2. *If X is recursive, then $\mathcal{C}(X)$ is in co-RE, that is its complement is recursively enumerable.*

Proof. As we noticed above, all approximations X_i are recursive, and moreover effectively findable. Now the result follows from the identity

$$\overline{\mathcal{C}(X)} = \bigcup_{i \geq 0} \overline{X_i},$$

where bar is used for the complementation. Indeed, a method to algorithmically list all the elements of $\overline{\mathcal{C}(X)}$ is as follows: Enumerate all words $w_1, w_2, w_3, \ldots$ and test for all i and j whether $w_i \in \overline{X_j}$. Whenever a positive answer is obtained output w_i.

For regular languages X we have the following result.

Theorem 3. *Let $X \in A^+$ be regular. If $\mathcal{C}(X)$ is regular, even noneffectively, then $\mathcal{C}(X)$ is recursive.*

Proof. We assume that X is regular, and effectively given, while $\mathcal{C}(X)$ is regular but potentially nonconstructively. We have to show how to decide the membership problem for $\mathcal{C}(X)$. This is obtained as a combination of two semialgorithms.

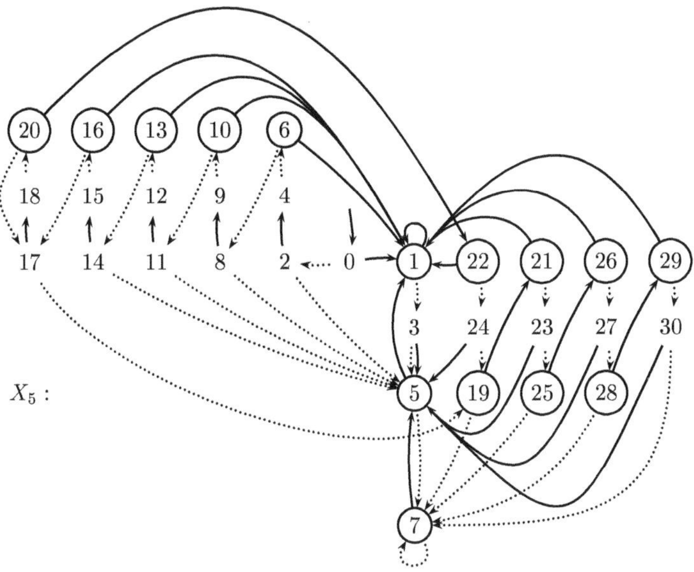

Fig. 5. The finite automaton recognizing X_5

A semialgorithm for the question "$X \in \overline{\mathcal{C}(X)}$?" is obtained as in the proof of Theorem 2. A semialgorithm for the complementary question is as follows: Given x, enumerate all regular languages $L_1, L_2, \ldots$ and test whether or not

$$(i) \quad L_i X = X L_i$$

and

$$(ii) \qquad x \in L_i.$$

Whenever the answers to both of the questions is affirmative output the input x. The correctness of the procedure follows since $\mathcal{C}(X)$ is assumed to be regular. Note also that the tests in (i) and (ii) can be done since L_i's and X are effectively known regular languages.

Theorem 3 is a bit amusing since it gives a meaningful example of the case where the regularity implies the recursiveness. Note also that a weaker assumption that $\mathcal{C}(X)$ is only context-free would not allow the proof, since the question in (i) is undecidble for contex-free languages, see [4].

We conclude with more practical comments. Although we have not been able to use our fixed point approach to answer Conway's Problem even in some new special cases, we can use it for concrete examples. Indeed, as shown by experiments, in most cases the iteration terminates in a finite number of steps. Typically in these cases the centralizer is of one of the following forms:

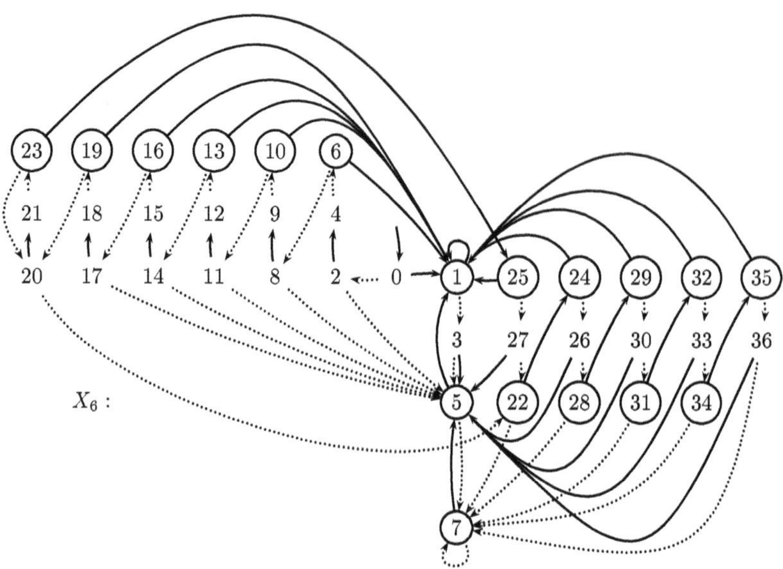

Fig. 6. The finite automaton recognizing X_6

(i) A^+,
(ii) X^+, or
(iii) $\{w \in A^+ \mid wX, Xw \subseteq XX^+\}$.

However, in some cases, as shown in Section 4, an infinite number of iterations are needed – and consequently some ad hoc methods are required to compute the centralizer.

The four element set of example 1, is an instance where the centralizer is not of any of the forms (i)–(iii). This, as well as some other such examples, can be verified by the fixed point approach using a computer, see [16].

References

1. Berstel, J., Karhumäki, J.: Combinatorics on words – A tutorial, *Bull. EATCS* 79 (2003), 178–229.
2. Choffrut, C., Karhumäki, J.: Combinatorics of Words. In Rozenberg, G., Salomaa, A. (eds.), *Handbook of Formal Languages*, Vol. 1, Springer-Verlag (1997) 329–438.
3. Choffrut, C., Karhumäki, J., Ollinger, N.: The commutation of finite sets: a challenging problem, *Theoret. Comput. Sci.*, 273 (1–2) (2002) 69–79.
4. Harju, T., Ibarra, O., Karhumäki, J., Salomaa, A., Decision questions concerning semilinearity, morphisms and commutation of languages, *J. Comput. System Sci.* 65 (2002) 278–294.

5. Hopcroft, J., Ullman, J., *Introduction to Automata Theory, Languages and Computation*, Addison-Wesley (1979).
6. Karhumäki, J.: Challenges of commutation: an advertisement, in *Proc of FCT 2001*, LNCS 2138, Springer (2001) 15–23
7. Karhumäki, J., Latteux, M., Petre, I.: The commutation with codes and ternary sets of words, in *Proc. of STACS 2003*, LNCS 2607, Springer (2003), 74–84; final version to appear.
8. Karhumäki, J. and Petre, I.: Conway's problem for three-word sets, *Theoret. Comput. Sci.*, 289/1 (2002) 705–725.
9. Karhumäki, J. and Petre, I., Two problems on commutation of languages, in: G. Păun, G. Rozenberg and A. Salomaa (eds.), *Current Trends in Theoretical Computer Science*, World Scientific, to appear.
10. Karhumäki, J., Plandowski, W., Rytter, W.: On the complexity of decidable cases of the commutation problem of languages, LNCS 2138, Springer (2001) 193–203; final version to appear.
11. Lothaire, M.: *Combinatorics on Words* (Addison-Wesley, Reading, MA.), (1983).
12. Lothaire, M.: *Algebraic Combinatorics on Words* (Cambridge University Press), (2002).
13. Păun, G., Rozenberg, G. and Salomaa, A., *DNA Computing. New Computing Paradigms*, Texts in Theoretical Computer Science. An EATCS series. Springer (1998)
14. Petre, I.: *Commutation Problems on Sets of Words and Formal Power Series*, PhD Thesis, University of Turku (2002)
15. Ratoandromanana, B.: Codes et motifs, *RAIRO Inform. Theor.*, 23(4) (1989) 425–444.
16. Salmela, P., *Rationaalisen kielen sentralisaattorista ja sen määrittämisestä kiintopistemetodilla*, Master's theses, University of Turku (2002)
17. Salomaa, A., *Formal Languages*, Academic Press (1973).
18. Grail+ 3.0 – software package, Department of Computer Science, University of Western Ontario, Canada, `http://www.csd.uwo.ca/research/grail/grail.html`.

Remarks on Relativisations and DNA Encodings

Claudio Ferretti and Giancarlo Mauri

Dipartimento di Informatica, Sistemistica e Comunicazione
Università degli Studi di Milano-Bicocca
via Bicocca degli Arcimboldi 8, 20136 Milano, Italy
{ferretti, mauri}@disco.unimib.it

Abstract. Biomolecular computing requires the use of carefully crafted DNA words. We compare and integrate two known different theoretical approaches ([2,3]) to the encoding problem for DNA languages: one is based on a relativization of concepts related to comma free codes, the other is a generalization to any given mapping θ of the notion of words appearing as substrings of others. Our first results show which parts of these two formalisms are in fact equivalent. The remaining remarks suggest how to define, to the benefits of laboratory experiments, properties of DNA encodings which merge relativization from [2] and the generality of θ from [3].

1 Introduction

A notion which has always been interesting for the problem of DNA encoding is that of *comma free codes* ([1]): in comma free codes, no word will appear as a substring across the concatenation of two words, i.e., somehow joining them together. This means that when concatenating many words from a comma free code, there will be no mistake in looking for borders between adjacent words, and thus, for instance, no mistake in parsing the message by starting from the middle of the string.

This property fits nicely with what we want in a test tube. If we perform a detection experiment with short probes on a long single stranded DNA molecule, each probe could hybridize independently to any corresponding pattern, each acting as a small parser working inside the longer string.

Head, in [2], described a formalism where it was proved how even codes which are not comma free could be used for biocomputing experiments. In fact, they can be split in sub-codes which are comma free, and whose corresponding sets of DNA probes can be put separately in the test tube, so to avoid unwanted hybridizations.

But one difference between strings and DNA molecules is that DNA probes could also hybridize among them, or the longer molecule could fold and self-hybridize. For this reason, theoretical models of biomolecular computation and DNA encoding try to consider how to avoid Watson-Crick reverse complementarity, which is required by self-hybridization.

In particular, Hussini et al. [3] define a formalism which further generalize these considerations. They study languages that avoid, in strings resulting from

N. Jonoska et al. (Eds.): Molecular Computing (Head Festschrift), LNCS 2950, pp. 132–138, 2004.

the concatenation of codewords, the undesired appearance of substrings which are mapping of other codewords, according to some generic morphism, or anti-morphism, θ.

In this paper we will formally present some relationships between the above mentioned two formalisms. In Section 3 we show that the first one is a relativising generalization of the second one, for θ being the identity mapping. On the other hand, in Section 4 we also suggest a way to combine the relativization of [2] with the generality of θ of [3].

As background references, we can point for some theory of codes to [1], and for a presentation of DNA computing to [7]. Further results related to the approach of [3] are presented in [6] and [5], and a software implementing some ideas from such formalism is discussed in [4].

2 Definitions and Known Results

For a finite alphabet X, X^* is the free monoid generated by X. Elements of X^* will be called indifferently *strings* or *words*.

Any subset of X^* is a *language*. In particular, a subset C of X^* is a *code* if it is *uniquely decipherable*, that is, for each word $w \in C^*$, there is only one non-negative integer n and one finite sequence $c_1, c_2, \ldots, c_{n-1}, c_n$ of words in C for which $w = c_1 c_2 \ldots c_{n-1} c_n$. Please note that the choice of considering a code as being always uniquely decipherable is the same as in [3], while in [2] any language was also considered a code.

A string w in X^* is a factor of a language L if there are strings $x, y \in X^*$ for which $xwy \in L$.

Among the definitions from [2], we are interested in those dealing with languages which avoid some ways of having common patterns in pairs of words.

Definition 1. *A language C is called* solid *iff the following two properties are satisfied:*

1. *u and puq in C can hold only if pq is null,*
2. *pu and uq in C and u non-null can hold only if pq is null.*

Moreover, if a language satisfies property 1 above, then it is called solid/1.

Property 1 forbids having a words *inside* another word, while the second required property forbids having the suffix of a word *overlapping* the prefix of another word.

The following definition introduces the notion of *relativization*, where the same properties above are verified only on words appearing in strings of a (potentially) different language.

Definition 2. *A language C is called* solid relative to a language L *iff the following two properties are satisfied:*

1. *$w = ypuqz$ in L with u and puq in C can hold only if pq is null,*
2. *$w = ypuqz$ in L with pu and uq in C and u non-null can hold only if pq is null.*

Moreover, if a language satisfies property 1 above, then it is called solid/1 relative to L.

One last definition from [2] considers another relativised property:

Definition 3. *A string w in A^* is a* join relative to a language L *iff w is a factor of L and for every u, v in A^* for which uwv in L, both u and v are also in L. A word w in a language L is a* join in L *iff w is a join relative to L^*.*

$J(L)$ will denote the set of all joins in the language L.

In [2] it is proved that:

- if a language on alphabet A is solid, then it is solid relative to every language in A^*,
- if a language is solid relative to a language L, then it is solid relative to every subset of L,
- if C is a code, then $J(C)$ is solid relative to C^* (Proposition 4 in [2]).

The definitions from [3] deal with involutions. An *involution* $\theta : A \rightarrow A$ is a mapping such that θ^2 equals to the identity mapping I: $I(x) = \theta(\theta(x)) = x$ for all $x \in A$. The following definitions, presented here with notations made uniform with those of [2], describe languages which avoid having a word mapped to a substring of another word, or to a substring of the concatenation of two words.

Definition 4. *If $\theta : A^* \rightarrow A^*$ is an involution, then a language $L \subseteq A^*$ is said to be* θ-compliant *iff u and $x\theta(u)y$ in L can hold only if xy is null.*

Moreover, if $L \cap \theta(L) = \emptyset$, then L is called strictly θ*-compliant.*

In [5] it is proved that a language is strictly θ-compliant iff u and $x\theta(u)y$ in L never holds.

Definition 5. *If $\theta : A^* \rightarrow A^*$ is an involution, then a language $L \subseteq A^*$ is said to be* θ-free *iff u in L with $x\theta(u)y$ in L^2 can hold only if x or y are null.*

Moreover, if $L \cap \theta(L) = \emptyset$, then L is called strictly θ*-free.*

In [3] it is proved that a language is strictly θ-free iff u in L with $x\theta(u)y$ in L^2 never holds.

In [3] it is also proved that if a language L is θ-free, then both L and $\theta(L)$ are θ-compliant.

3 Relationships Between the Two Formalisms

If we restrict ourselves to the case of $\theta = I$, with I being the identity mapping, then we can discuss the similarities between the properties of a DNA code as defined in [2] compared to those defined in [3].

First of all we observe that I-compliance and I-freedom cannot be strict, since $L \cap I(L) = L$ (unless we are in the case of $L = \emptyset$).

Moreover, I-compliance is equivalent to enjoying the first property required by solidity, which we called solidity/1. Therefore, if a language L is solid, or L is solid relative to a language in which all strings in L are factors, then L is also I-compliant.

Finally, it is obvious that if L is solid/1 then it is solid/1 relative to any language.

The following simple lemma considers another fact about solidity/1 properties, and will also be used later. It applies, for instance, to the case of $M = L$, or $M = L^2$, or $M = L^*$.

Lemma 1. *A language L is solid/1 iff L is solid/1 relative to a language M in which all strings in L are factors.*

Proof. Solidity/1 is obviously stronger than any relative solidity/1; on the other hand, relative solidity/1 of M verifies that no word of L is proper substring of any other.

Proposition 1. *A language L is solid relative to L^* iff L is solid relative to L^2.*

Proof. It is immediate to verify that solidity relative to L^* implies solidity relative to L^2, since L^2 is a subset of L^*. We prove the other direction by contradiction: if L is not solid relative to L^* then either L falsifies condition 1, and the same would be relatively to L^2, or a word in L^2 falsifies relative solidity, and this itself would be the contradiction, or a word $w = ypuqz \in L^{n+1}$, with pq non null and $n > 1$, would contradict solidity. This last case would falsify solidity relative to L^n in one of the ways that we will see, and so lead to contradiction for L^2. If w has a parsing in L^{n+1}according to which either yp has a prefix or qz has a suffix in L, then we could build from w, by dropping such prefix or suffix, a word which would falsify solidity in L^n. If no such parsing exists, then pu or uq falsify condition 1 of relative solidity, by spanning on at least one word of L in the string w.

Proposition 2. *A language L is I-free iff L is solid relative to L^*.*

Proof. Proposition 1 allows us to reduce the statement to consider only equivalence with solidity relative to L^2. We know that I-freedom implies I-compliance, which in turns implies solidity/1 and, by Lemma 1, solidity/1 relative to $M = L^2$. By absurd, if L is I-free but would not enjoy second property required by the relative solidity, then we would have a string $w \in L^2$ such that $w = ypuqz$ with u non null and $pu, uq \in L$. This would contradict solidity/1 relative to L^2, in case yz, or p or q, would be null, or I-freedom in other cases; for instance, if y and q are non null, word $pu \in L$ would falsify I-freedom.

In the other direction, the solidity relative to L^2 implies the I-freedom, otherwise we would have a word $st = yuz \in L^2$ such that $u \in L$ and both y and z are non null; u would contradict property 1 of relative solidity, if it is a substring of s or t, or property 2 otherwise.

We could say that these links between the two formalisms come at no surprise, since both [2] and [3] chose their definitions in order to model the same reactions, between DNA molecules, which have to be avoided.

To complete the picture, we could define the following language families:

- SOL, which contains all languages which are solid,
- IF, which contains all languages which are I-free,
- RS_n and RS_*, which contain all languages L which are solid relative to L^n and all languages L which are solid relative to L^*, respectively.

We can show the existence of a language E_1 which is in RS_2 but not in SOL, and of a language E_2 which is in RS_1 but not in RS_2: $E_1 = \{adb, bda\}$, $E_2 = \{ab, c, ba\}$ (E_2 was suggested for a similar purpose in [2]).

The above results allow to state that, with $n > 2$:

$$SOL \subset RS_2 = RS_n = RS_* = IF \subset RS_1.$$

We observe that this can be considered as an alternative way of proving the correctness of the two choices made in [2] and [3], which let *by definition* the family of comma free codes be in RS_* and IF, respectively. It appears that both formalisms extend the notion of comma free codes: the first toward languages which are not comma free, but which can contain comma free subsets $J(C)$, and the second toward the more general θ-freedom property.

4 Further Relativization

Relativization allowed in [2] to show how to scale an experiment on DNA molecules on a code C which is not comma free but that can be split in a number of sub-codes $J(C_k)$ which are comma free. This concept can be illustrated by this informal notation: $C = \bigcup_k J(C_k)$, where $C_0 = C, C_{i+1} = C_i \backslash J(C_i)$. In fact, we could use large codes, even when they are not comma free, by performing the hybridization step of a DNA experiment as a short series of sub-steps. In each sub-step we mix in the test tube, for instance, a long single stranded DNA molecule with short probes which are words of a single C_k.

Here we want to move from comma free codes, i.e., solid w.r.t. Kleene closure or, equivalently, I-free, to a definition similar to that of $J(C)$ but relative to θ-freedom, where involution θ is different from I. For instance, we could be interested in the antimorphic involution τ on the four letter alphabet $\Delta = \{A, G, C, T\}$ representing DNA nucleotides; τ will be defined so to model the reverse complementarity between DNA words.

We first need to define the *complement involution* $\gamma : \Delta^* \to \Delta^*$ as mapping each A to T, T to A, G to C, and C to G. The *mirror involution* $\mu : \Delta^* \to \Delta^*$ maps a word u to a word $\mu(u) = v$ defined as follows:

$$u = a_1 a_2 \ldots a_k, v = a_k \ldots a_2 a_1, a_i \in \Delta, 1 \leq i \leq k.$$

Now we can define the involution $\tau = \mu\gamma = \gamma\mu$ on Δ^*, modeling the Watson-Crick mapping between complementary single-stranded DNA molecules.

Results from Section 3, and the fact that $J(C) \subseteq C$, allow us to restate Proposition 4 from [2] (quoted in our Section 2) as follows: if C is a code, then $J(C)$ is I-free. In a similar way, we want a definition of $J_\tau(C)$, subset of C, so that $J_\tau(C)$ is τ-free.

The experimental motivation of this goal is that τ-freedom, and not I-freedom or relative solidity, would protect from the effects of unwanted reverse complementarity. I-freedom guarantees that if we put a set of probes from $J(C)$ in the test tube, they will correctly hybridize to the hypothetical longer DNA molecule, even if this is built from the larger set C. Nonetheless, among words in $J(C)$ there could still be words which can hybridize with words from the same set $J(C)$, interfering with the desired reactions. This would be formally equivalent to say that $J(C)$ is not τ-free.

We conjecture that we could define $J_\tau(C)$ as being related to $J(C)$, as follows:

$$J_\tau(C) = J(C) \backslash \tau(J(C)).$$

Such a subset of C would be I-free, since it is a subset of $J(C)$, and would avoid self hybridization among words of $J\tau(C)$. Further, it would induce a splitting in each $J(C)$, which would still allow to operate detection, i.e., matching between probes and segments on a longer molecule, in a relativised, step by step way, even if with a greater number of steps. Further theoretical studies on this issue is being carried on.

5 Final Remarks

In order to make these theoretical results actually useful for laboratory DNA experiments, more work should be done on example codes.

It would be interesting to generate codes which enjoy the properties defined in Section 4, and to compare them to codes generated as in Proposition 16 of [3] or as in Proposition 4 of [5].

On the other hand, algorithms could be studied that would decide whether a given code enjoys properties defined in Section 4, and whether there is a way of splitting it into a finite number of $J_\tau(C_k)$ sub-codes.

References

1. J. Berstel, D. Perrin. *Theory of Codes*, Academic Press, 1985.
2. T. Head, "Relativised code concepts and multi-tube DNA dictionaries", submitted, 2002.
3. S. Hussini, L. Kari, S. Konstantinidis. "Coding properties of DNA languages". *Theoretical Computer Science*, 290, pp.1557–1579, 2003.
4. N. Jonoska, D. Kephart, K. Mahaligam. "Generating DNA code words", *Congressum Numeratium*, 156, pp.99-110, 2002.

5. N. Jonoska, K. Mahalingam. "Languages of DNA code words". *Preliminary Proceedings of* DNA 9 *Conference*, pp.58–68, 2003
6. L. Kari, R. Kitto, G. Thierrin. "Codes, involutions and DNA encoding", *Formal and Natural Computing*, Lecture Notes in Computer Science 2300, pp.376–393, 2002.
7. Gh. Păun, G. Rozenberg, A. Salomaa. *DNA Computing. New Computing Paradigms*, Springer Verlag, 1998.

Splicing Test Tube Systems and Their Relation to Splicing Membrane Systems

Franziska Freund[1], Rudolf Freund[2], and Marion Oswald[2]

[1] Gymnasium der Schulbrüder, Strebersdorf
Anton Böck-Gasse 37, A-1215 Wien, Austria
franziska@emcc.at
[2] Department of Computer Science
Technical University Wien
Favoritenstrasse 9, A-1040 Wien Austria
{rudi, marion}@emcc.at

Abstract. We consider a variant of test tube systems communicating by applying splicing rules, yet without allowing operations in the tubes themselves. These test tube systems communicating by applying splicing rules of some restricted type are shown to be equivalent to a variant of splicing test tube systems using a corresponding restricted type of splicing rules. Both variants of test tube systems using splicing rules are proved to be equivalent to membrane systems with splicing rules assigned to membranes, too. In all the systems considered in this paper, for the application of rules leading from one configuration to the succeeding configuration we use a sequential model, where only one splicing rule is applied in one step; moreover, all these systems have universal computational power.

1 Introduction

Already in 1987, (a specific variant of) the *splicing* operation was introduced by Tom Head (see [10]). Since then, (formal variants of) splicing rules have been used in many models in the area of *DNA computing*, which rapidly evolved after Adleman in [1] had described how to solve an instance of the Hamiltonian path problem (an NP-complete problem) in a laboratory using DNA. The universality of various models of splicing systems (then also called *H systems*) was shown in [8] and [14].

Distributed models using splicing rules theoretically simulating possible lab implementations were called *test tube systems*. The universality of test tube systems with splicing rules first was proved in [2]; optimal results with respect to the number of tubes showing that two tubes are enough were proved in [7] (Pixton's results in [16] show that only regular sets can be generated using only one tube) . A first overview on the area of DNA computing was given in [15], another comprehensive overview can be found in [17].

In this paper we introduce a variant of test tube systems communicating by applying splicing rules to end-marked strings, yet without allowing operations in

N. Jonoska et al. (Eds.): Molecular Computing (Head Festschrift), LNCS 2950, pp. 139–151, 2004.

the tubes themselves. At each computation step, i.e., at each communication step from tube i to tube j, only one splicing rule is applied using one axiom (which we assume to be provided by the communication rule) and one other string from tube i (which is no axiom); the result of the application of the splicing rule then is moved to tube j. We suppose that this variant of test tube systems communicating by splicing may be interesting for lab applications, too. For all other systems considered in this paper, we shall also use this special variant of splicing rules where exactly one axiom is involved.

In 1998, Gheorghe Păun introduced *membrane systems* (see [11]), and since then the field of membrane systems (soon called *P systems*) has been growing rapidly. In the original model, the rules responsible for the evolution of the systems were placed inside the region surrounded by a membrane and had to be applied in a maximally parallel way. In most variants of P systems considered so far, the membrane structure consisted of membranes hierarchically embedded in the outermost skin membrane, and every membrane enclosed a region possibly containing other membranes and a finite set of evolution rules. Moreover, in the regions multisets of objects evolved according to the evolution rules assigned to the regions, and by applying these evolution rules in a non-deterministic, maximally parallel way, the system passed from one configuration to another one, in that way performing a computation.

Sequential variants of membrane systems were introduced in [5]. Various models of membrane systems were investigated in further papers, e.g., see [3], [11], [12]; for a comprehensive overview see [13], and recent results and developments in the area can be looked up in the web at [19].

A combination of both ideas, i.e., using splicing rules in P systems, was already considered from the beginning in the area of membrane systems, e.g., see [3]; for other variants, e.g., splicing P systems with immediate communication, and results the reader is referred to [13]. Non-extended splicing P systems and splicing P systems with immediate communication with only two membranes can already generate any recursively enumerable language as is shown in [18]. In [6], sequential P systems with only two membranes as well as splicing rules and conditions for the objects to move between the two regions (in some sense corresponding to the filter conditions of test tube systems) were shown to be universal. In [9] the model of P systems with splicing rules assigned to membranes was introduced; using strings from inside or outside the membrane, a splicing rule assigned to the membrane is applied and the resulting string is sent inside or outside the membrane. As the main results in this paper we show that these P systems with splicing rules assigned to membranes are equivalent to splicing test tube systems as well as to test tube systems communicating by splicing, which are introduced in this paper.

In the following section we first give some preliminary definitions and recall some definitions for splicing systems and splicing test tube systems; moreover, we introduce our new variant of test tube systems communicating by applying splicing rules (not using splicing rules in the test tubes themselves); finally, we recall the definition of P systems with splicing rules assigned to membranes. In the third section we show that both variants of test tube systems considered in this paper are equivalent and therefore have the same computational power. In the fourth section we show that these variants of test tube systems using splicing rules and P systems with splicing rules assigned to membranes are equivalent, too. An outlook to related results and a short summary conclude the paper.

2 Definitions

In this section we define some notions from formal language theory and recall the definitions of splicing schemes (H-schemes, e.g., see [2], [14]) and splicing test tube systems (HTTS, e.g., see [2], [7]). Moreover, we define the new variant of test tube systems with communication by splicing (HTTCS). Finally, we recall the definition of membrane systems with splicing rules assigned to membranes as introduced in [9].

2.1 Preliminaries

An *alphabet* V is a finite non-empty set of abstract *symbols*. Given V, the free monoid generated by V under the operation of concatenation is denoted by V^*; the *empty string* is denoted by λ, and $V^* \setminus \{\lambda\}$ is denoted by V^+. By $\mid x \mid$ we denote the length of the word x over V. For more notions from the theory of formal languages, the reader is referred to [4] and [17].

2.2 Splicing Schemes and Splicing Systems

A *molecular scheme* is a pair $\sigma = (B, P)$, where B is a set of *objects* and P is a set of *productions*. A production p in P in general is a partial recursive relation $\subseteq B^k \times B^m$ for some $k, m \geq 1$, where we also demand that for all $w \in B^k$ the range $p(w)$ is finite, and moreover, there exists a recursive procedure listing all $v \in B^m$ with $(w, v) \in p$. For any two sets L and L' over B, we say that L' is computable from L by a production p if and only if for some $(w_1, ..., w_k) \in B^k$ and $(v_1, ..., v_m) \in B^m$ with $(w_1, ..., w_k, v_1, ..., v_m) \in p$ we have $\{w_1, ..., w_k\} \subseteq L$ and $L' = L \cup \{v_1, ..., v_m\}$; we also write $L \Longrightarrow_p L'$ and $L \Longrightarrow_\sigma L'$. A computation in σ is a sequence $L_0, ..., L_n$ such that $L_i \subseteq B$, $0 \leq i \leq n$, $n \geq 0$, as well as $L_i \Longrightarrow_\sigma L_{i+1}$, $0 \leq i < n$; in this case we also write $L_0 \Longrightarrow_\sigma^n L_n$, and moreover, we write $L_0 \Longrightarrow_\sigma^* L_n$ if $L_0 \Longrightarrow_\sigma^n L_n$ for some $n \geq 0$. A *molecular system* is a triple $\sigma = (B, P, A)$ where (B, P) is a molecular scheme and A is a set of axioms from B. An *extended molecular system* is a quadruple $\sigma = (B, B_T, P, A)$, where (B, P, A) is a molecular system and B_T is a set of *terminal objects* with $B_T \subseteq B$. The *language generated by* σ is

$$L(\sigma) = \{w \mid A \Longrightarrow_\sigma^* L,\ w \in L \cap B_T\}.$$

Throughout this paper we shall consider end-marked strings as objects in a molecular scheme or system as well as in membrane systems. *End-marked strings* are objects of the form mwn, where $m, n \in M$, M a set of markers, and $w \in W^*$ for an alphabet W with $W \cap M = \emptyset$.

A *splicing scheme* (over end-marked strings) is a pair σ, $\sigma = (MW^*M, R)$, where M is a set of markers, W is an alphabet with $W \cap M = \emptyset$, and

$$R \subseteq (M \cup \{\lambda\}) W^* \# W^* (M \cup \{\lambda\}) \$ (M \cup \{\lambda\}) W^* \# W^* (M \cup \{\lambda\});$$

$\#, \$$ are special symbols not in $M \cup W$; R is the set of *splicing rules*. For $x, y, z \in MW^*M$ and a splicing rule $r = u_1 \# u_2 \$ u_3 \# u_4$ in R we define $(x, y) \Longrightarrow_r z$ if and only if $x = x_1 u_1 u_2 x_2$, $y = y_1 u_3 u_4 y_2$, and $z = x_1 u_1 u_4 y_2$ for some $x_1, y_1 \in MW^* \cup \{\lambda\}$, $x_2, y_2 \in W^*M \cup \{\lambda\}$. By this definition, we obtain the derivation relation $\Longrightarrow_\sigma$ for the splicing scheme σ in the sense of a molecular scheme as defined above. An *extended H-system* (or *extended splicing system*) γ is an extended molecular system of the form $\gamma = (MW^*M, M_T W_T^* M_T, R, A)$, where $M_T \subseteq M$ is the set of terminal markers, $V_T \subseteq V$ is the set of terminal symbols, and A is the set of axioms.

2.3 Splicing Test Tube Systems

A *splicing test tube system* (*HTTS* for short) with n test tubes is a construct σ,

$$\sigma = (MW^*M, M_T W_T^* M_T, A_1, ..., A_n, I_1, ..., I_n, R_1, ..., R_n, D)$$

where

1. M is a set of markers, W is an alphabet with $W \cap M = \emptyset$;
2. $M_T \subseteq M$ is the set of terminal markers and $W_T \subseteq W$ is the set of terminal symbols;
3. $A_1, ..., A_n$ are the sets of *axioms* assigned to the test tubes $1, ..., n$, where $A_i \subseteq MW^*M$, $1 \leq i \leq n$; moreover, we define $A := \bigcup_{i=1}^{n} A_i$;
4. $I_1, ..., I_n$ are the sets of *initial objects* assigned to the test tubes $1, ..., n$, where $I_i \subseteq MW^*M$, $1 \leq i \leq n$; moreover, we define $I := \bigcup_{i=1}^{n} I_i$ and claim $A \cap I = \emptyset$;
5. $R_1, ..., R_n$ are the sets of *splicing rules* over MW^*M assigned to the test tubes $1, ..., n$, $1 \leq i \leq n$; moreover, we define $R := \bigcup_{i=1}^{n} R_i$; every splicing rule in R_i has to contain exactly one axiom from A_i (which, for better readability, will be underlined in the following) as well as to involve another end-marked string from $MW^*M \setminus A$;
6. D is a (finite) set of *communication relations* between the test tubes in σ of the form (i, F, j), where $1 \leq i \leq n$, $1 \leq j \leq n$, and F is a filter of the form $\{A\} W^* \{B\}$ with $A, B \in M$ (for any i, j, there may be any finite number of such communication relations).

In the interpretation used in this section, a computation step in the system σ run as follows: In one of the n test tubes, a splicing rule from R_i is applied to an object in the test tube (which is not an axiom in A) together with an axiom from A_i. If the resulting object can pass a filter F for some $(i, F, j) \in D$, then this object may either move from test tube i to test tube j or else remain in test tube i, otherwise it has to remain in test tube i. The final result of the computations in σ consists of all terminal objects from $M_T W_T^* M_T$ that can be extracted from any of the n tubes of σ.

We should like to emphasize that for a specific computation we assume all axioms and all initial objects not to be available in an infinite number of copies, but only in a number of copies sufficiently large for obtaining the desired result.

2.4 Test Tube Systems Communicating by Splicing

A *test tube system communicating by splicing* (*HTTCS* for short) with n test tubes is a construct σ,

$$\sigma = (MW^*M, M_T W_T^* M_T, A, I_1, ..., I_n, C)$$

where

1. M is a set of markers, W is an alphabet with $W \cap M = \emptyset$;
2. $M_T \subseteq M$ is the set of terminal markers and $W_T \subseteq W$ is the set of terminal symbols;
3. A is a (finite) set of axioms, $A \subseteq MW^*M$;
4. $I_1, ..., I_n$ are the sets of *initial objects* assigned to the test tubes $1, ..., n$, where $I_i \subseteq MW^*M$, $1 \leq i \leq n$; moreover, we define $I := \bigcup_{i=1}^{n} I_i$ and claim $A \cap I = \emptyset$;
5. C is a (finite) set of *communication rules* of the form (i, r, j), where $1 \leq i \leq n$, $1 \leq j \leq n$, and r is a splicing rule over MW^*M which has to contain exactly one axiom from A as well as to involve another end-marked string from $MW^*M \setminus A$; moreover, we define $R := \bigcup_{(i,r,j) \in C} \{r\}$.

In the model defined above, no rules are applied in the test tubes themselves; an end-marked string in test tube i is only affected when being communicated to another tube j if the splicing rule r in a communication rule (i, r, j) can be applied to it; the result of the application of the splicing rule r to the end-marked string together with the axiom from A which occurs in r is communicated to tube j. The final result of the computations in σ consists of all terminal objects from $M_T W_T^* M_T$ that can be extracted from any of the n tubes of σ.

We should like to point out that again we avoid (target) conflicts by assuming all initial objects to be available in a sufficiently large number of copies, and, moreover, we then assume only one copy of an object to be affected by a communication rule, whereas the other copies remain in the original test tube. Obviously, we assume the axioms used in the communication rules to be available in a sufficiently large number.

Example 1. The HTTCS (see Figure 1)

$$(\{A,B,Z\}\{a,b\}^*\{A,B,Z\},\{A,B\}\{a,b\}^*\{A,B\},C,\{AaZ,ZbB\},\{AabB\},\emptyset)$$

with the communication rules

1. $(1,\underline{Aa\#Z}\$A\#a,2)$ (using the axiom AaZ) and
2. $(2,b\#B\$\underline{Z\#bB},1)$ (using the axiom ZbB)

generates the linear (but not regular) language

$$\left\{Aa^{n+1}b^nB,Aa^nb^nB \mid n\geq 1\right\}.$$

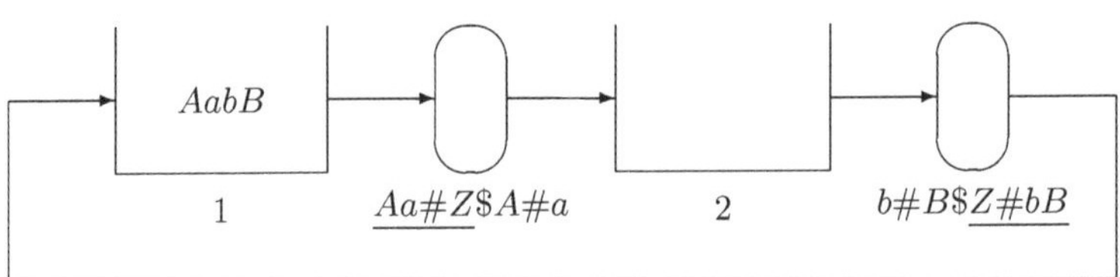

Fig. 1. Example of an HTTCS.

2.5 P Systems with Splicing Rules Assigned to Membranes

A *P system with splicing rules assigned to membranes* (PSSRAM for short) is a construct Π,

$$\Pi=(MW^*M,M_TW_T^*M_T,\mu,A_0,...,A_n,I_0,...,I_n,R_1,...,R_n)$$

where

1. M is a set of markers, W is an alphabet with $W\cap M=\emptyset$;
2. $M_T\subseteq M$ is the set of terminal markers and $W_T\subseteq W$ is the set of terminal symbols;
3. μ is a *membrane structure* (with the membranes labelled by non-negaitive integers $0,...,n$ in a one-to-one manner);
4. $A_0,...,A_n$ are the sets of *axioms* assigned to the membranes $0,...,n$, where $A_i\subseteq MW^*M$, $0\leq i\leq n$; moreover, we define $A:=\bigcup_{i=0}^{n}A_i$;
5. $I_0,...,I_n$ are the sets of *initial objects* assigned to the membranes $0,...,n$, where $I_i\subseteq MW^*M$, $0\leq i\leq n$; moreover, we define $I:=\bigcup_{i=0}^{n}I_i$ and claim $A\cap I=\emptyset$;

6. $R_1, ..., R_n$ are the sets of *rules* assigned to the membranes $1, ..., n$ which are of the form

$$(or_1, or_2; r; tar)$$

for a splicing rule r over MW^*M; every splicing rule occurring in a rule in R_i has to contain exactly one axiom from A as well as to involve another end-marked string from $MW^*M \setminus A$; $or_1, or_2 \in \{in, out\}$ indicate the origins of the objects involved in the application of the splicing rule r, whereas $tar \in \{in, out\}$ indicates the target region the resulting object has to be sent to, where in points to the region inside the membrane and out points to the region outside the membrane. Observe that we do not assign rules to the skin membrane labelled by 0.

A *computation* in Π starts with the initial configuration with the axioms from A_k as well as the initial objects from I_k, $1 \leq k \leq n$, being placed in region k. We assume all objects occurring in $A_k \cup I_k$, $1 \leq k \leq n$, to be available in an arbitrary (unbounded) number. A transition from one configuration to another one is performed by applying a rule from R_k, $1 \leq k \leq n$. The language generated by Π is the set of all terminal objects $w \in M_T W_T^* M_T$ obtained in any of the membrane regions by some computation in Π.

We should like to emphasize that we do not demand the axioms or initial objects really to appear in an unbounded number in any computation of Π. Instead we apply the more relaxed strategy to start with a limited but large enough number of copies of these objects such that a desired terminal object can be computed if it is computable by Π when applying the rules sequentially in a multiset sense; we do not demand things to evolve in a maximally parallel way. In that way we avoid target conflicts, i.e., the other copies of the strings involved in the application of a rule (being consumed or generated) just remain in their regions. This working mode of a P system with splicing rules assigned to membranes also reflects the sequential way of computations considered in the models of test tube systems defined above.

3 Splicing Test Tube Systems and Test Tube Systems Communicating by Splicing Are Equivalent

In this section we show that the splicing test tube systems and the test tube systems communicating by splicing newly introduced in this paper are equivalent models for generating end-marked strings.

Theorem 1. *For every HTTS we can construct an equivalent HTTCS.*

Proof. Let

$$\sigma = (MW^*M, M_T W_T^* M_T, A_1, ..., A_n, I_1, ..., I_n, R_1, ..., R_n, D)$$

be an HTTS.

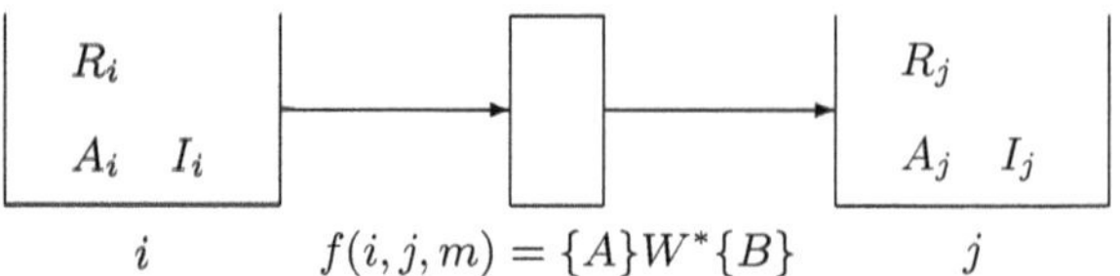

Fig. 2. Elements of HTTS σ.

Then we construct an equivalent HTTCS

$$\sigma' = (M'W^*M', M_T W_T^* M_T, A, I_1, ..., I_n, I'_1, ..., I'_l, C)$$

where

1. $M' = M \cup \{Z\}$;
2. $A = \left(\bigcup_{i=1}^{n} A_i\right) \cup \{XZ, ZX \mid X \in M\}$;
3. for every filter $f(i,j,m)$, $1 \leq m \leq n_{i,j}$, $n_{i,j} \geq 0$, of the form $\{A\} W^* \{B\}$ between the tubes i and j we introduce a new test tube (i,j,m) in σ'; these new test tubes altogether are the l additional test tubes we need in σ';
4. $I'_k = \emptyset$, $1 \leq k \leq l$;
5. the communication rules in C are defined in the following way:

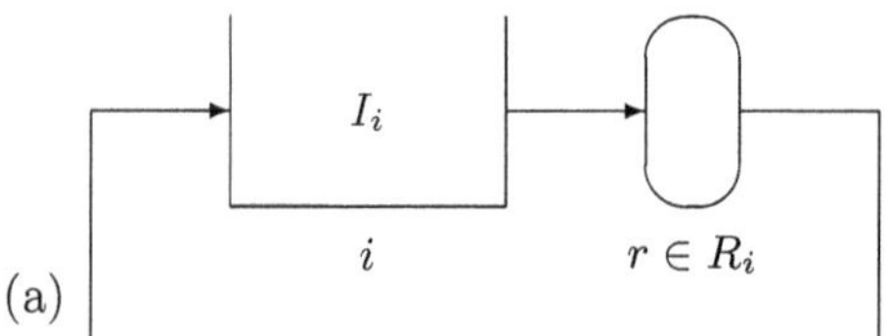

Fig. 3. Simulation of $r \in R_i$ in σ'.

(b) every splicing rule r in R_i is simulated by the communication rule (i, r, i) in C (here we use the axioms from A_i), see Figure 3;

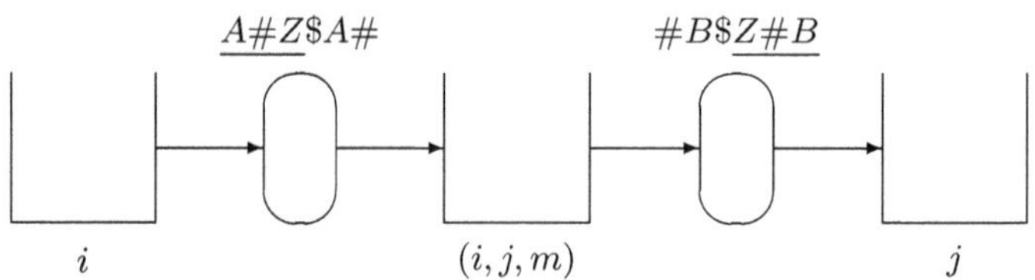

Fig. 4. Simulation of filter $\{A\}W^*\{B\}$ between test tubes i and j in σ'.

(c) every filter $f(i,j,m)$, $1 \leq m \leq n_{i,j}$, $n_{i,j} \geq 0$, of the form $\{A\}W^*\{B\}$ between the tubes i and j is simulated by the communication rules $(i, \underline{A\#Z}\$A\#, (i,j,m))$ and $((i,j,m), \#B\$\underline{Z\#B}, j)$ in C (here we use the additional axioms AZ and ZB), see Figure 4.

The definitions given above completely specify the HTTCS σ' generating the same language as the given HTTS σ. □

Theorem 2. *For every HTTCS we can construct an equivalent HTTS.*

Proof. We have to guarantee that in the HTTS every splicing rule is applied only once when simulating the splicing rule used at a communication step in the HTTCS (see Figure 5).

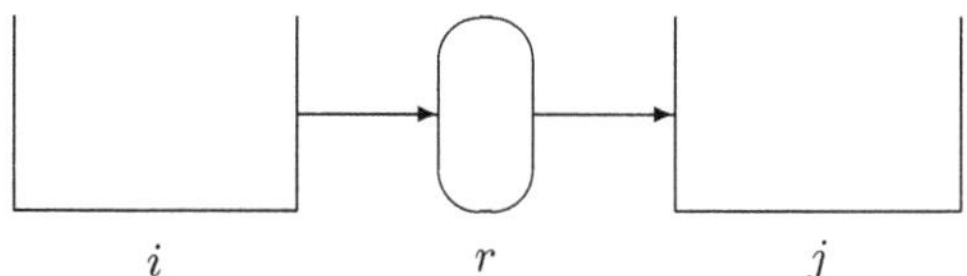

Fig. 5. Communication rule in HTTCS.

For this simulation, we have to consider two cases:

1. the splicing rule r in the communication rule (i,r,j) is of the form $r = (\underline{Au_1\#v_1B}\$Cu_2\#v_2)$; then for every $X \in M$ we take two new test tubes $((i,r,j), X, 1)$ and $((i,r,j), X, 2)$. In test tube $((i,r,j), X, 1)$, the splicing rule $\tilde{r} = \left(\underline{\tilde{A}u_1\#v_1B}\$Cu_2\#v_2\right)$ is applied using the axiom $\widetilde{ax} = \tilde{A}u_1v_1B$. The resulting end-marked string beginning with the marker $\tilde{A}$ then can pass the filter $\{\tilde{A}\}W^*\{X\}$ from tube $((i,r,j), X, 1)$ to tube $((i,r,j), X, 2)$, where the splicing rule $r' = \left(\underline{Au_1\#v_1B}\$\tilde{A}u_1\#\right)$ is applied using the (original) axiom $ax = Au_1v_1B$. After that, the resulting end-marked string can pass the filter $\{A\}W^*\{X\}$ to test tube j.
2. the splicing rule r in the communication rule (i,r,j) is of the form $r = (u_1\#v_1D\$\underline{Au_2\#v_2B})$; then for every $X \in M$ we take two new test tubes $((i,r,j), X, 1)$ and $((i,r,j), X, 2)$. In test tube $((i,r,j), X, 1)$, the splicing rule $\tilde{r} = \left(u_1\#v_1D\$\underline{Au_2\#v_2\tilde{B}}\right)$ is applied using the axiom $\widetilde{ax} = Au_2v_2\tilde{B}$. The resulting end-marked string ending with the marker $\tilde{B}$ then can pass the filter $\{X\}W^*\{\tilde{B}\}$ from tube $((i,r,j), X, 1)$ to tube $((i,r,j), X, 2)$, where the splicing rule $r' = \left(\#v_2\tilde{B}\$\underline{Au_2\#v_2B}\right)$ is applied using the (original) axiom $ax = Au_2v_2B$. After that, the resulting end-marked string can pass the filter $\{X\}W^*\{\tilde{B}\}$ to test tube j.

From these constructions described above, it is obvious how to obtain a complete description of the HTTS for a given HTTCS; the remaining details are left to the reader. □

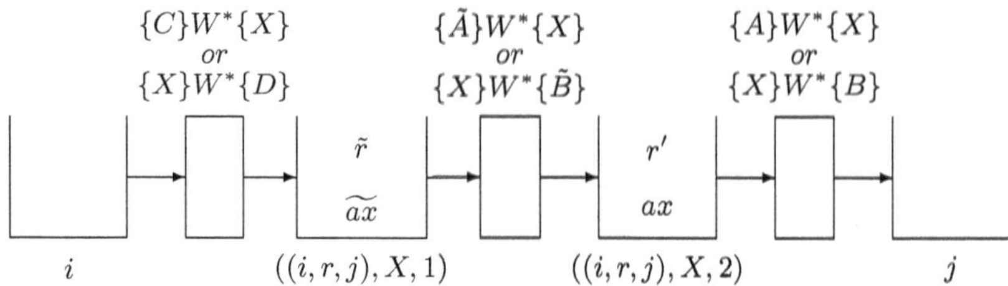

Fig. 6. Simulation of communication rule (i, r, j) in HTTS.

As we can derive from the results proved in [7], the results proved in this section show that test tube systems communicating by splicing have universal computational power; in contrast to splicing test tube systems, where according to the results proved in [7] already systems with two test tubes are universal, we cannot bound the number of membranes in the case of test tube systems communicating by splicing.

4 Test Tube Systems Using Splicing Rules and Membrane Systems with Splicing Rules Assigned to Membranes Are Equivalent

In this section we show that when equipped with splicing rules assigned to membranes, (sequential) P systems are equivalent to test tube systems using splicing rules.

Theorem 3. *For every HTTS we can construct an equivalent PSSRAM.*

Proof. Let

$$\sigma = (M'W^*M', M'_TW^*_TM'_T, A'_1, ..., A'_n, I'_1, ..., I'_n, R'_1, ..., R'_n, D)$$

be an HTTS.

Then we construct an equivalent PSSRAM (see Figure 7)

$$\Pi = (MW^*M, M_TW^*_TM_T, \mu, A_0, ..., A_{n+l}, I_0, ..., I_{n+l}, R_1, ..., R_{n+l})$$

where

1. $M = \{X_i \mid X \in M', 1 \leq i \leq n\} \cup \{Z\}$; $M_T = \{X_i \mid X \in M'_T, 1 \leq i \leq n\}$;
2. for every test tube i we use the membrane i in Π; moreover, for every filter $f(i,j,m) \in D$, $1 \leq m \leq n_{i,j}$, $n_{i,j} \geq 0$, of the form $\{A\} W^* \{B\}$ between the tubes i and j we take a membrane (i,j,m) in Π; these additional membranes altogether are l membranes within the skin membrane;
3. $A_0 = \emptyset$, $A_i = \{C_iwD_i \mid CwD \in A_i\}$, $1 \leq i \leq n$;
$A_{(i,j,m)} = \{A_jZ, ZB_j \mid f(i,j,m) = \{A\} W^* \{B\}, (i, f(i,j,m), j) \in D\}$;

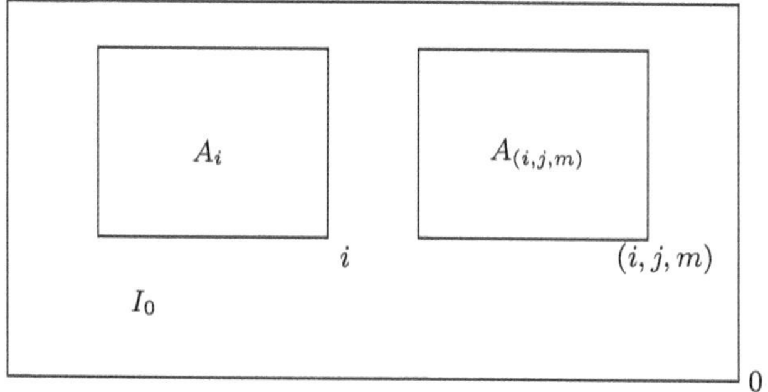

Fig. 7. PSSRAM simulating HTTS.

4. $I_0 = \{A_i w B_i \mid AwB \in I'_i, 1 \leq i \leq n\}$; $I_k = \emptyset$, $1 \leq k \leq l$;
5. $R_i = \{(in, out; \underline{C_i u_1 \# v_1 D_i} \$ A_i u_2 \# v_2; out) \mid (C u_1 \# u_2 D \$ A u_2 \# v_2) \in R'_i\} \cup$
$\{(out, in; u_1 \# v_1 B_i \$ \underline{C_i u_2 \# v_2 D_i}; out) \mid (u_1 \# v_1 B \$ C u_2 \# v_2 D) \in R'_i\}$,
$1 \leq i \leq n$;
moreover, for $i \neq j$ (which can be assumed without loss of generality) let $f(i,j,m) = \{A\} W^* \{B\}$ be a filter between tubes i and j; then we take
$R_{(i,j,m)} = \left\{ \left(in, out; \underline{A_j \# Z} \$ A_i \#; in \right), \left(in, in; \# B_i \$ \underline{Z \# B_j}; out \right) \right\}$.

The construction elaborated above proves that for every splicing test tube system we can effectively construct an equivalent membrane system with splicing rules assigned to membranes. □

Theorem 4. *For every PSSRAM we can construct an equivalent HTTCS.*

Proof. Let

$$\Pi = (MW^*M, M_T W_T^* M_T, \mu, A_0, ..., A_n, I_0, ..., I_n, R_1, ..., R_n)$$

be a PSSRAM. Without loss of generality, we assume that every axiom needed in the splicing rules of R_i, $1 \leq i \leq n$, is available in the corresponding given sets of axioms A_j, $0 \leq j \leq n$. Then we can easily construct an equivalent HTTCS

$$\sigma = (MW^*M, M_T W_T^* M_T, A, I_0, ..., I_n, C)$$

where $A = \left(\bigcup_{i=0}^{n} A_i \right)$ and the communication rules in C are defined in the following way:

Let k be a membrane, $1 \leq k \leq n$ (remember that there are no rules assigned to the skin membrane, which is labbelled by 0), and let the surrounding membrane be labelled by l, and let $(or_1, or_2; r; tar)$ be a rule assigned to membrane k for some splicing rule r. Then $(or_1, or_2; r; tar)$ can be simulated by a communication rule (i, r, j), where $j = k$ for $tar = in$ and $j = l$ for $tar = out$ as well as

1. for r being of the form $\underline{Au_1\#v_1B}\$u_2\#v_2$, $i = k$ for $or_2 = in$ and $i = l$ for $or_2 = out$,
2. for r being of the form $u_1\#v_1\$\underline{Au_2\#v_2B}$, $i = k$ for $or_1 = in$ and $i = l$ for $or_1 = out$.

The construction elaborated above proves that for every membrane system with splicing rules assigned to membranes we can effectively construct an equivalent test tube system communicating by splicing. □

As the results in this section show, membrane systems with splicing rules assigned to membranes are eqivalent to the variants of test tube systems using splicing rules that we considered in the previous section, and therefore they have universal computational power (also see [9]), too.

5 Summary and Related Results

In this paper we have shown that splicing test tube systems and the new variant of test tube systems communicating by splicing are equivalent to (sequential) P systems using splicing rules assigned to membranes. Although using quite restricted variants of splicing rules on end-marked strings, all the systems considered in this paper have universal computational power, which results already follow from results obtained previously.

For splicing test tube systems, the results proved in [7] show that two test tubes are enough for obtaining universal computational power, which result is optimal with respect to the number of tubes. For P systems using splicing rules assigned to membranes, the results proved in [9] (there the environment played the rôle of the region enclosed by the skin membrane) show that (with respect to the definitions used in this paper) two membranes are enough, which result then is optimal with respect to the number of membranes. On the other hand, the number of test tubes in the new variant of test tube systems communicating by splicing cannot be bounded.

Acknowledgements

We gratefully acknowledge many interesting and fruitful discussions on splicing and DNA computing with Tom Head.

References

1. L.M. Adleman, Molecular computation of solutions to combinatorial problems, *Science*, **226** (1994), 1021–1024.
2. E. Csuhaj-Varjú, L. Kari, Gh. Păun, Test tube distributed systems based on splicing, *Computers and Artificial Intelligence*, **15,** 2 (1996), 211–232.
3. J. Dassow, Gh. Păun, On the power of membrane computing, *Journal of Universal Computer Science* **5**, 2 (1999), 33–49 (`http://www.iicm.edu/jucs`).

4. J. Dassow, Gh. Păun, *Regulated Rewriting in Formal Language Theory*, Springer-Verlag (1989).
5. R. Freund, Generalized P-systems, *Proceedings of Fundamentals of Computation Theory Conf. (G.* Ciobanu, Gh. Păun, eds.), Lecture Notes in Computer Science **1684**, Springer-Verlag, Berlin (1999), 281–292.
6. F. Freund, R. Freund, Molecular computing with generalized homogenous P-systems, *DNA Computing. 6th International Workshop on DNA-Based Computers*, DNA 2000, Leiden, The Netherlands, June 2000, Revised Papers, Lecture Notes in Computer Science **2054**, Springer-Verlag, Berlin (2001), 130–144.
7. F. Freund, R. Freund, Test tube systems: When two tubes are enough, *Developments in Language Theory, Foundations, Applications and Perspectives* (G. Rozenberg, W. Thomas, eds.), World Scientific Publishing Co., Singapore (2000), 338–350.
8. R. Freund, L. Kari, Gh. Păun, DNA computing based on splicing: The existence of universal computers, *Theory of Computing Systems*, **32** (1999), 69–112.
9. F. Freund, R. Freund, M. Margenstern, M. Oswald, Yu. Rogozhin, S. Verlan, P systems with cutting/recombination rules or splicing rules assigned to membranes, *Pre-Proceedings of the Workshop on Membrane Computing, WMC-2003* (A. Alhazov, C. Martín-Vide, Gh. Păun, eds.), Tarragona, July 17–22 (2003), 241–251.
10. T. Head, Formal language theory and DNA: An analysis of the generative capacity of specific recombinant behaviors, *Bull. Math. Biology*, **49** (1987), 737–759.
11. Gh. Păun, Computing with membranes, *Journal of Computer and System Sciences* **61**, 1 (2000), 108–143 and *TUCS Research Report* 208 (1998) (`http://www.tucs.fi`).
12. Gh. Păun, Computing with Membranes: an Introduction, *Bulletin EATCS* **67** (1999), 139–152.
13. Gh. Păun, *Membrane Computing: An Introduction*, Springer-Verlag, Berlin (2002).
14. Gh. Păun, Regular extended H systems are computationally universal, *Journal of Automata, Languages and Combinatorics*, **1**, 1 (1996), 27–37.
15. Gh. Păun, G. Rozenberg, A. Salomaa, *DNA Computing. New Computing Paradigms*, Springer-Verlag, Berlin (1998).
16. D. Pixton, Splicing in abstract families of languages, *Theoretical Computer Science* **234** (2000), 135–166.
17. G. Rozenberg, A. Salomaa (Eds.), *Handbook of Formal Languages*, Springer-Verlag, Berlin, Heidelberg (1997).
18. S. Verlan, About splicing P systems with immediate communication and non-extended splicing P systems, *Pre-Proceedings of the Workshop on Membrane Computing, WMC-2003* (A. Alhazov, C. Martín-Vide, Gh. Păun, eds.), Tarragona, July 17–22 (2003), 461–473.
19. The P Systems Web Page: `http://psystems.disco.unimib.it`.

Digital Information Encoding on DNA

Max H. Garzon[1], Kiranchand V. Bobba[1], and Bryan P. Hyde[2]

[1] Computer Science, The University of Memphis
Memphis, TN 38152-3240, U.S.A.
{mgarzon, kbobba}@memphis.edu
[2] SAIC-Scientific Applications International Corporation
Huntsville, AL 35805, U.S.A.
brian.p.hyde@saic.com

Abstract. Novel approaches to information encoding with DNA are explored using a new Watson-Crick structure for binary strings more appropriate to model DNA hybridization. First, a Gibbs energy analysis of codeword sets is obtained by using a template and extant error-correcting codes. Template-based codes have too low Gibbs energies that allow cross-hybridization. Second, a new technique is presented to construct arbitrarily large sets of noncrosshybridizing codewords of high quality by two major criteria. They have a large minimum number of mismatches between arbitrary pairs of words and alignments; moreover, their pairwise Gibbs energies of hybridization remain bounded within a safe region according to a modified nearest-neighbor model that has been verified *in vitro*. The technique is scalable to long strands of up to 150-mers, is in principle implementable *in vitro*, and may be useful in further combinatorial analysis of DNA structures. Finally, a novel method to encode abiotic information in DNA arrays is defined and some preliminary experimental results are discussed. These new methods can be regarded as a different implementation of Tom Head's idea of writing on DNA molecules [22], although only through hybridization.

1 Introduction

Virtually every application of DNA computing [23,1,17] requires the use of appropriate sequences to achieve intended hybridizations, reaction products, and yields. The codeword design problem [4,19,3] requires producing sets of strands that are likely to bind in desirable hybridizations while minimizing the probability of erroneous hybridizations that may induce false positive outcomes. A fairly extensive literature now exists on various aspects and approaches of the problem (see [4] for a review). Approaches to this problem can be classified as evolutionary [7,15,9] and conventional design [6,19]. Both types of method require the use of a measure of the quality of the codewords obtained, through either a fitness function or a quantifiable measure of successful outcomes in test tubes.

Although some algorithms have been proposed for testing the quality of codeword sets in terms of being free of secondary structure [4,10], very few methods have been proposed to *systematically* produce codes of high enough quality to

N. Jonoska et al. (Eds.): Molecular Computing (Head Festschrift), LNCS 2950, pp. 152–166, 2004.

guarantee good performance in test tube protocols. Other than greedy "generate and filter" methods common in evolutionary algorithms [8], the only systematic procedure to obtain code sets for DNA computing by analytic methods is the *template method* developed in [2]. An application of the method requires the use of a binary word, so-called *template*, in combination with error-correcting codes from information theory [26], and produces codewords set designs with DNA molecules of size up to 32−mers (more below.)

This paper explores novel methods for encoding information in DNA strands. The obvious approach is to encode strings into DNA strands. They can be stored or used so that DNA molecules can self-assemble fault-tolerantly for biomolecular computation [9,11,19,18]. In Section 2, a binary analog of DNA is introduced as a framework for discussing encoding problems. Section 3.2 describes a new technique, analogous to the tensor product techniques used in quantum computing for error-correcting codes [29], to produce appropriate methods to encode information in DNA-like strands and define precisely what "appropriate" means. It is also shown how these error-preventing codes for binary DNA (BNA, for short) can be easily translated into codeword sets of comparable quality for DNA-based computations. Furthemore, two independent evaluations are discussed of the quality of these codes in ways directly related to their performance in test tube reactions for computational purposes with DNA. We also compare them to code sets obtained using the template method.

Direct encoding into DNA strands is not a very efficient method for storage or processing of massive amounts (over terabytes) of abiotic data because of the enormous implicit cost of DNA synthesis to produce the encoding sequences. A more indirect and more efficient approach is described in Section 4. Assuming the existence of a large basis of noncrosshybridizing DNA molecules, as obtained above, theoretical and experimental results are presented that allow a preliminary assesment of the reliability and potential capacity of this method. These new methods can be regarded as a different implementation of Tom Head's idea of aqueous computing for writing on DNA molecules [22,21], although only hybridization is involved. Section 5 summarizes the results and presents some preliminary conclusions about the technical feasibility of these methods.

2 Binary Models of DNA

DNA molecules can only process information by intermolecular reactions, usually hybridization in DNA-based computing. Due to the inherent uncertainty in biochemical processes, small variations in strand composition will not cause major changes in hybridization events, with consequent limitations on using similar molecules to encode different inputs. Input strands must be "far apart" from each other in hybridization affinity in order to ensure that only desirable hybridizations occur. The major difficulty is that the hybridization affinity between DNA strands is hard to quantify. Ideally, the Gibbs energy released in the process is the most appropriate criterion, but its exact calculation is difficult, even for pairwise interactions among small oligos (up to 60−mers), and using approximation models [8]. Hence an exhaustive search of strands sets of words maximally separated in a given coding space is infeasible, even for the small oligo-nucleotides useful in DNA computing.

To cope with this problem, a much simpler and computationally tractable model, the h-distance, was introduced in [18]. Here, we show how an abstraction of this concept to include binary strings can be used to produce code sets of high enough quality for use *in vitro*, i.e., how to encode symbolic information in biomolecules for robust and fault-tolerant computing. By introducing a DNA-like structure into binary data, one expects to introduce good properties of DNA into electronic information processing while retaining good properties of electronic data storage and processing (such as programmability and reliability). In this section it is shown how to associate basic DNA-like properties with binary strings, and in the remaining sections we show how to use good coding sets in binary with respect to this structure to obtain good codeword sets for computing with biomolecules *in vitro*.

2.1 Binary DNA

Basic features of DNA structure can be brought into data representations traditionally used in conventional computing as follows. Information is usually represented in binary strings, butthey are treated them as analogs of DNA strands, and refer to them as binary oligomers, or simply *biners*. The pseudo-base (or abstract base) 0 binds with 1, and *vice versa*, to create a complementary pair 0/1, in analogy with Watson-Crick bonding of natural nucleic bases. These concepts can be extended to biners recursively as follows

$$(xa)^R := y^R x^R, \quad \text{and} (xa)^{wc} := a^{wc} x^{wc}, \tag{1}$$

where the superscripts R and wc stand for the reversal operation and the Watson-Crick complementary operation, respectively. The resulting single and double strands are referred to as *binary DNA*, or simply *BNA*. Hybridization of single DNA strands is of crucial importance in biomolecular computing and it thus needs to be defined properly for BNA. The motivation is, of course, the analogous ensemble processes with DNA strands [30].

2.2 h-Distance

A measure of hybridization likelihood between two DNA strands has been introduced in [18]. A similar concept can be used in the context of BNA. The *h-measure* between two biners x, y is given by

$$|x, y| := \min_{-n<k<n} H(x, \sigma^k(y^{wc})) , \tag{2}$$

where σ^k is the (right-) left-shift by k positions (if $k < 0$, resp.), y^{wc} is the Watson-Crick complement of y obtained by reversing y and exchanging 0s and

1s, and $H(*,*)$ is the ordinary Hamming distance between binary strands. This h-measure takes the minimum of all Hamming distances obtained by successively shifting and lining up the reverse complement of y against x. If some shift of one strand perfectly matches a segment of the other, the measure is reduced by the length of the matching segment. Thus a small measure indicates that the two biners are likely to stick to each other one way or another. Measure 0 indicates perfect complementarity. A large measure indicates that in whatever relative position x finds itself in the proximity of y, they are far from containing a large number of complementary base pairs, and are therefore less likely to hybridize, i.e., they are more likely to avoid an error (unwanted hybridization). Strictly speaking, the h-measure is not a distance function in the mathematical sense, because it does not obey the usual properties of ordinary distances (*e.g.*, the distance between different strands may be 0 and, as a consequence, the triangle inequality fails). One obtains the h-distance, however, by grouping together complementary biners and passing to the quotient space. These pairs will continue to be referred to as biners. The distance between two such biners X, Y is now given by the minimum h-measure between all four possible pairs $|x, y|$, one x from X and one y from Y. There is the added advantage that in choosing a strand for encoding, its Watson-Crick complement is also chosen, which goes will the the fact that it is usually required for the required hybridization reactions.

One can use the h-distance to quantify the quality of a code set of biners. An optimal encoding maximizes the minimum h-distance among all pairs of its biners. For applications *in vitro*, a given set of reaction conditions is abstracted as a parameter $\tau \geq 0$ giving a bound on the maximum number of base pairs that are permitted to hybridize between two biners before an error (undesirable hybridization) is introduced in the computation. The analogy of the problem of finding good codes for DNA-based computing and finding good error-correcting codes in information theory was pointed out in [18,9]. The theory of error-correcting codes based on the ordinary Hamming distance is a well established field in information theory [26]. A *t-error-correcting code* [26] is a set of binary strands with a minimum Hamming distance $2t + 1$ between any pair of codewords. A good encoding set of biners has an analogous property. The difficulty with biomolecules is that the hybridization affinity, even as abstracted by the h-distance, is essentially different from the Hamming distance, and so the plethora of known error-correcting codes [26] doesnt translate immediately into good DNA code sets. New techniques are required to handle the essentially different situation, as further evidenced below.

In this framework, the codeword design problem becomes to produce encoding sets that permit them to maintain the integrity of information while enabling only the desirable interactions between appropriate biners. Using these codes does not really require the usual procedure of encoding and decoding information with error-correcting codes since there are no errors to correct. In fact, there is no natural way to enforce the redundancy equations characteristic of error-correcting codes *in vitro*. This is not the only basic difference with information theory. All bits in code sets are information bits, so the codes have rate 1, and are therefore more efficient for a given encoding length. The result of the computation only needs to re-interpreted in terms of the assigned mapping used to transduce the original problem's input into DNA strands. Therefore, given the nature of the inspiring biochemistry, analogous codes for hybridization distance are more appropriately called *error-preventing* codes, because the minimum separation in terms of hybridization distance prevents errors, instead of enabling error detection and correction once they have occurred.

2.3 Gibbs Energy

Ideally, the Gibbs energy released in the hybridization process between strand pairs is the most appropriate criterion of quality for a code set for experiments *in vitro*. Although hybridization reactions *in vitro* are governed by well established rules of local interaction between base pairings, difficulties arise in trying to extend these rules to even short oligonucleotides (less than 150-mers) in a variety of conditions [28]. Hence an exhaustive search of strand sets of words maximally separated in a given coding space is infeasible, even for the small size of oligonucleotides useful in DNA computing.

Computation of the Gibbs energy thus relies on approximations based on various assumptions about the type of interactions between neighboring bonds. Various models have been proposed for the range of oligonucleotides used in DNA-based computing, major among which are the nearest-neighbor model and the staggered-zipper model [28]. We use an extension of the nearest neighbor model proposed by [8] that computes optimal alignments between DNA oligonucleotides using a dynamic programming algorithm. There is evidence that this calculation of the Gibbs energy, while not as computationally efficient as the h-distance, is a good predictor of actual hybridization in the range of 20- to 60-mers in PCR reactions *in vitro* [5].

3 Error-Preventing Codes

Codes designs can now be described based on the tensor product operation and compared with codes obtained using the template method [2]. Both types of codes are obtained by first finding good biner sets (a template or a seed set) which is then used to generate a set of BNA strings. These codes are then transformed into code sets for DNA computation in real test tubes, so the basic question becomes how they map to codeword sets in DNA and how their quality (as measured by the parameter τ) maps to the corresponding reaction conditions and DNA ensemble behavior.

3.1 The Template Method

The template method requires the use of a template n-biner T which is as far from itself as possible in terms of h-measure, i.e., $|T,T| >> 0$, and an error-correcting code C with words of length n. The bits are mapped to DNA words by grouping (see below) DNA bases into two groups, the 0-group (say a, t) and the 1-group (say c, g), corresponding to the bits of T, and then rewriting the codewords in C so that their bits are interpreted as one (a or c) or the other (g or t). Codewords are generated in such a way that the distances would be essentially preserved (in proportion of $1:1$), as shown in Table 1(b).

The quality of the code set produced naturally depends on the quality of self-distance of T and the error-correcting capability of C. If a minimum self-distance is assured, the method can be used to produce codewords with appropriate c/g content to have nearly uniform melting temperatures. In [2], an exhaustive search was conducted of templates of length $l \leq 32$ and templates with minimum self-distance about $l/3$. However, the thermodynamical analysis and lab performance are still open problems. To produce a point of comparison, the template method was used, although with a different seed code (32-bit BCH codes [26]). The pairwise Gibbs energy profiles of the series of template codeword sets obtained using some templates listed in [2] are shown below in Fig. 2(a).

3.2 Tensor Products

A new iterative technique is now introduced to systematically construct large code sets with high values for the minimum h-distance τ between codewords. The technique requires a base code of short biners (good set of short oligos) to seed the process. It is based on a new operation between strands, called the tensor product between words from two given codes of s- and t-biners, to produce a code of $s+t$-biners. The codewords are produced as follows. Given two biners $x = ab$, and y from two coding sets C, D of s- and t-biners, respectively, where a, b are the corresponding halves of x, new codewords of length $s+t$ are given by $x \oslash y = a'y'b'$, where the prime indicates a cyclic permutation of the word obtained by bringing one or few of the last symbols of each word to the front. The tensor product of two sets $C \oslash D$ is the set of all such products between pairs of words x from C and y from D, where the cyclic permutation is performed once again on the previously used word so that no word from C or D is used more than once in the concatenation $a'y'b'$. The product code $C \oslash D$ therefore contains at least $|C||D|$ codewords. In one application below, the construction is used without applying cyclic permutations to the factor codewords with comparable, if lesser, results.

Size/τ	Best codes
5/1	10000,10100,11000
6/2	100010.100100,110000
7/2	1000010,1001000,1100000
8/2	11000100,11010000,11100000

BNA	DNA base
000,010/111,101	a/t
011,100/001,110	c/g

Table 1. (a) Best seed codes, and (b) mapping to convert BNA to a DNA strand.

To seed the process, we conducted exhaustive searches of the best codes of 5- and 6-biners possible. The results are shown in Table 1(a). No 3-element sets of 7- or 8-biners exist with h-distance $\tau > 2$. It is not hard to see that if n, m are the minimum h-distances between any two codewords in C and D, respectively, then the minimum distance between codewords in $C \oslash D$ may well be below $n + m$. By taking the two factors $C = D$ to be equal to the same seed n-biner code C from Table 1, we obtain a code D'. The operation is iterated to obtain codes of $2n$-, $4n$-, $8n$-biners, etc. Because of the cyclic shifts, the new codewords in D are not necessarily at distance $s + s = 2s$, double the minimum distance s between words in the original words. We thus further pruned the biners whose h-distances to others are lowest and most common, until the minimum distance between codewords that remain is at least 2s. In about a third of the seeds, at least one code set survives the procedure and systematically produces codes of high minimum h-distance between codewords. Since the size of the code set grows exponentially, we can obtain large code sets of relatively short length by starting with seeds of small size. Fig 1(a) shows the growth of the minimum h-distance over in the first few iterations of the product with the seeds shown in Table 1(a) in various combinations. We will discuss the tube performance of these codes in the following section, after showing how to use them to produce codeword sets of DNA strands.

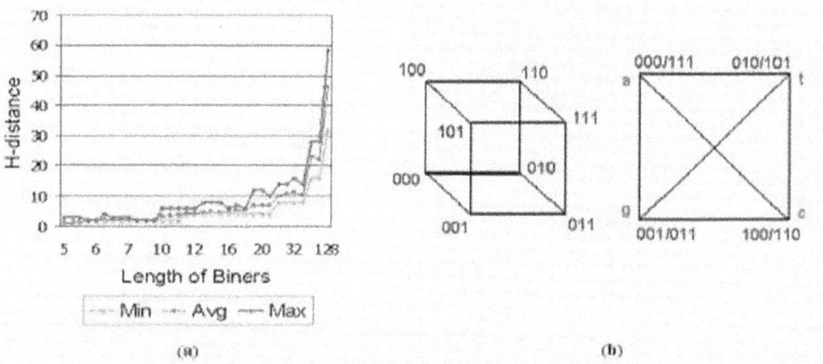

Fig. 1. (a) Minimum h-distance in tensor product BNA sets; and (b) two methods to convert biners to DNA oligonucleotides.

3.3 From BNA to DNA

Two different methods were used. The first method (*grouping*) associates to every one of the eight 3-biners a DNA base a, c, g, or t as shown in Table 1(b). Note that Watson-Crick complementarity is preserved, so that the hybridization distances should be roughly preserved (in proportion $3:1$).

The second method (*position*) assigns to each position in a biner set a fixed DNA base a, c, g, or t. The assignment is in principle arbitrary, and in particular can be done so that the composition of the strands follows a predetermined pattern (e.g., 60% c/g's and 40% a/t's). All biners are then uniformly mapped to a DNA codeword by the same assignment. The resulting code thus has minimum h-distance between DNA codewords no smaller than in the original BNA seed. Fig. 2(b) shows the quality of the codes obtained by this method, to be discussed next.

3.4 Evaluation of Code Quality

For biner codes, a first criterion is the distribution of h-distance between codewords. Fig. 1(a) shows this distribution for the various codes generated by tensor products. The average minimun pairwise distance is, on the average, about 33% of the strand length. This is comparable with the minimum distance of $l/3$ in template codes.

As mentioned in [2], however, the primary criterion of quality for codeword sets is performance in the test tube. A reliable predictor of such performance is the Gibbs energy of hybridization. A computational method to compute the Gibbs energy valid for short oligonucleotides (up to 60 bps) is given in [8]. Code sets obtained by generate-and-filter methods using this model have been tested in the tube with good experimental results [8,10]. The same Gibbs energy calculation was used to evaluate and compare the quality of code sets generated by the template and tensor product methods. According to this method, the highest Gibbs energy allowable between two codewords must be, at the very least, -6 Kcal/mole if they are to remain noncrosshybridizing [8].

Statistics on the Gibbs energy of hybridization are shown in Fig. 2. The template codes of very short lengths in Fig. 2(a) remain within a safe region (above -6 kCal/Mole), but they have too low Gibbs energies that permit crosshybridization for longer lengths. On the other hand, although the tensor product operation initially decreases the Gibbs energy, the Gibbs energies eventually stabilize and remain bounded within a safe region that does not permit cross hybridization, both on the average and minimum energies.

Fig. 3 shows a more detailed comparison of the frequency of codewords in each of the two types of codes with a Gibbs energy prone to cross hybridization (third bar from top). Over 80% of the pairwise energies are noncrosshybridizing for template codes (top and middle bars), but they dip below the limit in a significant proportion. On the other hand, all pairwise energies predict no crosshybridization among tensor product codewords, which also exhibit more positive energies. Finally, the number of codewords produced is larger with tensor products, and increases rapidly when the lengths of the codewords reaches 64– and 128–mers (for which no template codes are available.)

4 Encoding Abiotic Information in DNA Spaces

This section explores the broader problem of encoding abiotic information in DNA for storage and processing. The methods presented in the previous section

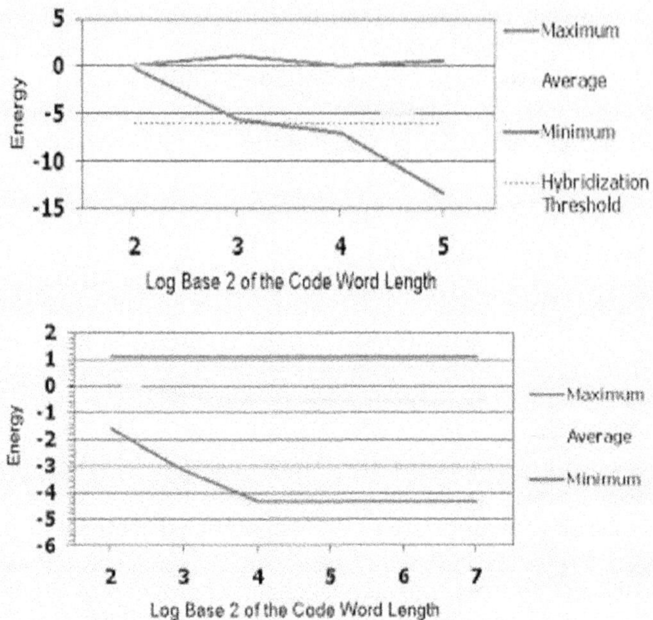

Fig. 2. Gibbs energies of (a = above) template codes; and (b = below) tensor product codes.

could possibly be used to encode symbolic information (strings) in DNA single strands. A fundamental problem is that the abiotic nature of the information would appear to require massive synthesis of DNA strands proportional to the amount to be encoded. Current methods produce massive amounts of DNA copies of the same species, but not of too many different species. Here, some theoretical foundation and experimental results are provided for a new method, described below. Theis method can be regarded as a new implementation of Tom Head's idea of aqueous computing for writing on DNA molecules [22,?], although in through simpler operations (only hybridization.)

4.1 A Representation Using a Non-crosshybridizing Basis

Let B be a set of DNA molecules (the encoding basis, or "stations" in Head's terminology [22], here not necessarily bistable), which is assumed to be finite and noncrosshybridizying according to some parameter τ (for example, the Gibbs energy, or the h-distance mentioned above.) For simplicity, it is also assumed that the length of the strands in B is a fixed integer n, and that B contains no hairpins. If $\tau = 0$ and the h-distance is the hybridization criterion, a maximal such set B can be obtained by selectiong one strand from every (non-palindromic) pair of Watson-Crick complementary strands. If $\tau = n$, a maximal set B consists of only one strand of length n, to which every other strand will hybridize under

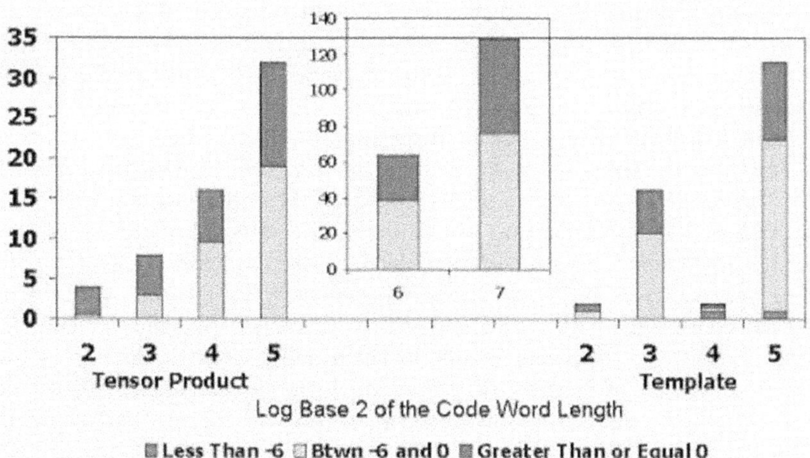

Fig. 3. Code size in error-preventing DNA Codes. Also shown are the percentages of pairwise Gibbs energies in three regions for each code (crosshybridizing: third bar from top, and noncrosshybrizing: top two.)

the mildest hybridization condition represented by such large τ. Let $N = |B|$ be the cardinality of B. Without loss of generality, it will be assumed that strings to be encoded are written in a four letter alphabet $\{a, c, g, t\}$.

Given a string x (ordinarily much larger than n), x is said to be h-dependent of B is there some concatenation c of elements of B such that x will hybridize to c under stringency τ, i.e., $|c, x| \leq \tau$. Shredding x to the corresponding fragments according to the components of c in B leads to the following slightly weaker but more manageable definition. The *signature* of x with respect to B is a vector V of size N obtained as follows. Shredding x to $|X|/n$ fragments of size n or less, V_i is the number of fragments f of x that are within threshold τ of strand i in B, i.e., such that $|f, i| < \tau/2$. The vector V can be visualized as a 1D signature, or as 2D matrix shown in Fig. 4(a), where the bases strands have been arranged on as they would be on a 2D DNA-chip, with one basis strand per spot.

Under realistic reaction conditions, the vector V may appear not to be well-defined, since it is clear that its calculation depends on the concentration of basis as well the strands x used in an experiment to compute it. To avoid these difficulties, this situation will be idealized to the case where only the same number of copies (say, one copy) of each strand is present in the tube, and that the tube is small enough that all possible hybridizations occur within reasonable time. This idealization is based on the results of the following experiments. The experiments were performed in simulation on a virtual test tube that is known to provide highly reliable predictions of wet tube reactions (see [13] for more details).

Six large plasmids of lengths varying between $3K$ and 4.5Kbps were shredded to fragments of size 40 or less and thrown into a tube containing a basis B of three different qualities (values of τ) consisting of 36−mers. The first set Nxh consisted of a set of non-crosshybridizing 40-mers, the length of the longest fragment in the shredded plasmids x; the second set $Randxh$ was a randomly chosen set of 40-mers; and the third set, a set Mxh of maximally crosshybrizing 40-mer set consisting of 18 copies of the same strand and 18 copies of its Watson-Crick complement. The hybridization reactions were determined by the h-distance for various thresholds τ between 0% and 50% of the basis strand length n, which varied between $n = 10, 20, 40$. Each experiment was repeated 10 times under identical conditions. The typical signatures of the six plasmids are illustrated in Fig. 4, averaged pixelwise for one of them, together with the corresponding standard deviation of the same. Fig. 5(a) shows the signal-to-noise ratio (SNR) in effect size units, *i.e.*, given by the pixel signals divided by the standard deviation for the same plasmid, and Fig. 5(b) shows the chipwise SNR comparison for all plasmids, in the same scale as shown in Fig. 4.

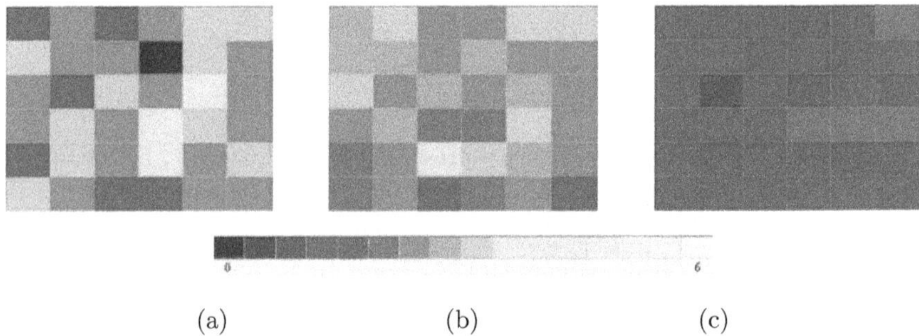

Fig. 4. Signatures of a plasmid genome in (*a*) one run (left); (*b*) average over 10 runs; and (*c*) noise (right, as standard deviation), on a set on non-crosshybridizing probes.

Examination of these results indicates that the signal obtained from the Mxh set is the weakest of all, producing results indistinguishable from noise, certainly due to the random wandering of the molecules in the test tube and their enountering different base strands (probes.) This is in clear contrast with the signals provided by the other sets, where the noise can be shown to be about 10% of the entire signal. The random set Randxh shows signals in between, and the Nxh shows the strongest signal. The conclusion is that, under the assumptions leading to the definition above, an appropriate base set and a corresponding hybridization stringency do provide an appropriate definition of the signature of a given set.

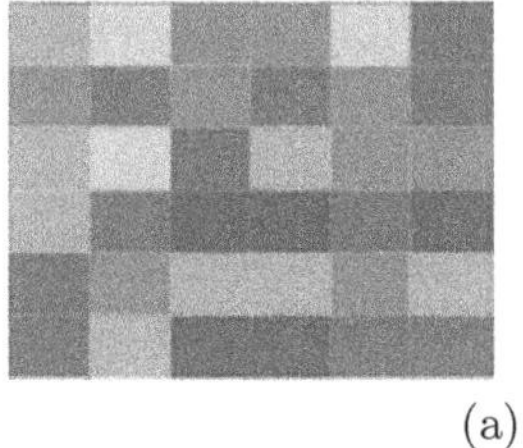

(a)

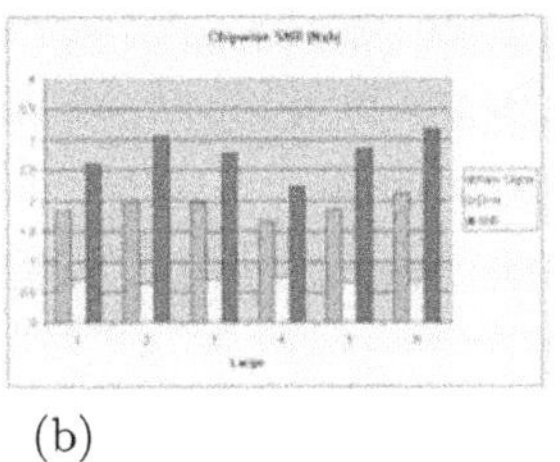

(b)

Fig. 5. Signal-to-noise ratio (a) pixelwise and (b) chipwise for all 6 plasmids.

4.2 The Dimension of DNA Spaces

Under the conditions of the definition of h-dependence, the signature of a single base strands is a delta function (a single pixel), while it is perfectly possible that an h-independent strand from B may have a signal-less signature. Thus there remain two questions, First, how to ensure the completeness of signatures for arbitrary strings x (of length n or larger) on a basis B of given $n-$mers, i.e., to ensure that the original strand x can be somewhat re-constructed from its signature V. And second, how to ensure universal representation for every strand, at least of length n. These problems are similar to, although distinct, of T. Head's *Common Algoritmic Problem* (CAP) [22], where a subest of largest cardinality is called for that fails to contain every one of the subsets in a given a familiy of forbidden subsets of DNA space of $n-$mers. Here, every strand $i \in B$ and a given τ determine a subset of strands S_i that hybridize i under stringency τ. We now want the smallest B that maximizes the "intersection" with every subset S_x in order to provide the strongest signal to represent an input strand x. (An instance of CAP with the complementary sets only asks for the smallest cardinality subset that just "hits", with a nontrivial intersection, all the permissible sets, i.e., the complements of the forbidden subsets.)

It is clear that completeness is ensured by choosing a maximal set of h-independent strands for which every strand of length n provides a nonempty signature. Such a set exists assuming that the first problem is solved. The *h-dimension* of a set of $n-$mer strands is defined as a complete basis for the space. The computation of the dimension of the full set of $n-$mer is thus tightly connected to its structure in terms of hybridization and it appears to be a difficult problem.

The first problem is more difficult to formulate. It is clear that if one insists in being able to recover the composition of strand x from its signature $V(x)$, the threshold must be set to $\tau = 0$ and the solution becomes very simple, if exceedingly large (exponential in the base strand length.) However, if one is more interested in the possibility of storing and processing information in terms of DNA chip signatures, the question arises whether the signature is good enough a representation to process x from the point of view of higher-level semantic relationships. Evidence in [14] shows that there is a real possibility that signatures for relatively short bases are complete for simple discriminating tasks that may suffice for many applications.

5 Summary and Conclusions

A new approach (tensor product) has been introduced for the systematic generation of sets of codewords to encode information in DNA strands for computational protocols so that undesirable crosshybridizations are prevented. This is a hybrid method that combines analytic techniques with a combinatorial fitness function (h-distance [18]) based on an abstract model of complementarity in binary strings. A new way to represent digital information has also been introduced using such sets of noncrosshybrizing codewords (possibly in solution, or affixed to a DNA-chip) and some properties of these representations have been explored.

An analysis of the energetics of codeword sets obtained by this method show that their thermodynamic properties are good enough for acceptable performance in the test tube, as determined the Gibbs energy model of Deaton et al. [8]. These results also confirm that the h-distance is not only a computationally tractable model, but also a reliable model for codeword design and analysis. The final judge of the quality of these code sets is, of course, the performance in the test tube. Preliminary evidence *in vitro* [5] shows that these codes are likely to perform well in wet tubes.

A related measure of quality is what can be termed the *coverage* of the code, i.e., how well spread the code is over the entire DNA space to provide representation for every strand. This measure of quality is clearly inversely related to the error-preventing quality of the code. it is also directly related to the capacity of DNA of a given length to encode information, which is given by a quantity that could be termed *dimension* of the space of DNA strands of length n, although properties analogous to the concept of dimension in euclidean spaces are yet to be assessed. Finally, determining how close tensor product codes come to optimal size is also a question worthy of further study.

Acknowledgements

Much of the work presented here has been done in close collaboration with the molecular computing consortium that includes Russell Deaton and Jin Wu (The U. of Arkansas), Junghuei Chen and David Wood (The U. Delaware). Support from the National Science Foundation grant QuBic/EIA-0130385 is gratefully acknowledged.

References

1. L. Adleman (1994), Molecular computation of solutions of combinatorial problems. Science 266, 1021-1024.

2. M. Arita, S. Kobayashi (2002), DNA Sequence Design Using Templates. New Generation Computing 20:3, 263-277. See also [20], 205-214.
3. E. Baum (1995), Building an Associative Memory Vastly larger than the Brain. Science 268, 583-585.
4. B. Brenneman, A. Condon (2001), Sequence Design for Biomolecular Computation. In press. Available at http://www.cs.ubc.edu/~condon/papers/wordsurvey.ps.
5. H. Bi, J. Chen, R. Deaton, M. Garzon, H. Rubin, D. Wood (2003). A PCR-based Protocol for In Vitro Selection of Non-Crosshybridizing Oligonucleotides. In [20], 196-204. J. of Natural Computing, in press.
6. A. Condon, G. Rozenberg, eds. (2000), Proc. 6th Int. Workshop on DNA-Based Computers DNA 2000 (Revised Papers), Leiden, The Netherlands. Lecture Notes in Computer Science LNCS 2054, Springer-Verlag.
7. R.J. Deaton, J.Chen, H. Bi, M. Garzon, H. Rubin, D.H. Wood (2002), A PCR-based protocol for In Vitro Selection of Non-crosshybridizing Oligonucleotides. In: [20], 105-114.
8. R.J. Deaton, J. Chen, H. Bi, J.A. Rose (2002b), A Software Tool for Generating Non-crosshybridizing Libraries of DNA Oligonucleotides. In: [20], 211-220.
9. R. Deaton, M. Garzon, R.E. Murphy, J.A. Rose, D.R. Franceschetti, S.E. Stevens, Jr. (1998), The Reliability and Efficiency of a DNA Computation. Phys. Rev. Lett. 80, 417.
10. U. Feldkamp, H. Ruhe, W, Banzhaf (2003), Sofware Tools for DNA Sequence Design. J. Genetic Programming and Evolvable Machines 4, 153-171.
11. A.G. Frutos, A. Condon, R. Corn (1997), Demonstration of a Word Design Strategy for DNA Computing on Surface. Nucleic Acids Research 25, 4748-4757.
12. M. Garzon, ed. (2003), Biomolecular Machines and Artificial Evolution. Special Issue of the Journal of Genetic Programming and Evolvable Machines 4:2, Kluwer Academic Publishers.
13. M. Garzon, D. Blain, K. Bobba, A. Neel, M. West. (2003), Self-Assembly of DNA-like structures In-Silico. In: [12], 185-200.
14. M. Garzon, A. Neel, K. Bobba (2003), Efficiency and Reliability of Semantic Retrieval in DNA-based Memories. In: J. Reif and J. Chen (eds), Proc. DNA9-2003, Int. Workshop on DNA-based Computers. Springer-Verlag Lecture Notes in Computer Science, in press.
15. M. Garzon, C. Oehmen (2001b), Biomolecular Computing in Virtual Test Tubes. In: [24], 117-128.
16. M. Garzon, R.J. Deaton (2000), Biomolecular Computing: A Definition. Kunstliche Intelligenz 1/00 (2000), 63-72.
17. M. Garzon, R.J. Deaton (1999), Biomolecular Computing and Programming. IEEE Trans. on Evolutionary Comp. 3:2, 36-50.
18. M. Garzon, P.I. Neathery, R. Deaton, R.C. Murphy, D.R. Franceschetti, S.E. Stevens, Jr. (1997), A New Metric for DNA Computing. In: [25], pp. 472-478.
19. M. Garzon, R. Deaton, P. Neathery, R.C. Murphy, D.R. Franceschetti, E. Stevens Jr. (1997), On the Encoding Problem for DNA Computing. Poster at The Third DIMACS Workshop on DNA-based Computing, U of Pennsylvania. Preliminary Proceedings, 230-237.
20. M. Hagiya, A. Ohuchi, eds. (2002), Proc. 8th Int. Meeting on DNA-Based Computers. Springer-Verlag Lecture Notes in Computer Science LNCS 2568. Springer-Verlag.
21. T. Head, M. Yamamura, S. Gal (2001), Relativized code concepts and multi-tube DNA dictionaries, in press.

22. T. Head, M. Yamamura, S. Gal (1999), Aqueous Computing: Writing on Molecules. Proceedings of the Congress on Evolutionary Computing (CEC'99).
23. T. Head (1986), Formal Language Thery and DNA; An Analysis of the Generative Capacity of Specific Recombinant Behaviors. Bull. of Mathematical Biology 49:6, 737-759.
24. N. Jonoska, N. Seeman, eds. (2001), Proc. 7th Int. Workshop on DNA-Based Computers DNA 2000, Tampa, Florida. Lecture Notes in Computer Science LNCS 2340, Springer-Verlag, 2002.
25. J.R. Koza, K. Deb, M. Dorigo, D.B. Fogel, M. Garzon, H. Iba, R.L. Riolo, eds. (1997). Proc. 2nd Annual Genetic Programming Conference. Morgan Kaufmann, San Mateo, California.
26. J. Roman (1995), The Theory of Error-Correcting Codes. Springer-Verlag, Berlin.
27. H. Rubin. D. Wood, eds. (1997), Third DIMACS Workshop on DNA-Based Computers, The University of Pennsylvania, 1997. DIMACS Series in Discrete Mathematics and Theoretical Computer Science, Providence, RI: American Mathematical Society 48 (1999).
28. J. SantaLucia, Jr., H.T. Allawi, P.A. Seneviratne (1990), Improved Nearest Neighbor Paramemeters for Predicting Duplex Stability. Biochemistry 35, 3555-3562.
29. P.W. Shor (1996), Fault-Tolerant Quantum Computation. Proc. 37th FOCS, 56-65.
30. J.G. Wetmur (1997), Physical Chemistry of Nucleic Acid Hybridization. In: [27], 1-23.

DNA-based Cryptography

Ashish Gehani, Thomas LaBean, and John Reif

Department of Computer Science, Duke University
Box 90129, Durham, NC 27708-0129, USA
{geha,thl,reif}@cs.duke.edu

Abstract. Recent research has considered DNA as a medium for ultra-scale computation and for ultra-compact information storage. One potential key application is DNA-based, molecular cryptography systems. We present some procedures for DNA-based cryptography based on one-time-pads that are in principle unbreakable. Practical applications of cryptographic systems based on one-time-pads are limited in conventional electronic media by the size of the one-time-pad; however DNA provides a much more compact storage medium, and an extremely small amount of DNA suffices even for huge one-time-pads. We detail procedures for two DNA one-time-pad encryption schemes: (i) a substitution method using libraries of distinct pads, each of which defines a specific, randomly generated, pair-wise mapping; and (ii) an XOR scheme utilizing molecular computation and indexed, random key strings. These methods can be applied either for the encryption of natural DNA or for artificial DNA encoding binary data. In the latter case, we also present a novel use of chip-based DNA micro-array technology for 2D data input and output. Finally, we examine a class of DNA steganography systems, which secretly tag the input DNA and then hide it within collections of other DNA. We consider potential limitations of these steganographic techniques, proving that in theory the message hidden with such a method can be recovered by an adversary. We also discuss various modified DNA steganography methods which appear to have improved security.

1 Introduction

1.1 Biomolecular Computation

Recombinant DNA techniques have been developed for a wide class of operations on DNA and RNA strands. There has recently arisen a new area of research known as DNA computing, which makes use of recombinant DNA techniques for doing computation, surveyed in [37]. Recombinant DNA operations were shown to be theoretically sufficient for universal computation [19]. Biomolecular computing (BMC) methods have been proposed to solve difficult combinatorial search problems such as the Hamiltonian path problem [1], using the vast parallelism available to do the combinatorial search among a large number of possible solutions represented by DNA strands. For example, [5] and [41] propose BMC

N. Jonoska et al. (Eds.): Molecular Computing (Head Festschrift), LNCS 2950, pp. 167–188, 2004.

methods for breaking the Data Encryption Standard (DES). While these methods for solving hard combinatorial search problems may succeed for fixed sized problems, they are ultimately limited by their volume requirements, which may grow exponentially with input size. However, BMC has many exciting further applications beyond pure combinatorial search. For example, DNA and RNA are appealing media for data storage due to the very large amounts of data that can be stored in compact volume. They vastly exceed the storage capacities of conventional electronic, magnetic, optical media. A gram of DNA contains about 10^{21} DNA bases, or about 10^8 tera-bytes. Hence, a few grams of DNA may have the potential of storing all the data stored in the world. Engineered DNA might be useful as a database medium for storing at least two broad classes of data: (i) processed, biological sequences, and (ii) conventional data from binary, electronic sources. Baum [3] has discussed methods for fast associative searches within DNA databases using hybridization. Other BMC techniques [38] might perform more sophisticated database operations on DNA data such as database join operations and various massively parallel operations on the DNA data.

1.2 Cryptography

Data security and cryptography are critical aspects of conventional computing and may also be important to possible DNA database applications. Here we provide basic terminology used in cryptography [42]. The goal is to transmit a message between a sender and receiver such that an eavesdropper is unable to understand it. Plaintext refers to a sequence of characters drawn from a finite alphabet, such as that of a natural language. Encryption is the process of scrambling the plaintext using a known algorithm and a secret key. The output is a sequence of characters known as the ciphertext. Decryption is the reverse process, which transforms the encrypted message back to the original form using a key. The goal of encryption is to prevent decryption by an adversary who does not know the secret key. An unbreakable cryptosystem is one for which successful cryptanalysis is not possible. Such a system is the one-time-pad cipher. It gets its name from the fact that the sender and receiver each possess identical notepads filled with random data. Each piece of data is used once to encrypt a message by the sender and to decrypt it by the receiver, after which it is destroyed.

1.3 Our Results

This paper investigates a variety of biomolecular methods for encrypting and decrypting data that is stored as DNA. In Section 2, we present a class of DNA cryptography techniques that are in principle unbreakable. We propose the secret assembly of a library of one-time-pads in the form of DNA strands, followed by a number of methods to use such one-time-pads to encrypt large numbers of short message sequences. The use of such encryption with conventional electronic media is limited by the large amount of one-time-pad data which must be created and transmitted securely. Since DNA can store a significant amount of information in a limited physical volume, the use of DNA could mitigate this concern. In Section 3, we present an interesting concrete example of a DNA cryptosystem in which a two-dimensional image input is encrypted as a solution of DNA strands. We detail how these strands are then decrypted using fluorescent DNA-on-a-chip technology. Section 4 discusses steganographic techniques in which the DNA encoding of the plaintext is hidden among other unrelated strands rather than actually being encrypted. We analyze a recently published genomic steganographic technique [45], where DNA plaintext strands were appended with secret keys and then hidden among many other irrelevant strands. While the described system is appealing for its simplicity, our entropy based analysis allows extraction of the message without knowledge of the secret key. We then propose improvements that increase the security of DNA steganography.

2 DNA Cryptosystems Using Random One-Time-Pads

One-time-pad encryption uses a codebook of random data to convert plaintext to ciphertext. Since the codebook serves as the key, if it were predictable (i.e., not random), then an adversary could guess the algorithm that generates the codebook, allowing decryption of the message. No piece of data from the codebook should ever be used more than once. If it was, then it would leak information about the probability distribution of the plaintext, increasing the efficiency of an attempt to guess the message. These two principles, true randomness and single use of pads, dictate certain features of the DNA sequences and of sequence libraries, which will be discussed further below. This class of cryptosystems using a secret random one-time-pad are the only cryptosystems known to be absolutely unbreakable [42].

We will first assemble a large one-time-pad in the form of a DNA strand, which is randomly assembled from short oligonucleotide sequences, then isolated and cloned. These one-time-pads will be assumed to be constructed in secret, and we further assume that specific one-time-pads are shared in advance by both the sender and receiver of the secret message. This assumption requires initial communication of the one-time-pad between sender and receiver, which is facilitated by the compact nature of DNA.

We propose two methods whereby a large number of short message sequences can be encrypted: (i) the use of substitution, where we encrypt each message sequence using an associatively matched piece from the DNA pad; or (ii) the use of bit-wise XOR computation using a biomolecular computing technique. The decryption is done by similar methods.

It is imperative that the DNA ciphertext is not contaminated with any of the plaintext. In order for this to be effected, the languages used to represent each should be mutually exclusive. The simplest way to create mutually exclusive languages is to use disjoint plain and ciphertext alphabets. This would facilitate the physical separation of plaintext strands from the ciphertext using a set of complementary probes. If the ciphertext remains contaminated with residual plaintext strands, further purification steps can be utilized, such as the use of the DNA-SIEVE described in Section 4.4.

2.1 DNA Cryptosystem Using Substitution

A substitution one-time-pad system uses a plaintext binary message and a table defining a random mapping to ciphertext. The input strand is of length n and is partitioned into plaintext words of fixed length. The table maps all possible plaintext strings of a fixed length to corresponding ciphertext strings, such that there is a unique reverse mapping.

Encryption occurs by substituting each plaintext DNA word with a corresponding DNA cipher word. The mapping is implemented using a long DNA pad that consisting of many segments, each of which specifies a single plaintext word to cipher word mapping. The plaintext word acts as a hybridization site for the binding of a primer, which is then elongated. This results in the formation of a plaintext-ciphertext word-pair. Further, cleavage of the word-pairs and removal of the plaintext portion must occur. A potential application is detailed in Section 3.

An ideal one-time-pad library would contain a huge number of pads and each would provide a perfectly unique, random mapping from plaintext words to cipher words. Our construction procedure approaches these goals. The structure of an example pad is given in Figure 1.

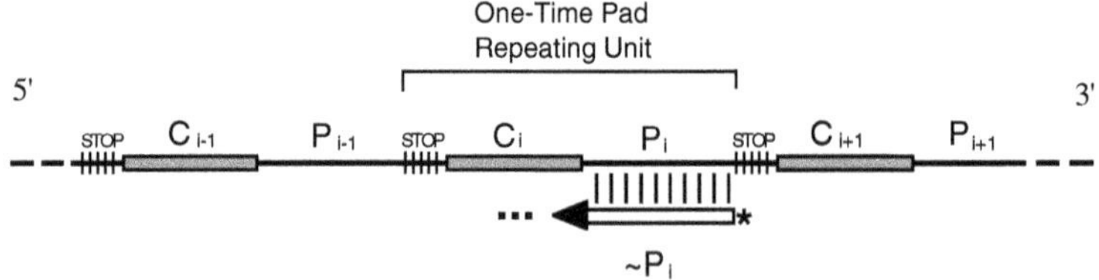

Fig. 1. One-time-pad Codebook DNA Sequences

The repeating unit is made up of: (i) one sequence word, C_i , from the set of cipher or codebook-matching words; (ii) one sequence word, P_i , from the set of plaintext words; and (iii) a polymerase "stopper" sequence. We note that each P_i includes a unique subsequence, which prevents frequency analysis attacks by mapping multiple instances of the same message plaintext to different ciphertext words. Further, this prefix could optionally be used to encode the position of the word in the message.

Each sequence pair i, uniquely associates a plaintext word with a cipher word. Oligo with sequence $\overline{P_i}$, corresponding to the Watson-Crick complement of plaintext word P_i , can be used as polymerase primer and be extended by specific attachment of the complement of cipher word C_i. The stopper sequence prohibits extension of the growing DNA strand beyond the boundary of the paired cipher word. A library of unique codebook strands is constructed using this theme. Each individual strand from this codebook library specifies a particular, unique set of word pairings.

The one-time-pad consists of a DNA strand of length n containing $d = \frac{n}{L_1+L_2+L_3}$ copies of the repeating pattern: a cipher word of length L_2, a plaintext word of length L_1, and stopper sequence of length L_3. We note that word length grows logarithmically in the total pad length; specifically $L_1 = c_1 \log_2 n, L_2 = c_2 \log_2 n$, and $L_3 = c_3$, for fixed integer constants $c_1, c_2, c_3 > 1$. Each repeat unit specifies a single mapping pair, and no codebook word or plaintext word will be used more than once on any pad. Therefore, given a cipher word C_i we are assured that it maps to only a single plaintext word P_i and vice versa. The stopper sequence acts as "punctuation" between repeat units so that DNA polymerase will not be able to continue copying down the template (pad) strand. Stopper sequences consist of a run of identical nucleotides which act to halt strand copying by DNA polymerase given a lack of complementary nucleotide triphosphate in the test tube. For example, the sequence TTTT will act as a stop point if the polymerization mixture lacks its base-pairing complement, A. Stopper sequences of this variety have been prototyped previously [18]. Given this structure, we can anneal primers and extend with polymerase in order to generate a set of oligonucleotides corresponding to the plaintext/cipher lexical pairings.

The experimental feasibility depends upon the following factors: (i) the size of the lexicon, which is the number of plaintext-ciphertext word-pairs, (ii) the size of each word, (iii) the number of DNA one-time-pads that can be constructed in a synthesis cycle, and (iv) the length of each message that is to be encrypted. If the lexicon used consisted of words of the English language, its size would be in the range of 10,000 to 25,000 word-pairs. If for experimental reasons, a smaller lexicon is required, then the words used could represent a more basic set such as ASCII characters, resulting in a lexicon size of 128. The implicit tradeoff is that this would increase message length. We estimate that in a single cloning procedure, we can produce 10^6 to 10^8 different one-time-pad DNA sequences. Choice of word encodings must guarantee an acceptable Hamming distance between sequences such that the fidelity of annealing is maximized. When generating sequences that will represent words, the space of all possibilities is much larger than the set that is actually needed for the implementation of the words in the lexicon. We also note that if the lexicon is to be split among multiple DNA one-time-pads, then care should be taken during pad construction to prevent a single word from being mapped to multiple targets.

If long-PCR with high fidelity enzymes introduces errors and the data in question is from an electronic source, we can pre-process it using suitable error-correction coding. If instead we are dealing with a wet database, the DNA one-time-pad's size can be restricted. This is done by splitting the single long one-time-pad into multiple shorter one-time-pads. In this case each cipher word would be modified to include a subsequence prefix that would denote which shorter one-time-pad should be used for its decryption. This increases the difficulty of cloning the entire set of pads.

The entire construction process can be repeated to prepare greater numbers of unique pads. Construction of the libraries of codebook pads can be approached using segmental assembly procedures used successfully in previous gene library construction projects [25,24] and DNA word encoding methods used in DNA computation [10,11,40,13,14,32]. One methodology is chemical synthesis of a diverse segment set followed by random assembly of pads in solution. An issue to consider with this methodology is the difficulty of achieving full coverage while avoiding possible conflicts due to repetition of plaintext or cipher words. We can set the constants c_1 and c_2 large enough so that the probability of getting repeated words on a pad of length n is acceptably small.

Another methodology would be to use a commercially available DNA chip [12,35,8,6,44]. See [31] for previous use of DNA chips for I/O. The DNA chip has an array of immobilized DNA strands, so that multiple copies of a single sequence are grouped together in a microscopic pixel. The microscopic arrays of the DNA chip are optically addressable, and there is known technology for growing distinct DNA sequences at each optically addressable site of the array. Light-directed synthesis allows the chemistry of DNA synthesis to be conducted in parallel at thousands of locations, i.e., it is a combinatorial synthesis. Therefore, the number of sequences prepared far exceeds the number of chemical reactions required. For preparation of oligonucleotides of length L, the 4^L sequences are synthesized in $4L$ chemical reactions. For example, the $\sim 65,000$ sequences of length 8 require 32 synthesis cycles, and the 1.67×10^7 sequences of length 12 require only 48 cycles. Each pixel location on the chip comprises 10 microns square, so the complete array of 12-mer sequences could be contained within a ~ 4 cm square.

2.2 DNA XOR One-Time-Pad Cryptosystem

The Vernam cipher [21] uses a sequence, S, of R uniformly distributed random bits known as a one-time-pad. A copy of S is stored at both the sender's and receiver's locations. L is the number of bits of S that have not yet been used for encrypting a message. Initially $L = R$. XOR operates on two binary inputs, yielding 0 if they are the same and 1 otherwise. When a plaintext binary message M which is $n < L$ bits long needs to be sent, each bit M_i is XOR'ed with the bit $K_i = S_{R-L+i}$ to produce the encrypted bit $C_i = M_i \oplus K_i$ for $i = 1, \ldots, n$. The n bits of S that have been consumed are then destroyed at the source and the encrypted sequence $C = (C_1, C_2, \ldots, C_n)$ is dispatched to the destination. At the destination the identical process is repeated - that is the sequence C is used in the place of M, performing bitwise XOR with bits from S, destroying the bits of S after they are consumed. The self-inverse property of binary XOR results in the initial message being reproduced since $C_i \oplus K_i = M_i$ and $M_i \oplus K_i \oplus K_i = M_i$.

To implement this algorithm with DNA, we need methods to (i) encode a plaintext message, (ii) create a DNA one-time-pad and (iii) effect bitwise XOR in DNA. Several methods exist to effect binary addition and XOR with DNA. In 1996, [16] prototyped single bit addition. Subsequent proposals [34,17] allowed for chaining outputs with inputs, and parallel operations. [22] experimentally demonstrated a logically reversible conditional XOR that required $O(n)$ recombinant DNA operations to act on n bit data. [26] described a specific DNA tiling implementation of XOR and addition, based on previous work on self-assembly of DNA tilings [47,46,48,49,50,36]. An example of cumulative XOR using self-assembled DNA tilings has recently been published [30].

DNA tiles are multi-strand complexes containing two or more double helical domains such that individual oligonucleotide chains might base-pair in one helix then cross-over and base-pair in another helix. Complexes involving crossovers (strand exchange points) create tiles which are multivalent and can have quite high thermal stability. Many helix arrangements and strand topologies are possible and several have been experimentally tested [28,27]. Tiles with specific uncomplemented sticky ends at the corners were constructed, with the purpose of effecting self-assembly.

A binary input string can be encoded using a single tile for each bit. The tiles are designed such that they assemble linearly to represent the binary string. The use of special connector tiles allow two such linear tile assemblies representing two binary input strings respectively, to come together and create a closed framework within which specially designed output tiles can fit. This process allows for unmediated parallel binary addition or XOR. As a result of the special design of these tiles, at the end of the process, there exists a single strand that runs through the entire assembly which will contain the two inputs and the output [27,26]. By using this property, we are able to effect the Vernam cipher in DNA.

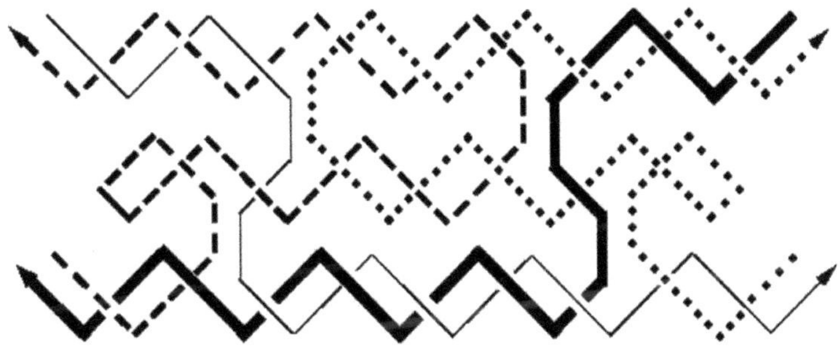

Fig. 2. TAO triple-helix tile

In particular, we use TAO triple-helix tiles (see Figure 2). The tile is formed by the complementary hybridization of four oligonucleotide strands (shown as different line types with arrowheads on their $3'$ ends). The three double-helical domains are shown with horizontal helix axes where the upper and lower helices end with bare strand ends and the central helix is capped at both ends with hairpins. Paired vertical lines represent crossover points where two strands exchange between two helices. Uncomplemented sticky ends can be appended to the four corners of the tile and encode neighbour rules for assembly of larger structures including computational complexes. For more details see [27,30].

We outline below how the bit-wise XOR operation may be done (see Figure 3). For each bit M_i of the message, we construct a sequence a_i that will represent the bit in a longer input sequence. By using suitable linking sequences, we can assemble the message M's n bits into the sequence $a_1 a_2 \ldots a_n$, which serves as the scaffold strand for one binary input to the XOR. The further portion of the scaffold strand $a'_1 a'_2 \ldots a'_n$ is created based on random inputs and serves as the one-time-pad. It is assumed that many scaffolds of the form $a'_1 a'_2 \ldots a'_n$ have been initially created, cloned using PCR [2,39] or an appropriate technique, and then separated and stored at both the source and destination points in advance. When encryption needs to occur at the source, the particular scaffold used is noted and communicated using a prefix index tag that both sender and destination know corresponds to a particular scaffold.

By introducing the scaffold for the message, the scaffold for the one-time-pad, connector tiles and the various sequences needed to complete the tiles, the input portion of the structure shown in Figure 3 forms. We call this the input assembly. This process of creating input scaffolds and assembling tiles on the scaffold has been carried out successfully [26]. Each pair of tiles (corresponding to a pair of binary inputs) in the input assembly creates a slot for the binding of a single output tile. When output tiles are introduced, they bind into appropriate binding slots by the matching of sticky ends.

Finally, adding ligase enzyme results in a continuous reporter strand R that runs through the entire assembly. If $b_i = a_i \oplus a'_i$, for $i = 1, \ldots, n$, then the reporter $R = a_1 a_2 \ldots a_n . a'_1 a'_2 \ldots a'_n . b_1 b_2 \ldots b_n$. The reporter strand is shown as a dotted line in Figure 3. This strand may be extracted by first melting apart the hydrogen bonding between strands and then purifying by polyacrylamide gel electrophoresis. It contains the input message, the encryption key, and the ciphertext all linearly concatenated. The ciphertext can be excised using a restriction endonuclease if a cleavage site is encoded between the a_0 and b_1 tiles. Alternatively the reporter strand could incorporate a nick at that point by using an unphosphorylated oligo between those tiles. The ciphertext could then be gel purified since its length would be half that of the remaining sequence. This may then be stored in a compact form and sent to a destination.

Since XOR is its own inverse, the decryption of a Vernam cipher is effected by applying the identical process as encryption with the same key. Specifically, the $b_1 b_2 \ldots b_n$ is used as one input scaffold, the other is chosen from the stored $a'_1 a'_2 \ldots a'_n$ according to the index indicating which sequence was used as the encryption key. The sequences for tile reconstitution, the connector tiles, and ligase are added. After self-assembly, the reporter strand is excised, purified, cut at the marker and the plain text is extracted.

We need to guard against loss of synchronization between the message and the key, which would occur when a bit is spuriously introduced or deleted from either sequence. Some fault tolerance is provided by the use of several nucleotides to represent each bit in the the tiles' construction. This reduces the probabiliity of such errors.

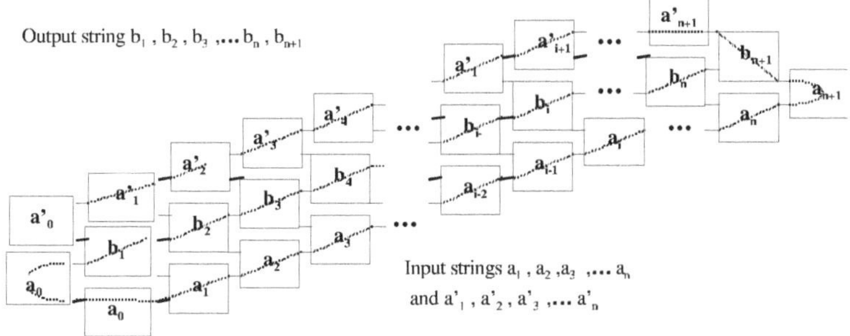

Fig. 3. XOR computation by the use of DNA tiles

3 Encrypting Images with DNA Chips and DNA One-Time-Pads

3.1 Overview of Method

In this section we outline a system capable of encryption and decryption of input and output data in the form of 2D images recorded using microscopic arrays of DNA on a chip. The system we describe here consists of: a data set to be encrypted, a chip bearing immobilized DNA strands, and a library of one-time-pads encoded on long DNA strands as described in Section 2.1. The data set for encryption in this specific example is a 2-dimensional image, but variations on the method may be useful for encoding and encrypting other forms of data or types of information. The DNA chip contains an addressable array of nucleotide sequences immobilized such that multiple copies of a single sequence are grouped together in a microscopic pixel. Such DNA chips are currently commercially available and chemical methods for construction of custom variants are well developed. Further chip details will be given below.

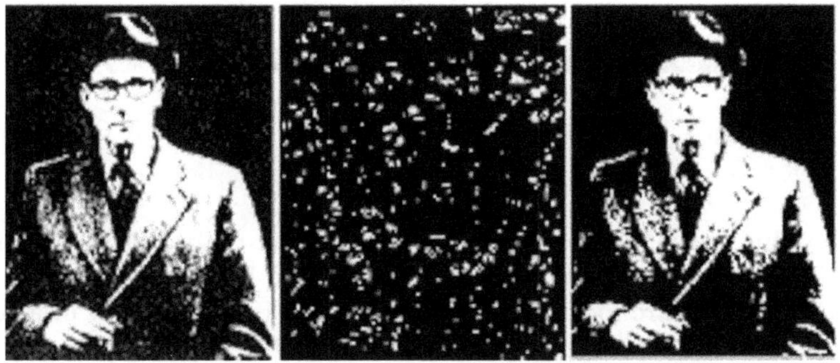

Fig. 4. DNA Chip Input/Output: Panel A: Message, Panel B: Encrypted Message, Panel C: Decrypted Message

Figure 4 gives a coarse grained outline of the I/O method. Fluorescently labeled, word-pair DNA strands are prepared from a substitution pad codebook as described in Section 2.1. These are annealed to their sequence complements at unique sites (pixels) on the DNA chip. The message information (Panel A) is transferred to a photo mask with transparent (white) and opaque (black) regions. Following a light-flash of the mask-protected chip, the annealed oligonucleotides beneath the transparent regions are cleaved at a photo-labile position. Their $5'$ sections dissociate from the annealed $3'$ section and are collected in solution. This test tube of fluorescently labeled strands is the encrypted message. Annealed oligos beneath the opaque regions are unaffected by the light-flash and can be subsequently washed off the chip and discarded. If the encrypted message oligos are reannealed onto a different DNA chip, they would anneal to different locations and the message information would be unreadable (Panel B). Note that if one used a chip identical to the encoding chip, and if the sequence lexicons for $5'$ segment (cipher word) and $3'$ segment (plaintext word) are disjoint, no binding would occur and the chip in Panel B would be completely black. Decrypting is accomplished by using the fluorescently labeled oligos as primers in asymmetric PCR with the same one-time codebook which was used to prepare the initial word-pair oligos. When the word-pair PCR product is bound to the same DNA chip, the decrypted message is revealed (Panel C).

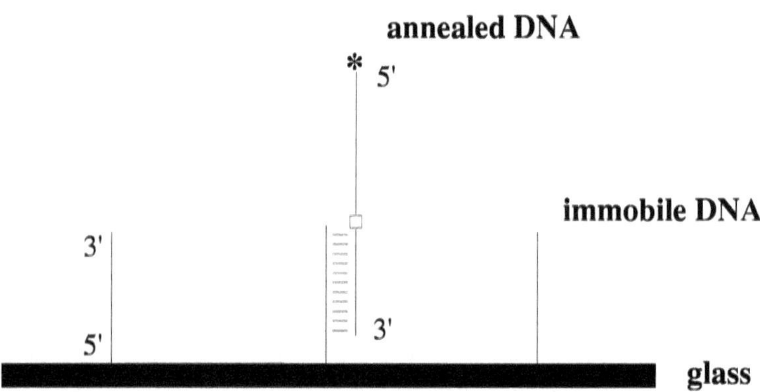

Fig. 5. Components and Organization of the DNA Chip

The annealed DNA in Figure 5 corresponds to the word-pair strands prepared from a random substitution pad as described in Section 2.1 above. Immobile DNA strands are located on the glass substrate of the chip in a sequence addressable grid according to currently used techniques. Ciphertext-plaintext word-pair strands anneal to the immobile ones. The annealed strand contains a fluorescent label on its $5'$ end (denoted with an asterisk in the figure). This is followed by the complement of a plaintext word (uncomplemented section) and the complement of a cipher word (complemented section). Located between the two words is a photo-cleavable base analog (white box in the figure) capable of cleaving the backbone of the oligonucleotide.

Figure 6 gives step by step procedures for encryption and decryption. For encryption, we start with a DNA chip displaying the sequences drawn from the cipher lexicon. In step one, the fluorescently labeled word-pair strands prepared from a one-time-pad are annealed to the chip at the pixel bearing the complement to their 3′ end. In the next step, the mask (heavy black bar) protects some pixels from a light-flash. At unprotected regions, the DNA backbone is cleaved at a site between the plaintext and cipher words. In the final step, the 5′ segments, still labeled with fluorophore at their 5′ ends, are collected and transmitted as the encrypted message.

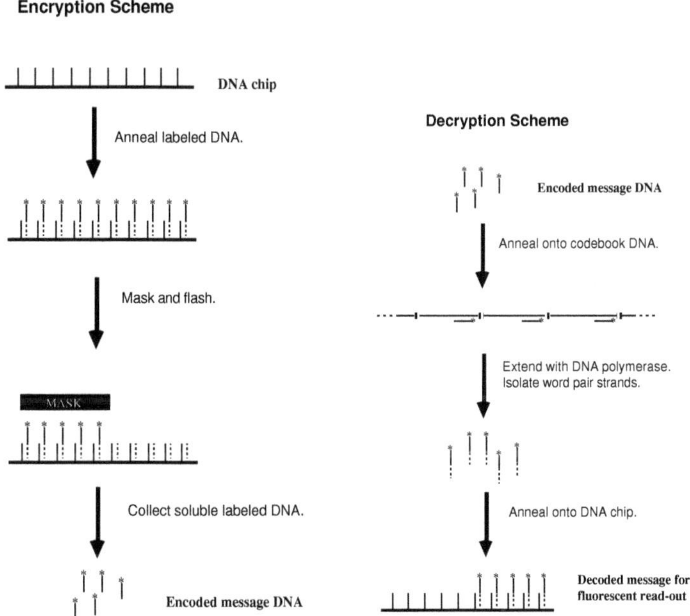

Fig. 6. Step by step procedures for encryption and decryption

A message can be decrypted only by using the one-time-pad and DNA chip identical to those used in the encryption process. First, the word-pair strands must be reconstructed by appending the proper cipher word onto each plaintext word. This is accomplished by primer extension or asymmetric PCR using transmitted strands as primer and one-time-pad as template. The strands bind to their specific locations on the pad and are appended with their proper cipher partner. Note that in the decrypting process the fluorescent label is still required, but the photo-labile base is unnecessary and not present. The final step of decryption involves binding the reformed word-pair strands to the DNA chip and reading the message by fluorescent microscopy.

3.2 Additional Technical Considerations

Some details concerning the configuration of the DNA chip should be mentioned. In the current incarnation of the method, reverse synthesis of oligos directly on the chip or "spot attachment" would be required. Chemical reagents for reverse synthesis are commercially available although not as widely used as those for 3'-to-5' methods. Spot attachment of oligos onto the chip results in decreased pixel density and increased work. However, recent chip improvements, including etching with hydrophobic gridlines, may alleviate this drawback.

One potential photo-cleavable base analog is 7−nitroindole nucleoside. It has previously been shown to produce a chemical lesion in the effected DNA strand which causes backbone cleavage. Use of 7−nitroindole nucleoside for strand cleavage has been shown to produce useable $5'$ and $3'$ ends [23]. Production of 'clean' $3'$ ends is critical for decrypting the message, since the cipher strands must hybridize to the one-time-pad and act as primers for the polymerase mediated strand elongation (PCR). Primers and templates containing the 7−nitroindole nucleoside have been shown to function properly in PCR and other enzymatic reactions.

4 DNA Steganography Analysis

Steganography using DNA is appealling due to its simplicity. One method proposed involves taking "plaintext" input DNA strands, tagging each with "secret key" strands, and then hiding them among random "distracter" strands. The plaintext is retrieved by hybridization with the complement of the secret key strands. It has been postulated that in the absence of knowledge of the secret key, it would be necessary to examine all the strands including the distracters to retrieve the plaintext. Based on the likely difference in entropy of the distracters and the plaintext, we argue that the message can be retrieved without the key.

4.1 Relevant Data Compression Result

The *compression ratio* is the quotient of the length of the compressed data divided by the length of the source data. For example, many images may be losslessly compressed to a 1/4 of their original size; English text and computer programs have compression ratios of about 1/3; most DNA has a compression ratio between 1/2 and 1/1.2 [15,29]. Protein coding sequences make efficient use of amino acid coding and have larger compression ratios [33,20]. The Shannon information theoretic *entropy rate* is denoted by $H_S \leq 1$. It is defined to be the rate that the entropy increases per symbol, for a sequence of length n with $n \to \infty$ [9]. It provides a measure of the asymptotic rate at which a source can be compressed without loss of information. Random sequences can not be compressed and therefore have an entropy rate of 1.

Lossless data compression with an algorithm such as Lempel-Ziv [51], is theoretically asymptotically optimal. For sequences whose length n is large, the compression ratio approaches the entropy rate of the source. In particular, it is of the form $(1 + \epsilon(n))H_S$, where $\epsilon(n) \to 0$ for $n \to \infty$. Algorithms such as Lempel-Ziv build an indexed dictionary of all subsequences parsed that can not be constructed as a catenation of current dictionary entries. Compression is performed by sequentially parsing the input text, finding maximal length subsequences already present in the dictionary, and outputting their index number instead. When a subsequence is not found in the dictionary, it is added to it (including the case of single symbols). Algorithms can achieve better compression by making assumptions about the distribution of the data [4]. It is possible to use a dictionary of bounded size, consisting of the most frequent subsequences. Experimentation on a wide variety of text sources shows that this method can be used to achieve compression within a small percentage of the ideal [43]. In particular, the compression ratio is of the form $(1 + \epsilon)H_S$, for a small constant $\epsilon > 0$ typically of at most $1/10$ if the source length is large.

Lemma 1. *The expected length of a parsed word is between* $\frac{L}{1+\epsilon}$ *and* L*, where* $L = \frac{\log_b n}{H_S}$.

Proof. Assume the source data has an alphabet of size b. An alphabet of the same size can be used for the compressed data. The dictionary can be limited to a constant size. We can choose an algorithm that achieves a compression ratio within $1 + \epsilon$ of the asymptotic optimal, for a small $\epsilon > 0$. Therefore, for large n, we can expect the compression ratio to approach $(1 + \epsilon)H_S$.

Each parsed word is represented by an index into the dictionary, and so its size would be $\log_b n$ if the source had no redundancy. By the choice of compression algorithm, the length of the compressed data is between H_S and $H_S(1+\epsilon)$ times the length of the original data. From these two facts, it follows that the expected length of a code word will be between $\frac{\log_b n}{(1+\epsilon)H_S}$ and $\frac{\log_b n}{H_S}$.

Lemma 2. *A parsed word has length* $\leq \frac{L}{p}$ *with probability* $\geq 1 - p$.

Proof. The probability of a parsed word having length $> \frac{L}{p}$ is $< p$, for all $p \in (0, 1)$, by the Markov inequality. The lemma follows from this.

Lemma 3. *A parsed word has length* $\geq c'L$ *with probability* $\geq 1 - p$*, if* $p > 1 - \frac{1}{c(1+\epsilon)}$ *and* $c' = c - \frac{c - \frac{1}{1+\epsilon}}{p} > 0$.

Proof. The maximum length of a parsed word has an upper bound in practice. We assume that this is cL for a small constant $c > 1$. We use Δ to denote the difference between the maximum possible and the actual length of a parsed word, and $\bar{\Delta}$ to denote the difference's expected value. Applying Lemma 1,

$0 < \bar{\Delta} < cL - \frac{L}{1+\epsilon} (= (c - \frac{1}{1+\epsilon})L)$. The probability that $\Delta > \frac{\bar{\Delta}}{p} (= \frac{(c - \frac{1}{1+\epsilon})L}{p})$, is $< p$, by the Markov inequality. Therefore, with probability $< p$, a parsed word has length $< cL - \frac{\bar{\Delta}}{p} = c'L$, where $c' = c - \frac{c - \frac{1}{1+\epsilon}}{p}$. We choose $p > 1 - \frac{1}{c(1+\epsilon)}$ so that $0 < c' \leq c$, since parsed words must have positive length that does not exceed the maximum postulated.

Lemma 4. *A parsed word has length between $c'L$ and $\frac{L}{p}$ with probability $\geq (1-p)^2$, if $p > 1 - \frac{1}{c(1+\epsilon)}$ and $c' > H_S$.*

Proof. This follows from Lemmas 2 and 3.

4.2 Analysis Assumptions

We assume all the following. The alphabet in question is that of DNA and therefore has size 4. The probability distribution of the "plaintext" DNA source S is known - for example, that it is generated by a stationary ergodic process. The "distracter" DNA strands have a random uniform distribution over the 4 DNA bases. Both "plaintext" and "distracter" DNA strands have the same length since they may otherwise be distinguished by length. A Lempel-Ziv algorithm variant that meets the criteria of Lemma 4 is known. p is fixed just above $1 - \frac{1}{c(1+\epsilon)}$. $f(n) \approx g(n)$ if $\frac{f(n)}{g(n)} \to 1$ and $(1 - \frac{1}{n})^n \approx \frac{1}{e}$, for large n.

4.3 Constructing a Dictionary

Let $L = H_S \log_4 n$. D is the set of d most frequently occurring words of the source, where d is the size of the dictionary. D' is the subset of D that consists of words that meet the following two criteria. The first is that the word's length must be between $c'L$ and $\frac{L}{p}$. The second is that the word's frequency in the source S must be $> \frac{1}{n'}$, where $n' = (1-p)^2 \frac{n}{L}$.

Lemma 5. *The probability that a word w in D' is a parsed word of the "plaintext" DNA sequence is $> 1 - \frac{1}{e}$.*

Proof. Let X be a "plaintext" DNA sequence of length n. Consider D'', the subset of D containing words of length between $c'L$ and $\frac{L}{p}$. D'' contains at least $(1-p)^2$ of the parsed words of X by Lemma 4. D' is the subset of D'' which consists of only words that have frequency $> \frac{1}{n'}$. Consider a word v parsed from X. The probability that a word w from D' is not v is $< 1 - \frac{1}{n'}$ by construction. X has an expected number $\frac{n}{L}$ parsed words. By Lemma 1, there are an expected number $(1-p)^2 \frac{n}{L}$ words with length in the range between $c'L$ and $\frac{L}{p}$. The probability that w is none of these words is therefore $< (1 - \frac{1}{n'})^{(1-p)^2 \frac{n}{L}} \approx \frac{1}{e}$. Thus, a word w in D' is some parsed word of X with probability $> 1 - \frac{1}{e}$.

4.4 DNA-SIEVE: Plaintext Extraction Without the Key

Strand separation operates on a test tube of DNA, using recombinant DNA technology to split the contents into two tubes, T and T' with separation error rates $\sigma^-, \sigma^+ : 0 < \sigma^-, \sigma^+ < 1$. The goal is to transfer all the strands that contain the subsequence w into the tube T', leaving all the rest in tube T. A fraction $< \sigma^-$ of the strands without subsequence w enter T'. A fraction $> 1 - \sigma^+$ of the strands containing w are left in T. We assume that $\rho = \frac{\sigma^-}{(1-\sigma^+)(1-\frac{1}{e})}, 0 < \rho < 1$. Modest expectations for separation technology yield $0 < \sigma^+ < 0.2$ and $0 < \sigma^+ < 0.05$. Using $\sigma^- = \sigma^+ = 0.2$ suffices to obtain ρ in the desired range.

DNA-SIEVE is to be used to extract the "plaintext" DNA strands from the mix in which there are many "distracters". It begins with a tube T. The separate operation is iteratively applied. In each round, a previously unused word w from the set D' is chosen. All strands that contain it are retained by using hybridization with the complement of w. We use $r(T)$ to denote the ratio of the distracters to the plaintext, and $F(T)$ to denote the tube from which the strands with subsequence w were removed.

4.5 DNA-SIEVE Analysis

The success of DNA-SIEVE rests on the fact that a word in D' is likely to occur in plaintext X with probability $1 - \frac{1}{e}$, while it is expected to occur in a random text R with probability close to 0.

Lemma 6. *The probability that a word in D' is a subsequence of R is* $\approx n4^{-c'L} = \frac{1}{n^{\frac{c'}{H_S}-1}}$.

Proof. Let R denote a random "distracter" sequence of length n over the alphabet of the 4 DNA bases. Since all sequences are equiprobable, one of length $c'L = c'\frac{\log_4 n}{H_S}$ is likely to occur with probability $4^{-c'L} = 4^{\log_4 n^{\frac{-c'}{H_S}}} = \frac{1}{n^{\frac{c'}{H_S}}}$. Since it can occur at any of $\approx n$ locations in R, the probability of it occurring in R is $n4^{-c'L} = \frac{1}{n^{\frac{c'}{H_S}-1}}$. By assumption in Lemma 4, $c' > H_S$, so $\frac{c'}{H_S} - 1 > 0$.

Lemma 7. *If DNA-SIEVE operates on tube T and results in tube $F(T)$, then at the most $\approx \sigma^-$ of the distracters in T are in $F(T)$, while at least $\approx (1-\sigma^+)(1-\frac{1}{e})$ of the plaintext strands of T are in $F(T)$.*

Proof. The probability that a distracter strand in T is present in $F(T)$ is limited by $\sigma^- + n4^{-c'L}$, the sum of the error and the theoretical chance. By assumption, the error rate is $< \sigma^-$. By the Lemma 6 the chance is $< \frac{1}{n^{\frac{c'}{H_S}-1}}$. Since n is large, this is ≈ 0. Therefore at most σ^- of the distracters in T reach $F(T)$. Similarly, by Lemma 5, at least $1 - \frac{1}{e}$ of the plaintext strands in T are expected to be in $F(T)$. By assumption, at most σ^+ of the strands that should reach $F(T)$ are left behind due to separation error. Therefore, $\approx (1 - \sigma^+)(1 - \frac{1}{e})$ of the plaintext strands actually reach $F(T)$.

Lemma 8. *The probability distribution of the distracter strands returns to the original one after an expected number of $\frac{2L}{p}$ iterations of DNA-SIEVE.*

Proof. Each iteration of DNA-SIEVE uses a word w from D' that has not been previously used. Assume w is a prefix or suffix of another word w' in D'. Once DNA-SIEVE has been using w, the probability distribution of the distracters complementary to w' will be altered. The distribution of the rest will remain unaffected. There are $\leq \frac{2L}{p}$ words that can overlap with a given word from D'. Therefore, a particular separation affects $\leq \frac{2L}{p}$ other iterations. If w is chosen randomly from D', then after an expected number of $\frac{2L}{p}$ iterations, all the strands will be equally effected and hence the probability distribution of the distracters will be the same as before the sequence of iterations. Such a number of iterations is termed "independent".

Theorem 1. *To reduce the ratio of distracter to plaintext strands by a factor r, it suffices to apply DNA-SIEVE an expected number of $O(\log n) \log r$ times.*

Proof. Denote the ratio of the distracter strands to the plaintext strands in test tube T with r'. Now consider a tube $F(T)$ that results from applying DNA-SIEVE t times till this ratio has been reduced by a factor r. Denote the ratio for tube $F(T)$ by r''. By Lemma 8, after an expected number of $\frac{2L}{p}$ iterations, a test tube $G(T)$ is produced with the same distribution of distracters as in T. Applying Lemma 7, we expect that after every set of $\frac{2L}{p}$ iterations, the ratio will change by at least $\rho = \frac{\sigma^-}{(1-\sigma^+)(1-\frac{1}{e})}$. We expect a decrease in the concentration after t iterations by a factor of $\rho^{\frac{t}{\frac{2L}{p}}}$. To attain a decrease of $\frac{r''}{r'}$, we need $t = \frac{2L}{p} \log \frac{r''}{r'} \log \rho$. Since $L = O(\log n)$ and $\rho = O(1)$, $t = O(\log n) \log r$.

4.6 DNA-SIEVE Implementation Considerations

The theoretical analysis of DNA-SIEVE was used to justify the expected geoemetric decrease in the conentration of the distracter strands. It also provides two further insights. The number of "plaintext" DNA strands may decrease by a factor of $(1-\sigma^+)(1-1/e)$ after each iteration. It is therefore prudent to increase the absolute number of copies periodically (by ligating all the strands with primer sequences, PCR cycling, and then digesting the primers that were added). The number of iterations that can be performed is limited to n' due to the fact that a distinct word from D' must be used each time. This limits the procedure to operation on a population where the number of distracter strands is $< 4^{n'}$.

4.7 Empirical Analysis

We performed an empirical analysis of DNA-SIEVE. We assumed that the test tube would start with 10^8 distracter strands and 10^3 message strands. The first

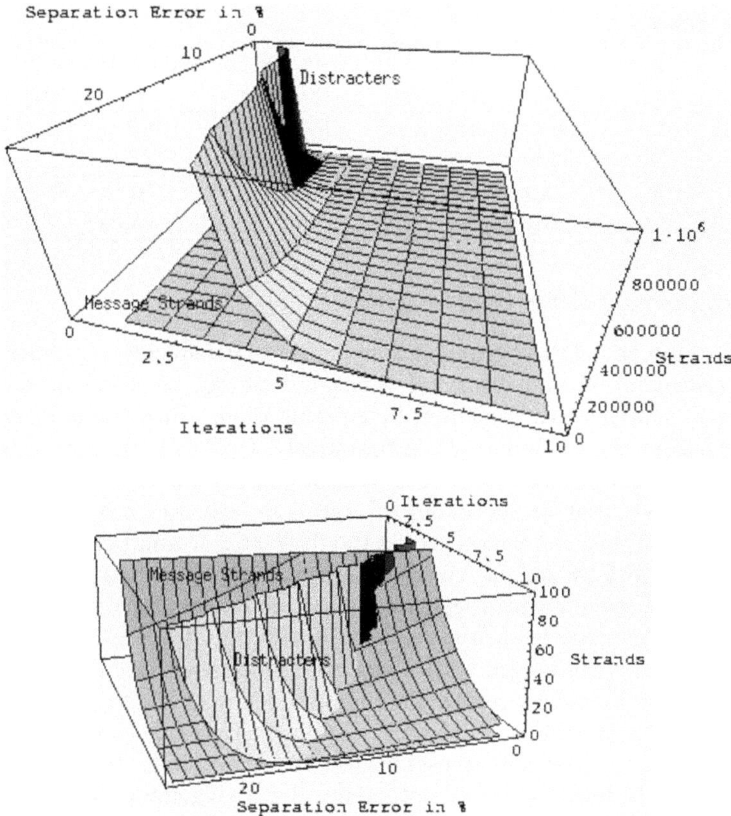

Fig. 7. Simulation of DNA-SIEVE : Distracters have a sharper drop-off in concentration

parameter varied was the number of “indpendent” iterations of DNA-SIEVE - from 1 to 10. The second parameter varied was the separation error rate - from 0.01 to 0.25 in multiples of 0.01. Here we do not assume a difference in the error rate for false positives and false negatives that occur during the separation. In each case, the number of distracters and message strands remaining was computed. The results are plotted in Figure 7. From this we can see that 5 to 10 iterations of DNA-SIEVE suffice to reduce the distracter population's size to below that of the message strands when the separation error is < 0.18. The table illustrates the number of distracters and message strands left after 3, 6 and 9 iterations with varying separation error rates. If the error rate is reasonable, it can be seen from the table that there remain enough message strands for the plaintext to be found. If the separation error rate is high, the number of strands used must be increased to allow enough iterations of DNA-SIEVE to occur before the message strands are all lost.

Iterations	Separation Error	0.05	0.1	0.15	0.2	0.25
3	Distracters	12500	100000	337500	800000	1562500
	Messages	217	184	155	129	107
6	Distracters	2	100	1139	6400	24414
	Messages	47	34	24	17	11
9	Distracters	0	0	4	51	381
	Messages	10	6	4	2	1

4.8 Improving DNA Steganography

We can improve a DNA steganography system by reducing the difference between the plaintext and distracter strands. This can be done by making the distracters similar to the plaintext by creating them using random assembly of elements from the dictionary D. Alternatively, DNA-SIEVE can be employed on a set of random distracters to shape the population into one whose distribution matches that of the plaintext. We note, however, that if the relative entropy [9] between the plaintext and the distracter strand populations is large enough, DNA-SIEVE can be employed as previously described. An attacker can use a larger dictionary, which provides a better model of the plaintext source, to increase the relative entropy re-enabling the use of DNA-SIEVE. If the M plaintext strands are tagged with a sequence that allows them to be extracted, then they may be recognized by the high frequency of the tag sequence in the population. To guard against this, N sets of M strands each can be mixed in. This results in a system that uses $V = O(MN)$ volume. To prevent a brute force attack, N must be large, potentially detracting from the practicality of using using the DNA steganographic system.

The other approach to reduce the distinguishability of the plaintext from the distracters is to make the former mimic the latter. By compressing the plaintext with a lossless algorithm, such as Lempel-Ziv [51], the relative entropy of the message and the distracter populations can be reduced. If the plaintext is derived from an electronic source, it can be compressed in a preprocessing step. If the source is natural DNA, it can be recoded using a substitution method similar to the one described in Section 2. However, the security of such a recoding remains unknown. In the case of natural DNA, for equivalent effort, DNA cryptography offers a more secure alternative to DNA steganography.

5 Conclusion

This paper presented an initial investigation into the use of DNA-based information security. We discussed two classes of methods: (i) DNA cyptography methods based on DNA one-time-pads, and (ii) DNA steganography methods. Our DNA substitution and XOR methods are based on one-time-pads, which are in principle unbreakable. We described our implementation of DNA cyptography with 2D input/output. We showed that a class of DNA steganography methods offer limited security and can be broken with a reasonable assumption about the entropy of the plaintext messages. We considered modified DNA steganographic systems with improved security. Steganographic techniques rest on the assumption that the adversary is unaware of the existence of the data. When this does not hold, DNA cryptography must be relied upon.

Acknowledgments

Work supported by Grants NSF/DARPA CCR-9725021, CCR-96-33567, NSF IRI-9619647, ARO contract DAAH-04-96-1-0448, and ONR contractN00014-99-1-0406. A preliminary version appears in DIMACS DNA Based Computers V, American Mathematical Society, 2000.

References

1. L.M. Adleman, Molecular computation of solutions to combinatorial problems, *Science*, 266 (1994), 1021–1024.
2. W.M. Barnes, PCR amplification of up to 35-kb DNA with high fidelity and high yield from bacteriophage templates, *Proc. Natl. Acad. Sci.*, 91 (1994), 2216–2220.
3. E.B. Baum, Building an associative memory vastly larger than the brain, *Science*, 268 (1995), 583–585.
4. T. Bell, I.H. Witten, and J.G. Cleary, Modeling for Text Compression, *ACM Computing Surveys*, 21, 4 (1989), 557–592.
5. D. Boneh, C. Dunworth, and R.J. Lipton, Breaking DES Using a Molecular Computer, *DNA Based Computers* (E.B. Baum and R.J. Lipton, eds.), American Mathematical Society, 1996, *DIMACS: Series in Discrete Mathematics and Theoretical Computer Science.*
6. A.P. Blanchard, R.J. Kaiser, and L E. Hood, High-density oligonucleotide arrays, *Biosens. Bioelec.*, 11 (1996), 687–690.
7. D. Boneh, C. Dunworth, R.J. Lipton, and J. Sgall, Making DNA computers error resistant, *DNA based computer II* (Editors: L. Landwaber, E. Baum) DIMACS series in Discrete Math. and Theoretical Comp. Sci. **44** (1999).
8. M. Chee, R. Yang, E. Hubbell, A. Berno, X.C. Huang, D. Stern, J. Winkler, D.J. Lockhart, M.S. Morris, S.P.A. Fodor, Accessing genetic information with high-density DNA arrays, *Science*, 274 (1996), 610–614.
9. Th.M. Cover and J.A. Thomas, *Elements of Information Theory*, New York, NY, USA, John Wiley & Sons, 1991.
10. R. Deaton, R.C. Murphy, M. Garzon, D.R. Franceschetti, and S.E. Stevens, Jr., Good Encodings for DNA-based Solutions to Combinatorial Problems, *DNA based computer II* (Editors: L. Landwaber, E. Baum) DIMACS series in Discrete Math. and Theoretical Comp. Sci. **44** (1999). *Proceedings of the Second Annual Meeting on DNA Based Computers*, 1996, American Mathematical Society, *DIMACS: Series in Discrete Mathematics and Theoretical Computer Science*, 1052–1798.
11. R. Deaton, R.C. Murphy, M. Garzon, D.R. Franceschetti, and S.E.Stevens, Jr., Reliability and efficiency of a DNA-based computation, *Phys. Rev.Lett.*, 80 (1998), 417–420.

12. S. Fodor, J.L. Read, M.C. Pirrung, L. Stryer, A. Tsai Lu, and D. Solas, Light-directed spatially addressable parallel chemical synthesis, *Science*, 251 (1991), 767–773.
13. A.G. Frutos, A.J. Thiel, A.E. Condon, L.M. Smith, and R.M. Corn, DNA Computing at Surfaces: 4 Base Mismatch Word Design, *DNA based computer III* (Editors: H. Rubin, D. Wood) DIMACS series in Discrete Math. and Theoretical Comp. Sci. **vol 48** (1999) 238.
14. J.M. Gray, T.G. Frutos, A. Michael Berman, A.E. Condon, M.G. Lagally, L.M. Smith, and R.M. Corn, Reducing Errors in DNA Computing by Appropriate Word Design, November, 1996.
15. S. Grumbach and F. Tahi, A new challenge for compression algorithms: genetic sequences, *Inf. Proc. and Management*, 30, 6 (1994), 875–886.
16. F. Guarnieri, M. Fliss, and C. Bancroft, Making DNA Add, *Science*, 273 (1996), 220–223.
17. V. Gupta, S. Parthasarathy, and M.J. Zaki, Arithmetic and Logic Operations with DNA, *DNA based computer III* (Editors: H. Rubin, D. Wood) DIMACS series in Discrete Math. and Theoretical Comp. Sci. **vol 48** (1999) 212–220.
18. M. Hagiya, M. Arita, D. Kiga, K. Sakamoto, and S. Yokoyama, Towards Parallel Evaluation and Learning of Boolean μ-Formulas with Molecules, *DNA based computer III* (Editors: H. Rubin, D. Wood) DIMACS series in Discrete Math. and Theoretical Comp. Sci. **vol 48** (1999) 105–114.
19. T. Head, Splicing schemes and DNA, *Lindenmayer Systems; Impact on Theoretical computer science and developmental biology* (G. Rozenberg and A. Salomaa, eds.), Springer Verlag, Berlin, 1992, 371–383.
20. S. Henikoff and J. G. Henikoff, Amino acid substitution matrices from protein blocks, *Proc. Natl. Acad. Sci.*, 89 (1992), 10915–10919.
21. D. Kahn, *The Codebreakers*, Macmillan, NY, 1967.
22. J.P. Klein, T.H. Leete, and H. Rubin, A biomolecular implementation of logical reversible computation with minimal energy dissipation, *Proceedings 4th DIMACS Workshop on DNA Based Computers*, University of Pennysylvania, Philadelphia, 1998 (L. Kari, H. Rubin, and D.H. Wood, eds.), 15–23.
23. M. Kotera, A.G. Bourdat, E. Defrancq, and J. Lhomme, A highly efficient synthesis of oligodeoxyribonucleotides containing the 2'-deoxyribonolactone lesion, *J. Am. Chem. Soc.*, 120 (1998), 11810–11811.
24. T.H. LaBean and T.R. Butt, Methods and materials for producing gene libraries, U.S. Patent Number 5,656,467, 1997.
25. T. LaBean and S.A. Kauffman, Design of synthetic gene libraries encoding random sequence proteins with desired ensemble characteristics, *Protein Science*, 2 (1993), 1249–1254.
26. T.H. LaBean, E. Winfree, and J.H. Reif, Experimental Progress in Computation by Self-Assembly of DNA Tilings, *DNA Based Computers V*, 1999.
27. T.H. LaBean, H. Yan, J. Kopatsch, F. Liu, E. Winfree, H.J. Reif, and N C. Seeman, The construction, analysis, ligation and self-assembly of DNA triple crossover complexes, *J. Am. Chem. Soc.*, 122 (2000), 1848–1860.
28. X. Li, X. Yang, J. Qi, and N.C. Seeman, Antiparallel DNA double crossover molecules as components for nanoconstruction, *J. Amer. Chem. Soc.*, 118 (1996), 6131–6140.
29. Loewenstern and Yianilos, Significantly Lower Entropy Estimates for Natural DNA Sequences, *DCC: Data Compression Conference*, IEEE Computer Society TCC, (1997), 151–161.

30. C. Mao, T.H. LaBean, J.H. Reif, and N C. Seeman, Logical computation using algorithmic self-assembly of DNA triple-crossover molecules, *Nature,* 407 (2000), 493–496.
31. A.P. Mills Jr., B. Yurke, and P.M. Platzman, Article for analog vector algebra computation, *Proceedings 4th DIMACS Workshop on DNA Based Computers,* University of Pennysylvania, Philadelphia, 1998 (L. Kari, H. Rubin, and D.H. Wood, eds.), 175–180.
32. K.U. Mir, A Restricted Genetic Alphabet for DNA Computing, *Proceedings of the Second Annual Meeting on DNA Based Computers*, Princeton University, 1996, in *DNA based computer II* (Editors: L. Landwaber, E. Baum) DIMACS series in Discrete Math. and Theoretical Comp. Sci. **44** (1999).
33. C.G. Nevill-Manning and I.H. Witten, Protein is Incompressible, *IEEE Data Compression Conference*, IEEE Computer Society TCC, 1999, 257–266.
34. M. Orlian, F. Guarnieri, and C. Bancroft, Parallel Primer Extension Horizontal Chain Reactions as a Paradigm of Parallel DNA-Based Computation, *DNA based computer III* (Editors: H. Rubin, D. Wood) DIMACS series in Discrete Math. and Theoretical Comp. Sci. **vol 48** (1999) 142–158.
35. A.C. Pease, D. Solas, E.J. Sullivan, M.T. Cronin, C.P. Holmes, and S. P. Fodor, Light-generated oligonucleotide arrays for rapid DNA sequence analysis, *Proc. Natl Acad. Sci. USA,* 91 (1994), 5022–5026.
36. J.H. Reif, Local Parallel Biomolecular Computing, *DNA based computer III* (Editors: H. Rubin, D. Wood) DIMACS series in Discrete Math. and Theoretical Comp. Sci. **vol 48** (1999) 243–264.
37. J.H. Reif, Paradigms for Biomolecular Computation, *Unconventional Models of Computation* (C.S. Calude, J. Casti, and M.J. Dinneen, eds.), Springer, 1998.
38. J.H. Reif, Parallel Molecular Computation: Models and Simulations, *Algorithmica, Special Issue on Computational Biology*, 1998.
39. S.S. Roberts, Turbocharged PCR, *Jour. of N.I.H. Research*, 6 (1994), 46–82.
40. J.A. Rose, R. Deaton, M. Garzon, R.C. Murphy, D.R. Franceschetti, and S.E. Stevens Jr., The effect of uniform melting temperatures on the efficiency of DNA computing, *DNA based computer III* (Editors: H. Rubin, D. Wood) DIMACS series in Discrete Math. and Theoretical Comp. Sci. **vol 48** (1999) 35–42.
41. S. Roweis, E. Winfree, R. Burgoyne, N.V. Chelyapov, M.F. Goodman, P.W.K. Rothemund, and L.M. Adleman, A Sticker Based Architecture for DNA Computation, *DNA based computer II* (Editors: L. Landwaber, E. Baum) DIMACS series in Discrete Math. and Theoretical Comp. Sci. **44** (1999) 1-29.
42. B. Schneier, *Applied Cryptography: Protocols, Algorithms, and Source Code in C*, John Wiley & Sons, Inc., 1996.
43. J.A. Storer, *Data Compression: Methods and Theory*, Computer Science Press, 1988.
44. A. Suyama, DNA chips - Integrated Chemical Circuits for DNA Diagnosis and DNA computers, 1998.
45. C.T. Taylor, V. Risca, and C. Bancroft, Hiding messages in DNA microdots, *Nature*, 399 (1999), 533–534.
46. E. Winfree, On the Computational Power of DNA Annealing and Ligation, *DNA Based Computers* (E.B. Baum and R.J. Lipton, eds.), American Mathematical Society, *DIMACS: Series in Discrete Mathematics and Theoretical Computer Science*, 1995, 187–198.
47. E. Winfree, Complexity of Restricted and Unrestricted Models of Molecular Computation, *DNA Based Computers* (E.B. Baum and R.J. Lipton, eds.), American

Mathematical Society, 27, *DIMACS: Series in Discrete Mathematics and Theoretical Computer Science*, 1995, 187–198.

48. E. Winfree, Simulations of Computing by Self-Assembly, *Proceedings of the Fourth DIMACS Meeting on DNA Based Computing*, 1998, 213–242.
49. E. Winfree, F. Liu, L.A. Wenzler, and N.C. Seeman, Design and Self-Assembly of Two Dimensional DNA Crystals, *Nature*, 394 (1998), 539–544.
50. E. Winfree, X. Yang, and N.C. Seeman, Universal Computation via Self-assembly of DNA: Some Theory and Experiments, *DNA based computer II* (Editors: L. Landwaber, E. Baum) DIMACS series in Discrete Math. and Theoretical Comp. Sci. **44** (1999) 191–214.
51. J. Ziv and A. Lempel, A universal algorithm for sequential data compression, *IEEE Trans. Inf. Theory*, IT-23 (1977), 337–343.

Splicing to the Limit

Elizabeth Goode[1] and Dennis Pixton[2]

[1] Mathematics Department, Towson University
Towson, MD 21252, USA
egoode@towson.edu

[2] Department of Mathematical Sciences, Binghamton University
Binghamton, NY 13902-6000, USA
dennis@math.binghamton.edu

Abstract. We consider the result of a wet splicing procedure after the reaction has run to its completion, or limit, and we try to describe the molecules that will be present at this final stage. In language theoretic terms the splicing procedure is modeled as an H system, and the molecules that we want to consider correspond to a subset of the splicing language which we call the *limit language.* We give a number of examples, including one based on differential equations, and we propose a definition for the limit language. With this definition we prove that a language is regular if and only if it is the limit language of a reflexive and symmetric splicing system.

1 Introduction

Tom Head [5] invented the notion of a *splicing system* in 1987 in an attempt to model certain biochemical reactions involving cutting and reconnecting strands of double-sided DNA. Since then there have been many developments and extensions of the basic theory, but some of the basic questions have remained unanswered. For example, there is still no simple characterization of the languages defined by splicing systems.

The language defined by a splicing system is called the *splicing language*, and this corresponds to the set of molecular types that are created during the evolution of the system. In discussions several years ago Tom proposed analyzing the outcome of a splicing system in a way that is closer to what is actually observed in the laboratory. The idea is that certain molecules which appear during the biochemical reactions are *transient* in the sense that they are used up by the time the splicing experiment has "run to completion." The molecules that are left at this stage can be termed the *limit molecules* of the system, and the corresponding formal language defined by the splicing system is called the *limit language.*

The first major problem is to properly define this limit language. One possibility is to try to use differential equations to model the set of molecules, and then to define the limit language in terms of the solutions. Although it is reasonably clear how to set up such equations, solving them is another matter, since

N. Jonoska et al. (Eds.): Molecular Computing (Head Festschrift), LNCS 2950, pp. 189–201, 2004.

the system of differential equations is non-linear and, in non-trivial cases, has infinitely many variables.

Another possibility is to try to avoid quantitative issues and to analyze the splicing process qualitatively, in the context of formal languages, but guided by our understanding of the actual behavior of the biochemical systems.

In this paper we present a number of examples from the second viewpoint, illustrating some of the features that we expect in a limit language. We also give some of the details of a simple example from the differential equations viewpoint, to illustrate the general approach.

We then propose a definition of limit language in the formal language setting. It is not clear that our definition covers all phenomena of limit behavior, but it is sufficient for the range of examples we present, and leads to some interesting questions. It is well known that splicing languages (based on finite splicing systems) are regular, and it is natural to ask whether the same is true for limit languages. We do not know the answer in general, but in the special case of a reflexive splicing system we can give a satisfying answer: The limit language is regular; and, moreover, every regular language occurs as a limit language.

We present this paper not as a definitive statement, but as a first step toward understanding the limit notion.

2 Motivation

We review briefly the basic definitions; for more details see [4]. A *splicing rule* is a quadruple $r = (u_1, v_1;\, u_2, v_2)$ of strings, and it may be used to *splice* strings $x_1u_1v_1y_1$ and $x_2u_2v_2y_2$ to produce $x_1u_1v_2y_2$. A *splicing scheme* or *H scheme* is a pair $\sigma = (A, R)$ where A is the alphabet and R is a finite set of rules over A. A *splicing system* or *H system* is a pair (σ, I) where σ is a splicing scheme and $I \subset A^*$ is a finite set, the *initial language.* The *splicing language* defined by such a splicing system is the smallest language L that contains I and is closed under splicing using rules of σ. Usually L is written as $\sigma^*(I)$.

Associated to a rule $r = (u_1, v_1;\, u_2, v_2)$ are three other rules: its *symmetric twin* $\bar{r} = (u_2, v_2;\, u_1, v_1)$ and its *reflexive restrictions* $\dot{r} = (u_1, v_1;\, u_1, v_1)$ and $\ddot{r} = (u_2, v_2;\, u_2, v_2)$. A splicing scheme σ is called *symmetric* if it contains the symmetric twins of its rules, and it is called *reflexive* if it contains the reflexive restrictions of its rules.

We consider I as modeling an initial set of DNA molecule types, with σ modeling a test-tube environment in which several restriction enzymes and ligases act simultaneously. In this case the language L models the set of all molecule types which appear at any stage in this reaction. In this interpretation we can see that, in fact, representatives of all molecule types in L appear very quickly (although perhaps in very small concentrations) after the beginning of the reactions.

Here we want to consider a somewhat different scenario: We ask not for the set of all molecules that are created in this reaction, but for the set of molecules that will be present after the system has "reached equilibrium". In other words, we expect that some of the molecular types will be used up in constructing different molecular types, so they will eventually disappear from the solution. We call the subset L_∞ of L corresponding to the remaining "final-state" molecules the *limit set* of the splicing system. In this section we consider several examples of such limit sets, without giving a formal definition.

Example 1. Initial language $I = \{ ab, cd \}$, rule $r = (a, b;\ c, d)$, together with its symmetric and reflexive versions. Then the splicing language is $L = \{ab, cd, ad, cb\}$ but the limit language is $L_\infty = \{ad, cb\}$.

Discussion. One of the first wet-lab experiments validating the splicing model was described in [7]. This example is just the symbolic version of that experiment. The molecules ab and cd are spliced by rules r and $\bar{r}$ to produce ad and bc; they are also spliced to themselves by rules $\dot{r}$ and $\ddot{r}$ to regenerate themselves. The words ad and bc are "inert" in the sense that they do not participate in any further splicings, and any inert words will obviously appear in the limit language. However, initial stocks of the initial strings ab and cd will eventually be used up (if they are present in equal numbers), so they will not appear in the limit language, and the limit language should be just $\{ ad, cb \}$. We call the words that eventually disappear "transient". Of course, if there are more molecules of ab than of cd present at the start then we should expect that there will also be molecules of ab left in the final state of the reaction.

Our definition of limit language will ignore this possibility of unbalanced numbers of molecules available for various reactions. Put simply, we will define the limit language to be the words that are predicted to appear if some amount of each initial molecule is present, and sufficient time has passed for the reaction to reach its equilibrium state, regardless of the balance of the reactants in a particular experimental run of the reaction. In the example above, one particular run of the reaction may yield molecules of ab at equilibrium, while another run may not yield molecules of ab at equilibrium, but rather molecules of cd. Both reactions, however, are predicted to yield molecules of ad and bc, and hence these molecules constitute the limit language.

Limit languages would be easy to analyze if they consisted only of inert strings. The following example involves just a slight modification to the previous example and provides a limit language without inert words.

Example 2. Initial language $I = \{ ab, cd \}$, rules $r_1 = (a, b;\ c, d)$ and $r_2 = (a, d;\ c, b)$, together with their symmetric twins and reflexive restrictions. Then there are no inert words and the splicing language and limit language are the same, $L = L_\infty = \{ ab, cd, ad, cb \}$.

Discussion. The molecules ab and cd are spliced by rules r_1 and its symmetric twin as in Example 1 producing ad and cb. These products in turn can be spliced using r_2 and its symmetric twin to produce ab and cd. As in Example 1 each molecule can regenerate itself using one of the reflexive rules. None of the strings are inert, nor do any of them disappear. Thus each molecule in L_∞ is considered to be in a reactive "steady-state" at equilibrium. Their distribution can be calculated using differential equations that model the chemical reactions for the restriction enzymes, ligase, and initial molecules involved.

The next example demonstrates the possibility of an infinite number of reactive steady state molecules at equilibrium.

Example 3. Initial language $I = \{ aa, aaa \}$, rule $r = (a, a;\ a, a)$. Then there are no inert words and the splicing language and the limit language are the same, $L = L_\infty = aa^+$.

Discussion. It is easy to see that $L = aa^+$. Moreover, each string a^k in L is the result of splicing two other strings of L; in fact, there are infinitely many pairs of strings in L which can be spliced to regenerate a^k. In effect, all copies of the symbol a are shuffled between the various molecular types a^k in L. We consider each word a^k of L to be in the limit language. This interpretation is buttressed by the example in Section 3, which calculates the limiting concentrations of the molecules in a very similar system and finds that all limiting concentrations are positive.

Our definition of limit language will avoid the detailed calculation of limiting distribution; in cases like this we will be content to note that any molecule a^k will be present in the limit.

The following illustrates a very different phenomenon: Molecules disappear by growing too big.

Example 4. Initial language $I = \{ abc \}$, splicing rule $r = (b, c;\ a, b)$. Then the splicing language is $L = ab^+c$ but the limit language is empty.

Discussion. The calculation of the splicing language is straightforward; note that splicing ab^kc and ab^jc produces $ab^{k+j}c$. Hence abc is not the result of any splicing operation although it is used in constructing other molecules. Therefore all molecules of type abc will eventually be used up, so abc cannot appear in the limit language. But now, once all molecules of type abc have disappeared, then there is no way to recreate ab^2c and ab^3c by splicing the remaining strings ab^kc, $k \geq 2$. Hence all molecules of types ab^2c or ab^3c will eventually be used up, so they cannot appear in the limit language. The remainder of L is analyzed similarly, using induction.

The H scheme in Example 4 is neither reflexive nor symmetric, so it is hard to justify this example as a model of an actual chemical process. In fact, the next example shows that this phenomenon, in which molecules disappear by "converging to infinitely long molecules", can also happen in reflexive and symmetric H systems. However, we shall see later (Corollary 1) that in the reflexive case the limit language is not empty unless the initial language is empty.

Example 5. Initial language $I = \{ abc \}$, splicing rule $r = (b, c;\ a, b)$ together with its symmetric twin and its reflexive restrictions. Then the splicing language is $L = ab^*c$ and the limit language is ac.

Discussion. The calculation of the splicing language follows as in Example 4, and all molecules of type ab^kc for $k > 0$ will again eventually disappear under splicing using r. The symmetric splice using $\bar{r}$ generates the single inert string ac, which will eventually increase in quantity until it consumes virtually all of the a and c pieces. Thus this system involves an infinite set of transient strings and a finite limit language consisting of a single inert string.

This system could run as a dynamic wet lab experiment, and one would expect to be able to detect by gel electrophoresis an increasingly dark band indicating the presence of more and more ac over time. Over the same time frame we predict the strings of the type ab^kc would get very long, and this would eventually cause these molecules to remain at the top of the gel. In theory, at the final state of the reaction, there would be one very long such molecule that consumed all of the b pieces. We might expect the dynamics of the system to make it increasingly difficult for molecules of ab^kc to be formed as k gets very large. A wet splicing experiment designed to demonstrate the actual dynamics of such a system would be a logical step to test the accuracy of this model of limit languages.

It is essential for the dynamic models that are introduced in Section 3 that we use a more detailed version of splicing. The action of splicing two molecules represented as $w_1 = x_1u_1v_1y_1$ and $w_2 = x_2u_2v_2y_2$ using the splicing rule $(u_1, v_1; u_2, v_2)$ to produce $z = x_1u_1v_2y_2$ is actually a three step process. A restriction enzyme first acts on w_1, cutting it between u_1 and v_1. The resulting fragments have "sticky ends" corresponding to the newly cut ends, in the sense that a number of the bonds between base pairs at the cut site are broken, so part of the cut end on one side is actually single stranded, with the complementary bases on the other fragment. Similarly the molecule w_2 is cut between u_2 and v_3, generating two fragments with sticky ends. Finally, if the fragments corresponding to x_1u_1 and v_2y_2 meet in the presence of a ligase, they will join at the sticky ends to form z. It is clear that any dynamic analysis of the situation must account for the various fragments, keeping track of sticky ends.

There are two primary techniques for this: cut-and-paste systems as defined by Pixton [8], and cutting/recombination systems as defined by Freund and his coworkers [2]. The two approaches are essentially equivalent and, in this context, differ only in how they handle the "sticky ends".

We shall use cut-and-paste. In this formulation there are special symbols called "end markers" which appear at the ends of all strings in the system and are designed to encode the sticky ends, as well as the ends of "complete" molecules. The cutting and pasting rules are themselves encoded as strings with end markers. A cutting rule $\alpha z\beta$ encodes the cutting action $xzy \Longrightarrow x\alpha + \beta y$, while a pasting rule $\alpha w\beta$ corresponds to the pasting action $x\alpha + \beta y \Longrightarrow xwy$. Thus a splicing rule $(u_1, v_1; u_2, v_2)$ can be represented by cutting rules $\alpha_1u_1v_1\beta_1$ and $\alpha_2u_2v_2\beta_2$ and a pasting rule $\alpha_1u_1v_2\beta_2$. In this case α_1 encodes the sticky end on the left half after cutting between u_1 and v_1 and β_2 corresponds to the sticky end on the right half after cutting between u_2 and v_2, so the pasting rule reconstitutes u_1v_2 when the sticky ends α_1 and β_2 are reattached. For details see [8].

Example 6. This is a cut-and-paste version of Example 5. The set of endmarkers is $\{\alpha, \beta_1, \beta_2, \gamma, \delta\}$, the set of cutting rules is $C = \{\alpha ab\beta_1, \beta_2 bc\gamma\}$ and the set of pasting rules is $P = C \cup \{\beta_2 bb\beta_1, \alpha ac\gamma\}$, and the initial set is $\{\delta abc\delta\}$. Then the full splicing language is $L = \delta ab^*c\delta + \delta\alpha + \gamma\delta + \delta ab^*\beta_2 + \beta_1 b^*c\gamma + \beta_1 b^*\beta_2$, and the limit language is $L_\infty = \{\delta ac\delta\}$.

Discussion. $\delta ac\delta$ is inert and is created from $\delta\alpha$ and $\gamma\delta$ by pasting. Thus the strings $\delta\alpha$ and $\gamma\delta$ will be eventually used up. These strings are obtained by cutting operations on the strings in $\delta ab^+c\delta + \delta ab^*\beta_2 + \beta_1 b^*c\gamma$, so these in turn are transient. The remaining strings are in $\beta_1 b^*\beta_2$, and these are never cut but grow in length under pasting using the rule $x\beta_2 + \beta_1 y \Longrightarrow xbby$. Hence these strings disappear in the same fashion as in Example 4.

One possibility that we have not yet discussed is that a DNA fragment may produce a circular string by *self pasting*, if both of its ends are sticky and there is an appropriate pasting rule. We use $\hat{}w$ to indicate the circular version of the string w.

Example 7. If we permit circular splicing in Example 6, then the limit language is $L_\infty = \{\delta ac\delta\} \cup \{\hat{}b^k : k \geq 2\}$.

Discussion. The only strings that can circularize are in $\beta_1 b^+\beta_2$, and if we self paste $\beta_1 b^j \beta_2$ using the pasting rule $\beta_2 bb\beta_1$, then we obtain $\hat{}b^{j+2}$. These circular strings cannot be cut, so they are inert.

Our final examples illustrate somewhat more complex dynamics.

Example 8. Initial language $I = \{abb, cdd\}$ and splicing rules $r_1 = (ab, b;\ a, b)$, $r_2 = (c, d;\ a, b)$, $r_3 = (ad, d;\ a, d)$, together with their symmetric and reflexive counterparts. The splicing language is $L = ab^+ + ad^+ + cb^+ + cdd$ and the limit language is $L_\infty = ad^+ + cb^+$.

Discussion. Rule r_1 is used to "pump" ab^+ from abb while rule $\bar{r}_1$ produces ab from two copies of abb. Rules r_2 and $\bar{r}_2$ splice copies of cdd and ab^k to produce cb^k and add. Rule r_3 generates add^+ from add, while $\bar{r}_3$ generates ad from two copies of add. The transient words cdd and ab^k are used up to produce add and cb^+, and the inert words cb^+ are limit words. However all words of ad^+ are continually recreated from other words in ad^+ using rules r_3, $\bar{r}_3$ and their reflexive counterparts (in the same manner as in Example 3), so these are also limit words. Thus, in this example there are infinitely many inert strings, infinitely many steady-state strings, and infinitely many transient strings.

It is not hard to extend this example to have several "levels" of transient behavior.

Example 9. The splicing rules are

$$r_1 = (a, b;\ c, d), \quad r_2 = (a, d;\ f, b), \quad r_3 = (a, e;\ a, e), \quad r_4 = (a, f;\ a, f)$$

together with their symmetric twins and reflexive restrictions, and the initial language I consists of

$$w_1 = cd, \quad w_2 = ad, \quad w_3 = afd, \quad w_4 = afbae, \quad w_5 = cbae, \quad w_6 = abae$$

Then the splicing language L is the same as I, but the limit language is $\{w_3, w_4, w_5, w_6\}$.

Discussion. We shall use a notation that will be made more precise later: For w and z in L we write $w \to z$ to mean that a splicing operation involving w produces z. It is easy to check that there is a cycle $w_2 \to w_3 \to w_4 \to w_5 \to w_6 \to w_2$ so all these words would seem to be "steady-state" words. On the other hand, w_1 can be generated only by splicing w_1 and w_1 using $\ddot{r}_1 = (c, d;\ c, d)$, but w_1 also participates in splicing operations that lead to the w_2–w_6 cycle. For example, w_6 and w_1 splice using r_1 to produce w_2, so $w_1 \to w_2$.

Clearly then w_1 will be eventually used up, and so it should not be part of the limit language. We let $L_1 = L \setminus \{ w_1 \}$, and reconsider the limit possibilities. There are only two ways to produce w_2 by splicing: from w_2 and w_2 using $\dot{r}_1 = (a, b;\ a, b)$ or from w_6 and w_1 using r_1. Since w_1 is not available in L_1 we see that the w_2–w_6 cycle is broken. In fact, since w_2 is used in the splicing $w_2 \to w_3$, it must eventually disappear, so it is not part of the limit language. The remaining words do not form a cycle but they are connected as follows: $w_3 \to w_4 \to w_3$ using r_4 and $w_4 \to w_5 \to w_6 \to w_4$ using r_3. Thus these words appear to be limit words; and, since the splicings in these cycles do not involve w_2, they remain limit words after w_2 is removed.

With somewhat more work we can construct a splicing language in which the words w_k are replaced by infinite sublanguages W_k, but demonstrating the same phenomenon: The disappearance of a set of transients like W_1 can change the apparent limit behavior of the remaining strings.

3 Dynamics

We present here a dynamical model based on a very simple cut-and-paste system. The molecules all have the form F^j for $j > 0$, where F is a fixed fragment of DNA. We suppose that our system contains a restriction enzyme which cuts between two copies of F and a ligase which joins any two molecules at their ends. That is, the simplified cut-and-paste rules are just $(F^i, F^j) \iff F^{i+j}$. For simplicity we do not allow circular molecules.

The dynamics of our model will be parameterized by two rates, α and β. We interpret α as the probability that a given ordered pair A, B of molecules will paste to yield AB in a unit of time, and we interpret β as the probability that a cut operation will occur at a given site on a given molecule A in a unit of time.

Let $S_k = S_k(t)$ be the set of molecules of type F^k at time t, let $N_k = N_k(t)$ be the number of such molecules (or the concentration – i.e., the number per unit volume), and let $N = N(t) = \sum_{j=0}^{\infty} N_k$ be the total number of molecules. Consider the following four contributions to the rate of change of N_k in a small time interval of length Δt:

First, a certain number will be pasted to other molecules. A single molecule A of type F^k can be pasted to any other molecule B in two ways (yielding either AB or BA), so the probability of pasting A and B is $2\alpha\Delta t$. Since there are $N-1$ other molecules the probability that A will be pasted to some other molecule is $2\alpha(N-1)\Delta t$, and, since this operation removes A from S_k, the expected change in N_k due to pastings with other molecules is $-2\alpha(N-1)N_k\Delta t$. Since N is very large we shall approximate this as $-2\alpha N N_k$. Note that this may seem to overestimate the decrease due to pastings when both A and B are in S_k, since both A and B will be considered as candidates for pasting; but this is correct, since if two elements of S_k are pasted, then N_k decreases by 2, not by 1.

Of course, pasting operations involving smaller molecules may produce new elements of S_k. For molecules in S_i and S_j where $i+j=k$ the same reasoning as above produces $\alpha N_i N_j \Delta t$ new molecules in S_k. There is no factor of 2 here since the molecule from S_i is considered to be pasted on the left of the one from S_j. Also, if $i = j$, then we actually have $\alpha(N_i - 1)N_i\Delta t$ because a molecule cannot paste to itself, and we approximate this as $\alpha N_i^2\Delta t$. This is not as easy to justify as above, since N_i is not necessarily large; however, this approximation seems to be harmless. The total corresponding change in N_k is $\alpha\sum_{i+j=k} N_i N_j \Delta t$.

Third, a certain number of the molecules in S_k will be cut. Each molecule in S_k has $k-1$ cutting sites, so there are $(k-1)N_k$ cutting sites on the molecules of S_k, so we expect N_k to change by $-\beta(k-1)N_k\Delta t$.

Finally, new molecules appear in S_k as a result of cutting operations on longer molecules, and, since Δt is a very small time interval, we do not consider multiple cuts on the same molecule. If A is a molecule in S_m, then there are $m-1$ different cutting sites on A, and the result of cutting at the jth site is two fragments, one in S_j and the other in S_{m-j}. Hence, if $m > k$, exactly two molecules in S_k can be generated from A by cutting, and the total expected change in N_k due to cutting molecules in S_m is $2\beta N_m \Delta t$. Summing these gives a total expected change in N_k of $2\beta\sum_{m>k} N_m \Delta t$.

So we have the following basic system of equations:

$$N_k' = -2\alpha N N_k - \beta(k-1)N_k + \alpha \sum_{i+j=k} N_i N_j + 2\beta \sum_{m>k} N_m. \tag{1}$$

Define $M = \sum_k kN_k$. If μ is the mass of a single molecule in S_1, then $M\mu$ represents the total mass of DNA, so M should be a constant. We verify this from (1) as a consistency check:

We will ignore all convergence questions. We have $M' = \sum_k kN_k'$. Plugging in (1), we have four sums to consider, as follows:

$$-2\alpha \sum_k kNN_k = -2\alpha MN,$$

$$\alpha \sum_k \sum_{i+j=k} kN_iN_j = \alpha \sum_{i,j}(i+j)N_iN_j = 2\alpha MN,$$

$$-\beta \sum_k k(k-1)N_k = -\beta \sum_m m(m-1)N_m,$$

$$2\beta \sum_k \sum_{m>k} kN_m = \beta \sum_m \left(2 \sum_{k<m} k\right) N_m = \beta \sum_m m(m-1)N_m.$$

Adding these gives $M' = 0$.

Summing the equations (1) to get an equation for N yields four sums, which we treat similarly:

$$-2\alpha \sum_k NN_k = -2\alpha N^2,$$

$$\alpha \sum_k \sum_{i+j=k} N_iN_j = \alpha \sum_{i,j} N_iN_j = \alpha N^2,$$

$$-\beta \sum_k (k-1)N_k = -\beta(M-N),$$

$$2\beta \sum_k \sum_{m>k} N_m = 2\beta \sum_m \left(\sum_{k<m} 1\right) N_m = 2\beta(M-N).$$

Hence

$$N' = -\alpha N^2 - \beta N + \beta M. \tag{2}$$

Equations of this form are solved explicitly in elementary differential equation texts; in this case the solution is

$$N = \bar{N} + \frac{\gamma C e^{-\gamma t}}{\alpha(1 - Ce^{-\gamma t})}, \tag{3}$$

where $\gamma = \sqrt{\beta^2 + 4\alpha\beta M}$, $\bar{N}$ is the positive solution of

$$-\alpha\bar{N}^2 - \beta\bar{N} + \beta M = 0, \tag{4}$$

and C is determined by the initial conditions. From the solution (3) and a consideration of the direction field for the differential equation (2) it is clear that if $N(0) > 0$, then $N(t)$ is defined for all $t \geq 0$ and $N \to \bar{N}$ as $t \to \infty$.

We want to find the limiting values of the quantities N_k. First we have a simple result in differential equations.

Lemma 1. *Suppose a and b are continuous functions on $[t_0, \infty)$ and $a(t) \to \bar{a} > 0$ and $b(t) \to \bar{b}$ as $t \to \infty$. Then any solution of $Y' = -aY + b$ satisfies $\lim_{t\to\infty} Y(t) = \bar{b}/\bar{a}$.*

We omit the proof of the lemma, which uses standard comparison techniques for solutions of ordinary differential equations.

Now we replace $\sum_{m>k} N_m$ with $N - \sum_{i=1}^{k-1} N_i - N_k$ in (1) and rearrange to get

$$\begin{aligned} N_k' &= -2\alpha N N_k - \beta(k-1)N_k + \alpha \sum_{i+j=k} N_i N_j + 2\beta \left(N - \sum_{i=1}^{k-1} N_i - N_k \right) \\ &= -(2\alpha N + \beta(k+1))N_k + \alpha \sum_{i+j=k} N_i N_j + 2\beta \left(N - \sum_{i=1}^{k-1} N_i \right) \\ &= -a_k N_k + b_k . \end{aligned}$$

The point of this rearrangement is that the coefficients a_k and b_k only depend on the functions N_j for $j < k$ and on the known function N. Hence, if we solve the equations in sequence, then the coefficients a_k and b_k can be treated as known functions of t.

It is convenient to write our results in terms of the asymptotic average molecular size $\bar{W} = M/\bar{N}$; using (4) we have $\beta\bar{W} = \alpha\bar{N} + \beta$. Clearly the coefficient a_k has the limit $\bar{a}_k = 2\alpha\bar{N} + \beta(k+1)$. We can find the limit $\bar{N}_k$ of N_k from Lemma 1 once we know the limit $\bar{b}_k$ of b_k. Using a routine but messy induction (which we omit) we show that $\bar{b}_k = \bar{a}_k \bar{N}_k$, where

$$\bar{N}_k = \lim_{t\to\infty} N_k(t) = \frac{\bar{N}}{\bar{W}} \left(\frac{\bar{W}-1}{\bar{W}} \right)^{k-1} .$$

4 Definitions

We suppose we have a finite H scheme σ and an initial language I, defining the splicing language $L = \sigma^*(I)$. Given two words w, z in L we write $w \to_L z$ to mean that there is some word w' in L (possibly the same as w or z) so that either w, w' or w', w splices, using a rule of σ, to produce z. Then $\to_L$ is a binary relation on L. As usual we define the transitive closure $\to_L^+$ and the reflexive transitive closure $\to_L^*$ of $\to_L$. Precisely, $w \to_L^+ z$ means there is some finite sequence $w_0, w_1, \ldots, w_n$ of words of L so that $n \geq 1$, $w_0 = w$, $w_n = z$, and $w_k \to_L w_{k+1}$ for $0 \leq k < n$, and $w \to_L^* z$ means $w \to_L^+ z$ or $w = z$.

We say a word $w \in L$ is a *first-order limit* of L iff for any z in L for which $w \to_L^+ z$ we have $z \to_L^* w$. A word which is not a first-order limit is called *transient* in L; in other words, w is transient in L iff there is a word z in L so that $w \to_L^+ z$ but $z \to_L^* w$ is false. This notion of transience is meant to model the following: The splicing operations $w \to_L^+ z$ contributes some of the material of the molecule w to the molecule z, and this material is never reassembled in a molecule of type w; hence the material of w will eventually be used up.

We now define L_1 to be the set of first-order limit words of L, and we continue recursively to define the set L_k of kth-order limits of L to be the set of first order limits of L_{k-1}. That is, we obtain L_k from L_{k-1} by deleting the words that are transient in L_{k-1}.

Finally we define the *limit language* as

$$L_\infty = \bigcap_{k=1}^{\infty} L_k.$$

The limit languages described informally in the examples in Section 2 all satisfy this definition. Example 9 shows the difference between limits of order 1 and order 2.

We will often use the following interpretation of the relation $\rightarrow_L$: We consider a directed graph G_L in which the vertices are the words of L, so that there is an edge from w to z if and only if $w \rightarrow_L z$. We call this the *splicing graph* of L, although it is not determined just by L but by the pair (L, σ).

In this interpretation we can describe the limit language as follows: We start by determining the *strongly connected components* of G_L; these are the maximal subgraphs C of the graph so that, for any vertices w, z of C, we have $w \rightarrow_L^* z$. Such a component is called a *terminal* component if there is no edge $w \rightarrow_L z$ with w in C and z not in C. The first order limit language L_1 consists of the vertices which lie in the terminal components, and the transient words in L are the vertices of the non-terminal components.

We then define a subgraph G_L^1 of G_L whose vertices are the vertices of the terminal components, so that there is an edge from w to z if and only if there is a splicing operation $(w, w') \Longrightarrow z$ or $(w', w) \Longrightarrow z$ using a rule of σ in which w' is in L_1. Then, of course, L_∞ is the set of vertices of the intersection G_L^∞ of the chain of subgraphs constructed recursively by this process.

5 Regularity

We continue with the terminology of Section 4. It is well-known that the splicing language L is regular ([1], [9]), and it is natural to ask whether the limit language L_∞ is regular. We do not know the answer in general. However, we can give a satisfactory answer in the important special case of a *reflexive* splicing system. In Head's original formulation of splicing [5] both symmetry and reflexivity were understood, and there are good biochemical reasons to require that any H system that purports to model actual chemical reactions must be both symmetric and reflexive.

Theorem 1. *Suppose σ is a finite reflexive H scheme and I is a finite initial language. Then the limit language L_∞ is regular.*

Proof. First, let S be the collection of all sites of rules in σ. That is, the string s is in S if and only if there is a rule $(u_1, v_1; u_2, v_2)$ in σ with either $s = u_1 v_1$ or $s = u_2 v_2$. For each $s \in S$ we let $L_s = L \cap A^* s A^*$; that is, L_s consists of the words of L which contain s as a factor. We shall refer to a set of the form L_s as a *site class*. We also define $L_I = L \setminus A^* S A^*$. This is the set of *inert* words of L; that is, the set of words which do not contain any sites.

Suppose $x, y \in L_s$. By reflexivity we may find a rule $r = (u, v;\, u, v)$ of σ so that $s = uv$. Write $x = x_1uvx_2$ and $y = y_1uvy_2$. Then splicing x and y using r produces $z = x_1uvy_2$, and splicing y and z using r produces y_1uvy_2, or y. Hence we have $x \to_L z \to_L y$, and z is in L_s. We conclude that the subgraph of G_L with the elements of L_s as vertices forms a strongly connected subgraph of the vertices of G_L, and so it lies in one strongly connected component of G_L.

We conclude that each strongly connected component of G_L is either a singleton $\{\, w \,\}$ with $w \in L_I$ or a union of a collection of site classes. The first type of component is clearly terminal. Thus the set L_1 of first-order limit words is the union of L_I and some collection of site classes. Moreover, any site class L_s which appears in L_1 is still a strongly connected subset of the vertex set of G_L^1. Hence we can use induction to extend this decomposition to see that each L_k is the union of L_I and a collection of site classes.

Since the sets $L_{k+1} \subset L_k$ and there are only finitely many site classes we see that the sequence L_k is eventually constant, so $L_\infty = L_k$ for all sufficiently large k. Hence L_∞ is the union of L_I and a collection of site classes. This is a finite union of sets, each of which is regular (using the regularity of L), so L_∞ is regular.

Corollary 1. *If σ is a finite reflexive H scheme and I is not empty, then L_∞ is not empty.*

Proof. This is clear from the proof, since there are a finite number of strongly connected components (aside from the inert words) at each stage, and so there are terminal components at each stage; and the recursive construction stops after finitely many steps.

It has been a difficult problem to determine the class of splicing languages as a subclass of the regular class, even if we restrict attention to reflexive splicing schemes; see [3]. Hence the following is rather unexpected.

Theorem 2. *Given any regular language K there is a finite reflexive and symmetric splicing scheme σ and a finite initial language I so that the limit language L_∞ is K.*

Proof. This is an easy consequence of a theorem of Head [6], which provides a finite reflexive and symmetric splicing scheme σ_0 and a finite initial language I so that $\sigma_0^*(I) = cK$, where c is some symbol not in the alphabet of K. In fact, every rule of σ_0 has the form $(cu, 1;\, cv, 1)$ for strings u and v.

Our only modification involves adding the rules $r_0 = (c, 1;\, 1, c)$, $\bar{r}_0 = (1, c;\, c, 1)$, $\dot{r}_0 = (c, 1;\, c, 1)$ and $\ddot{r}_0 = (1, c;\, 1, c)$ to σ_0 to define σ. This changes the splicing language to $L = c^*K$. To see this, note that if j and k are positive, then $c^j w$ and $c^k z$ splice, using any of the new rules to produce $c^m z$, where, depending on the rule used, m ranges over the integers from 0 to $j + k$.

Now all words of $L \setminus K$ have the form $c^k w$ with $k > 0$, and these are transient since any such word splices with itself using $\bar{r}_0$ to produce the inert word $w \in K$. Hence the limit language is just K, the set of inert words.

6 Conclusion

We wrote this paper to introduce the notion of the limit language. This notion can be supported by qualitative reasoning, as in the examples in Section 2. However, the splicing operation is not a realistic model of the actual chemical reactions; for this we need something like cut and paste operations to model the separate chemical actions of restriction enzymes and ligases. Thus Examples 6 and 7 give a better account of the molecules involved during the chemical reactions than Example 5.

We believe that a complete understanding of the limit language must wait until a suitable dynamical model has been developed and supported by experiment. It is not hard to formulate a generalization of the example in Section 3 to apply to arbitrary cut-and-paste systems, but it is much more difficult to solve such a generalization. We would hope eventually to derive regularity of the splicing language from the dynamical system, and then to define and analyze the limit language directly from the dynamical system.

Our preliminary definition of the limit language in Section 4 is simply an attempt to circumvent the considerable difficulties of the dynamical systems approach and see what we might expect for the structure of the limit language. We cannot yet prove that this definition coincides with the natural definition based on dynamical systems, but we believe that it will, probably with minor restrictions on the splicing scheme.

References

1. K. Culik II and T. Harju, Splicing Semigroups of Dominoes and DNA. *Discrete Applied Mathematics*, 31 (1991), 261–277.
2. R. Freund, E. Csuhaj-Varjú, and F. Wachtler, Test Tube Systems with Cutting/Recombination Operations. *Pacific Symposium on Biocomputing*, 1997 (`http://WWW-SMI.Stanford.EDU/people/altman/psb97/index.html`).
3. E. Goode and D. Pixton, Recognizing Splicing Languages: Syntactic Monoids and Simultaneous Pumping. Submitted 2003.
4. T. Head, Gh. Păun, and D. Pixton, Generative Mechanisms Suggested by DNA Recombination. In *Handbook of Formal Languages* (G. Rozenberg and A. Salomaa, Eds.), Springer, Berlin, 1997, chapter 7 in vol. 2, 295–360.
5. T. Head, Formal Language Theory and DNA: an Analysis of the Generative Capacity of Specific Recombinant Behaviors. *Bulletin of Mathematical Biology*, 49 (1987), 737–759 (`http://math.binghamton.edu/tom/index.html`).
6. T. Head, Splicing Languages Generated with One Sided Context. In *Computing with Bio-Molecules. Theory and Experiments* (Gh. Păun, Ed.), Springer, Singapore, (1998), 158–181.
7. E. Laun and K.J. Reddy, Wet Splicing Systems. *Proceedings of the 3rd DIMACS Workshop on DNA Based Computers, held at the University of Pennsylvania*, June 23–25, 1997 (D.H. Wood, Ed.), 115–126.
8. D. Pixton, Splicing in Abstract Families of Languages. *Theoretical Computer Science*, 234 (2000), 135–166.
9. D. Pixton, Regularity of Splicing Languages. *Discrete Applied Mathematics*, 69 (1996), 101–124.

Formal Properties of Gene Assembly: Equivalence Problem for Overlap Graphs

Tero Harju[1], Ion Petre[2], and Grzegorz Rozenberg[3]

[1] Department of Mathematics, University of Turku
FIN-20014 Turku, Finland
harju@utu.fi

[2] Department of Computer Science, Åbo Akademi University
FIN-20520 Turku, Finland
ipetre@abo.fi

[3] Leiden Institute for Advanced Computer Science, Leiden University
Niels Bohrweg 1, 2333 CA Leiden, the Netherlands
and
Department of Computer Science, University of Colorado, Boulder
Co 80309-0347, USA
rozenber@liacs.nl

Abstract. Gene assembly in ciliates is a life process fascinating from both the biological and the computational points of view. Several formal models of this process have been formulated and investigated, among them a model based on (legal) strings and a model based on (overlap) graphs. The latter is more abstract because the translation of legal strings into overlap graphs is not injective. In this paper we consider and solve the overlap equivalence problem for realistic strings: when do two different realistic legal strings translate into the same overlap graph? Realistic legal strings are legal strings that "really" correspond to genes generated during the gene assembly process.

1 Introduction

Ciliates (ciliated protozoa) is a group of single-celled eukaryotic organisms. It is an ancient group - about two billion years old, which is very diverse - some 8000 different species are currently known. A unique feature of ciliates is nuclear dualism - they have two kinds of functionally different nuclei in the same cell, a micronucleus and a macronucleus (and each of them is present in multiple copies), see, e.g., [15] for basic information on ciliates. The macronucleus is the "standard household nucleus" that provides RNA transcripts for producing proteins, while the micronucleus is a dormant nucleus that gets activated only during sexual reproduction. At some stage of sexual reproduction the genome of a micronucleus gets transformed into the genome of a macronucleus in the process called *gene assembly*. The process of gene assembly is fascinating from both the biological (see, e.g., [14], [15], [16], and [18]) and from the computational (see, e.g., [4], [8], [10], [12], [13], [17]) points of view. A number of formal

N. Jonoska et al. (Eds.): Molecular Computing (Head Festschrift), LNCS 2950, pp. 202–212, 2004.

models of gene assembly were studied, including both the *intermolecular* models (see, e.g., [12], [13]) and the *intramolecular* models (see, e.g., [4] – [11], [17]). This paper continues research on intramolecular models. We consider here formal representations of genes generated during the process of gene assembly in two such models: one based on (legal) strings (see, e.g., [8]) and one based on (overlap) graphs (see, e.g., [4]). The latter model is more abstract than the former because the translation of legal strings into overlap graphs is not injective. This fact leads naturally to the following *overlap equivalence problem* for legal strings: when do two different legal strings yield the same overlap graph?

In this paper we consider the overlap equivalence problem and solve it for realistic legal strings (i.e., those legal strings that "really" correspond to genes generated during the process of gene assembly, see, e.g., [4]).

2 Preliminaries

For positive integers k and n, we denote by $[k,n] = \{k, k+1, \dots, n\}$ the interval of integers between k and n.

Let $\Sigma = \{a_1, a_2, \dots\}$ be an alphabet. The set of all (finite) strings over Σ is denoted by Σ^* - it includes the empty string, denoted by Λ. For strings $u, v \in \Sigma^*$, we say that v is a *substring* of u, if $u = w_1 v w_2$ for some $w_1, w_2 \in \Sigma^*$. Also, v is a *conjugate* of u, if $u = w_1 w_2$ and $v = w_2 w_1$ for some strings w_1 and w_2.

For an alphabet Σ, let $\overline{\Sigma} = \{\overline{a} \mid a \in \Sigma\}$ be a signed copy of Σ (disjoint from Σ). Then the total alphabet $\Sigma \cup \overline{\Sigma}$ is called a *signed alphabet.* The set of all strings over $\Sigma \cup \overline{\Sigma}$ is denoted by $\Sigma^{\maltese}$, that is,

$$\Sigma^{\maltese} = (\Sigma \cup \overline{\Sigma})^* .$$

A string $v \in \Sigma^{\maltese}$ is called a *signed string over* Σ. We adopt the convention that $\overline{\overline{a}} = a$ for each $a \in \Sigma$.

Let $v \in \Sigma^{\maltese}$ be a signed string over Σ. We say that a letter $a \in \Sigma \cup \overline{\Sigma}$ *occurs* in v, if either a or $\overline{a}$ is a substring of v. Let $\mathrm{dom}(v) = \{a \mid a \in \Sigma \text{ and } a \text{ occurs in } v\}$ be the *domain* of v. Thus the domain of a signed string consists of the unsigned letters that occur in it.

Example 1. Let $\Sigma = \{a, b\}$, and thus $\overline{\Sigma} = \{\overline{a}, \overline{b}\}$. The letters a and b occur in the signed string $v = a\overline{b}\overline{a}a$, although b is not a substring of v (its signed copy $\overline{b}$ of b is a substring of v). Also, $\mathrm{dom}(v) = \{a, b\}$. □

The signing $a \mapsto \overline{a}$ can be extended to longer strings as follows. For a nonempty signed string $u = a_1 a_2 \dots a_n \in \Sigma^{\maltese}$, where $a_i \in \Sigma \cup \overline{\Sigma}$ for each i, the *inversion* $\overline{u}$ of u is defined by

$$\overline{u} = \overline{a}_n \overline{a}_{n-1} \dots \overline{a}_1 \in \Sigma^{\maltese} .$$

Also, let $u^R = a_n a_{n-1} \dots a_1$ and $u^C = \overline{a}_1 \overline{a}_2 \dots \overline{a}_n$ be the *reversal* and the *complement* of u, respectively. It is then clear that $\overline{u} = (u^R)^C = (u^C)^R$.

Let Σ and Γ be two alphabets. A function $\varphi\colon \Sigma^* \to \Gamma^*$ is a *morphism*, if $\varphi(uv) = \varphi(u)\varphi(v)$ for all strings $u, v \in \Sigma^*$. A morphism $\varphi\colon \Sigma^{\maltese} \to \Gamma^{\maltese}$ between signed strings is required to satisfy $\overline{\varphi(u)} = \varphi(\overline{u})$ for all $u, v \in \Sigma^{\maltese}$. Note that, for a morphism φ, the images $\varphi(a)$ of the letters $a \in \Sigma$ determine φ, i.e., if the images of the letters $a \in \Sigma$ are given, then the image of a string over Σ is determined.

Example 2. Let $\Sigma = \{a, b\}$ and $\Gamma = \{0, 1, 2\}$, and let $\varphi\colon \Sigma^{\maltese} \to \Gamma^{\maltese}$ be the morphism defined by $\varphi(a) = 0$ and $\varphi(b) = 2$. Then for $u = a\overline{b}b$, we have $\varphi(u) = 0\overline{2}2$ and $\varphi(\overline{u}) = \varphi(\overline{b}b\overline{a}) = \overline{2}2\overline{0} = \overline{\varphi(u)}$. □

Signed strings $u \in \Sigma^{\maltese}$ and $v \in \Gamma^{\maltese}$ are *isomorphic*, if there exists an injective morphism $\varphi\colon \Sigma^{\maltese} \to \Gamma^{\maltese}$ such that $\varphi(\Sigma) \subseteq \Gamma$ and $\varphi(u) = v$. Hence two strings are isomorphic if each of the strings can be obtained from the other by renaming letters.

A signed string $v = a_1 a_2 \dots a_n$ is a *signed permutation* of another signed string $u = b_1 b_2 \dots b_n$, if there exists a permutation $i_1, i_2, \dots, i_n$ of $[1, n]$ such that $a_j \in \{b_{i_j}, \overline{b}_{i_j}\}$ for each $j \in [1, n]$.

Example 3. Let $u = a\,\overline{a}\,\overline{b}\,c\,b \in \{a, b, c\}^{\maltese}$. Then u is a signed permutation of $a\,c\,a\,b\,b$, as well as of $\overline{a}\,\overline{b}\,a\,b\,\overline{c}$ (and many other signed strings). Also, u is isomorphic to the signed string $v = a\,\overline{a}\,\overline{c}\,b\,c$. The isomorphism φ in question is defined by: $\varphi(a) = a$, $\varphi(c) = b$, and $\varphi(b) = c$. Hence also $\varphi(\overline{a}) = \overline{a}$, $\varphi(\overline{c}) = \overline{b}$, and $\varphi(\overline{b}) = \overline{c}$. □

3 MDS Arrangements and Legal Strings

The structural information about the micronuclear or an intermediate precursor of a macronuclear gene can be given by the sequence of MDSs forming the gene. The whole assembly process can be then thought of as a process of assembling MDSs through splicing, to obtain the MDSs in the orthodox order $M_1, M_2, \dots, M_\kappa$. Thus one can represent a macronuclear gene, as well as its micronuclear or an intermediate precursor, by its sequence of MDSs only.

Example 4. The actin I gene of *Sterkiella nova* has the following MDS/IES micronuclear structure:

$$M_3 I_1 M_4 I_2 M_6 I_3 M_5 I_4 M_7 I_5 M_9 I_6 \overline{M}_2 I_7 M_1 I_8 M_8 \tag{1}$$

(see Fig. 1(a)). The 'micronuclear pattern' is obtained by removing the IESs I_i. For our purposes, the so obtained string

$$\alpha = M_3 M_4 M_6 M_5 M_7 M_9 \overline{M}_2 M_1 M_8$$

has the same information as the structure given in the representation (1).

The macronuclear version of this gene is given in Fig. 1(b). There the IESs have been excised and the MDSs have been spliced (by overlapping of their ends) in the orthodox order $M_1, M_2, \dots, M_9$. □

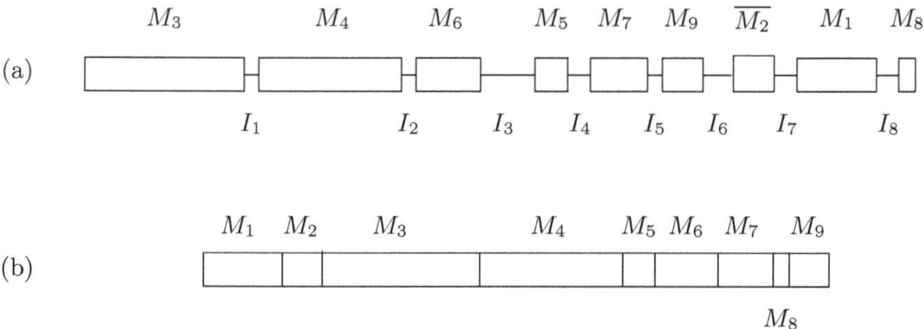

Fig. 1. (a) The micronuclear version of the actin I gene in *Sterkiella nova.* (b) The macronuclear version of the actin I gene of *Sterkiella nova.* The vertical lines describe the positions where the MDSs have been spliced together (by overlapping of their ends)

For each $\kappa \geq 1$, let

$$\Theta_\kappa = \{ M_i \mid 1 \leq i \leq \kappa \}$$

be an alphabet representing *elementary* MDSs, i.e., each M_i denotes an MDSs that is present in the micronucleus. The signed strings in $\Theta_\kappa^{\maltese}$ are *MDS arrangements (of size κ)*. An MDS arrangement $\alpha \in \Theta_\kappa^*$ is *orthodox*, if it is of the form $M_1 M_2 \ldots M_\kappa$. Note that an orthodox MDS arrangement does not contain any inversions of MDSs, and the MDSs are in their orthodox order. A signed permutation of an orthodox MDS arrangement α is a *realistic MDS arrangement.*

A nonempty signed string $v \in \Sigma^{\maltese}$ is a *legal string* over Δ if every letter $a \in \mathrm{dom}(v)$ occurs exactly twice in v. (Recall that an occurrence of a letter can be signed.)

A letter a is *positive* in a legal string v, if both a and $\overline{a}$ are substrings of v, otherwise, a is *negative* in v.

Let $\Delta = \{2, 3, \ldots\}$ be a designated alphabet of *pointers.*

Example 5. Let $v = 2\,4\,3\,\overline{2}\,\overline{5}\,3\,4\,5$ be a legal string over Δ. Pointers 2 and 5 are positive in u, while 3 and 4 are negative in v. On the other hand, the string $w = 2\,4\,3\,\overline{2}\,\overline{5}\,3\,5$ is not legal, since 4 has only one occurrence in w. □

We shall now represent the MDS arrangements by legal strings over the pointer alphabet Δ. In this way legal strings become a formalism for describing the sequences of pointers present in the micronuclear and the intermediate molecules.

Let $\varrho_\kappa \colon \Theta_\kappa^{\maltese} \to \Delta^{\maltese}$ be the morphism defined by:

$$\varrho_\kappa(M_1) = 2, \quad \varrho_\kappa(M_\kappa) = \kappa, \quad \varrho_\kappa(M_i) = i\, i+1 \quad \text{for } 2 < i < \kappa\,,$$

and $\varrho_\kappa(\overline{M_i}) = \overline{\varrho_\kappa(M_i)}$ for $1 \leq i \leq \kappa$.

We say then that a legal string u is *realistic* if there exists a realistic MDS arrangement α such that $u = \varrho_\kappa(\alpha)$.

Example 6. Consider the micronuclear MDS arrangement of the actin I gene of *Sterkiella nova*: $\alpha = M_3 M_4 M_6 M_5 M_7 M_9 \overline{M}_2 M_1 M_8$. We have

$$\varrho_9(\alpha) = 34\ 45\ 67\ 56\ 78\ 9\ \overline{3}\,\overline{2}\ 2\ 89\,.$$

Since α is a realistic MDS arrangement, $\varrho_9(\alpha)$ is a realistic legal string. □

The following example shows that there exist legal strings that are not realistic.

Example 7. The string $u = 2\,3\,4\,3\,2\,4$ is legal, but it is not realistic, since it has no 'realistic parsing'. Indeed, suppose that there exists a realistic MDS arrangement α such that $u = \varrho_4(\alpha)$. Clearly, α must end with the MDS M_4, since $\varrho_4(M) \neq 24$ for all MDSs M. Similarly, since $\varrho_4(M) \neq 32$ for all MDSs M, α must end with $M_2 M_4$. Now, however, $u = 2\,3\,4\,3\,\varrho_4(M_2 M_4)$ gives a contradiction, since 3 and 43 are not images of any MDSs M. □

4 Overlap Graphs of Legal Strings

Let $u = a_1 a_2 \dots a_n \in \Sigma^{\maltese}$ be a legal string over Σ, where $a_i \in \Sigma \cup \overline{\Sigma}$ for each i. Let for each letter $a \in \mathrm{dom}(u)$, i and j with $1 \le i < j \le n$ be indices such that $a_i, a_j \in \{a, \overline{a}\}$. Then the substring

$$u_{(a)} = a_i a_{i+1} \dots a_j$$

is the *a-interval* of u. Two different letters $a, b \in \Sigma$ are said to *overlap in* u, if the a-interval and the b-interval of u overlap: if $u_{(a)} = a_{i_1} \dots a_{j_1}$ and $u_{(b)} = a_{i_2} \dots a_{j_2}$, then either $i_1 < i_2 < j_1 < j_2$ or $i_2 < i_1 < j_2 < j_1$.

Example 8. Let $u = 2\,4\,3\,5\,3\,\overline{2}\,\overline{6}\,\overline{5}\,7\,4\,6\,7$ be a string of pointers. The 2-interval of u is the substring $u_{(2)} = 2\,4\,3\,5\,3\,\overline{2}$, which contains only one occurrence of pointer 4 and pointer 5, but either two or no occurrences of $3, 6$ and 7. Hence pointer 2 overlaps with 4 and 5, but not with $3, 6$ and 7. □

For a finite set V, let

$$E(V) = \{\{x, y\} \mid x, y \in V,\ x \neq y\}$$

be the set of all unordered pairs of different elements of V. A *graph* is a pair (V, E), where V is a finite set of *vertices* and $E \subseteq E(V)$ is a set of *edges*. A *signed graph* $\gamma = (V, E, \sigma)$ consists of a graph (V, E) together with a labelling $\sigma\colon V \to \{-, +\}$ of the vertices. Vertices labelled by $+$ are *positive*, and those labelled by $-$ are *negative*. We use x^+ and x^- to indicate that $\sigma(x) = +$ and $\sigma(x) = -$, respectively.

We shall use signed graphs to represent the structure of overlaps of pointers occurring in a legal string as follows.

For a pointer $p \in \Delta$, let $\mathbf{p} = \{p, \overline{p}\}$. Let $v \in \Delta^{\maltese}$ be a legal string of pointers, and let

$$\mathbf{P}_v = \{\mathbf{p} \mid p \text{ occurs in } v\}\,.$$

Then the *overlap graph* of v is the signed graph $\gamma_v = (\mathbf{P}_v, E, \sigma)$ such that

$$\sigma(\mathbf{p}) = \begin{cases} + , & \text{if } p \in \Delta \text{ is positive in } v\,, \\ - , & \text{if } p \in \Delta \text{ is negative in } v\,, \end{cases}$$

and

$$\{\mathbf{p}, \mathbf{q}\} \in E \iff p \text{ and } q \text{ overlap in } v\,.$$

Example 9. Consider the legal string $v = 3\,4\,\overline{5}\,2\,3\,\overline{3}\,\overline{2}\,4\,5 \in \Delta^{\maltese}$. Then its overlap graph is given in Fig. 2. Indeed, pointers 2, 3 and 5 are positive in v (and hence the vertices **2**, **3** and **5** have sign + in the overlap graph γ_v), while pointer 4 is negative in v (and the vertex **4** has sign − in γ_v). The intervals of the pointers in v are

$$v_{(2)} = 2\,3\,\overline{3}\,\overline{2}, \qquad v_{(3)} = 3\,4\,\overline{5}\,2\,3\,\overline{3}\,,$$
$$v_{(4)} = 4\,\overline{5}\,2\,3\,\overline{3}\,\overline{2}\,4, \qquad v_{(5)} = \overline{5}\,2\,3\,\overline{3}\,\overline{2}\,4\,5\,,$$

The overlappings of these intervals yield the edges of the overlap graph. □

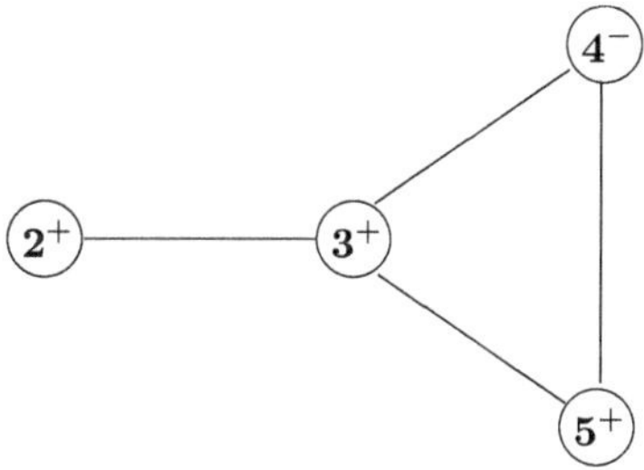

Fig. 2. The overlap graph of the signed string $v = 3\,4\,\overline{5}\,2\,3\,\overline{3}\,\overline{2}\,4\,5$. An edge is present between the signed vertices **p** and **q** if and only if p and q overlap in v

Remark 1. Overlap graphs of unsigned legal strings are also known as *circle graphs* (see [1,2,3]) . These graphs have a geometric interpretation as chord diagrams, where a chord diagram $(\mathcal{C}, V)$ consists of a circle $\mathcal{C}$ in the plane together with a set V of chords. A graph $\gamma = (V, E)$ is a circle graph, if there exists a chord diagram $(\mathcal{C}, V)$ such that $\{x, y\} \in E$ for different x and y if and only if the chords x and y intersect each other. □

5 The Overlap Equivalence Problem

The mapping $w \mapsto \gamma_w$ of legal strings to overlap graphs is not injective. Indeed, for each legal string $w = w_1 w_2$, we have

$$\gamma_{w_1 w_2} = \gamma_{w_2 w_1} \quad \text{and} \quad \gamma_w = \gamma_{c(w)}\,,$$

where c is any morphism that selects one element $c(p)$ from $\mathbf{p} = \{p, \overline{p}\}$ for each p (and then obviously $c(\overline{p}) = \overline{c(p)}$). In particular, all conjugates of a legal string w have the same overlap graph. Also, the reversal w^R and the complementation w^C of a legal string w define the same overlap graph as w does.

Example 10. The following eight legal strings of pointers (in $\Delta^{\maltese}$) have the same overlap graph: $2\overline{3}23$, $\overline{3}232$, $232\overline{3}$, $32\overline{3}2$, $\overline{232}3$, $\overline{32}3\overline{2}$, $\overline{2}3\overline{23}$, $3\overline{232}$. □

Example 11. All string representations of an overlap graph γ need not be obtained from a fixed representation string w by using the operations of conjugation, inversion and reversing (and by considering isomorphic strings equal). For instance, the strings $v_1 = 23342554$ and $v_2 = 35242453$ define the same overlap graph (see Fig. 3), while v_1 is not isomorphic to any signed string obtained from v_2 by the above operations. However, we will demonstrate that if strings u and v are realistic, then $\gamma_u = \gamma_v$ implies that u and v can be obtained from each other by using the operations of conjugation, inversion and reversing. □

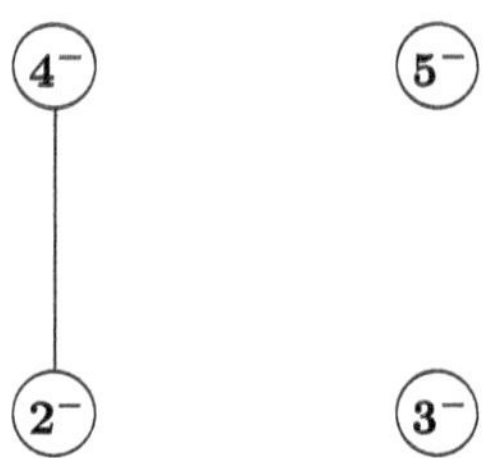

Fig. 3. The common overlap graph of legal strings $v_1 = 23342554$ and $v_2 = 35242453$. The signed string v_1 and v_2 are not obtainable from each other by the operations of conjugation, reversal and complementation

Examples 10 and 11 lead naturally to the *overlap equivalence problem for realistic legal strings*: when do two *realistic* legal strings u and v have the same overlap graph, i.e., when $\gamma_u = \gamma_v$? To solve this problem, we begin by characterizing legal substrings of realistic legal strings.

For a signed string $v \in \Delta^{\maltese}$, let

$$v_{\min} = \min(\mathrm{dom}(v)) \quad \text{and} \quad v_{\max} = \max(\mathrm{dom}(v))\,.$$

Thus, $\mathrm{dom}(v) \subseteq [v_{\min}, v_{\max}]$.

Example 12. The string $v = \overline{4}\,8\,3\,4\,\overline{8}\,3$ is legal. We have $\mathrm{dom}(v) = \{3, 4, 8\}$, and hence $v_{\mathrm{min}} = 3$ and $v_{\mathrm{max}} = 8$. □

Lemma 1. *Let u be a realistic legal string with $\mathrm{dom}(u) = [2, \kappa]$, and let v be a legal substring of u. Then either $\mathrm{dom}(v) = [p, q]$ or $\mathrm{dom}(v) = [2, \kappa] \setminus [p, q]$, for some p and q with $p \leq q$.*

Proof. Let $u = u_1 v u_2$ - without loss of generality we can assume that u_1 and u_2 are not both empty strings.

(a) Assume that there exists a pointer p with $v_{\mathrm{min}} < p < v_{\mathrm{max}}$ such that $p \notin \mathrm{dom}(v)$ and $p - 1 \in \mathrm{dom}(v)$. Then $p > 2$. Let $w = (p-1)p$ $(= \varrho_\kappa(M_{p-1}))$. Since u contains w or $\overline{w}$, it follows that either

$$u = u_{11}\overline{p}\,(\overline{p-1})v_1 u_2, \text{ where } v = (\overline{p-1})v_1, \text{ or}$$
$$u = u_1 v_1 (p-1) p u_{21}, \text{ where } v = v_1(p-1).$$

(b) Similarly, if there exists q with $v_{\mathrm{min}} < q < v_{\mathrm{max}}$ such that $q \notin \mathrm{dom}(v)$ and $q + 1 \in \mathrm{dom}(v)$, then either

$$u = u_{12} q (q+1) v_2 u_2, \text{ where } v = (q+1)v_2, \text{ or}$$
$$u = u_1 v_2\,(\overline{q+1})\overline{q} u_{22}, \text{ where } v = v_2(\overline{q+1}).$$

The following three cases can hold now and each of them yields the conclusion from the statement of the lemma.

1. There are no pointers p as in (a). In this case, $\mathrm{dom}(v) = [q+1, \kappa]$, where $q > 2$ is a unique pointer for which (b) holds.
2. There are no pointers q as in (b). In this case, $\mathrm{dom}(v) = [2, p-1]$, where $p \leq \kappa$ is a unique pointer for which (a) holds.
3. There exists a pointer p as in (a) and a pointer q as in (b). By the above, p is unique for (a) and q is unique for (b). Hence either $\mathrm{dom}(v) = [q+1, p-1]$, if $p - 1 \geq q + 1$ or $\mathrm{dom}(v) = [2, p-1] \cup [q+1, \kappa]$, if $p - 1 < q + 1$.

□

Example 13. Let

$$u = \overline{7}\,\overline{6}\,7\,8\,2\,\overline{9}\,\overline{8}\,\overline{3}\,\overline{2}\,9\,3\,4\,\overline{5}\,\overline{4}\,5\,6\,.$$

Then u is realistic, since $u = \varrho_9(\overline{M}_6 M_7 M_1 \overline{M}_8 \overline{M}_2 M_9 M_3 \overline{M}_4 M_5)$. The string u has the following legal substrings (in addition to u and Λ): $v_1 = 4\,\overline{5}\,\overline{4}\,5$ with $\mathrm{dom}(v_1) = [4, 5]$, $v_2 = 8\,2\,\overline{9}\,\overline{8}\,\overline{3}\,\overline{2}\,9\,3$ with $\mathrm{dom}(v) = [2, 3] \cup [8, 9]$, and $v_3 = 8\,2\,\overline{9}\,\overline{8}\,\overline{3}\,\overline{2}\,9\,3\,4\,\overline{5}\,\overline{4}\,5$ with $\mathrm{dom}(v_3) = [2, 5] \cup [8, 9]$. □

The following result provides more details of the structure of legal substrings of realistic legal strings.

Lemma 2. *Let u be a realistic string with* $\text{dom}(u) = [2, \kappa]$, *and let v be a legal substring of u.*

(i) If $2 \notin \text{dom}(v)$, *then either* $v = v_{\min}v'$ *or* $v = v'\overline{v}_{\min}$.
(ii) If $\kappa \notin \text{dom}(v)$, *then either* $v = \overline{v}_{\max}v'$ *or* $v = v'v_{\max}$.
(iii) If $2, \kappa \in \text{dom}(v)$, *then* $\text{dom}(v) = [2,p] \cup [q,\kappa]$ *and either* $v = qv'p$ *or* $v = \overline{p}v'\overline{q}$.

Proof. Assume that $v \neq u$, and suppose first that $2 \notin \text{dom}(v)$, and let $p = v_{\min}$. By Lemma 1, $\text{dom}(v) = [p, q]$ for some pointer q with $p \leq q \leq \kappa$. Now since $p - 1 \notin \text{dom}(v)$ and either $(p-1)p$ or $\overline{p}(\overline{p-1})$ is a substring of u, it must be that either v begins or v ends with an occurrence of p. But this is possible only when $v = pv'$ or $v = v'\overline{p}$, as required.

The claim for the case $\kappa \notin \text{dom}(v)$ is similar to the above case.

If $2, \kappa \in \text{dom}(v)$, then $\text{dom}(v) = [2,p] \cup [q,\kappa]$ for some p and q with $p < q$ by Lemma 1. Since either $p(p+1)$ or $\overline{p}(\overline{p-1})$ is a substring of u, either $v = v_1p$ or $v = \overline{p}v_1$. Similarly, either $v = qv_2$ or $v = v_2\overline{q}$, and so the last claim from the statement of the lemma holds. □

The following theorem gives a characterization for those pairs of realistic legal strings that yield the same overlap graph.

Recall that for a string $v = p_1p_2\dots p_n \in \Delta^{\maltese}$, the reversal of v is defined by $v^R = p_np_{n-1}\dots p_1$, and the complementation by $v^C = \overline{p}_1\overline{p}_2\dots\overline{p}_n$. For two signed strings $u, v \in \Delta^{\maltese}$, let $u \approx v$ if u is obtained from a conjugate of v by a composition of the operations of reversal and complementation. Thus, $u \approx v$ if and only if there exist strings v_1 and v_2 such that $v = v_1v_2$ and u is one of the strings v_2v_1, $(v_2v_1)^R$, $(v_2v_1)^C$, or $((v_2v_1)^R)^C$.

Theorem 1. *Let u and v be two realistic legal strings such that* $\text{dom}(u) = [2, \kappa] = \text{dom}(v)$. *Then* $\gamma_u = \gamma_v$ *if and only if* $u \approx v$.

Proof. Assume that $\gamma_u = \gamma_v$. We prove the theorem by induction on κ. If $\kappa = 2$, then clearly $u \approx v$. Suppose then that the claim holds for strings over $[2, \kappa - 1]$.

Since we now consider strings up to equivalence with respect to conjugation, reversal and complementation, we can assume without loss of generality that both u and v end with $\varrho_\kappa(M_{k-1}) = (\kappa-1)\kappa$. Now

$$u = u_1\,\kappa'\,u_2(\kappa-1)\kappa \quad \text{and} \quad v = v_1\,\kappa'\,v_2(\kappa-1)\kappa\,, \tag{2}$$

where $\kappa' = \kappa$ or $\kappa' = \overline{\kappa}$. The strings $u' = u_1u_2(\kappa-1)$ and $v' = v_1v_2(\kappa-1)$, that are obtained from u and v by erasing the occurrences of κ, are realistic, and since $\gamma_{u'} = \gamma_{v'}$, we have $u' \approx v'$ by the induction hypothesis. Since the sign of the last occurrence of $(\kappa-1)$ is the same in u' and v', complementation is not used in the above equivalence $u' \approx v'$. Therefore either u' is a conjugate of v' or of $(v')^R$. Moreover, the last occurrences $(\kappa-1)$ in u' and in v' both correspond to the end marker, which implies that $u' = v'$. If $u_2 = v_2$ then also $u = v$, and the claim holds. Suppose then that $u \neq v$. Now, $u_2 \neq v_2$, and either u_2 is a suffix of v_2 or v_2 is a suffix of u_2. By symmetry, we can assume that the first alternative holds: $v_2 = wu_2$, and then also $u_1 = v_1 w$. Since $\gamma_u = \gamma_v$, the substring w must be a legal string (if only one occurrence of a pointer p is in w, then κ would overlap with p either in u or v, but not in both). Now

$$u = v_1 w \kappa' u_2(\kappa - 1)\kappa \quad \text{and} \quad v = v_1 \kappa' w u_2(\kappa - 1)\kappa .$$

By the form of the substrings $w\kappa'$ in u and $\kappa' w$ in v in the above, w must be an image $\varrho_\kappa(\alpha) = w$ for some $\alpha \in \Theta^{\maltese}_{\kappa-2}$. By Lemma 1, $\mathrm{dom}(w) = [p, q]$ for some pointers $p \leq q$, and by Lemma 2(ii), either $w = w'q$ or $w = \overline{q}w'$ for a substring w'. Moreover, as shown in the proof of Lemma 2, in the former case $w(q+1)$ is a substring of both u and v and in the latter case, $(\overline{q+1})w$ is a substring of both u and v. It then follows that $q = \kappa - 1$, and so $\varrho_\kappa(M_{\kappa-1})$ occurs twice in u and v - this is impossible since u and v are realistic. □

Acknowledgements. Research supported partly by the MolCoNet project, IST-2001-32008. T. Harju gratefully acknowledges the support of the Academy of Finland under project 39802. G. Rozenberg gratefully acknowledges partial support by NSF grant 0121422.

References

1. Bouchet, A., Circle graphs. *Combinatorica* **7** (1987), 243–254.
2. Bouchet, A., Circle graph obstructions. *J. Combin. Theory Ser. B* **60** (1994), 107–144.
3. de Fraysseix, H., A characterization of circle graphs. *European J. Combin.* **5** (1984), 223–238.
4. Ehrenfeucht, A., Harju, T., Petre, I., Prescott, D. M., and Rozenberg, G., Formal systems for gene assembly in ciliates. *Theoret. Comput. Sci.* **292** (2003), 199–219.
5. Ehrenfeucht, A., Harju, T., Petre, I., and Rozenberg, G., Characterizing the micronuclear gene patterns in ciliates. *Theory of Computation and Systems* **35** (2002), 501–519.
6. Ehrenfeucht, A., Harju, T., and Rozenberg, G., Gene assembly through cyclic graph decomposition. *Theoretic Comput. Syst.* **281** (2002), 325–349.
7. Ehrenfeucht, A., Petre, I., Prescott, D. M., and Rozenberg, G., Universal and simple operations for gene assembly in ciliates. In *Words, Sequences, Languages: Where Computer Science, Biology and Linguistics Meet*, V. Mitrana, C. Martin-Vide (eds.), Kluwer Academic Publishers, Dortrecht/Boston, 329–342, 2001.
8. Ehrenfeucht, A., Petre, I., Prescott, D. M., and Rozenberg, G., String and graph reduction systems for gene assembly in ciliates. *Math. Structures Comput. Sci.* **12** (2001), 113–134.
9. Ehrenfeucht, A., Petre, I., Prescott, D. M., and Rozenberg, G., Circularity and other invariants of gene assembly in cliates. In *Words, Semigroups, and Transductions*, M. Ito, Gh. Păun, S. Yu (eds.), World Scientific, Singapore, 81–97, 2001.
10. Ehrenfeucht, A., Prescott, D. M., and Rozenberg, G., Computational aspects of gene (un)scrambling in ciliates. In *Evolution as Computation*, L. Landweber, E. Winfree (eds.), 45–86, Springer-Verlag, Berlin, Heidelberg, 2001.
11. Harju, T. and Rozenberg, G., Computational processes in living cells: gene assembly in ciliates. *Lecure Notes in Comput. Sci.*, to appear.

12. Landweber, L. F., and Kari, L., The evolution of cellular computing: nature's solution to a computational problem. In *Proceedings of the 4th DIMACS Meeting on DNA Based Computers*, Philadelphia, PA, 3–15 (1998).
13. Landweber, L. F., and Kari, L., Universal molecular computation in ciliates. In *Evolution as Computation*, L. Landweber, E. Winfree (eds.), Springer-Verlag, Berlin, Heidelberg, 2002.
14. Prescott, D. M., The unusual organization and processing of genomic DNA in Hypotrichous ciliates. *Trends in Genet.* **8** (1992), 439–445.
15. Prescott, D. M., The DNA of ciliated protozoa. *Microbiol Rev.* **58**(2) (1994), 233–267.
16. Prescott, D. M., Genome gymnastics: unique modes of DNA evolution and processing in ciliates. *Nat Rev Genet.* 1(3) (2000), 191–198.
17. Prescott, D. M., Ehrenfeucht, A., and Rozenberg, G., Molecular operations for DNA processing in hypotrichous ciliates. *European Journal of Protistology* **37** (2001), 241–260.
18. Prescott, D. M., and Rozenberg, G., How ciliates manipulate their own DNA – A splendid example of natural computing. *Natural Computing* **1** (2002), 165–183.

n-Insertion on Languages

Masami Ito and Ryo Sugiura

Faculty of Science, Kyoto Sangyo University
Kyoto 603, Japan
ito@ksuvx0.kyoto-su.ac.jp

Abstract. In this paper, we define the n-insertion $A \rhd^{[n]} B$ of a language A into a language B and provide some properties of n-insertions. For instance, the n-insertion of a regular language into a regular language is regular but the n-insertion of a context-free language into a context-free language is not always context-free. However, it can be shown that the n-insertion of a regular (context-free) language into a context-free (regular) language is context-free. We also consider the decomposition of regular languages under n-insertion.

1 Introduction

Insertion and deletion operations are well-known in DNA computing – see, e.g., [5] and [6]. In this paper we investigate a generalized insertion operation, namely the shuffled insertion of (substrings of) a string in another string.

Formally, we deal with the following operation.

Let $u, v \in X^*$ and let n be a positive integer. Then the *n-insertion* of u into v, i.e., $u \rhd^{[n]} v$, is defined as $\{v_1u_1v_2u_2 \cdots v_nu_nv_{n+1} \mid u = u_1u_2 \cdots u_n, u_1, u_2, \ldots, u_n \in X^*, v = v_1v_2 \cdots v_nv_{n+1}, v_1, v_2, \ldots, v_n, v_{n+1} \in X^*\}$. For languages $A, B \subseteq X^*$, the *n-insertion* $A \rhd^{[n]} B$ of A into B is defined as $\bigcup_{u \in A, v \in B} u \rhd^{[n]} v$. The shuffle product $A \diamond B$ of A and B is defined as $\bigcup_{n \geq 1} A \rhd^{[n]} B$.

In Section 2, we provide some properties of n-insertions. For instance, the n-insertion of a regular language into a regular language is regular but the n-insertion of a context-free language into a context-free language is not always context-free. However, it can be shown that the n-insertion of a regular (context-free) language into a context-free (regular) language is context-free. In Section 3, we prove that, for a given regular language $L \subseteq X^*$ and a positive integr n, it is decidable whether $L = A \rhd^{[n]} B$ for some nontrivial regular languages $A, B \subseteq X^*$. Here a language $C \subseteq X^*$ is said to be *nontrivial* if $C \neq \{\epsilon\}$, where ϵ is the empty word.

Regarding definitions and notations concerning formal languages and automata, not defined in this paper, refer, for instance, to [2].

2 Shuffle Product and n-Insertion

First, we consider the shuffle product of languages.

N. Jonoska et al. (Eds.): Molecular Computing (Head Festschrift), LNCS 2950, pp. 213–218, 2004.

Lemma 21 *Let $A, B \subseteq X^*$ be regular languages. Then $A \diamond B$ is a regular language.*

Proof. By $\overline{X}$ we denote the new alphabet $\{\overline{a} \mid a \in X\}$. Let $\mathcal{A} = (S, X, \delta, s_0, F)$ be a finite deterministic automaton with $\mathcal{L}(\mathcal{A}) = A$ and let $\mathcal{B} = (T, X, \theta, t_0, G)$ be a finite deterministic automaton with $\mathcal{L}(\mathcal{B}) = B$. Define the automaton $\overline{\mathcal{B}} = (T, \overline{X}, \overline{\theta}, t_0, G)$ as $\overline{\theta}(t, \overline{a}) = \theta(t, a)$ for any $t \in T$ and $a \in X$. Let ρ be the homomorphism of $(X \cup \overline{X})^*$ onto X^* defined as $\rho(a) = \rho(\overline{a}) = a$ for any $a \in X$. Moreover, let $\mathcal{L}(\overline{\mathcal{B}}) = \overline{B}$. Then $\rho(\overline{B}) = \{\rho(\overline{u}) \mid \overline{u} \in \overline{B}\} = B$ and $\rho(A \diamond \overline{B}) = A \diamond B$. Hence, to prove the lemma, it is enough to show that $A \diamond \overline{B}$ is a regular language over $X \cup \overline{X}$. Consider the automaton $\mathcal{A} \diamond \overline{\mathcal{B}} = (S \times T, X \cup \overline{X}, \delta \diamond \overline{\theta}, (s_0, t_0), F \times G)$ where $\delta \diamond \overline{\theta}((s,t), a) = (\delta(s, a), t)$ and $\delta \diamond \overline{\theta}((s,t), \overline{a}) = (s, \theta(t, a))$ for any $(s,t) \in S \times T$ and $a \in X$. Then it is easy to see that $w \in \mathcal{L}(\mathcal{A} \diamond \overline{\mathcal{B}})$ if and only if $w \in A \diamond \overline{B}$, i.e., $A \diamond \overline{B}$ is regular. This completes the proof of the lemma.

Proposition 21 *Let $A, B \subseteq X^*$ be regular languages and let n be a positive integer. Then $A \rhd^{[n]} B$ is a regular language.*

Proof. Let the notations of $\overline{X}$, $\overline{B}$ and ρ be the same as above. Notice that $A \rhd^{[n]} \overline{B} = (A \diamond \overline{B}) \cap (\overline{X}^* X^*)^n \overline{X}^*$. Since $(\overline{X}^* X^*)^n \overline{X}^*$ is regular, $A \rhd^{[n]} \overline{B}$ is regular. Consequently, $A \rhd^{[n]} B = \rho(A \rhd^{[n]} \overline{B})$ is regular.

Remark 21 The n-insertion of a context-free language into a context-free language is not always context-free. For instance, it is well known that $A = \{a^n b^n \mid n \geq 1\}$ and $B = \{c^n d^n \mid n \geq 1\}$ are context-free languages over $\{a, b\}$ and $\{c, d\}$, respectively. Since $(A \rhd^{[2]} B) \cap a^+ c^+ b^+ d^+ = \{a^n c^m b^n d^m \mid n, m \geq 1\}$ is not context-free, $A \rhd^{[2]} B$ is not context-free. Therefore, for any $n \geq 2$, n-insertion of a context-free language into a context-free language is not always context-free. However, $A \rhd^{[1]} B$ is a context-free language for any context-free languages A and B (see [4]). Usually, a 1-insertion is called an *insertion*.

Now consider the n-insertion of a regular (context-free) language into a context-free (regular) language.

Lemma 22 *Let $A \subseteq X^*$ be a regular language and let $B \subseteq X^*$ be a context-free language. Then $A \diamond B$ is a context-free language.*

Proof. The notations which we will use for the proof are assumed to be the same as above. Let $\mathcal{A} = (S, X, \delta, s_0, F)$ be a finite deterministic automaton with $\mathcal{L}(\mathcal{A}) = A$ and let $\mathcal{B} = (T, X, \Gamma, \theta, t_0, \epsilon)$ be a pushdown automaton with $\mathcal{N}(\mathcal{B}) = B$. Let $\overline{\mathcal{B}} = (T, \overline{X}, \Gamma, \overline{\theta}, t_0, \gamma_0, \epsilon)$ be a pushdown automaton such that $\overline{\theta}(t, \overline{a}, \gamma) = \theta(t, a, \gamma)$ for any $t \in T, a \in X \cup \{\epsilon\}$ and $\gamma \in \Gamma$. Then $\rho(\mathcal{N}(\overline{\mathcal{B}})) = B$. Now define the pushdown automaton $\mathcal{A} \diamond \overline{\mathcal{B}} = (S \times T, X \cup \overline{X}, \Gamma \cup \{\#\}, \delta \diamond \overline{\theta}, (s_0, t_0), \gamma_0, \epsilon)$ as follows:

1. $\forall a \in X, \delta \diamond \overline{\theta}((s_0, t_0), a, \gamma_0) = \{((\delta(s_0, a), t_0), \#\gamma_0)\}$,
   $\delta \diamond \overline{\theta}((s_0, t_0), \overline{a}, \gamma_0) = \{((s_0, t'), \#\gamma') \mid (t', \gamma') \in \overline{\theta}(t_0, \overline{a}, \gamma_0)\}$.
2. $\forall a \in X, \forall (s,t) \in S \times T, \forall \gamma \in \Gamma \cup \{\#\}, \delta \diamond \overline{\theta}((s,t), a, \gamma) = \{((\delta(s, a), t), \gamma)\}$.

3. $\forall a \in X, \forall (s,t) \in S \times T, \forall \gamma \in \Gamma, \delta \diamond \overline{\theta}((s,t),\overline{a},\gamma) = \{((s,t'),\gamma') \mid (t',\gamma') \in \overline{\theta}(t,\overline{a},\gamma)\}$.
4. $\forall (s,t) \in F \times T, \delta \diamond \overline{\theta}((s,t),\epsilon,\#) = \{((s,t),\epsilon)\}$.

Let $w = \overline{v}_1 u_1 \overline{v}_2 u_2 \ldots \overline{v}_n u_n \overline{v}_{n+1}$ where $u_1, u_2, \ldots, u_n \in X^*$ and $\overline{v}_1, \overline{v}_2, \ldots, \overline{v}_{n+1} \in \overline{X}^*$. Assume $\delta \diamond \overline{\theta}((s_0,t_0),w,\gamma_0) \neq \emptyset$. Then we have the following configuration: $((s_0,t_0),w,\gamma_0) \vdash^*_{\mathcal{A}\diamond\overline{\mathcal{B}}} ((\delta(s_0,u_1u_2\cdots u_n),t'),\epsilon,\#\cdots\#\gamma')$ where $(t',\gamma') \in \overline{\theta}(t_0,\overline{v}_1\overline{v}_2\cdots\overline{v}_{n+1},\gamma_0)$. If $w \in \mathcal{N}(\mathcal{A}\diamond\overline{\mathcal{B}})$, then $(\delta(s_0,u_1u_2\cdots u_n),t')$, $\epsilon, \#\cdots\#\gamma') \vdash^*_{\mathcal{A}\diamond\overline{\mathcal{B}}} (\delta(s_0,u_1u_2\cdots u_n),t'),\epsilon,\epsilon)$. Hence $(\delta(s_0,u_1u_2\cdots u_n),t') \in F \times T$ and $\gamma' = \epsilon$. This means that $u_1u_2\cdots u_n \in A$ and $\overline{v}_1,\overline{v}_2,\ldots,\overline{v}_{n+1} \in \overline{B}$. Hence $w \in A \times \overline{B}$. Now let $w \in A \times \overline{B}$. Then, by the above configuration, we have $((s_0,t_0),w,\gamma_0) \vdash^*_{\mathcal{A}\diamond\overline{\mathcal{B}}} ((\delta(s_0,u_1u_2\cdots u_n),t'),\epsilon,\#\cdots\#) \vdash^*_{\mathcal{A}\diamond\overline{\mathcal{B}}} ((\delta(s_0,u_1u_2\cdots u_n),t'),\epsilon,\epsilon)$ and $w \in \mathcal{N}(\mathcal{A}\diamond\overline{\mathcal{B}})$. Thus $A \diamond \overline{B} = \mathcal{N}(\mathcal{A}\diamond\overline{\mathcal{B}})$ and $A \diamond \overline{B}$ is context-free. Since $\rho(A \diamond \overline{B}) = A \diamond B$, $A \diamond B$ is context-free.

Proposition 22 *Let $A \subseteq X^*$ be a regular (context-free) language and let $B \subseteq X^*$ be a context-free (regular) language. Then $A \triangleright^{[n]} B$ is a context-free language.*

Proof. We consider the case that $A \subseteq X^*$ is regular and $B \subseteq X^*$ is context-free. Since $A \triangleright^{[n]} \overline{B} = (A \diamond \overline{B}) \cap (\overline{X}^* X^*)^n \overline{X}^*$ and $(\overline{X}^* X^*)^n \overline{X}^*$ is regular, $A \triangleright^{[n]} \overline{B}$ is context-free. Consequently, $A \triangleright^{[n]} B = \rho(A \triangleright^{[n]} \overline{B})$ is context-free.

3 Decomposition

Let $L \subseteq X^*$ be a regular language and let $\mathcal{A} = (S, X, \delta, s_0, F)$ be a finite deterministic automaton accepting the language L, i.e., $\mathcal{L}(\mathcal{A}) = L$. For $u, v \in X^*$, by $u \sim v$ we denote the equivalence relation of finite index on X^* such that $\delta(s,u) = \delta(s,v)$ for any $s \in S$. Then it is well known that for any $x, y \in X^*$, $xuy \in L \Leftrightarrow xvy \in L$ if $u \sim v$. Let $[u] = \{v \in X^* \mid u \sim v\}$ for $u \in X^*$. It is easy to see that $[u]$ can be effectively constructed using $\mathcal{A}$ for any $u \in X^*$. Now let n be a positive integer. We consider the decomposition $L = A \triangleright^{[n]} B$. Let $K_n = \{([u_1],[u_2],\ldots,[u_n]) \mid u_1,u_2,\ldots,u_n \in X^*\}$. Notice that K_n is a finite set.

Lemma 31 *There is an algorithm to construct K_n.*

Proof. Obvious from the fact that $[u]$ can be effectively constructed for any $u \in X^*$ and $\{[u] \mid u \in X^*\} = \{[u] \mid u \in X^*, |u| \leq |S|^{|S|}\}$. Here $|u|$ and $|S|$ denote the length of u and the cardinality of S, respectively.

Let $u \in X^*$. By $\rho_n(u)$, we denote $\{([u_1],[u_2],\ldots,[u_n]) \mid u = u_1u_2\cdots u_n, u_1, u_2,\ldots,u_n \in X^*\}$. Let $\mu = ([u_1],[u_2],\ldots,[u_n]) \in K_n$ and let $B_\mu = \{v \in X^* \mid \{v_1\}[u_1]\{v_2\}[u_2]\cdots\{v_n\}[u_n]\{v_{n+1}\} \subseteq L$ for any $v = v_1v_2\cdots v_nv_{n+1}, v_1, v_2, \ldots, v_n, v_{n+1} \in X^*\}$.

Lemma 32 *$B_\mu \subseteq X^*$ is a regular language and it can be effectively constructed.*

Proof. Let $S^{(i)} = \{s^{(i)} \mid s \in S\}, 0 \leq i \leq n$, and let $\tilde{S} = \bigcup_{0 \leq i \leq n} S^{(i)}$. We define the following nondeterministic automaton $\tilde{\mathcal{A}}' = (\tilde{S}, X, \tilde{\delta}, \{s_0^{(0)}\}, S^{(n)} \setminus F^{(n)})$ with ϵ-moves, where $F^{(n)} = \{s^{(n)} \mid s \in F\}$. The state transition relation $\tilde{\delta}$ is defined as follows:

$\tilde{\delta}(s^{(i)}, a) = \{\delta(s, a)^{(i)}, \delta(s, au_{i+1})^{(i+1)}\}$, for any $a \in X \cup \{\epsilon\}$ and $i = 0, 1, \ldots, n-1$, and $\tilde{\delta}(s^{(n)}, a) = \{\delta(s, a)^{(n)}\}$, for any $a \in X$.

Let $v \in \mathcal{L}(\tilde{\mathcal{A}}')$. Then $\delta(s_0, v_1u_1v_2u_2 \cdots v_nu_nv_{n+1})^{(n)} \in \tilde{\delta}(s_0^{(0)}, v_1v_2 \cdots v_nv_{n+1}) \cap (S^{(n)} \setminus F^{(n)})$ for some $v = v_1v_2 \cdots v_nv_{n+1}, v_1, v_2, \ldots, v_n, v_{n+1} \in X^*$. Hence $v_1u_1v_2u_2 \cdots v_nu_nv_{n+1} \notin L$, i.e., $v \in X^* \setminus B_\mu$. Now let $v \in X^* \setminus B_\mu$. Then there exists $v = v_1v_2 \cdots v_nv_{n+1}, v_1, v_2, \ldots, v_n.v_{n+1} \in X^*$ such that $v_1u_1v_2u_1 \cdots v_nu_nv_{n+1} \notin L$. Therefore, $\tilde{\delta}(s_0^{(0)}, v_1v_2 \cdots v_nv_{n+1}) \in S^{(n)} \setminus F^{(n)}$, i.e., $v = v_1v_2 \cdots v_nv_{n+1} \in \mathcal{L}(\tilde{\mathcal{A}}')$. Consequently, $B_\mu = X^* \setminus \mathcal{L}(\tilde{\mathcal{A}}')$ and B_μ is regular. Notice that $X^* \setminus \mathcal{L}(\tilde{\mathcal{A}}')$ can be effectively constructed.

Symmetrically, consider $\nu = ([v_1], [v_2], \ldots, [v_n], [v_{n+1}]) \in K_{n+1}$ and $A_\nu = \{u \in X^* \mid \forall u = u_1u_2 \cdots u_n, u_1, u_2, \ldots, u_n \in X^*, [v_1]\{u_1\}[v_2]\{u_2\} \cdots [v_n]\{u_n\} [v_{n+1}] \subseteq L\}$.

Lemma 33 *$A_\nu \subseteq X^*$ is a regular language and it can be effectively constructed.*

Proof. Let $S^{(i)} = \{s^{(i)} \mid s \in S\}, 1 \leq i \leq n+1$, and let $\overline{S} = \bigcup_{1 \leq i \leq n+1} S^{(i)}$. We define the nondeterministic automaton $\overline{\mathcal{B}}' = (\overline{S}, X, \overline{\delta}, \{\delta(s_0, v_1)^{(1)}\}, S^{(n+1)} \setminus F^{(n+1)})$ with ϵ-move where $F^{(n+1)} = \{s^{(n+1)} \mid s \in F\}$. The state transition relation $\overline{\delta}$ is defined as follows:

$\overline{\delta}(s^{(i)}, a) = \{\delta(s, a)^{(i)}, \delta(s, au_{i+1})^{(i+1)}\}$, for any $a \in X \cup \{\epsilon\}$ and $i = 1, 2, \ldots, n$.

By the same way as in the proof of Lemma 6, we can prove that $A_\nu = X^* \setminus \mathcal{L}(\overline{\mathcal{B}'})$. Therefore, A_ν is regular. Notice that $X^* \setminus \mathcal{L}(\overline{\mathcal{B}'})$ can be effectively constructed.

Proposition 31 *Let $A, B \subseteq X^*$ and let $L \subseteq X^*$ be a regular language. If $L = A \rhd^{[n]} B$, then there exist regular languages $A', B' \subseteq X^*$ such that $A \subseteq A', B \subseteq B'$ and $L = A' \rhd^{[n]} B'$.*

Proof. Put $B' = \bigcap_{\mu \in \rho_n(A)} B_\mu$. Let $v \in B$ and let $\mu \in \rho_n(A)$. Since $\mu \in \rho_n(A)$, there exists $u \in A$ such that $\mu = ([u_1], [u_2], \ldots, [u_n])$ and $u = u_1u_2 \cdots u_n, u_1, u_2, \ldots, u_n \in X^*$. By $u \rhd^{[n]} v \subseteq L$, we have $\{v_1\}[u_1]\{v_2\}[u_2] \cdots \{v_n\}[u_n]\{v_{n+1}\} \subseteq L$ for any $v = v_1v_2 \ldots v_nv_{n+1}, v_1, v_2, \ldots, v_n, v_{n+1} \in X^*$. This means that $v \in B_\mu$. Thus $B \subseteq \bigcap_{\mu \in \rho_n(A)} B_\mu = B'$. Now assume that $u \in A$ and $v \in B'$. Let $u = u_1u_2 \cdots u_n, u_1, u_2, \ldots, u_n \in X^*$ and let $\mu = ([u_1], [u_2], \ldots, [u_n]) \in \rho_n(u) \subseteq \rho_n(A)$. By $v \in B' \subseteq B_\mu$, $v_1u_1v_2u_2 \cdots v_nu_nv_{n+1} \in \{v_1\}[u_1]\{v_2\}[u_2] \cdots \{v_n\}[u_n]\{v_{n+1}\} \subseteq L$ for any $v = v_1v_2 \cdots v_nv_{n+1}, v_1, v_2, \ldots, v_n, v_{n+1} \in X^*$. Hence $u \rhd^{[n]} v \subseteq L$ and $A \rhd^{[n]} B' \subseteq L$. On the other hand, since $B \subseteq B'$ and $A \rhd^{[n]} B = L$, we have $A \rhd^{[n]} B' = L$. Symmetrically, put $A' = \bigcap_{\nu \in \rho_{n+1}(B')} A_\nu$. By the same way as the above, we can prove that $A \subseteq A'$ and $L = A' \rhd^{[n]} B'$.

Theorem 1. *For any regular language $L \subseteq X^*$ and a positive integer n, it is decidable whether $L = A \triangleright^{[n]} B$ for some nontrivial regular languages $A, B \subseteq X^*$.*

Proof. Let $\mathbf{A} = \{A_\nu \mid \nu \in K_{n+1}\}$ and $\mathbf{B} = \{B_\mu \mid \mu \in K_n\}$. By the preceding lemmata, $\mathbf{A}$, $\mathbf{B}$ are finite sets of regular languages which can be effectively constructed. Assume that $L = A \triangleright^{[n]} B$ for some nontrivial regular languages $A, B \subseteq X^*$. In this case, by Proposition 8, there exist regular languages $A \subseteq A'$ and $B \subseteq B'$ which are an intersection of languages in $\mathbf{A}$ and an intersection of languages in $\mathbf{B}$, respectively. It is obvious that A', B' are nontrivial languages. Thus we have the following algorithm:

1. Take any languages from $\mathbf{A}$ and let A' be their intersection.
2. Take any languages from $\mathbf{B}$ and let B' be their intersection.
3. Calculate $A' \triangleright^{[n]} B'$.
4. If $A' \triangleright^{[n]} B' = L$, then the output is "YES".
5. If the output is "NO", search another pair of $\{A', B'\}$ until obtaining the output "YES".
6. This procedure terminates after a finite-step trial.
7. Once we get the output "YES", then $L = A \triangleright^{[n]} B$ for some nontrivial regular languages $A, B \subseteq X^*$.
8. Otherwise, there are no such decompositions.

Let n be a positive integer. By $\mathcal{F}(n, X)$, we denote the class of finite languages $\{L \subseteq X^* \mid max\{|u| \mid u \in L\} \leq n\}$. Then the following result by C. Câmpeanu et al. ([1]) can be obtained as a corollary of Theorem 9.

Corollary 31 *For a given positive integer n and a regular language $A \subseteq X^*$, the problem whether $A = B \diamond C$ for a nontrivial language $B \in \mathcal{F}(n, X)$ and a nontrivial regular language $C \subseteq X^*$ is decidable.*

Proof. Obvious from the following fact: If $u, v \in X^*$ and $|u| \leq n$, then $u \diamond v = u \triangleright^{[n]} v$.

The proof of the above corollary was given in a different way in [3] using the following result: *Let $A, L \subseteq X^*$ be regular languages. Then it is decidable whether there exists a regular languages $B \subseteq X^*$ such that $L = A \diamond B$.*

References

1. C. Câmpeanu, K. Salomaa and S. Vágvölgyi, Shuffle quotient and decompositions, *Lecture Notes in Computer Science* 2295, Springer, 2002, 186–196.
2. J.E. Hopcroft and J.D. Ullman, *Introduction to Automata Theory, Languages and Computation*, Addison-Wesley, Reading MA,1979.
3. M. Ito, Shuffle decomposition of regular languages, *Journal of Universal Computer Science*, 8 (2002), 257–259.
4. L. Kari, *On insertion and deletion in formal languages*, PhD thesis, University of Turku, 1991.

5. L. Kari, Gh. Păun, G. Thierrin, S. Yu, At the crossroads of DNA computing and formal languages: Characterizing RE using insertion-deletion systems, *Prof. 3rd DIMACS Workshop on DNA Based Computers*, Philadelphia, 1997, 318–333.
6. Gh. Păun, G. Rozenberg, A. Salomaa, *DNA Computing. New Computing Paradigms*, Springer, Berlin, 1998.

Transducers with Programmable Input by DNA Self-assembly

Nataša Jonoska[1], Shiping Liao[2], and Nadrian C. Seeman[2]

[1] University of South Florida, Department of Mathematics
Tampa, FL 33620, USA
jonoska@math.usf.edu
[2] New York University, Department of Chemistry
New York, NY 1003, USA
sl558@nyu.edu, ned.seeman@nyu.edu

Abstract. Notions of Wang tiles, finite state machines and recursive functions are tied together. We show that there is a natural way to simulate finite state machines with output (transducers) with Wang tiles and we show that recursive (computable) functions can be obtained as composition of transducers through employing Wang tiles. We also show how a programmable transducer can be self-assembled using TX DNA molecules simulating Wang tiles and a linear array of DNA PX-JX_2 nanodevices.

1 Introduction

In recent years there have been several major developments both experimentally and theoretically, that use DNA for obtaining three dimensional nanostructures, computation and as a material for nano-devices.

Nanostructures. The inherently informational character of DNA makes it an attractive molecule for use in applications that entail targeted assembly. Genetic engineers have used the specificity of sticky-ended cohesion to direct the construction of plasmids and other vectors. Naturally-occurring DNA is a linear molecule in the sense that its helix axis is a line, although that line is typically not straight. Linear DNA molecules are not well-suited to serve as components of complex nanomaterials, but it is easy to construct DNA molecules with stable branch points [20]. Synthetic molecules have been designed and shown to assemble into branched species [12,23], and more complex species that entail the lateral fusion of DNA double helices [21], such as DNA double crossover (DX) molecules [7], triple crossover (TX) molecules [13] or paranemic crossover (PX) molecules. Double and triple cross-over molecules have been used as tiles and building blocks for large nanoscale arrays [24,25]. In addition, three dimensional structures such as a cube [4], a truncated octahedron [29] and arbitrary graphs [10,19] have been constructed from DNA duplex and junction molecules.

Computation. Theoretically, it has been shown that two dimensional arrays can simulate the dynamics of a bounded one dimensional cellular automaton and so are capable of potentially performing computations as a Universal Turing

N. Jonoska et al. (Eds.): Molecular Computing (Head Festschrift), LNCS 2950, pp. 219–240, 2004.

machine [24]. Several successful experiments performing computation have been reported, most notably the initial successful experiment by Adleman [1] and the recent one from the same group solving an instance of SAT with 20 variables [3]. Successful experiments that have confirmed computation such as the binary addition (simulation of XOR) using triple cross-over molecules (tiles) have been reported in [16]. In [6] a 9-bit instance of the "knight problem" has been solved using RNA and in [8] a small instance of the maximal clique problem has been solved using plasmids. Theoretically it has been shown that by self-assembly of three dimensional graph structure many hard computational problems can be solved in one (constant) biostep operation [10,11].

Nano Devices. Based on the B-Z transition of DNA, a nano-mechanical device was introduced in [17]. Soon after, "DNA fuel" strands were used to produce devices whose activity is controlled by DNA strands [26,28]. The PX-JX_2 device introduced in [26] has two distinct structural states, differing by a half-rotation; each state is obtained by addition of a pair of DNA strands that hybridizes with the device such that the molecule is in either the JX_2 state or in the PX state.

We consider it a natural step at this point to use the PX-JX_2 device and DNA self-assembly to develop a programmable finite state machine. A simulation of a finite state automaton that uses duplex DNA molecules and a restriction endonuclease to recognize a sequence of DNA was reported in [2]. Unfortunately this model is such that the DNA representing the input string is "eaten up" during the process of computation and no output (besides "accept-reject") is produced.

In this paper we exploit the idea of using DNA tiles (TX molecules) that correspond to Wang tiles to simulate a finite state machine with output, i.e., a transducer. Theoretically we show that by composition of transducers we can obtain all recursive functions and hence all computable functions (see Sections 3, and 4). This result is not surprising considering the fact that iterations of a generalized sequential machine can simulate type-0 grammars [14,15]. However, the explicit connection between recursive functions and Wang tiles has not been described before. Also, although the dynamics of transducers have been considered [22], the composition of transducers as means to obtain class of recursive functions represents a new way to tie together recursive functions, tiles and transducers. Moreover, in this paper, the purpose of Wang tiles is not to tile the plane, as they usually appear in literature, but merely to facilitate our description of the use of DNA TX molecules to simulate transducers and their composition. In particular, we assume the existence of "boundary colors" at some of the tiles which do not allow tiling extension beyond the appearance of such a color. The input tiles contain input "colors" on one side and a boundary color on the opposite side, such that the computation is performed in the direction of the input color. This is facilitated with composition colors on the sides (Section 4).

We go beyond two dimensions in this paper. We show how PX-JX_2 DNA devices (whose rotation uses three dimensional space) can be incorporated into the boundary such that the input of the transducer can be programmed and potentially the whole finite-state nano-machine can be reusable. The paper is organized as follows. Section 2 describes the relationship between Wang tiles and transducers. It describes the basic prototiles that can be used to simulate a transducer. The composition of transducers obtained through Wang tiles is described in Section 3. Here we also show how primitive recursive functions can be obtained through composition of transducers. The general recursion (computable functions) simulated with Wang tiles is presented in Section 4. DNA implementation of transducers with TX molecules is described in Section 5 and the use of the PX-JX_2 devices for setting up the input sequence is described in Section 5.2. We end with few concluding remarks.

2 Finite State Machines with Output: Basic Model

In this section we show the general idea of performing a computation by a finite state machine using tiles. The actual assembly by DNA will be described in Section 5. We briefly recall the definition of a transducer. This notion is well known in automata theory and an introduction to transducers (Mealy machines) can be found in [9].

A finite state machine with output or *a transducer* is $\mathcal{T} = (\Sigma, \Sigma', Q, \delta, s_0, F)$ where Σ and Σ' are finite alphabets, Q is a finite set of states, δ is the transition function, $s_0 \in Q$ is the initial state, and $F \subseteq Q$ is the set of final or terminal states. The alphabet Σ is the input alphabet and the alphabet Σ' is the output alphabet. We denote with Σ^* the set of all words over the alphabet Σ. This includes the word with "no symbols", the empty word denoted with λ. For a word $w = a_1 \cdots a_k$ where $a_i \in \Sigma$, the length of w denoted with $|w|$ is k. For the empty word λ, we have $|\lambda| = 0$.

The transition operation δ is a subset of $Q \times \Sigma \times \Sigma' \times Q$. The elements of δ are denoted with (q, a, a', q') or $(q, a) \stackrel{\delta}{\mapsto} (a', q')$ meaning that when $\mathcal{T}$ is in state q and scans input symbol a, then $\mathcal{T}$ changes into state q' and gives output symbol a'. In the case of *deterministic* transducers, δ is a function $\delta : Q \times \Sigma \to \Sigma' \times Q$, i.e., at a given state reading a given input symbol, there is a unique output state and an output symbol. Usually the states of the transducer are presented as vertices of a graph and the transitions defined with δ are presented as directed edges with input/output symbols as labels. If there is no edge from a vertex q in the graph that has input label a, we assume that there is an additional "junk" state $\bar{q}$ where all such transitions end. This state is usually omitted from the graph since it is not essential for the computation. The transducer is said to recognize a string (or a word) w over alphabet Σ if there is a path in the graph from the initial state s_0 to a terminal state in F with input label w. The set of words recognized by a transducer $\mathcal{T}$ is denoted with $L(\mathcal{T})$ and is called a *language* recognized by $\mathcal{T}$. It is well known that finite state transducers recognize the class of regular languages.

We concentrate on deterministic transducers. In this case the transition function δ maps the input word $w \in \Sigma^*$ to a word $w' \in (\Sigma')^*$. So the transducer $\mathcal{T}$ can be considered to be a function from $L(\mathcal{T})$ to $(\Sigma')^*$, i.e., $\mathcal{T} : L(\mathcal{T}) \to (\Sigma')^*$.

Examples:

1. The transducer $\mathcal{T}_1$ presented in Figure 1 (a) has initial and terminal state s_0. The input alphabet is $\{0,1\}$ and the output alphabet is $\Sigma' = \emptyset$. It recognizes the set of binary strings that represent numbers divisible by 3. The states s_0, s_1, s_2 represent the remainder of the division of the input string with 3.

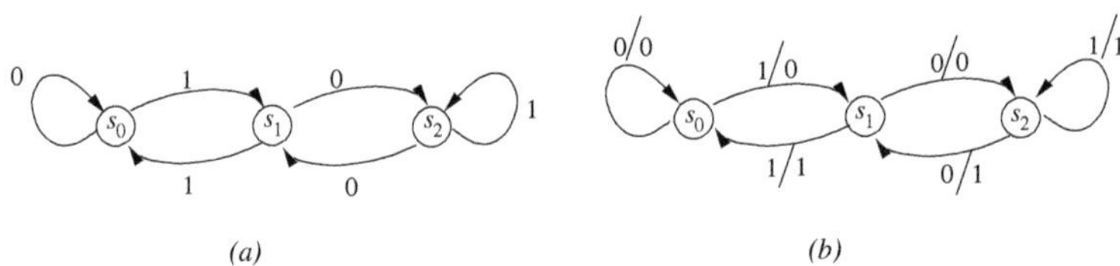

Fig. 1. A finite state machine that accepts binary strings that are divisible by 3. The machine in figure (b) outputs the result of dividing the binary string with 3 in binary

2. The transducer $\mathcal{T}_2$ presented in Figure 1 (b) is essentially the same as $\mathcal{T}_1$ except that now the output alphabet is also $\{0,1\}$. The output in this case is the result of the division of the binary string with three. On input 10101 (21 in decimal) the transducer gives the output 00111 (7 in decimal).
3. Our next example refers to encoders. Due to manufacturing constraints of magnetic storage devices, the binary data cannot be stored verbatim on the disk drive. One method of storing binary data on a disk drive is by using the modified frequency modulation (MFM) scheme currently used on many disk drives. The MFM scheme inserts a 0 between each pair of data bits, unless both data bits are 0, in which case it inserts a 1. The finite state machine, transducer, that provides this simple scheme is represented in Figure 2 (a). In this case the output alphabet is $\Sigma' = \{00, 01, 10\}$. If we consider rewriting of the symbols with $00 \mapsto \alpha$, $01 \mapsto \beta$ and $10 \mapsto \gamma$ we have the transducer in Figure 2 (b).
4. A transducer that performs binary addition is presented in Figure 2 (c). The input alphabet is $\Sigma = \{00, 01, 10, 11\}$ representing a pair of digits to be added, i.e., if $x = x_1 \cdots x_k$ and $y = y_1 \cdots y_k$ are two numbers written in binary ($x_i, y_i = \{0,1\}$), the input for the transducer is written in form $[x_k y_k]$ $[x_{k-1} y_{k-1}]$ $\cdots$ $[x_1 y_1]$. The output of the transducer is the sum of those numbers. The state s_1 is the "carry", s_0 is the initial state and all states are terminal. In [16] essentially the same transducer was simultaed by gradually connecting TX molecules.

For a given $\mathcal{T} = (\Sigma, \Sigma', Q, \delta, s_0, F)$ the transition $(q, a) \overset{\delta}{\mapsto} (a', q')$ schematically can be represented with a square as shown in Figure 3. Such a square can be considered as a Wang tile with colored edges, such that left and right we have the state colors encoding the input and output states of the transition and down

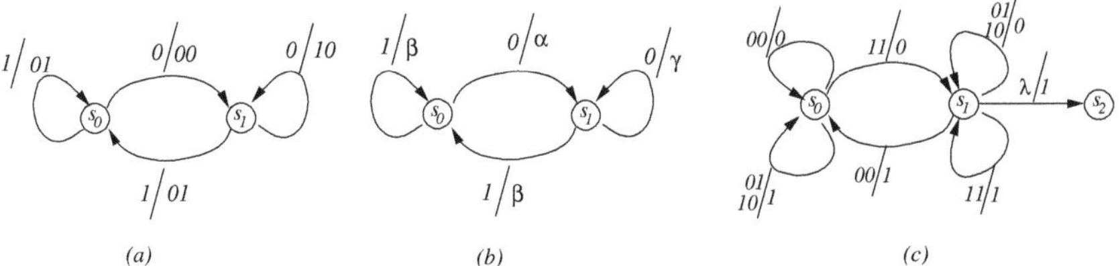

Fig. 2. An encoder that produces the modified frequency modulation code.

and up we have colors encoding input and output symbols. Then a computation with $\mathcal{T}$ is obtained by a process of assembling the tiles such that the abutted edges are colored with the same color. Only translation, and no rotations of the tiles are allowed. We describe this process in a better detail below.

2.1 Finite State Machines with Tile Assembly

Tiles: A Wang tile is a unit square with colored edges. A finite set of distinct unit squares with colored edges are called *Wang prototiles.* We assume that from each prototile there are arbitrarily large number of copies that we call *tiles.* A tile τ with left edge colored l, bottom edge colored b, top edge colored t and right edge colored r is denoted with $\tau = [l, b, t, r]$. No rotation of the tiles is allowed. Two tiles $\tau = [l, b, t, r]$ and $\tau' = [l', b', t', r']$ can be placed next to each other, τ to the left of τ' iff $r = l'$, and τ' on top of τ iff $t = b'$.

- *Computational tiles.* For a transducer $\mathcal{T}$ with a transition of form $(q, a) \mapsto (a', q')$ we associate a prototile $[q, a, a', q']$ as presented in Figure 3. If there are m transitions in the transducer, we associate m such prototiles. These tiles will be called *computational* tiles.

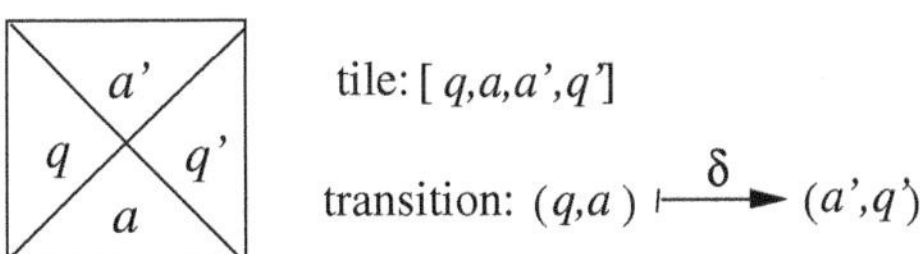

Fig. 3. A computational tile for a transducer.

- *Input and output tiles.* Additional colors called *border* are added to the set of colors. These colors will not be part of the assembly, but will represent the boundary of the computational assembly. Hence the left border is distinct from the right border. We denote these with $\beta_l, \beta_b, \beta_r, \beta_t$ for left, bottom, right and top border. We assume that each input word is from $\Sigma^*\alpha$ where α is a new symbol "end of input" that does not belong to Σ. For

$a \in \Sigma$ an *input* prototile $\tau_a = [c, \beta_b, a, c]$ is added. There are $|\Sigma|$ input prototiles, each representing an input symbol. The left and right sides are colored with the same color, *connect* color c. For the "end of input" symbol α a prototile $\tau_\alpha = [c, \beta_b, \alpha, \beta_r]$ is constructed. The output tiles are essentially the same as the input tiles, except they have the "top" border: $\tau'_a = [c', a, \beta_t, c']$ where c' is another connect color. The output row of tiles starts with $\tau_{left} = [\beta_l, \beta'_t, \beta_t, c']$ and ends with $\tau_{right} = [c'\beta'_t, \beta_t, \beta_r]$. For DNA implementation, β_t may be represented with a set of different motiffs that will fascilitate the "read out" of the result. With these sets of input and output tiles, every computation with $\mathcal{T}$ is obtained as a tiled rectangle surrounded by boundary colors.

- *Start tiles and accepting (end) tiles.* Start of the input for the transducer $\mathcal{T}$ is indicated with the start prototile $S_{\mathcal{T}} = [\beta_l, \beta_b, T, c]$ where T is a special color specifying the transducer $\mathcal{T}$. The input tiles can then be placed next to such a starting tile. For the starting state s_0 a starting prototile $\tau_0 = [\beta_l, T, \eta, s_0]$ is constructed. Then τ_0 can be placed on top of $S_{\mathcal{T}}$. The color η can be equal to T if we want to iterate the same transducer or it can indicate another transducer that should be applied after $\mathcal{T}$. If the computation is to be ended, η is equal to β'_t, indicating the start of "top boundary". For each terminal state $f \in F$ we associate a terminal prototile $\tau_f = [f, \alpha, \alpha, \beta_r]$ if another computation is required, otherwise, $\tau_f = [f, \alpha, \beta'_t, \beta_r]$ which stops the computation.

The set of tiles for executing a computation for transducer $\mathcal{T}_2$ that performs division by 3 (see Figure 1 (b)) is depicted in Figure 4 (a).

Computation: The computation is performed by first assembling the input starting with tile S and a sequence of input tiles ending with τ_α. The computation of the transducer starts by assembling the computation tiles according to the input state (to the left) and the input symbol (at the bottom). The computation ends by assembling the end tile τ_f which can lie next to both the last input tile and the last computational tile iff it ends with a terminal state. The output result will be read from the sequence of the output colors assembled with the second row of tiles and application of the output tiles. In this way one computation with $\mathcal{T}$ is obtained with a tiled $3 \times n$ rectangle ($n > 2$) such that the sides of the rectangle are colored with boundary colors. Denote all such $3 \times n$ rectangles with $D(\mathcal{T})$. By construction, the four boundary colors are different and since no rotation of Wang tiles is allowed, each boundary color can appear at only one side of such a rectangle. For a rectangle $\rho \in D(\mathcal{T})$ we denote w_ρ the sequence of colors opposite of boundary β_b and w'_ρ the sequence of colors opposite of boundary β_t. Then we have the following

Proposition 21 *For a transducer $\mathcal{T}$ with input alphabet Σ, output alphabet Σ', and any $\rho \in D(\mathcal{T})$ the following hold:*

(i) $w_\rho \in L(\mathcal{T})$ *and* $w'_\rho \in \Sigma'$.
(ii) $\mathcal{T}(w_\rho) = w'_\rho$

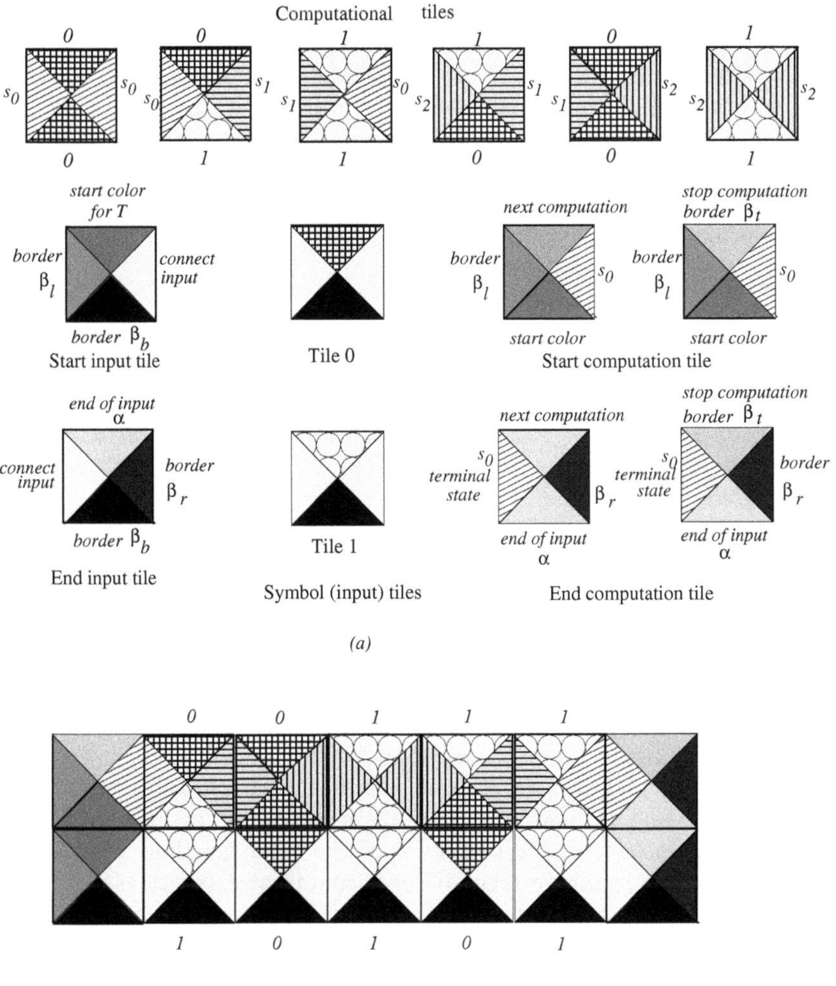

Fig. 4. All prototiles needed for performing the computation of the transducer presented in Figure 1 and a simple computation over the string 10101 with result 00111.

Moreover, if $w \in L(\mathcal{T})$ *and* $\mathcal{T}(w) = w'$, *then there is a* $\rho \in D(\mathcal{T})$ *such that* $w = w_\rho$ *and* $w' = w'_\rho$.

The tile computation of $\mathcal{T}_2$ from Figure 1 (b) for the input string 10101 is shown in Figure 4 (b). The output tiles are not included.

3 Primitive Recursive Functions from Transducers

In this section we show how the basic idea of modeling finite transducers by Wang tiles, with a small modification to allow composition of such computations, can perform computations of any primitive recursive function over the set of natural numbers. There is extensive literature on (primitive) recursive functions, and some introduction to the topic can be found in [5].

The primitive recursive functions are defined inductively as follows:

Definition 31

(i) Three initial functions: "zero" function $z(x) = 0$, "add one" function $s(x) = x + 1$ and the "i-th projection" function $\pi_i^n(x_1, \ldots, x_n) = x_i$ are primitive recursive.

(ii) Let $\mathbf{x}$ denote $(x_1, \ldots, x_n)$. If $f(x_1, \ldots, x_k)$ and $g_1(\mathbf{x}), \ldots, g_k(\mathbf{x})$ are primitive recursive, then the composition $h(\mathbf{x}) = f(g_1(\mathbf{x}), \ldots, g_k(\mathbf{x}))$ is primitive recursive too.

(iii) Let $f(x_1, \ldots, x_n)$ and $g(y, x_1, \ldots, x_n, z)$ be two primitive recursive functions, then the function h defined recursively (recursion):

$$h(x_1, \ldots, x_n, 0) = f(x_1, \ldots, x_n)$$
$$h(x_1, \ldots, x_n, t+1) = g(h(x_1, \ldots, x_n, t), x_1, \ldots, x_n, t)$$

is also primitive recursive.

Primitive recursive functions are total functions (defined for every input of natural numbers) and their range belongs to the class of recursive sets of numbers (see Section [5] and 4). Some of the primitive recursive functions include: $x + y$, $x \cdot y$, $x!$, x^y... In fact, in practice, most of the total functions (defined for all initial values) are primitive recursive. All computable functions are recursive (see Section 4), and there are examples of such functions that are not primitive recursive (see for ex. [5]).

In order to show that the previous tiling model can simulate every primitive recursive function we need to show that the initial functions can be modeled in this manner, as well as their composition and recursion.

3.1 Initial Functions

The input and the output alphabets for the transducers are equal: $\Sigma = \Sigma' = \{0, 1\}$. An input representing the number n is given with a non-empty list of 0's followed with $n+1$ ones and a non-empty list of 0's, i.e., $0 \ldots 01^{n+1}0 \ldots 0$. That means that the string 01110 represents the number 2. The input as an n-tuple $\mathbf{x} = (x_1, \ldots, x_n)$ is represented with

$$0 \cdots 01^{x_1+1}0^{s_1}1^{x_2+1}0^{s_2} \cdots 0^{s_n}1^{x_n+1}0 \cdots 0$$

where $s_i > 0$. This is known as the *unary representation* of $\mathbf{x}$. In Figure 5, transducers corresponding to the initial primitive recursive functions $z(x)$, $s(x)$, $U_i^n(\mathbf{x})$

are depicted. Note that the only input accepted by these functions is the correct representation of x or $\mathbf{x}$. The zero transducer has three states, q_0 is initial and q_2 is terminal. The transducer for the function $s(x)$ contains a transition that adds a symbol 0 to the input if the last 0 is changed into 1. This transition utilizes the "end of input" symbol α. The state q_0 is the initial and q_3 and ! are terminal states for this transducer. The increase of space needed as a result of computation is provided with the end tile (denoted !) for q_3. This tile is of the form $! = !_\alpha = [q_3, \beta_b, \alpha, \beta_r]$ which corresponds to the transition $(q_3, \lambda) \mapsto (\alpha, !)$. (In DNA implementation, this corresponds to a tile whose bottom side has no sticky ends.) The i-th projection function U_i^n accepts as inputs strings $0\cdots01^{x_1+1}0^{s_1}1^{x_2+1}0\cdots01^{x_n+1}0\cdots0$, i.e, unary representations of $\mathbf{x}$ and it has $2n+1$ states. Transitions starting from every state change 1's into 0's except at the state q_i where the i-th entry of 1's is copied.

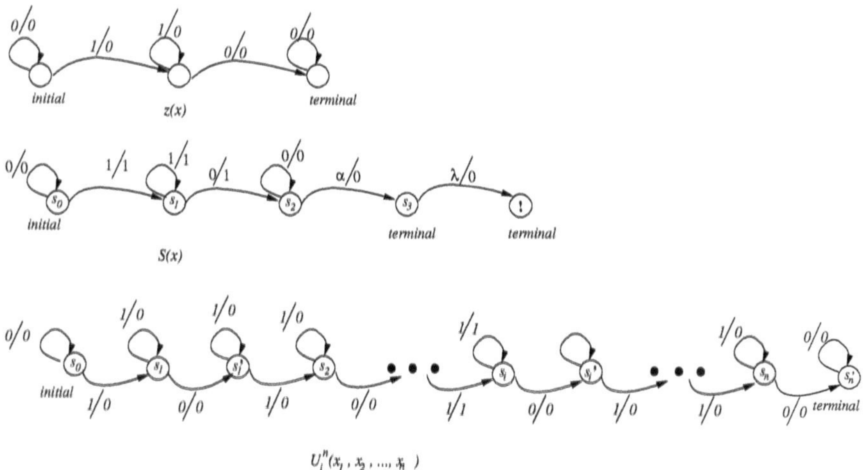

Fig. 5. The three initial functions of the primitive recursive functions

3.2 Composition

Let h be the composition $h(\mathbf{x}) = f(g_1(\mathbf{x}), \ldots, g_k(\mathbf{x}))$ where $\mathbf{x} = (x_1, \ldots, x_n)$. Since the input of h is $\mathbf{x}$ and there are k functions that need to be simultaneously computed on that input, we use a new alphabet $\bar{\Sigma} = \{(a_1, \ldots, a_k) \mid a_i \in \Sigma\}$ as an aid. By the inductive hypothesis, there are sets of prototiles $P(g_i)$ that perform computations for g_i with input $\mathbf{x}$ for each $i = 1, \ldots, k$. Also there is a set of prototiles $P(f)$ that simulates the computation for f. The composition function h is obtained with the following steps. Each step is explained below.

1. Check for correct input.
2. Translate the input symbol a into k-tuples $(a, a, \ldots, a)$ for $a \in \Sigma$.

3. For $i = 1, \ldots, k$ apply $\bar{P}(g_i)$. The start prototiles are adjusted as follows: If the current transducer $\mathcal{T}_i$ is to be followed by transducer $\mathcal{T}_j$, then the start tile for $\mathcal{T}_i$ is $[\beta_l, \mathcal{T}_i, \mathcal{T}_j, s_0]$ where s_0 is the initial state for $\mathcal{T}_i$.
4. For all $i = 1, \ldots, k-1$ in the i-th coordinate of the input mark the end of the last 1.
5. For all $i = 2, \ldots, k$ shift all 1's in the i-th coordinate to the right of all 1's in the $i-1$ coordinate. This ensures a proper input for the tiles that compute f. After this, the k-tuples contain at most one 1.
6. Translate back k-tuples into 1 if they contain a symbol 1, otherwise into 0.
7. Apply the tiles that perform the computation for function f.

Step 1 and 2 The input tiles are the same as described in Section 2. The translation of the input from a symbol to a k-tuple is obtained with the transducer T_k depicted in Figure 6. Note that this transducer is also checking the correctness of the input.

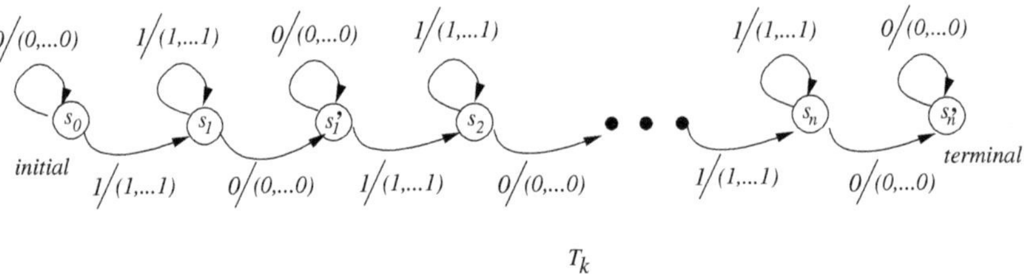

Fig. 6. The transducer T_k that translates symbols into k-tuples

Step 3 Each computational prototile for g_i of the form $[q, a, a', q']$ is substituted with a set of prototiles $[q, (a_1, \ldots, a_{i-1}, a, \ldots, a_k), (a_1, \ldots, a_{i-1}, a', \ldots, a_k), q']$ where a_i is any symbol in the alphabet Σ_{g_i} used by transducers for g_i. The end tile $!_\alpha = [q_i, \beta_b, \alpha, \beta_r]$ that increases the computational space is substituted with $!_{(\alpha,\ldots,\alpha)} = [q_i, \beta_b, (\alpha, \ldots, \alpha), \beta_r]$. We call this set of prototiles $\bar{P}(g_i)$. The idea is to use the tiles $\bar{P}(g_i)$ to compute the function g_i on coordinate i and leave the rest of the coordinates unchanged.

Step 4 After the application of the tiles $\bar{P}(g_1)$ to $\bar{P}(g_k)$, the k-tuples have the property that for each i the sequence of symbols that appears in the i-th coordinate represents a number $m_i = g(\mathbf{x})$ of the form $w_i = 0 \cdots 01^{m_i}0 \cdots 0\alpha^{s_i}$ for some $s_i \geq 1$. The end of input is obtained with the symbol $(\alpha, \ldots, \alpha)$. In order to prepare an input for the function f, all 1's in the i-th coordinate have to appear to the right of the first 0 after the 1's in the $i-1$ coordinate for all $i = 2, \ldots, k$. Hence, first we mark the end of 1's in each coordinate with an additional symbol γ. The transducer M_i substitutes γ instead of the first 0 after the sequence of 1's in the i-th coordinate. It has the following transitions (all states have superscripts i which are avoided to simplify the reading, we denote $\mathbf{y} = (a_1, \ldots, a_k)$ and $\mathbf{y}' = (a'_1, \ldots, a'_k)$)

$$(s_0, \mathbf{y}) \mapsto \begin{cases} (\mathbf{y}, s_0) & \text{for } a_i = 0 \\ (\mathbf{y}, s_1) & \text{for } a_i = 1 \end{cases}$$

when 1 is found in the i-th coordinate, change to s_1

$$(s_1, \mathbf{y}) \mapsto \begin{cases} (\mathbf{y}, s_1) & \text{for } a_i = 1 \\ (\mathbf{y}', s_2) & \text{for } a_i = 0, a'_i = \gamma \end{cases}$$

when all 1's are scanned, change the first 0 into γ

$$(s_2, \mathbf{y}) \mapsto (\mathbf{y}, s_2)$$

Note that the starting tile for each M_i $(i \neq k)$ is $[\beta_l, M_i, M_{i+1}, s_0^i]$. For $i = k$, M_{i+1} is substituted with σ_2 which is the function that start shifting with the second coordinate. Similarly, for $i = k$ the end of input $(\alpha, \ldots, \alpha)$ in the end tile is substituted with $(\sigma_2, \sigma_2, \ldots, \sigma_2)$.

Step 5 Now the shifting of the i-th input to the right of the $i-1$ input is achieved with the function σ_i. This function corresponds to a transducer with the following transitions (all states have superscripts i which are avoided to simplify the reading, hence we denote $\mathbf{y} = (a_1, \ldots, a_k)$ and $\mathbf{y}' = (a'_1, \ldots, a'_k)$):

$$(s_0, \mathbf{y}) \mapsto \begin{cases} (\mathbf{y}, s_0) & \text{for } a_i = 0, a_{i-1} \neq \gamma \\ (\mathbf{y}, D) & \text{for } a_i = 0, a_{i-1} = \gamma \\ (\mathbf{y}', s_1) & \text{for } a_i = 1, a_j = a'_j, a'_i = 0 \ (j \neq i) \end{cases}$$

start the shift with s_1 unless there is a γ in the $i-1$ coordinate before there is 1 in the i-th coordinate, in that case go to the final state D

$$(s_1, \mathbf{y}) \mapsto \begin{cases} (\mathbf{y}, s_1) & \text{for } a_i = 1 \\ (\mathbf{y}, s_2) & \text{for } a_i = \gamma, a'_i = 1, a_j = a'_j (j \neq i) \end{cases}$$

copy the 1s, change the end γ into 1 and go to s_2

$$(s_2, \mathbf{y}) \mapsto (\mathbf{y}', s_3) \text{ for } a_i = 0, \alpha, a'_i = \gamma$$

record the end symbol γ and change to s_3, note that a_i cannot be 1

$$(s_3, \mathbf{y}) \mapsto \begin{cases} (\mathbf{y}, s_3) & \text{for } a_i \neq \sigma_i, \mathbf{y} \neq \lambda \\ (\mathbf{y}', !_{\mathbf{y}'}) & \text{for } \mathbf{y} = \lambda, a'_j = \sigma_i \ (j = 1, \ldots, k) \end{cases}$$

copy the input while there are input symbols, if no input is to be found indicate the end of input and go to the next shifting, expand space if necessary

if $i < k$ then

$$(D, \mathbf{y}) \mapsto (\mathbf{y}, D)$$

go to the end of input and change the shift to the next coordinate using the ending tile $\tau_D = [D, (\sigma_i, \ldots, \sigma_i), (\sigma_{i+1}, \ldots, \sigma_{i+1}), \beta_r]$

if $i = k$ then

$(D, \mathbf{y}) \mapsto (\mathbf{y}, D)$ with ending tile $\tau_D = [D, (\sigma_k, \ldots, \sigma_k), T_R, \beta_r]$

Since the shifting is performed one at a time, σ_i potentially has to be repeated many times. This is obtained by setting the "end of input" of the form $(\sigma_i, \ldots, \sigma_i, \ldots, \sigma_i)$ which indicates that the next shift is the function σ_i. The terminal states D and s_3 make distinction between whether the shift σ_i has to be repeated (state s_3) or the shifting of the i'th coordinates is completed (state D). When $i = k$ the end tile at state D changes such that $(\sigma_k, \ldots, \sigma_k)$ is substituted with T_R which is the transducer that translates symbols from k-tuples back into symbols from Σ.

Step 6 The transition from k-tuples of symbols to a single symbol is done with the following simple transitions and only one state (note that after shifting, each n-tuple contains at most one 1):

$$(q_0, \mathbf{y}) \mapsto \begin{cases} 0 & \text{if } \mathbf{y} \text{ contains no 1's} \\ 1 & \text{if } \mathbf{y} \text{ contains a 1} \end{cases}$$

The start tile for T_R is $[\beta_l, T_R, f, q_0]$ indicating that the next computation has to be performed with tiles for function f and the end tile for T_R is $[q_0, T_R, \alpha, \beta_r]$ fixing the input for f in the standard form. After the application of T_R, the input reads the unary representation of $(g_1(\mathbf{x}), \ldots, g_k(\mathbf{x}))$.

Step 7 The composition is completed using tiles for f. These tiles exist by the inductive hypothesis.

3.3 Recursion

In order to perform the recursion, we observe that the composition and translation into n-tuples is sufficient. Denote with $\mathbf{x}$ the n-tuple $(x_1, \ldots, x_n)$. Then we have $h(\mathbf{x}, 0) = f(\mathbf{x})$ for some primitive recursive function f. By induction we assume that there is a set of tiles that performs the computation for the function f. For $h(\mathbf{x}, t+1) = g(h(\mathbf{x}, t), \mathbf{x}, t)$, also by the inductive hypothesis we assume that there is a set of tiles that performs the computation of $g(y, \mathbf{x}, t)$.

Set up the input. We use the following input $01^{t+1}\nu\bar{\mathbf{x}}010$ where $\bar{\mathbf{x}}$ denotes the unary representation of the n-tuple $\mathbf{x}$. We denote this with $(t+1 \,|\, \mathbf{x}, 0)$. The symbol ν is used to separate the computational input from the counter of the recursion. Since this input is not standard, we show that we can get to this input from a standard input for h which is given with an ordered $(n+1)$-tuple $(\mathbf{x}, y)$. First we add a 0 unary represented with string 010 at the end of the input to obtain $(\mathbf{x}, y, 0)$. This is obtained with the transducer $T^{n+1}_{add(0)}$ depicted in Figure 7. Note that for $k = 1$ the translation function T_k (Figure 6) is in fact the identity function, hence, there is a set of prototiles that represents the function $\mathrm{Id}(\mathbf{x}, y, 0)$. Consider the composition $\mathrm{Id}(U^{n+1}_{n+2}, U^1_{n+2}, \ldots, U^n_{n+2}, U^{n+2}_{n+2})$. This composition transforms $(\mathbf{x}, y, 0)$ into $(y, \mathbf{x}, 0)$. We obtain the desired entry for the recursion by marking the end of the first coordinate input (used as a counter for the recursion) with symbol ν. This is obtained by adding the following transitions to $\mathrm{Id}(\mathbf{x}, 0)$: $(i_0, 0) \mapsto (0, i_0)$, $(i_0, 1) \mapsto (1, i_1)$, $(i_1, 1) \mapsto (1, i_1)$ and

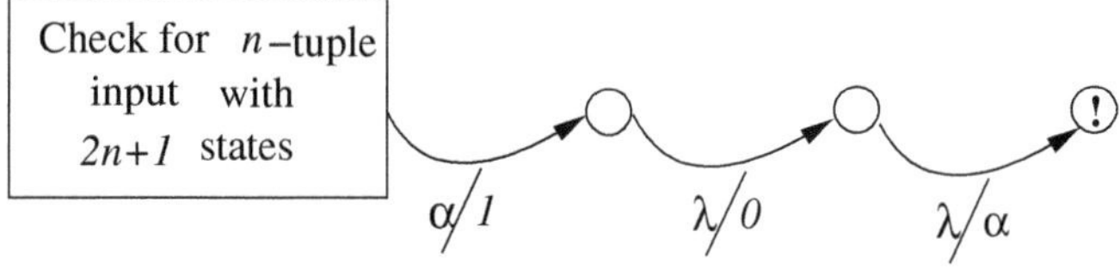

Fig. 7. The transducer $T^n_{add(0)}$ that increases the input from n-tuple $\mathbf{x}$ to $(n+1)$-tuple $(\mathbf{x}, 0)$.

$(i_1, 0) \mapsto (\nu, s_0)$ where i_0, i_1 are new states and s_0 is the starting state for $\mathrm{Id}(\mathbf{x}, 0)$. With this, our entry is in the desired form of $(y, x, 0) = (t+1 \,|\, \mathbf{x}, 0)$.

Execution of the recursion. Each transducer T that is used for computation of functions f and g is adjusted to T' which skips the input until symbol γ is obtained. This is obtained with one additional state $(c^T, i) \mapsto (i, c^T)$ for $i = 0, 1$ and $(c^T, \nu) \mapsto (\nu, s_0)$ where s_0 is the starting state for T. As in the case of composition we further adjust the prototiles for functions f and g into prototiles $\bar{P}(f)'$ and $\bar{P}(g)'$ which read/write pairs of symbols, but the computation of f and g is performed on the first coordinate, i.e., every prototile of the form $[q, a, a', q']$ in the set of prototiles for f and g is substituted with $[q, (a, a_2), (a', a_2), q']$ where a_2 are in $\{0, 1\}$. Second coordinates are kept to "remember" the input for f and g that is used in the current computation and will be used in the next computation as well.

In this case the recursion is obtained by the following procedure:

- Translate input $(t+1 \,|\, \mathbf{x}, 0)$ into $(t+1 \,|\, \bar{\mathbf{x}}, \bar{0})$ using the translation transducer T_2 as presented in Figure 6 and adjusted with the additional state to skip the initial $t+1$. Now each symbol a in the input portion $(\mathbf{x}, 0)$ is translated into a pair (a, a).
- Apply $\bar{P}(f)'$, hence the result $f(\mathbf{x})$ can be read from the first coordinates of the input symbols and the input $(\mathbf{x}, 0)$ is read in the second coordinates.
- Mark with M_1' the end of input from the first coordinate. This is the same transducer M_1 as used in the composition, except in this case there is an extra state c^{M_1} that skips the counter $t+1$ at the beginning.
- Shift coordinates using σ_2', i.e., the same transducer as σ_2 for the composition with the additional state c^{σ_2} to skip the counter.
- Translate back from pairs into $\{0, 1\}$ with T_R'. Now the result reads as $(t+1 \,|\, f(\mathbf{x}), \mathbf{x}, 0)$, i.e., $h(\mathbf{x}, 0) = f(\mathbf{x})$ is read right after symbol ν.
- For an input $(t+1 \,|\, f(\mathbf{x}), \mathbf{x}, 0)$ reduce $t+1$ for one. This is done with transducer presented in Figure 8 (a). The new input reads $(t \,|\, f(\mathbf{x}), \mathbf{x}, 0)$.
- Check for end of computation with transducer that accepts $0 \cdots 0 \nu w$ for any word w. Note that the language $0^* \nu (0+1)^*$ is regular and so accepted by a finite state transducer that has each transition with output symbols same as input symbols.

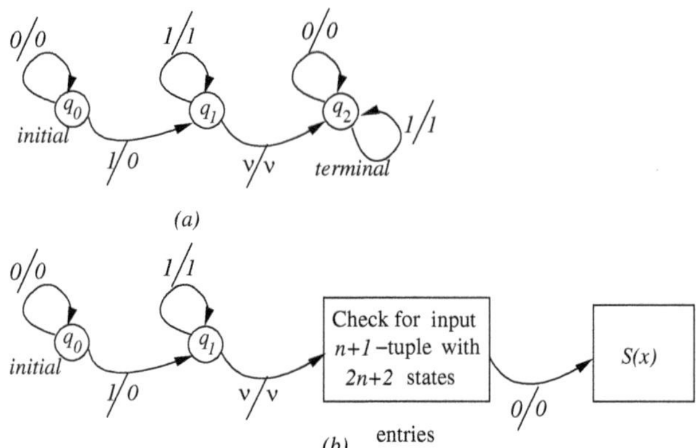

Fig. 8. The transducer that reduces the count of t and adds 1 to the input for the recursion. When count down is reduced to 0, the recursion has been applied $t+1$ times

- For $j = t, \ldots, 0$ perform the recursion on input $(j \mid y_j, \mathbf{x}, t-j)$:
  - Translate with T_2' leaving the first entry unchanged, and making all entries $2, \ldots, n+2$ into pairs.
  - Perform $\bar{P}(g)'$ on the first coordinate of these entries (skipping the first entry j). In other words, g is applied on the entry $(y_j, \mathbf{x}, t-j)$, and by the inductive hypothesis $y_j = h(\mathbf{x}, t-j)$.
  - Mark the end of first coordinate input in the pairs of symbols with M_1'.
  - Shift with σ_2'.
  - Translate back from pairs into single symbols with T_R'. Now the result reads as $(j \mid g(y_j, \mathbf{x}, t-j), \mathbf{x}, t-j)$. By definition, $g(y_j, \mathbf{x}, t-j) = g(h(\mathbf{x}, t-j), \mathbf{x}, t-j)$ is in fact $h(\mathbf{x}, t-j+1)$. Denote $y_{j-1} = g(y_j, \mathbf{x}, t-j)$.
  - For an input $(j \mid y_{j-1}, \mathbf{x}, t-j)$ reduce j for one and increase the last entry by 1. This is done with transducer presented in Figure 8 (b). The new input now reads $(j-1 \mid y_{j-1}, \mathbf{x}, t-j+1)$.
  - Check for end of the loop with transducer that accepts $0 \cdots 0 \nu w$ for any word w.
- The end of the loop reads the following $(0 \mid y_0, \mathbf{x}, t)$ and by construction $h(\mathbf{x}, t+1) = y_0$. Now change all symbols before and after y_0 into 0 (note that y_0 appears right after ν). This can be obtained with the following transitions:

$$\begin{array}{llll} (s_0,0) \mapsto (0,s_0) & (s_0,\nu) \mapsto (0,s_1) & (s_1,0) \mapsto (0,s_1) & (s_1,1) \mapsto (1,s_2) \\ (s_2,1) \mapsto (1,s_2) & (s_2,0) \mapsto (0,s_3) & (s_3,0) \mapsto (0,s_3) & (s_3,1) \mapsto (0,s_3) \end{array}$$

- Read out the result with upper border.

4 Recursive Functions as Composition of Transducers

Recursive functions are defined through the so called *minimization* operator. Given a function $g(x_1,\ldots,x_n,y) = g(\mathbf{x},y)$ the minimization $f(\mathbf{x}) = \mu_y g(\mathbf{x},y)$ is defined as:

$$f(\mathbf{x}) = \mu_y g(\mathbf{x},y) = \begin{cases} y & \text{if } y \text{ is the least natural number such that} \\ & \quad g(\mathbf{x},y)=0 \\ \text{undefined} & \text{if } g(\mathbf{x},y) \neq 0 \text{ for all } y \end{cases}$$

A function $f(\mathbf{x})$ is called *total* if it is defined for all $\mathbf{x}$. The class of *recursive* or *computable* functions is the class of total functions that can be obtained from the initial functions $z(x), s(x), U_n^i(\mathbf{x})$ with a finite number of applications of composition, recursion and minimization. The Kleene normal form theorem shows that every recursive function can be obtained with only one application of minimization (see for example [5]).

In order to show that all recursive (computable) functions can be obtained as a composition of transducers we are left to observe that the minimization operation can be obtained in this same way. As in the case of recursion we adjust the prototiles for function g into prototiles $\bar{P}(g)$ which contain pairs as symbols, but the computation of g is performed on the first coordinate i.e., every prototile of the form $[q,a,a',q']$ in the set of prototiles for g is substituted with $[q,(a,a_2),(a',a_2),q']$ where a_2 are in $\{0,1\}$. Second coordinates are kept to "remember" the input for g. Now the minimization is obtained in the following way (all representations of $\mathbf{x}$ are in unary).

- Fix the input $\mathbf{x}$ into $(\mathbf{x},0)$ by the transducer $T^n_{add(0)}$.
- Translate input $(\mathbf{x},0)$ into $(\bar{\mathbf{x}},\bar{0})$ using the translation transducer T_2 as presented in Figure 6. Now each symbol a in the input portion $(\mathbf{x},0)$ is translated into a pair (a,a).
- Apply $\bar{P}(g)$, hence the result $g(\mathbf{x})$ can be read from the first coordinates of the input symbols. The second coordinates read $(\mathbf{x},0)$.
- Until the first coordinate reads 0 continue:
  - (*) Check whether $g(\mathbf{x}) = 0$ in the first coordinate. This can be done with a transducer which "reads" only the first coordinate of the input pair and accepts strings of the form 0^+10^+ and rejects $0^+11^+0^+$.
  - If the transducer is in "accept" form, then
    - Change 1 from the first coordinate in the input into 0.
    - Translate back from pairs of symbols into a single symbol with T_R.
    - Apply U^{n+1}_{n+1}, stop.
  - If the transducer is in "reject" form,
    - Change all 1's from the first coordinate with 0's. (I.e., apply $z(x)$ on the first coordinate.)
    - Apply $T^{n+1}_{add(0)}$ to the second coordinate. Hence an input that reads from the second coordinate $(\mathbf{x},y)$ is changed into $(\mathbf{x},y+1)$.
    - Copy the second coordinates to the first with transitions: $(s,(a_1,a_2)) \mapsto ((a_2,a_2),s)$.

* Apply $\bar{P}(g)$, i.e., compute $g(\mathbf{x}, y+1)$ on the first coordinate, go back to (*).

5 DNA Implementation

5.1 The Basic Transducer Model

Winfree et. al [24] have shown that 2D arrays made of DX molecules can simulate dynamics of one-dimensional cellular automata (CA). This provides a way to simulate a Turing machine, since at one time step the configuration of a one-dimensional CA can represent one instance of the Turing machine. In the case of tiling representations of finite state machines (transducers) each tile represents a transition, such that the result of the computation with a transducer is obtained within one row of tiles. Transducers can be seen as Turing machines without left movement. Hence, the Wang tile simulation gives the result of a computation with such a machine within one row of tiles, which is not the case in the simulation of CA. A single tile simulating a transducer transition requires more information and DX molecules are not suitable for this simulation. Here we propose to use tiles made of triple crossover (TX) DNA molecules with sticky ends corresponding to one side of the Wang tile. An example of such molecule is presented in Figure 9 (b) with $3'$ends indicated with arrowheads. The connector is a sticky part that does not contain any information but it is necessary for the correct assembly of the tiles (see below). The TX tile representing transitions for a transducer have the middle duplex longer than the other two (see the schematic presentation in Figure 13). It has been shown that such TX molecules can self-assemble in an array [13], and they have been linearly assembled such that a cumulative XOR operation has been executed [16]. However, we have yet to demonstrate that they can be assembled reliably in a programmable way in a two dimensional array. Progress towards using double crossover molecules (DX) for assembly of a 2D array that generates the Sierpinski triangle was recently reported by the Winfree lab [18].

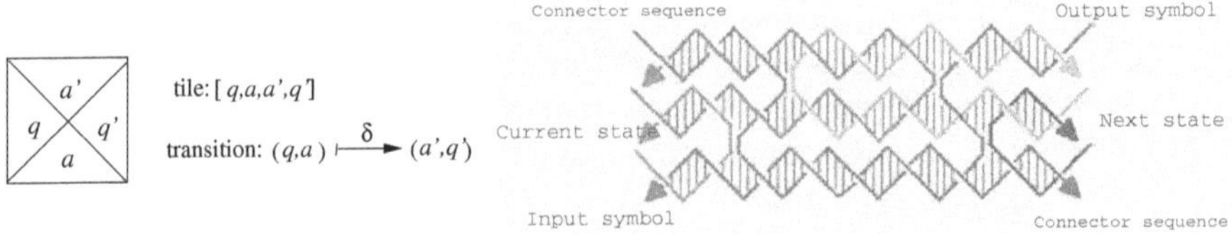

Fig. 9. A computational tile for a FSM and a proposed computational TX molecule tile.

Controlling the right assembly can be done by two approaches: (1) by regulating the temperature of the assembly such that only tiles that can bind three sticky ends at a time hybridize, and those with less than three don't, or (2) by including competitive imperfect hairpins that will extend to proper sticky ends only when all three sites are paired properly.

An iterated computation of the machine can be obtained by allowing third, fourth etc. rows of assembly and hence primitive recursive functions can be obtained. The input for this task is a combination of DX and TX molecules as presented in Figure 10. The top TX duplex (not connected to the neighboring DX) will have the right end sticky part encoding one of the input symbols and the left sticky end will be used as connector. The left (right) boundary of the assembly is obtained with TX molecules that have the left (right) sides of their duplexes ending with hairpins instead of sticky ends. The top boundary contains different motifs (such as the half hexagon in Figure 11 (b)) for different symbols. For a two symbol alphabet, the output tile for one symbol may contain a motif that acts as a topographic marker, and the other not. In this way the output can be detectable by atomic force microscopy.

Fig. 10. Input for the computational assembly.

5.2 Programable Computations with DNA Devices

Recent developments in DNA nanotechnology enable us to produce FSM's with variable and potentially programmable inputs. The first of these developments is the sequence-dependent PX-JX_2 2-state nanomechanical device [26]. This robust device, whose machine cycle is shown in Figure 11 to the left, is directed by the addition of set strands to the solution that forms its environment.

The set strands, drawn thick and thin, establish which of the two states the device will assume. They differ by rotation of a half-turn in the bottom parts of their structures. The sequence-driven nature of the device means that many different devices can be constructed, each of which is individually addressable; this is done by changing the sequences of the strands connecting the DX portion AB with the DX portion CD where the thick or thin strands pair with them. The thick and thin strands have short, unpaired extensions on them. The state of the device is changed by binding the full biotin-tailed complements of thick or thin strands, removing them from solution by magnetic streptavidin beads, and then adding the other strands. Figure 11 to the right shows that the device can change the orientation of large DNA trapezoids, as revealed by atomic force microscopy. The PX (thick strand) state leads to parallel trapezoids and the JX_2 (thin strand) state leads to a zig-zag pattern.

Linear arrays of a series of PX-JX_2 devices can be adapted to set the input of a FSM. This is presented in Figure 12 where we have replaced the trapezoids

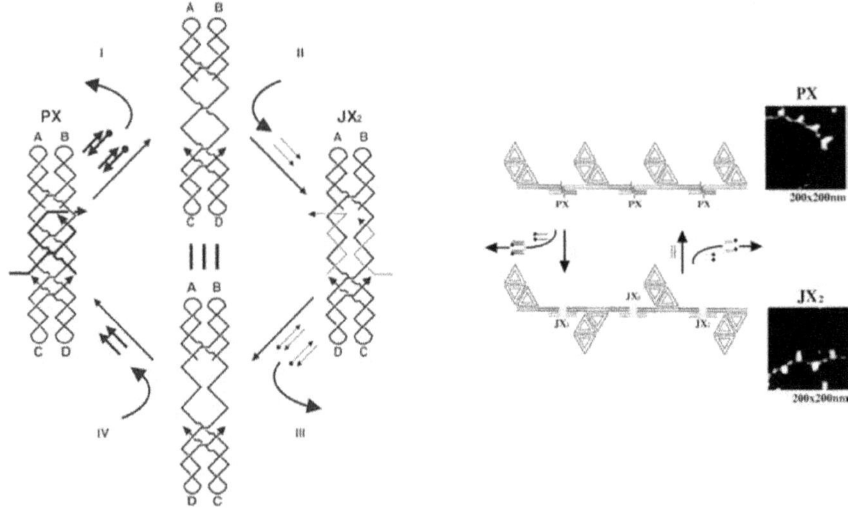

Fig. 11.

with double-triangle double diamonds. The computational set-up is illustrated schematically in Figure 13 with just two computational tiles made of DNA triple crossover (TX) molecules.

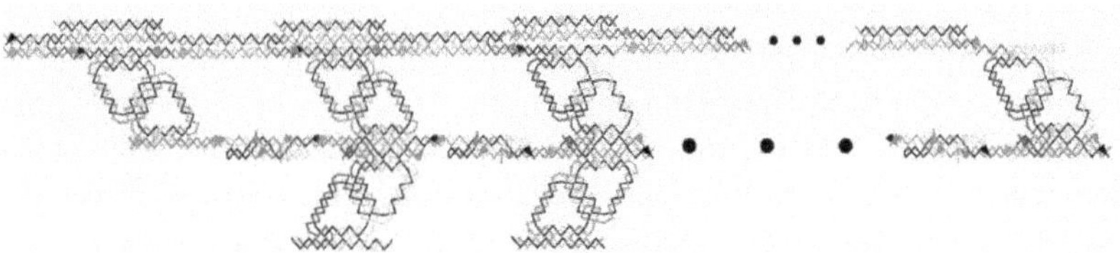

Fig. 12. Linear array of a series of devices to set-up the input.

Both diamonds and trapezoids are the result of edge-sharing motifs [27]. A key difference between this structure and the previous one is that the devices between the two illustrated double diamond structures differ from each other. Consequently, we can set the states of the two devices independently. The double-diamond structures contain domains that can bind to DX molecules; this design will set up the linear array of the input, see Figure 12. Depending on the state of the devices, one or the other of the double diamonds will be on the top side. The two DX molecules will then act similarly to bind with computational TX molecules, in the gap between them, as shown schematically in Figure 13. The system is designed to be asymmetric by allowing an extra diamond for the "start" and the "end" tiles (this is necessary in order to distinguish the top and

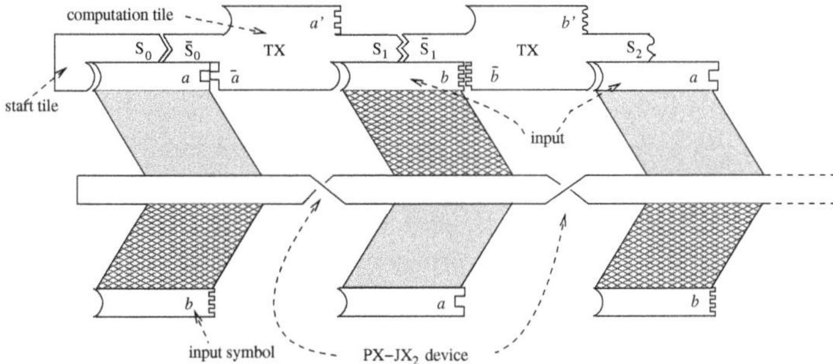

Fig. 13. Schematic view for a couple of first step computations. The linear array of DX and TX molecules that sets up the input is substituted only with a rectangle for each of the input symbol.

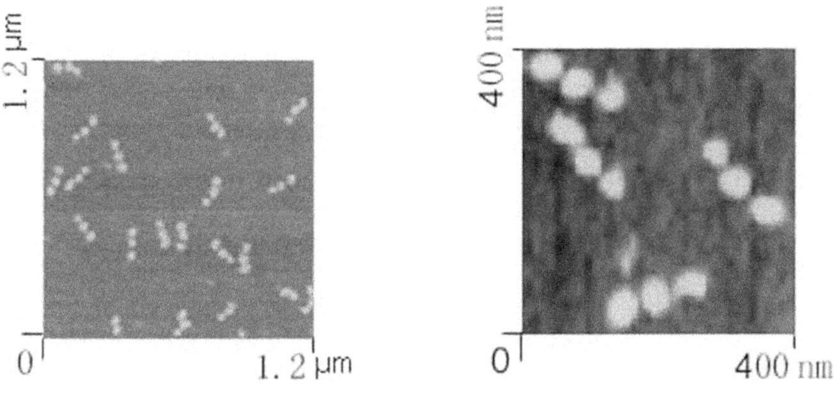

Fig. 14.

the bottom by in atomic force microscopy, (AFM) visualizations). The left side must contain an initiator (shown blunt at the top) and the right side will contain a terminator. The bottom assembly can be removed using the same Yurke-type techniques [28] that are used to control the state of the devices for removal of the set strands. Successive layers can be added to the device by binding to the initiator on the left, and to the terminator on the right.

For this task, we have prototyped the assembly of the central 4-diamond patterns, connecting them with sticky ends, rather than PX-JX_2 devices. AFM images of these strings are shown on the left in Figure 14. We have also built a 5-diamond system connected by devices. Its AFM pattern is shown on the right in Figure 14.

6 Discussion

What is the feasibility of this proposed system for a finite state machine? The TX tiles have already been developed and have been successfully self-assembled into a periodic 2D array [13]. Thus, we feel it is a reasonable extension of this previous work to extend this system to prototiles in a Wang tile system. Likewise, we have developed and purified a short double diamond-like system of connected PX-JX_2 devices. The experimental problems that we envision are the following: (i) Purifying a long linear system so that we do not have a mixture of input components. We expect that this can be done as done with the short system, using specifically biotinylated molecules on the end members of the input array. If the system is to be run similarly to the PX-JX_2 device previously constructed, we will need to remove the biotin groups from the system before operation. (ii) A second problem that we envision is that three sticky ends need to bind, but two sticky ends must be rejected. We have noted above two different methods by which this goal might be achieved. The most obvious form of competition is to make all of the sticky ends imperfect hairpins that require perfect mates to come undone. Experimentation will be required to identify the most effective protocol for this purpose.

An iterated computation (or a composition) of the machine can be obtained by allowing third, fourth etc. rows of assembly and hence recursive functions can be obtained. Although theoretically this has already been observed, the proposed way to simulate a transducer provides a much faster computational device than simulating a CA. As shown by examples in Figure 1 and 2 functions like addition and division as well as encoding devices can be simulated with a single row of tiles. However, the theoretical proofs for composition and recursion use transducers that perform a single rewriting step and are not much different from the computational steps simulated by CA. It is a theoretical question to characterize the functions that can be computed by a single application of a transducer. Knowing these functions, we can get a better understanding of the computational complexity of their iterations and compositions and with that a better understanding of the practical power of the proposed method. From another point, it is clear that errors can be minimized with minimal number of prototiles. It is known that the iterations of a four-state gsm provide universal computation [15], but what will be the minimal number of transitions in a transducer whose iterations provide universal computation remains to be determined.

With this paper we have presented ideas how to obtain a finite state nanomachine that is programmable, produces an output and is potentially reusable. Success in this objective relies on contributions of multiple disciplines, including biochemistry and information theory. Chemistry and biochemistry provide the infrastructure for dealing with biomolecules such as DNA and RNA. The development of programmable nano-devices provides us with a tool for building new (bio)molecules. Information theory provides the algorithms as well as the understanding of the potential and limitations of these devices as information processing machines. In turn, we expect that information technology will gain improved molecular based methods that ultimately will prove to be of value for difficult applications and computational problems.

Acknowledgements. This research has been supported by grants GM-29554 from the National Institute of General Medical Sciences, N00014-98-1-0093 from the Office of Naval Research, grants DMI-0210844, EIA-0086015, EIA-0074808, DMR-01138790, and CTS-0103002 from the National Science Foundation, and F30602-01-2-0561 from DARPA/AFSOR.

References

1. L. Adleman, *Molecular computation of solutions of combinatorial problems*, Science **266** (1994) 1021-1024.
2. Y. Benenson, T. Paz-Elizur, R. Adar, E. Keinan, E. Shapiro: *Programmable and autonomous computing machine made of biomolecules*, Nature **414** (2001) 430-434.
3. R. S. Braich, N. Chelyapov, C. Johnson, P.W.K. Rothemund, L. Adleman, *Solution of a 20-variable 3-SAT problem on a DNA Computer*, Science **296** (2002) 499-502.
4. J.H. Chen, N.C. Seeman, *Synthesis from DNA of a molecule with the connectivity of a cube,* Nature **350** (1991) 631-633.
5. N.J. Cutland, *Computability, an introduction to recursive function theory*, Cambridge University Press, Cambridge 1980.
6. D. Faulhammer, A.R. Curkas, R.J. Lipton, L.F. Landweber, *Molecular computation: RNA solution to chess problems*, PNAS **97** (2000) 1385-1389.
7. T.J. Fu, N.C. Seeman, *DNA double crossover structures*, Biochemistry **32** 3211-3220 (1993).
8. T. Head et.al, *Computing with DNA by operating on plasmids*, BioSystems **57** (2000) 87-93.
9. J.E. Hopcroft, J.D. Ullman, *Introduction to automata theory, languages and computation*, Addison-Wesley 1979.
10. N. Jonoska, P. Sa-Ardyen, N.C. Seeman, *Computation by self-assembly of DNA graphs*, Genetic Programming and Evolvable Machines **4** (2003) 123-137.
11. N. Jonoska, S. Karl, M. Saito, *Three dimensional DNA structures in computing,* BioSystems **52** (1999) 143-153.
12. N.R. Kallenbach, R.-I. Ma, N.C. Seeman, *An immobile nucleic acid junction constructed from oligonucleotides*, Nature **305** (1983) 829-831.
13. T.H. LaBean, H. Yan, J. Kopatsch, F. Liu, E. Winfree, J.H. Reif, N.C. Seeman, *The construction, analysis, ligation and self-assembly of DNA triple crossover complexes*, J. Am. Chem. Soc. **122** (2000) 1848-1860.
14. V. Manca, C. Martin-Vide, Gh. Păun, *New computing paradigms suggested by DNA computing: computing by carving*, BioSystems **52** (1999) 47-54.
15. V. Manca, C. Martin-Vide, Gh. Păun, *Iterated gsm mappings: a collapsing hierarchy*, in Jewels are forever, (J. Karhumaki, H. Maurer, Gh. Păun, G. Rozenberg eds.) Springer-Verlag 1999, 182-193.
16. C. Mao, T.H. LaBean, J.H. Reif, N.C. Seeman, *Logical computation using algorithmic self-assembly of DNA triple-crossover molecules*, Nature **407** (2000) 493-496.
17. C. Mao, W. Sun, Z. Shen, N.C. Seeman, *A nanomechanical device based on the B-Z transition of DNA*, Nature **397** (2000) 144-146.

18. P. Rothemund, N. Papadakis, E. Winfree, *Algorithmic self-assembly of DNA sierpinski triangles* (abstract) Preproceedings of the 9th International Meeting of DNA Based Computers, June 1-4, 2003.
19. P. Sa-Ardyen, N. Jonoska, N. Seeman, *Self-assembling DNA graphs*, Journal of Natural Computing (to appear).
20. N.C. Seeman, *DNA junctions and lattices*, J. Theor. Biol. **99** (1982) 237-247.
21. N.C. Seeman, *DNA nicks and nodes and nanotechnology*, NanoLetters **1** (2001) 22-26.
22. P. Siwak, *Soliton like dynamics of filtrons on cyclic automata*, Inverse Problems **17**, Institute for Physics Publishing (2001) 897-918.
23. Y. Wang, J.E. Mueller, B. Kemper, N.C. Seeman, *The assembly and characterization of 5-arm and 6-arm DNA junctions*, Biochemistry **30** (1991) 5667-5674.
24. E. Winfree, X. Yang, N.C. Seeman, *Universal computation via self-assembly of DNA: some theory and experiments*, *DNA computers II*, (L. Landweber, E. Baum editors), AMS DIMACS series **44** (1998) 191-214.
25. E. Winfree, F. Liu, L.A. Wenzler, N.C. Seeman, *Design and self-assembly of two-dimensional DNA crystals*, Nature **394** (1998) 539-544.
26. H. Yan, X. Zhang, Z. Shen and N.C. Seeman, *A robust DNA mechanical device controlled by hybridization topology*, Nature **415** (2002) 62-65.
27. H. Yan, N.C. Seeman, *Edge-sharing motifs in DNA nanotechnology*, Journal of Supramolecular Chemistry **1** (2003) 229-237.
28. B. Yurke, A.J. Turberfield, A.P. Mills, F.C. Simmel Jr., *A DNA fueled molecular machine made of DNA*, Nature **406** (2000) 605-608.
29. Y. Zhang, N.C. Seeman, *The construction of a DNA truncated octahedron*, J. Am. Chem. Soc. **160** (1994) 1661-1669.

Methods for Constructing Coded DNA Languages

Nataša Jonoska and Kalpana Mahalingam

Department of Mathematics
University of South Florida, Tampa, FL 33620, USA
jonoska@math.usf.edu, mahaling@helios.acomp.usf.edu

Abstract. The set of all sequences that are generated by a biomolecular protocol forms a language over the four letter alphabet $\Delta = \{A, G, C, T\}$. This alphabet is associated with a natural involution mapping θ, $A \mapsto T$ and $G \mapsto C$ which is an antimorphism of Δ^*. In order to avoid undesirable Watson-Crick bonds between the words (undesirable hybridization), the language has to satisfy certain coding properties. In this paper we build upon an earlier initiated study and give general methods for obtaining sets of code words with the same properties. We show that some of these code words have enough entropy to encode $\{0, 1\}^*$ in a symbol-to-symbol mapping.

1 Introduction

In bio-molecular computing and in particular DNA based computations and DNA nanotechnology, one of the main problems is associated with the design of the oligonucleotides such that mismatched pairing due to the Watson-Crick complementarity is minimized. In laboratory experiments non-specific hybridizations pose potential problems for the results of the experiment. Many authors have addressed this problem and proposed various solutions. Common approach has been to use the Hamming distance as a measure for uniqueness [3,8,9,11,19]. Deaton et al. [8,11] used genetic algorithms to generate a set of DNA sequences that satisfy predetermined Hamming distance. Marathe et al. [20] also used Hamming distance to analyze combinatorial properties of DNA sequences, and they used dynamic programing for design of the strands used in [19]. Seeman's program [23] generates sequences by testing overlapping subsequences to enforce uniqueness. This program is designed for producing sequences that are suitable for complex three-dimensional DNA structures, and the generation of suitable sequences is not as automatic as the other programs have proposed. Feldkamp et al. [10] also uses the test for uniqueness of subsequences and relies on tree structures in generating new sequences. Ruben at al. [22] use a random generator for initial sequence design, and afterwards check for unique subsequences with a predetermined properties based on Hamming distance. One of the first theoretical observations about number of DNA code words satisfying minimal Hamming distance properties was done by Baum [3]. Experimental separation of strands with "good" codes that avoid intermolecular cross hybridization was

N. Jonoska et al. (Eds.): Molecular Computing (Head Festschrift), LNCS 2950, pp. 241–253, 2004.

reported in [7]. One of the more significant attempt to design a method for generating good codes of uniform length is obtained by Arita and Kobayashi [2]. They design a single template DNA molecule, encode it binary and by taking a product with already known good binary code obtain a sequence of codes that satisfy certain minimal distance conditions. In this paper we do not consider the Hamming distance.

Every biomolecular protocol involving DNA or RNA generates molecules whose sequences of nucleotides form a language over the four letter alphabet $\Delta = \{A, G, C, T\}$. The Watson-Crick (WC) complementarity of the nucleotides defines a natural involution mapping θ, $A \mapsto T$ and $G \mapsto C$ which is an antimorphism of Δ^*. Undesirable WC bonds (undesirable hybridizations) can be avoided if the language satisfies certain coding properties. Tom Head considered comma-free levels of a language by identifying the maximal comma-free sublanguage of a given $L \subseteq \Delta^*$. This notion is closely related to the ones investigated in this paper by taking θ to be the identity map I. In [13], the authors introduce the theoretical approach followed here. Based on these ideas and code-theoretic properties, a computer program for generating code words is being developed [15] and additional properties were investigated in [14]. Another algorithm based on backtracking, for generating such code words is also developed by Li [18]. In particular for DNA code words it was considered that no involution of a word is a subword of another word, or no involution of a word is a subword of a composition of two words. These properties are called θ-compliance and θ-freedom respectively. The case when a DNA strand may form a hairpin, (i.e. when a word contains a reverse complement of a subword) was introduced in [15] and was called θ-subword compliance.

We start the paper with definitions of coding properties that avoid intermolecular and intramolecular cross hybridizations. The definitions of θ-compliant and θ-free languages are same as the ones introduced in [13]. Here we also consider intramolecular hybridizations and subword hybridizations. Hence, we have two additional coding properties: θ-subword compliance and θ-k-code. We make several observations about the closure properties of the code word languages. Section 3 provides several general methods for constructing code words with certain desired properties. For each of the resulting sets we compute the informational entropy showing that they can be used to encode $\{0,1\}^*$. We end with few concluding remarks.

2 Definitions and Closure Properties

An alphabet Σ is a finite non-empty set of symbols. We will denote by Δ the special case when the alphabet is $\{A, G, C, T\}$ representing the DNA nucleotides. A word u over Σ is a finite sequence of symbols in Σ. We denote by Σ^* the set of all words over Σ, including the empty word 1 and, by Σ^+, the set of all non-empty words over Σ. We note that with concatenation, Σ^* is the free monoid and Σ^+ is the free semigroup generated by Σ. The length of a word $u = a_1 \cdots a_n$ is n and is denoted with $|u|$.

For words representing DNA sequences we use the following convention. A word u over Δ denotes a DNA strand in its $5' \to 3'$ orientation. The Watson-Crick complement of the word u, also in orientation $5' \to 3'$ is denoted with $\overleftarrow{u}$. For example if $u = AGGC$ then $\overleftarrow{u} = GCCT$. There are two types of unwanted hybridizations: intramolecular and intermolecular. The intramolecular hybridization happens when two sequences, one being a reverse complement of the other appear within the same DNA strand (see Fig. 1). In this case the DNA strand forms a hairpin.

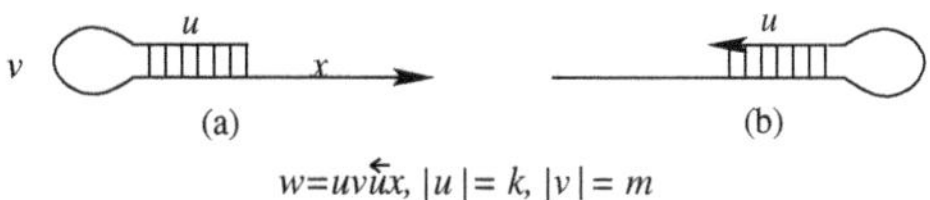

Fig. 1. Intramolecular hybridization (θ-subword compliance): (a) the reverse complement is at the beginning of the $5'$ end, (b) the reverse complement is at the end of the $3'$. The $3'$ end of the DNA strand is indicated with an arrow.

Two particular intermolecular hybridizations are of interest (see Fig. 2). In Fig. 2 (a) the strand labeled u is a reverse complement of a subsequence of the strand labeled v, and in the same figure (b) represents the case when u is the reverse complement of a portion of a concatenation of v and w.

Fig. 2. Two types of intermolecular hybridization: (a) (θ-compliant) one code word is a reverse complement of a subword of another code word, (b) (θ-free) a code word is a reverse complement of a subword of a concatenation of two other code words. The $3'$ end is indicated with an arrow.

Throughout the rest of the paper, we concentrate on finite sets $X \subseteq \Sigma^*$ that are *codes* such that every word in X^+ can be written uniquely as a product of words in X. In other words, X^* is a free monoid generated with X. For the background on codes we refer the reader to [5]. We will need the following definitions:

$$\begin{aligned}\text{Pref}(w) &= \{u \mid \exists v \in \Sigma^*, uv = w\}\\ \text{Suff}(w) &= \{u \mid \exists v \in \Sigma^*, vu = w\}\\ \text{Sub}(w) &= \{u \mid \exists v_1, v_2 \in \Sigma^*, v_1uv_2 = w\}\end{aligned}$$

We define the set of prefixes, suffixes and subwords of a set of words. Similarly, we have $\text{Suff}_k(w) = \text{Suff}(w) \cap \Sigma^k$, $\text{Pref}_k(w) = \text{Pref}(w) \cap \Sigma^k$ and $\text{Sub}_k(w) = \text{Sub}(w) \cap \Sigma^k$.

We follow the definitions initiated in [13] and used in [15,16].

An involution $\theta : \Sigma \to \Sigma$ of a set Σ is a mapping such that θ^2 equals the identity mapping, $\theta(\theta(x)) = x$, $\forall x \in \Sigma$.

The mapping $\nu : \Delta \to \Delta$ defined by $\nu(A) = T$, $\nu(T) = A$, $\nu(C) = G$, $\nu(G) = C$ is an involution on Δ and can be extended to a morphic involution of Δ^*. Since the Watson-Crick complementarity appears in a reverse orientation, we consider another involution $\rho : \Delta^* \to \Delta^*$ defined inductively, $\rho(s) = s$ for $s \in \Delta$ and $\rho(us) = \rho(s)\rho(u) = s\rho(u)$ for all $s \in \Delta$ and $u \in \Delta^*$. This involution is antimorphism such that $\rho(uv) = \rho(v)\rho(u)$. The Watson-Crick complementarity then is the antimorphic involution obtained with the composition $\nu\rho = \rho\nu$. Hence for a DNA strand u we have that $\rho\nu(u) = \nu\rho(u) = \overleftarrow{u}$. The involution ρ reverses the order of the letters in a word and as such is used in the rest of the paper.

For the general case, we concentrate on morphic and antimorphic involutions of Σ^* that we denote with θ. The notions of θ-free and θ-compliant in 2, 3 of Definition 21 below were initially introduced in [13]. Various other intermolecular possibilities for cross hybridizations were considered in [16] (see Fig. 3). All of these properties are included with θ-k-code introduced in [14] (4 of Definition 21).

Definition 21 *Let $\theta : \Sigma^* \to \Sigma^*$ be a morphic or antimorphic involution. Let $X \subseteq \Sigma^*$ be a finite set.*

1. *The set X is called* $\theta(k, m_1, m_2)$-subword compliant *if for all $u \in \Sigma^*$ such that for all $u \in \Sigma^k$ we have $\Sigma^* u \Sigma^m \theta(u) \Sigma^* \cap X = \emptyset$ for $m_1 \leq m \leq m_2$.*
2. *We say that X is called* θ-compliant *if $\Sigma^*\theta(X)\Sigma^+ \cap X = \emptyset$ and $\Sigma^+\theta(X)\Sigma^* \cap X = \emptyset$.*
3. *The set X is called* θ-free *if $X^2 \cap \Sigma^+\theta(X)\Sigma^+ = \emptyset$.*
4. *The set X is called* θ-k-code *for some $k > 0$ if $\mathrm{Sub}_k(X) \cap \mathrm{Sub}_k(\theta(X)) = \emptyset$.*
5. *The set X is called* strictly θ *if $X' \cap \theta(X') = \emptyset$ where $X' = X \setminus \{1\}$.*

The notions of prefix, suffix (subword) compliance can be defined naturally from the notions described above, but since this paper does not investigate these properties separately, we don't list the formal definitions here.

We have the following observations:

Observation 22 In the following we assume that $k \leq \min\{\,|x| \,:\, x \in X\}$.

1. X is strictly θ-compliant iff $\Sigma^*\theta(X)\Sigma^* \cap X = \emptyset$.
2. If X is strictly θ-free then X and $\theta(X)$ are strictly θ-compliant and $\theta(X)$ is θ-free.
3. If X is θ-k-code, then X is θ-k'-code for all $k' > k$.
4. X is a θ-k-code iff $\theta(X)$ is a θ-k-code.
5. If X is strictly θ such that X^2 is $\theta(k, 1, m)$-subword compliant, then X is strictly θ-k-code.
6. If X is a θ-k-code then both X and $\theta(X)$ are $\theta(k, 1, m)$-subword compliant, $\theta(k, 1, m)$ prefix and suffix compliant for any $m \geq 1$, θ-compliant. If $k \leq \frac{|x|}{2}$ for all $x \in X$ then X is θ-free and hence avoids the cross hybridizations as shown in Fig. 1 and 2.

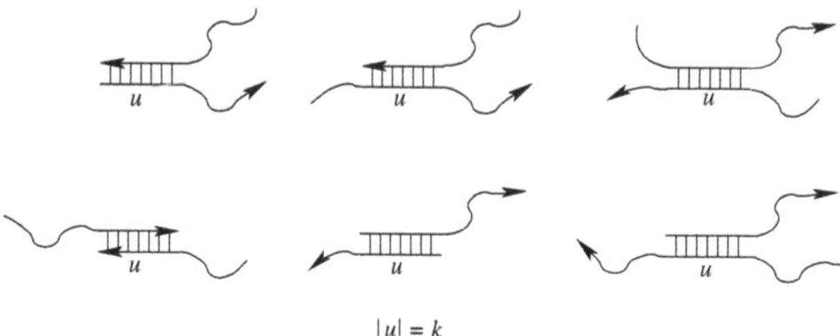

Fig. 3. Various cross hybridizations of molecules one of which contains subword of length k and the other its complement.

7. If X is a θ-k-code then X and $\theta(X)$ avoids all cross hybridizations of length k shown in Fig. 3 and so all cross hybridizations presented in Fig. 2 of [16].

It is clear that θ-subword compliance implies θ-prefix and θ-suffix compliance. We note that when $\theta = \rho\nu$, the $\theta(k, m_1, m_2)$-subword compliance of the code words $X \subseteq \Delta^*$ does not allow intramolecular hybridization as in Fig. 1 for a predetermined k and $m_1 \leq m \leq m_2$. The maximal length of a word that together with its reverse complement can appear as subwords of code words is limited with k. The length of the hairpin, i.e. "distance" between the word and its reversed complement is bounded between m_1 and m_2. The values of k and m_1, m_2 would depend on the laboratory conditions (ex. the melting temperature, the length of the code words and the particular application). In order to avoid intermolecular hybridizations as presented in Fig. 2, X has to satisfy θ-compliance and θ-freedom. Most applications would require X to be strictly θ. The most restricted and valuable properties are obtained with θ-k-code, and the analysis of this type of codes also seems to be more difficult. When X is θ-k-code, all intermolecular hybridizations presented in Fig. 3 are avoided. We include several observations in the next section.

2.1 Closure Properties

In this part of the paper we consider several closure properties of languages that satisfy some of the properties described with the Definition 21. We concentrate on closure properties of union and concatenation of "good" code words. From practical point of view, we would like to know under what conditions two sets of available "good" code words can be joined (union) such that the resulting set is still a "good" set of code words. Also, whenever ligation of strands is involved, we need to consider concatenation of code words. In this case it is useful to know under what conditions the "good" properties of the codewords will be preserved after arbitrary ligations. The following table shows closure properties of these languages under various operations.

Table 2.1 Closure Properties

	θ-subword compl.	θ-compl.	θ-free	θ-k-code
Union	Yes	No	No	No
Intersection	Yes	Yes	Yes	Yes
Complement	No	No	No	No
Concatenation ($XY, X \neq Y$)	No	Y/N	No	No
Kleene *	No	No	Yes	No

Most of the properties included in the table are straight forward. We add couple of notes for clarification.

- $\theta(k, m_1, m_2)$-subword compliant languages are not closed under concatenation and Kleene*. For example consider the set $X = \{aba^2\} \subseteq \{a,b\}^*$ with the morphic involution $\theta : a \mapsto b, b \mapsto a$. Then X is $\theta(2,2,4)$-subword compliant. But $X^2 = \{aba^3ba^2\}$ is not $\theta(2,2,4)$-subword compliant, i.e. X^2 contains $(ab)a^3\theta(ab)a$.
- When θ is a morphism θ-compliant languages are closed under concatenation, but when θ is antimorphism, they may not be. Consider the following example: $X_1 = \{a^2, ab\}$ and $X_2 = \{b^2, aba\}$ with the antimorphic involution $\theta : a \mapsto b, b \mapsto a$. Then $ab^3 \in X_1X_2$ and $bab^3 \in \theta(X_1X_2)$.

The next proposition which is a stronger version of Proposition 10 in [13], shows that for an antimorphic θ, concatenation of two distinct θ-compliant or θ-free languages is θ-compliant or θ-free whenever their union is θ-compliant i.e. θ-free respectively. The proof is not difficult and is omitted.

Proposition 23 *Let $X, Y \subseteq \Sigma^+$ be two θ-compliant (θ-free) languages for a morphic or antimorphic θ. If $X \cup Y$ is θ-compliant (θ-free) then XY is θ-compliant (θ-free).*

3 Methods to Generate Good Code Words

In the previous section closure properties of "good" code words were obtained. With the constructions in this section we show several ways to generate such codes. Many authors have realized that in the design of DNA strands it is helpful to consider three out of the four bases. This was the case with several successful experiments [4,9,19]. It turns out that this, or a variation of this technique can be generalized such that codes with some of the desired properties can be easily constructed. In this section we concentrate on providing methods to generate "good" code words X and for each code X the entropy of X^* is computed. The entropy measures the information capacity of the codes, i.e., the efficiency of these codes when used to represent information.

The standard definition of entropy of a code $X \subseteq \Sigma^*$ uses probability distribution over the symbols of the alphabet of X (see [5]). However, for a p-symbol alphabet, the maximal entropy is obtained when each symbol appears with the same probability $\frac{1}{p}$. In this case the entropy essentially counts the average number of words of a given length as subwords of the code words [17]. From the coding theorem, it follows that $\{0,1\}^*$ can be encoded by X^* with $\Sigma \mapsto \{0,1\}$ if the entropy of X^* is at least $\log 2$ ([1], see also Theorem 5.2.5 in [6]). The codes for θ-free, strictly θ-free, and θ-k-codes designed in this section have entropy larger than $\log 2$ when the alphabet has $p = 4$ symbols. Hence, such DNA codes can be used for encoding bit-strings.

We start with the entropy definition as defined in [6].

Definition 31 *Let X be a code. The entropy of X^* is defined by*

$$\hbar(X) = lim_{n \to \infty} \frac{1}{n} \log |\mathrm{Sub}_n(X^*)|.$$

If G is a deterministic automaton or an automaton with a delay that recognizes X^* and A_G is the adjacency matrix of G, then by Perron-Frobenius theory A_G has a positive eigen value $\bar{\mu}$ and the entropy of X^* is $\log \bar{\mu}$ (see Chapter 4 of [6]). We will use this fact in the following computations of the entropies of the designed codes. In [13], Proposition 16, authors designed a set of DNA code words that is strictly θ-free. The following propositions show that in a similar way we can construct codes with additional "good" propoerties.

In what follows we assume that Σ is a finite alphabet with $|\Sigma| \geq 3$ and $\theta : \Sigma \to \Sigma$ is an involution which is not identity. We denote with p the number of symbols in Σ. We also use the fact that X is (strictly) θ-free iff X^* is (strictly) θ-free, (Proposition 4 in [14]).

Proposition 32 *Let $a, b \in \Sigma$ be such that for all $c \in \Sigma \setminus \{a,b\}$, $\theta(c) \notin \{a,b\}$. Let $X = \bigcup_{i=1}^{\infty} a^m(\Sigma \setminus \{a,b\})^i b^m$ for a fixed integer $m \geq 1$.*

Then X and X^* are θ-free. The entropy of X^* is such that $\log(p-2) < \hbar(X^*)$.

Proof. Let $x_1, x_2, y \in X$ such that $x_1x_2 = s\theta(y)t$ for some $s, t \in \Sigma^+$ such that $x_1 = a^m p b^m$, $x_2 = a^m q b^m$ and $y = a^m r b^m$, for $p, q, r \in (\Sigma \setminus \{a,b\})^+$. Since θ is an involution, if $\theta(a) \neq a, b$, then there is a $c \in \Sigma \setminus \{a,b\}$ such that $\theta(c) = a$, which is excluded by assumption. Hence, either $\theta(a) = a$ or $\theta(a) = b$. When θ is morphic $\theta(y) = \theta(a^m)\theta(r)\theta(b^m)$ and when θ is antimorphic $\theta(y) = \theta(b^m)\theta(r)\theta(a^m)$. So, $\theta(y) = a^m\theta(r)b^m$ or $\theta(y) = b^m\theta(r)a^m$. Since $x_1x_2 = a^m p b^m a^m q b^m = s b^m \theta(r) a^m t$ or $x_1x_2 = a^m p b^m a^m q b^m = s a^m \theta(r) b^m t$ the only possibilities for r are $\theta(r) = p$ or $\theta(r) = q$. In the first case $s = 1$ and in the second case $t = 1$ which is a contradiction with the definition of θ-free. Hence X is θ-free.

Let $\mathcal{A} = (V, E, \lambda)$ be the automaton that recognizes X^* where $V = \{1, ..., 2m+1\}$ is the set of vertices, $E \subseteq V \times \Sigma \times V$ and $\lambda : E \to \Sigma$ (with $(i, s, j) \mapsto s$) is the labeling function.

An edge (i, s, j) is in E if and only if:

$$s = \begin{cases} a, & \text{for } 1 \leq i \leq m,\ j = i+1 \\ b, & \text{for } m+2 \leq i \leq 2m,\ j = i+1, \text{ and } i = 2m+1, j = 1 \\ s, & \text{for } i = m+1, m+2,\ j = m+2,\ s \in \Sigma \setminus \{a,b\} \end{cases}$$

Then the adjacency matrix for $\mathcal{A}$ is a $(2m+1) \times (2m+1)$ matrix with ijth entry equal to the number of edges from vertex i to vertex j. Then the characteristic polynomial can be computed to be $\det(A-\mu I) = (-\mu)^{2m}(p-2-\mu)+(-1)^{2m}(p-2)$. The eigen values are solutions of the equation $\mu^{2m}(p-2) - \mu^{2m+1} + p - 2 = 0$ which gives $p-2 = \mu - \frac{\mu}{\mu^{2m}+1}$. The largest real value for μ is positive. Hence $0 < \frac{\mu}{\mu^{2m}+1} < 1$, i.e., $p-2 < \mu < p-1$.□

In the case of the DNA alphabet, $p = 4$ and for $m = 1$ the above characteristic equation becomes $\mu^3 - 2\mu^2 - 2 = 0$. The largest real value of μ is approximately 2.3593 which means that the entropy of X^* is greater than $\log 2$.

Example 1. Consider the DNA alphabet Δ with $\theta = \rho\nu$. Let m=2 and choose A and T such that $X \subseteq \bigcup_{i=1}^{n} A^2\{G,C\}^i T^2$. Then X and so X^* is θ-free.

Proposition 33 *Choose distinct $a, b, c \in \Sigma$ such that $\theta(c) \neq a, b$ $\theta(a) \neq a$ and $\theta(b) \neq b$. Let $X = \bigcup_{i=1}^{\infty} a^m(\Sigma^{m-1}c)^i b^m$ for some $m \geq 2$.*

Then X, and so X^* is strictly θ-free. The entropy of X^* is such that $\log(p^{\frac{m-1}{m}}) < \hbar(X^*) < \log((p^{m-1}+1)^{\frac{1}{m}})$.

Proof. (Sketch) Suppose there are $x, x_1, x_2 \in X$ such that $s\theta(x)t = x_1x_2$ for some $s, t \in \Sigma^+$. Let $x = a^m s_1 c s_2 c..s_k c b^m$ then $\theta(x)$ is either $\theta(a^m)\theta(s_1c...s_kc)\theta(b^m)$ or $\theta(b^m)\theta(s_1c...s_kc)\theta(a^m)$ which cannot be a proper subword of x_1x_2 for any $x_1, x_2 \in X$. Hence X is θ-free.

Let $\mathcal{A} = (V, E, \lambda)$ be the automaton that recognizes X^* where $V = \{1, ..., 3m\}$ is the set of vertices, $E \subseteq V \times \Sigma \times V$ is the set of edges and $\lambda : E \to \Sigma$ (with $(i, s, j) \mapsto s$) is the labeling function.

An edge (i, s, j) is in E if and only if:

$$s = \begin{cases} a, & \text{for } 1 \leq i \leq m,\ j = i+1 \\ b, & \text{for } 2m+1 \leq i \leq 3m-1,\ j = i+1, \text{ and } i = 3m, j = i+1 \\ c, & \text{for } i = 2m,\ j = m+1, \text{ and } i = 2m,\ j = 2m+1 \\ t, & \text{for } m+1 \leq i \leq 2m-1\ j = i+1\ t \in \Sigma \end{cases}$$

Note that this automaton is not deterministic, but it has a delay 1, hence the entropy of X^* can be obtained form its adjacency matrix. Let A be the adjacency matrix of this automaton. The characteristic equation for A is $-(\mu)^{3m} + (\mu)^{2m}p^{m-1} + p^{m-1} = 0$. This implies $p^{m-1} = \frac{\mu^{3m}}{\mu^{2m}+1} = \mu^m - \frac{\mu^m}{\mu^{2m}+1}$. Since p is an integer and $0 < \frac{\mu^m}{\mu^{2m}+1} < 1$, we have $p^{\frac{m-1}{m}} < \mu$.□

For the DNA alphabet, $p = 4$ and for $m = 2$ the above characteristic equation becomes $\mu^6 - 4\mu^4 - 4 = 0$. Solving for μ, the largest real value of μ is 2.055278539. Hence the entropy of X^* is greater than $\log 2$.

Example 2. Consider Δ and $\theta = \rho\nu$ and let $m = 2$, $a = A, c = C, b = G$. Then $X = \bigcup_{i=1}^{\infty} AA(\Delta C)^i GG$ and X^* are strictly θ-free.

With the following propositions we consider ways to generate θ-subword compliant and θ-k-codes.

Proposition 34 *Choose $a, b \in \Sigma$ such that $\theta(a) = b$.*
Let $X = \cup_{i=1}^{\infty} a^{k-1}((\Sigma \setminus \{a,b\})^k b)^i$ for $k > 1$.
Then X is $\theta(k,k,k)$-subword compliant. The entropy of X^* is such that $\hbar(X^*) > \log((p-2)^{\frac{k}{k+1}})$.

Proof. We use induction on i. For $i = 1$, $X = a^{k-1}B^k b$, where $B = \Sigma - \{a, b\}$. Then for every $x \in X$, $|x| = 2k$. Hence X is $\theta(k,k,k)$-subword compliant. Assume that the statement is true for $i = n$. Let $x \in X \subseteq \cup_{i=1}^{n+1} a^{k-1}((\Sigma \setminus \{a,b\})^k b)^i = \cup_{i=1}^{n+1} a^{k-1}(B^k b)^i$. Suppose $x \in \Sigma^* u \Sigma^k \theta(u) \Sigma^*$ for some $u \in \Sigma^k$. Then there are several ways for $\theta(u)$ to be a subword of x. If $\theta(u)$ is a subword of $B^k b B^k$ then u is a subword of $B^k a B^k$ which is not possible. If $\theta(u)$ is a subword of bB^k then u is a subword of either $B^k a$ or aB^k. In the first case u would not be a subword of X and in the second case $x \in \Sigma^* u \Sigma^t \theta(u) \Sigma^*$ for some $t \neq k$ which is a contradiction. If $\theta(u) \in B^k$ then $u \in B^k$ which is a contradiction to our assumption that $k > 1$. Hence X is $\theta(k,k,k)$ subword compliant.

Let $\mathcal{A} = (V, E, \lambda)$ be the automaton that recognizes X^* where $V = \{1, ..., 2k\}$ is the set of vertices, $E \subseteq V \times \Sigma \times V$ and $\lambda : E \to \Sigma$ (with $(i,s,j) \mapsto s$) is the labeling function.

An edge (i, s, j) is in E if and only if:

$$s = \begin{cases} a, & \text{for } 1 \le i \le k-1,\ j = i+1 \\ b, & \text{for } i = 2k,\ j = k, \text{ and } i = 2k, j = 1 \\ s, & \text{for } k \le i \le 2k-1,\ j = i+1\ s \in \Sigma \setminus \{a,b\} \end{cases}$$

This automaton is with delay 1. Let A be the adjacency matrix of this automaton. The characteristic equation is $(\mu)^{2k} - (p-2)^k \mu^{k-1} - (p-2)^k = 0$. Then $(p-2)^k = \mu^{k+1} - \frac{\mu^{k+1}}{\mu^{k-1}+1}$. Since $p-2$ is an integer $\mu > 1$. Hence $\mu > (p-2)^{\frac{k}{k+1}}$. □

For the DNA alphabet, $p = 4$ and for $k = 2$ the above characteristic equation becomes $\mu^4 - 4\mu - 4 = 0$. When we solve for μ we get the largest real value of μ to be 1.83508 which is greater than the golden mean (1.618), but not quite gets to $\log 2$. In this case encoding bits with symbols from Σ is not possible with X^*, although as k approaches infinity, μ approaches 2.□

Example 3. Consider Δ with $\theta = \rho\nu$. Let $k = 2$ and choose A and T so that $X = \bigcup_{i=1}^{n} A(\{G,C\}^2 T)^i$. Clearly X is $\theta(2,2,2)$-subword compliant.

Proposition 35 *Let $a, b \in \Sigma$ such that $\theta(a) = b$ and let $X = \bigcup_{i=1}^{\infty} a^{k-1}((\Sigma \setminus \{a,b\})^{k-2} b)^i$.*
Then X is $\theta(k,1,m)$-subword compliant for $k \ge 3$ and $m \ge 1$. Moreover, when θ is morphic then X is a θ-k-code. The entropy of X^* is such that $\log((p-2)^{\frac{k-2}{k-1}}) < \hbar(X^*)$.

Proof. Suppose there exists $x \in X$ such that $x = rus\theta(u)t$ for some $r, t \in \Sigma^*$, $u, \theta(u) \in \Sigma^k$ and $s \in \Sigma^m$. Let $x = a^{k-1} s_1 b s_2 b ... s_n b$ where $s_i \in (\Sigma \setminus a, b)^{k-2}$. Then the following are all possible cases for u:

1. u is a subword of $a^{k-1} s_1$,
2. u is a subword of $a s_1 b$,

3. u is a subword of s_1bs_2,
4. u is a subword of bs_ib for some $i \leq n$.

In all these cases, since $\theta(a) = b$, $\theta(u)$ is not a subword of x . Hence X is $\theta(k,1,m)$ subowrd compliant.

Let $\mathcal{A} = (V, E, \lambda)$ be the automaton that recognizes X^* where $V = \{1, ..., 2k-2\}$ is the set of vertices, $E \subseteq V \times \Sigma \times V$ and $\lambda : E \to \Sigma$ (with $(i,s,j) \mapsto s$) is the labeling function.

An edge (i,s,j) is in E if and only if:

$$s = \begin{cases} a, & \text{for } 1 \leq i \leq k-1,\ j = i+1 \\ b, & \text{for } i = 2k-2,\ j = k,\ \text{and } i = 2k-2, j = 1 \\ s, & \text{for } k \leq i \leq 2k-3,\ j = i+1\ s \in \Sigma \setminus \{a,b\} \end{cases}$$

This automaton is with delay 1.

Let A be the adjacency matrix of this automaton. The characteristic equation is $(-\mu)^{2k-2} - (p-2)^{k-2}\mu^{k-1} - (p-2)^{k-2} = 0$. So $(p-2)^{k-2} = \mu^{k-1} - \frac{\mu^{k-1}}{\mu^{k-1}+1}$. Since $0 < \frac{\mu^{k-1}}{\mu^{k-1}+1} < 1$, $(p-2)^{\frac{k-2}{k-1}} < \mu$.□

For the DNA alphabet, $p = 4$ and for $k = 3$ the above characteristic equation becomes $\mu^4 - 2\mu^2 - 2 = 0$. Solving for μ, the largest real value is 1.6528 which is greater than the golden mean (1.618), but less than 2. Again as in the previous case, the asymptotic value for μ is 2 when k approaches infinity.

Example 4. Consider Δ with $\theta = \rho\nu$ and choose $k = 3$. Then $X = \bigcup_{i=1}^{\infty} AA(\{G,C\}T)^i$ is $\theta(3,1,m)$-subword compliant for any $m \geq 1$.

As other authors have observed, note that it is easy to get θ-k-code if one of the symbols in the alphabet is completely ignored in the construction of the code X.

Proposition 36 *Assume that $\theta(a) \neq a$ for all symbols $a \in \Sigma$. Let $b, c \in \Sigma$ such that $\theta(b) = c$ and let $X = \bigcup_{i=1}^{\infty} a^{k-1}((\Sigma \setminus \{c\})^{k-2}b)^i$ for $k \geq 3$.*

Then X and X^* are a θ-k-code. The entropy of X^* is such that $log((p-1)^{\frac{k-2}{k-1}}) < \hbar(X^*)$.

Proof. The fact that X^* is a θ-k-code is straight forward, since every subword of $x \in X$ of length k is either power of a or contains the symbol b.

Let $\mathcal{A} = (V, E, \lambda)$ be the automaton that recognizes X^* where $V = \{1, ..., 2k-2\}$ is the set of vertices, $E \subseteq V \times \Sigma \times V$ and $\lambda : E \to \Sigma$ (with $(i,s,j) \mapsto s$) is the labeling function and as in the previous propositions, the edges (i,s,j) are defined such that:

$$s = \begin{cases} a, & \text{for } 1 \leq i \leq k-1,\ j = i+1 \\ b, & \text{for } i = 2k-2,\ j = k,\ \text{and } i = 2k-2, j = 1 \\ s, & \text{for } k \leq i \leq 2k-3,\ j = i+1\ s \in \Sigma \setminus \{c\} \end{cases}$$

This automaton is with delay 1.

Let A be the adjacency matrix of this automaton with the characteristic equation $\mu^{2k-2} - \mu^{k-1}(p-1)^{k-2} - (p-1)^{k-2} = 0$. This implies $(p-1)^{k-2} = \mu^{k-1} - \frac{\mu^{k-1}}{\mu^{k-1}+1}$. We are interested in the largest real value for μ. Since $\mu > 0$, we have $0 < \frac{\mu^{k-1}}{\mu^{k-1}+1} < 1$ which implies $(p-1)^{\frac{k-2}{k-1}} < \mu$.□

For the DNA alphabet, $p = 4$ and for $k = 4$ the above estimate says that $\mu > 3^{\frac{2}{3}} > 2$. Hence the entropy of X^* in this case is greater than $\log 2$.

4 Concluding Remarks

In this paper we investigated theoretical properties of languages that consist of DNA based code words. In particular we concentrated on intermolecular and intramolecular cross hybridizations that can occur as a result that a Watson-Crick complement of a (sub)word of a code word is also a (sub)word of a code word. These conditions are necessary for a design of good codes, but certainly may not be sufficient. For example, the algorithms used in the programs developed by Seeman [23], Feldkamp [10] and Ruben [22], all check for uniqueness of k-length subsequences in the code words. Unfortunately, none of the properties from Definition 21 ensures uniqueness of k-length words. Such code word properties remain to be investigated. We hope that the general methods of designing such codewords will simplify the search for "good" codes. Better characterizations of good code words that are closed under Kleene $*$ operation may provide even faster ways for designing such codewords. Although the Proposition 36 provides a rather good design of code words, the potential repetition of certain subwords is not desirable. The most challenging questions of characterizing and designing good θ-k-codes that avoids numerous repetition of subwords remains to be developed.

Our approach to the question of designing "good" DNA codes has been from the formal language theory aspect. Many issues that are involved in designing such codes have not been considered. These include (and are not limited to) the free energy conditions, melting temperature as well as Hamming distance conditions. All these remain to be challenging problems and a procedure that includes all or majority of these aspects will be desirable in practice. It may be the case that regardless of the way the codes are designed, the ultimate test for the "goodness" of the codes will be in the laboratory.

Acknowledgment
This work has been partially supported by the grant EIA-0086015 from the National Science Foundation, USA.

References

1. R.L. Adler, D. Coppersmith, M. Hassner, *Algorithms for sliding block codes -an application of symbolic dynamics to information theory*, IEEE Trans. Inform. Theory **29** (1983), 5-22.

2. M. Arita and S. Kobayashi, *DNA sequence design using templates*, New Generation Comput. **20**(3), 263–277 (2002). (Available as a sample paper at http://www.ohmsha.co.jp/ngc/index.htm.)
3. E.B. Baum, *DNA Sequences useful for computation* unpublished article, available at: http://www.neci.nj.nec.com/homepages/eric/seq.ps (1996).
4. R.S. Braich, N. Chelyapov, C. Johnson, P.W.K. Rothemund, L. Adleman, *Solution of a 20-variable 3-SAT problem on a DNA computer*, Science **296** (2002) 499-502.
5. J. Berstel, D. Perrin, *Theory of codes,* Academis Press, Inc. Orlando Florida, 1985.
6. D. Lind, B. Marcus, *An introduction to Symbolic Dynamics and Coding,* Cambridge University Press, Inc. Cambridge United Kingdom, 1999.
7. R. Deaton, J. Chen, H. Bi, M. Garzon, H. Rubin, D.F. Wood, *A PCR-based protocol for in vitro selection of non-crosshybridizing oligonucleotides*, *DNA Computing: Proceedings of the 8th International Meeting on DNA Based Computers* (M. Hagiya, A. Ohuchi editors), Springer LNCS **2568** (2003) 196-204.
8. R. Deaton et. al, *A DNA based implementation of an evolutionary search for good encodings for DNA computation, Proc. IEEE Conference on Evolutionary Computation* ICEC-97 (1997) 267-271.
9. D. Faulhammer, A. R. Cukras, R. J. Lipton, L. F.Landweber, *Molecular Computation: RNA solutions to chess problems,* Proceedings of the National Academy of Sciences, USA **97** 4 (2000) 1385-1389.
10. U. Feldkamp, S. Saghafi, H. Rauhe, *DNASequenceGenerator - A program for the construction of DNA sequences*, *DNA Computing: Proceedings of the 7th International Meeting on DNA Based Computers* (N. Jonoska, N.C. Seeman editors), Springer LNCS **2340** (2002) 23-32.
11. M. Garzon, R. Deaton, D. Reanult, *Virtual test tubes: a new methodology for computing, Proc. 7th. Int. Symposium on String Processing and Information retrieval, A Coruña, Spain.* IEEE Computing Society Press (2000) 116-121.
12. T. Head, *Relativised code properties and multi-tube DNA dictionaries* in *Finite vs. Infinite* (Ed. by C. Calude, Gh. Paun) Springer, (2000) 175-186.
13. S. Hussini, L. Kari, S. Konstantinidis, *Coding properties of DNA languages*, *DNA Computing: Proceedings of the 7th International Meeting on DNA Based Computers* (N. Jonoska, N.C. Seeman editors), Springer LNCS **2340** (2002) 57-69.
14. N. Jonoska, K. Mahalingam *Languages of DNA based code words* Preliminary Proceedings of the 9th International Meeting on DNA Based Computers, Madison, Wisconsin June 1-4, (2003) 58-68.
15. N. Jonoska, D. Kephart, K. Mahalingam, *Generating DNA code words* Congressus Numernatium **156** (2002) 99-110.
16. L. Kari, S. Konstantinidis, E. Losseva, G. Wozniak, *Sticky-free and overhang-free DNA languages* preprint.
17. M.S. Keane, *Ergodic theory an subshifts of finite type*, in *Ergodic theory, symbolic dynamics and hyperbolic spaces* (ed. T. edford, et.al.) Oxford Univ. Press, Oxford 1991, pp.35-70.
18. Z. Li, *Construct DNA code words using backtrack algorithm*, preprint.
19. Q. Liu et al., *DNA computing on surfaces*, Nature **403** (2000) 175-179.
20. A. Marathe, A.E. Condon, R.M. Corn, *On combinatorial word design*, Preliminary Preproceedings of the 5th International Meeting on DNA Based Computers, Boston (1999) 75-88.
21. Gh. Paun, G. Rozenberg, A. Salomaa, *DNA Computing, new computing paradigms,* Springer Verlag 1998.

22. A.J. Ruben, S.J. Freeland, L.F. Landweber, *PUNCH: An evolutionary algorithm for optimizing bit set selection*, *DNA Computing: Proceedings of the 7th International Meeting on DNA Based Computers* (N. Jonoska, N.C. Seeman editors), Springer LNCS **2340** (2002) 150-160.
23. N.C. Seeman, *De Novo design of sequences for nucleic acid structural engineering* J. of Biomolecular Structure & Dynamics **8** (3) (1990) 573-581.

On the Universality of P Systems with Minimal Symport/Antiport Rules*

Lila Kari[1], Carlos Martín-Vide[2], and Andrei Păun[3]

[1] Department of Computer Science, University of Western Ontario
London, Ontario, Canada N6A 5B7
lila@csd.uwo.ca

[2] Research Group on Mathematical Linguistics
Rovira i Virgili University
Pl. Imperial Tàrraco 1, 43005 Tarragona, Spain
cmv@astor.urv.es

[3] Department of Computer Science
College of Engineering and Science
Louisiana Tech University, Ruston
P.O. Box 10348, Louisiana, LA-71272 USA
apaun@coes.latech.edu

Abstract. P systems with symport/antiport rules of a minimal size (only one object passes in any direction in a communication step) were recently proven to be computationally universal. The proof from [2] uses systems with nine membranes. In this paper we improve this results, by showing that six membranes suffice. The optimality of this result remains open (we believe that the number of membranes can be reduced by one).

1 Introduction

The present paper deals with a class of P systems which has recently received a considerable interest: the purely communicative ones, based on the biological phenomena of symport/antiport.

P systems are distributed parallel computing models which abstract from the structure and the functioning of the living cells. In short, we have a *membrane structure*, consisting of several membranes embedded in a main membrane (called the *skin*) and delimiting *regions* (Figure 1 illustrates these notions) where multisets of certain *objects* are placed. In the basic variant, the objects evolve according to given *evolution rules*, which are applied non-deterministically (the rules to be used and the objects to evolve are randomly chosen) in a maximally parallel manner (in each step, all objects which can evolve must evolve). The objects can also be communicated from one region to another one. In this way, we get *transitions* from a *configuration* of the system to the next one. A sequence of transitions constitutes a *computation*; with each *halting computation* we associate a *result*, the number of objects in an initially specified *output membrane*.

* Research supported by Natural Sciences and Engineering Research Council of Canada grants and the Canada Research Chair Program to L.K. and A.P.

N. Jonoska et al. (Eds.): Molecular Computing (Head Festschrift), LNCS 2950, pp. 254–265, 2004.

Since these computing devices were introduced ([8]) several different classes were considered. Many of them were proved to be computationally complete, able to compute all Turing computable sets of natural numbers. When membrane division, membrane creation (or string-object replication) is allowed, NP-complete problems are shown to be solved in polynomial time. Comprehensive details can be found in the monograph [9], while information about the state of the art of the domain can be found at the web address `http://psystems.disco.unimib.it`.

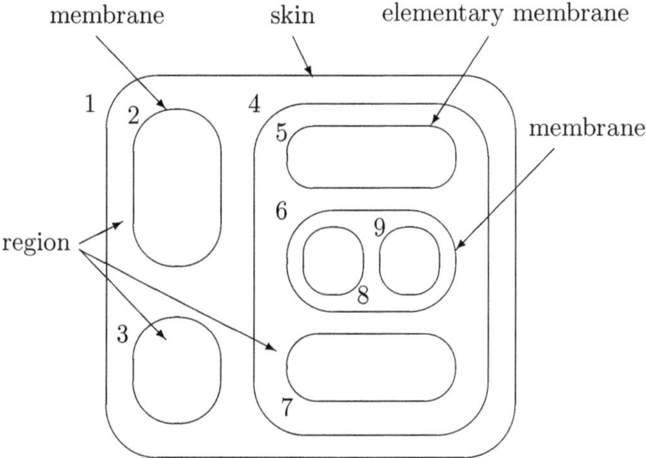

Figure 1: A membrane structure

A purely communicative variant of P systems was proposed in [7], modeling a real life phenomenon, that of membrane transport in pairs of chemicals – see [1]. When two chemicals pass through a membrane only together, in the same direction, we say that we have a process of *symport*. When the two chemicals pass only with the help of each other, but in opposite directions, the process is called *antiport*. For uniformity, when a single chemical passes through a membrane, one says that we have an *uniport*.

Technically, the rules modeling these biological phenomena and used in P systems are of the forms $(x, in), (x, out)$ (for symport), and $(x, out; y, in)$ (for antiport), where x, y are strings of symbols representing multisets of chemicals. Thus, the only used rules govern the passage of objects through membranes, the objects only change their places in the compartments of the membrane structure, they never transform/evolve.

Somewhat surprisingly, computing by communication only, in this "osmotic" manner, turned out to be computationally universal: by using only symport and antiport rules we can compute all Turing computable sets of numbers, [7]. The results from [7] were improved in several places – see, e.g., [3], [4], [6], [9] – in what concerns the number of membranes used and/or the size of symport/antiport rules.

Recently, a rather unexpected result was reported in [2]: in order to get the universality, minimal symport and antiport rules, that is of the forms $(a, in), (a, out), (a, out; b, in)$, where a, b are objects, are sufficient. The price was to use nine membranes, much more than in the results from [3] and [4], for example. The problem whether or not the number of membranes can be decreased was formulated as an open problem in [2]. We contribute here to this question, by improving the result from [2]: six membranes suffice. The proof uses the same techniques as the proofs from [2], [4]: simulating a counter automaton by a P system with minimal symport/antiport rules.

It is highly probable that our result is not optimal, but we conjecture that it cannot be significantly improved; we believe that at most one membrane can be saved.

2 Counter Automata

In this section we briefly recall the concept of counter automata, useful in the proof of our main theorem. We follow here the style of [2] and [4]. Informally speaking, a *counter automaton is a finite state machine* that has a finite number of *counters* able to store values represented by natural numbers; the machine runs a program consisting of instructions which can increase or decrease by one the contents of registers, changing at the same time the state of the automaton; starting with each counter empty, the machine performs a computation; if it reaches a terminal state, then the number stored in a specified counter is said to be generated during this computation. It is known (see, e.g., [5]) that counter automata (of various types) are computationally universal, they can generate exactly all Turing computable sets of natural numbers.

More formally, a *counter automaton* is a construct $M = (Q, F, p_0, C, c_{out}, S)$, where:

- Q is the set of the possible states,
- $F \subseteq Q$ is the set of the final states,
- $p_0 \in Q$ is the start state,
- C is the set of the counters,
- $c_{out} \in C$ is the output counter,
- S is a finite set of instructions of the following forms:
  $(p \to q, +c)$, with $p, q \in Q,\ c \in C$: add 1 to the value of the counter c and move from state p into state q;
  $(p \to q, -c)$, with $p, q \in Q,\ c \in C$: if the current value of the counter c is not zero, then subtract 1 from the value of the counter c and move from state p into state q; otherwise the computation is blocked in state p;
  $(p \to q, c = 0)$, with $p, q \in Q,\ c \in C$: if the current value of the counter c is zero, then move from state p into state q; otherwise the computation is blocked in state p.

A transition step in such a counter automaton consists in updating/checking the value of a counter according to an instruction of one of the types presented above and moving from a state to another one. Starting with the number zero stored in each counter, we say that the counter automaton computes the value n if and only if, starting from the initial state, the system reaches a final state after a finite sequence of transitions, with n being the value of the output counter c_{out} at that moment.

Without loss of generality, we may assume that in the end of the computation the automaton makes zero all the counters but the output counter; also, we may assume that there are no transitions possible that start from a final state (this is to avoid the automaton getting stuck in a final state).

As we have mentioned above, such counter automata are computationally equivalent to Turing machines, and we will make below an essential use of this result.

3 P Systems with Symport/Antiport Rules

The language theory notions we use here are standard, and can be found in any of the many monographs available, for instance, in [11].

A membrane structure is pictorially represented by a Venn diagram (like the one in Figure 1), and it will be represented here by a string of matching parentheses. For instance, the membrane structure from Figure 1 can be represented by $[_1[_2\ ]_2[_3\ ]_3[_4[_5\ ]_5[_6[_8\ ]_8[_9\ ]_9]_6[_7\ ]_7]_4]_1$.

A multiset over a set X is a mapping $M : X \longrightarrow \mathbf{N}$. Here we always use multisets over finite sets X (that is, X will be an alphabet). A multiset with a finite support can be represented by a string over X; the number of occurrences of a symbol $a \in X$ in a string $x \in X^*$ represents the multiplicity of a in the multiset represented by x. Clearly, all permutations of a string represent the same multiset, and the empty multiset is represented by the empty string, λ.

We start from the biological observation that there are many cases where two chemicals pass at the same time through a membrane, with the help of each other, either in the same direction, or in opposite directions; in the first case we say that we have a *symport*, in the second case we have an *antiport* (we refer to [1] for details).

Mathematically, we can capture the idea of symport by considering rules of the form (ab, in) and (ab, out) associated with a membrane, and stating that the objects a, b can enter, respectively, exit the membrane together. For antiport we consider rules of the form $(a, out; b, in)$, stating that a exits and at the same time b enters the membrane. Generalizing such kinds of rules, we can consider rules of the unrestricted forms (x, in), (x, out) (generalized symport) and $(x, out; y, in)$ (generalized antiport), where x, y are non-empty strings representing multisets of objects, without any restriction on the length of these strings.

Based on rules of this types, in [7] one introduces *P systems with symport/antiport* as constructs

$$\Pi = (V, \mu, w_1, \ldots, w_m, E, R_1, \ldots, R_m, i_o),$$

where:

- V is an alphabet (its elements are called *objects*);
- μ is a membrane structure consisting of m membranes, with the membranes (and hence the regions) injectively labeled with $1, 2, \ldots, m$; m is called the *degree* of Π;
- $w_i, 1 \leq i \leq m$, are strings over V representing multisets of objects associated with the regions $1, 2, \ldots, m$ of μ, present in the system at the beginning of a computation;
- $E \subseteq V$ is the set of objects which are supposed to continuously appear in the environment in arbitrarily many copies;
- $R_1, \ldots, R_m$ are finite sets of symport and antiport rules over the alphabet V associated with the membranes $1, 2, \ldots, m$ of μ;
- i_o is the label of an elementary membrane of μ (the *output membrane*).

For a symport rule (x, in) or (x, out), we say that $|x|$ is the *weight* of the rule. The *weight* of an antiport rule $(x, out; y, in)$ is $\max\{|x|, |y|\}$.

The rules from a set R_i are used with respect to membrane i as explained above. In the case of (x, in), the multiset of objects x enters the region defined by the membrane, from the surrounding region, which is the environment when the rule is associated with the skin membrane. In the case of (x, out), the objects specified by x are sent out of membrane i, into the surrounding region; in the case of the skin membrane, this is the environment. The use of a rule $(x, out; y, in)$ means expelling the objects specified by x from membrane i at the same time with bringing the objects specified by y into membrane i. The objects from E (in the environment) are supposed to appear in arbitrarily many copies; since we only move objects from a membrane to another membrane and do not create new objects in the system, we need a supply of objects in order to compute with arbitrarily large multisets. The rules are used in the non-deterministic maximally parallel manner specific to P systems with symbol objects: in each step, a maximal number of rules is used (all objects which can change the region should do it).

In this way, we obtain transitions between the configurations of the system. A configuration is described by the m-tuple of multisets of objects present in the m regions of the system, as well as the multiset of objects from $V - E$ which were sent out of the system during the computation; it is important to keep track of such objects because they appear only in a finite number of copies in the initial configuration and can enter the system again. On the other hand, it is not necessary to take care of the objects from E which leave the system because they appear in arbitrarily many copies in the environment as defined before (the environment is supposed to be inexhaustible, irrespective how many copies of an object from E are introduced into the system, still arbitrarily many remain in the environment). The initial configuration is $(w_1, \ldots, w_m, \lambda)$. A sequence of transitions is called a computation.

With any halting computation, we may associate an output represented by the number of objects from V present in membrane i_o in the halting configuration. The set of all such numbers computed by Π is denoted by $N(\Pi)$. The family of all sets $N(\Pi)$ computed by systems Π of degree at most $m \geq 1$, using symport rules of weight at most p and antiport rules of weight at most q, is denoted by $NOP_m(sym_p, anti_q)$ (we use here the notations from [9]).

Details about P systems with symport/antiport rules can be found in [9]; a complete formalization of the syntax and the semantics of these systems is provided in [10].

We recall from [3], [4] the best known results dealing with the power of P systems with symport/antiport.

Theorem 1. *$NRE = NOP_m(sym_r, anti_t)$, for $(m, r, t) \in \{(1, 1, 2), (3, 2, 0), (2, 3, 0)\}$.*

The optimality of these results is not known. In particular, it is an *open problem* whether or not also the families $NOP_m(sym_r, anti_t)$ with $(m, r, t) \in \{(2, 2, 0), (2, 2, 1)\}$ are equal to NRE.

Note that we do not have here a universality result for systems of type $(m, 1, 1)$. Recently, such a surprising result was proved in [2]:

Theorem 2. *$NRE = NOP_9(sym_1, anti_1)$.*

Thus, at the price of using nine membranes, uniport rules together with antiport rules as common in biology (one chemical exits in exchange with other chemical) suffice for obtaining the Turing computational level. The question whether or not the number of membranes can be decreased was formulated as an open problem in [2].

4 Universality with Six Membranes

We (partially) solve the problem from [2], by improving the result from Theorem 2: the number of membranes can be decreased to six – but we do not know whether this is an optimal bound or not.

Theorem 3. *$NRE = NOP_6(sym_1, anti_1)$.*

Proof. Let us consider a counter automaton $M = (Q, F, p_0, C, c_{out}, S)$ as specified in Section 2. We construct the symport/antiport P system

$$\Pi = (V, \mu, w_1, w_2, w_3, w_4, w_5, w_6, E, R_1, R_2, R_3, R_4, R_5, R_6, i_o),$$

where:

$$\begin{aligned}
V = {} & Q \cup \{c_q \mid c \in C,\ q \in Q,\ \text{and } (p \to q, +c) \in S\} \\
& \cup \{c'_q,\ d_{c,q} \mid c \in C,\ q \in Q,\ \text{and } (p \to q, -c) \in S\} \\
& \cup \{c''_q,\ d'_{c,q} \mid c \in C,\ q \in Q,\ \text{and } (p \to q, c = 0) \in S\} \\
& \cup \{a_1, a_2, a_3, a_4, b_1, b_2, i_1, i_2, i_3, i_4, i_5, h, h', h'', n_1, n_2, n_3, n_4, \#_1, \#_3\},
\end{aligned}$$

$$\mu = [_1 [_2 \, [_3 [_4 \,]_4]_3 \, [_5 [_6 \,]_6]_5 \,]_2]_1,$$

$$
\begin{aligned}
w_1 &= b_1 b_2 \#_3, \\
w_2 &= a_1 i_1 i_2 i_4 i_5 n_1 n_2 n_3 n_4 h'' \#_1, \\
w_3 &= a_2 a_3 i_3, \\
w_4 &= a_4 \#_3, \\
w_5 &= h', \\
w_6 &= \lambda, \\
E &= V - \{\#_1, b_1\}, \\
i_o &= 6, \\
R_i &= R'_i \cup R''_i \cup R'''_i, \text{ where } 1 \le i \le 6.
\end{aligned}
$$

Each computation in M will be simulated by Π in three main phases; the first phase will use rules from R'_i, $1 \le i \le 6$, R''_i contains the rules for the second phase, and R'''_i are the rules used for the third phase. These phases perform the following operations: (1) preparing the system for the simulation, (2) the actual simulation of the counter automaton, and (3) terminating the computation and moving the relevant objects into the output membrane.

We give now the rules from the sets R'_i, R''_i, R'''_i for each membrane together with explanations about their use in the simulation of the counter automaton.

Phase 1 performs the following operations: we bring in membrane 2 an arbitrary number of objects $q \in Q$ that represent the states of the automaton, then we also bring in membrane 4 an arbitrary number of objects $d_{c,q}$ and $d'_{c,q}$ that will be used in the simulation phase for simulating the rules $(p \to q, -c)$ and $(p \to q, c = 0)$, respectively. The rules used in this phase are as follows:

$$
\begin{aligned}
R'_1 &= \{(b_1, out; X, in) \mid X \in Q \cup \{d_{c,p}, d'_{c,p} \mid c \in C, p \in Q\}\} \cup \{(b_1, in)\}, \\
R'_2 &= \{(b_2, out; X, in) \mid X \in Q \cup \{d_{c,p}, d'_{c,p} \mid c \in C, p \in Q\}\} \cup \{(b_2, in)\} \\
&\quad \cup \{(a_1, out; b_1, in), (b_2, out; \#_3, in), (a_2, out; a_1, in), (a_2, out; \#_3, in)\}, \\
R'_3 &= \{(a_3, out; d, in) \mid d \in \{d_{c,p}, d'_{c,p} \mid c \in C, p \in Q\}\} \cup \{(a_3, in)\} \\
&\quad \cup \{(a_2, out; b_2, in), (a_4, out, b_1, in), (\#_3, in), (\#_3, out)\}, \\
R'_4 &= \{(a_4, out; d, in) \mid d \in \{d_{c,p}, d'_{c,p} \mid c \in C, p \in Q\}\} \cup \{(a_4, in)\} \\
&\quad \cup \{(a_4, out; b_2, in), (b_2, out; a_3, in)\}, \\
R'_5 &= \{(h', out; h'', in), (h'', out, h', in)\}, \\
R'_6 &= \emptyset.
\end{aligned}
$$

The special symbol b_1 brings from the environment the objects q, $d_{c,q}$, $d'_{c,q}$ by means of the rules $(b_1, out; X, in)$, and at the same time the symbol b_2 enters membrane 2 using the rule (b_2, in). At the next step b_1 comes back in the system, while b_2 moves the object that was introduced in membrane 1 in the previous step, q, $d_{c,q}$, or $d'_{c,q}$ into membrane 2 by means of the rules $(b_2, out; X, in)$. We can iterate these steps since we reach a configuration similar with the original configuration.

If the objects moved from the environment into membrane 2 are versions of d, then those objects are immediately moved into membrane 4 by the rules

$(a_3, out; d, in) \in R'_3$ and $(a_4, out; d, in) \in R'_4$. One can notice that in the simulation phase these special symbols that we bring in the system in this initial phase are used to simulate some specific rules from the counter automaton. A difficulty appears here because there are rules that will allow such a symbol to exit membrane 1 bringing in another such symbol (this leads to a partial simulation of rules from the counter automaton). To solve this problem we make sure that the symbols that we bring in the system do end up into one of membranes 2 or 4: if an object q, $d_{c,q}$, or $d'_{c,q}$ exits membrane 1 (using a rule from R''_1) immediately after it is brought in, then b_2 which is in membrane 2 now has to use the rule $(a_2, out; \#_3, in) \in R'_2$ and then the computation will never halt since $\#_3$ is present in membrane 2 and will move forever between membranes 2 and 3 by means of $(\#_3, in)$, $(\#_3, out)$ from R'_3.

After bringing in membrane 2 an arbitrary number of symbols q and in membrane 4 an arbitrary number of symbols $d_{c,q}, d'_{c,q}$ we pass to the second phase of the computation in Π, the actual simulation of rules from the counter automaton. Before that we have to stop the "influx" of special symbols from the environment: instead of going into environment, b_1 is interchanged with a_1 from membrane 2 by means of the rule $(a_1, out; b_1, in)$; at the same step b_2 enters the same membrane 2 by (b_2, in). Next b_2 is interchanged with a_2 by using the rule $(a_2, out; b_2, in) \in R'_3$, then in membrane 4 the same b_2 is interchanged with a_4 by means of $(a_4, out; b_2, in) \in R'_4$; simultaneously, for membrane 2 we apply $(a_2, out; a_1, in)$. At the next step we use the rules (a_4, out, b_1, in) and $(b_2, out; a_3, in)$ from R'_4.

There are two delicate points in this process. First, if b_2 instead of bringing in membrane 2 the objects from environment starts the finishing process by $(a_2, out; b_2, in) \in R'_3$, then at the next step the only rule possible is $(a_2, out; \#_3, in) \in R'_2$, since a_1 is still in membrane 2, and then the computation will never render a result. The second problem can be noticed by looking at the rules (a_4, in) and $(a_4, out; b_1, in)$ associated with membranes 4 and 3, respectively: if instead of applying the second rule in the finishing phase of this step, we apply the rule of membrane 4, then the computation stops in membranes 1 through 4, but for membrane 5 we will apply the rules from R'_5 continuously.

The second phase starts by introducing the start state of M in membrane 1, then we simulate all the rules from the counter automaton; to do this we use the following rules:

$$
\begin{aligned}
R''_1 = \{&(a_4, out; p_0, in), (\#_1, out), (\#_1, in)\} \cup \{(p, out; c_q, in), (p, out; c'_q, in), \\
&(d_{c,q}, out; q, in), (p, out; c''_q, in), (d'_{c,q}, out; q, in) \mid p, q \in Q, c \in C\}, \\
R''_2 = \{&(q, out; c_q, in), (\#_1, out; c_q, in), (n_1, out; c'_q, in), (d_{c,q}, out; n_4, in), \\
&(i_1, out; c''_q, in), (d_{c,q}, out; \#_3, in), (d'_{c,q}, out; i_5, in), (d'_{c,q}, out; \#_3, in) \mid \\
&q \in Q, c \in C\} \\
\cup \{&(a_4, out), (n_2, out; n_1, in), (n_3, out; n_2, in), (n_4, out; n_3, in), \\
&(\#_1, out; n_4, in), (i_2, out; i_1, in), (i_3, out; i_2, in), (i_4, out; i_3, in), \\
&(i_5, out; i_4, in)\},
\end{aligned}
$$

$$R''_3 = \{(c'_q, in), (d_{c,q}, out; c_\alpha, in), (i_3, out; c''_q, in), (d'_{c,q}, out; c_\alpha, in), (d'_{c,q}, out; i_3, in) \mid q, \alpha \in Q, c \in C\},$$
$$R''_4 = \{(d_{c,q}, out; c'_q, in), (\#_3, out; c'_q, in), (d'_{c,q}, out; c''_q, in), (\#_3, out; c''_q, in) \mid q \in Q, c \in C\},$$
$$R''_5 = \emptyset,$$
$$R''_6 = \emptyset.$$

We explain now the usage of these rules: we bring in the system the start state p_0 by using the rules $(a_4, out) \in R''_2$ and then $(a_4, out; p_0, in) \in R''_1$; a_4 should be in membrane 2 at the end of the step 1 if everything went well in the first phase.

We are now ready to simulate the transitions of the counter automaton.

The simulation of an instruction $(p \to q, +c)$ is done as follows. First p is exchanged with c_q by $(p, out; c_q, in) \in R''_1$, and then at the next step q is pushed in membrane 1 while c_q enters membrane 2 by means of the rule $(q, out; c_q, in) \in R''_2$. If there are no more copies of q in membrane 2, then we have to use the rule $(\#_1, out; c_q, in) \in R''_2$, which kills the computation. It is clear that the simulation is correct and can be iterated since we have again a state in membrane 1.

The simulation of an instruction $(p \to q, -c)$ is performed in the following manner. The state p is exchanged in this case with c'_q by the rule $(p, out; c'_q, in) \in R''_1$. The object c'_q is responsible of decreasing the counter c and then moving the automaton into state q. To do this c'_q will go through the membrane structure up to membrane 4 by using the rules $(n_1, out; c'_q, in) \in R''_2$, $(c'_q, in) \in R''_3$, and $(d_{a,q}, out; c'_q, in) \in R''_4$. When entering membrane 2, it starts a "timer" in membrane 1 and when entering membrane 4 it brings out the symbol $d_{c,q}$ which will perform the actual decrementing of the counter c.

The next step of the computation involves membrane 3, by means of the rule $(d_{c,q}, out; c_\alpha, in) \in R''_3$, which is effectively decreasing the content of counter c. If no more copies of c_α are present in membrane 2, then $d_{c,q}$ will sit in membrane 3 until the object n, the timer, reaches the subscript 4 and then $\#_3$ is brought in killing the computation, by means of the following rules from R''_2: $(n_2, out; n_1, in), (n_3, out; n_2, in), (n_4, out; n_3, in), (\#_1, out; n_4, in)$. If there is at least one copy of c_α in membrane 2, then we can apply $(d_{c,q}, out; n_4, in) \in R''_2$ and then we finish the simulation by bringing q in membrane 1 by means of $(d_{c,q}, out; q, in) \in R''_1$. If $d_{c,q}$ was not present in membrane 4, then $\#_3$ will be released from membrane 4 by $(\#_3, out; c'_q, in) \in R''_4$. It is clear that also these instructions are correctly simulated by our system, and also the process can be iterated.

It remains to discuss the case of rules $(p \to q, c = 0)$ from the counter automaton. The state p is replaced by c''_q by $(p, out; c''_q, in) \in R''_1$, then this symbol will start to increment the subscripts of i when entering membrane 2: $(i_1, out; c''_q, in) \in R''_2$, at the next step the subscript of i is incremented in membrane 1 and also i_3 is pushed in membrane 2 by means of $(i_3, out; c''_q, in) \in R''_3$. At the next step the special marker $d'_{c,q}$ is brought out of membrane 4 by means $(d'_{c,q}, out; c''_q, in) \in R''_4$ and the subscript of i is still incremented by $(i_3, out; i_2, in) \in R''_2$. Now $d'_{c,q}$ performs the checking for the counter c (whether it is zero or not): if there is at least one c_α present, then $d'_{c,q}$ will enter membrane 2, and at the next step will bring $\#_3$ from membrane 1 since the subscript of i did not reach position 5; on the other hand, if there are no copies of c in membrane 2, then $d'_{c,q}$ will sit unused in membrane 3 for one step until i_3 is brought from membrane 1 by $(i_4, out; i_3, in) \in R''_2$, then we apply the following rules: $(i_5, out; i_4, in) \in R''_2$ and $(d'_{c,q}, out; i_3, in) \in R''_3$. Next we can apply $(d'_{c,q}, out; i_5, in) \in R''_2$ and then in membrane 1 we finish the simulation by using $(d'_{c,q}, out; q, in) \in R''_1$. One can notice that all the symbols are in the same place as they were in the beginning of this simulation (i_3 is back in membrane 3, i_1, i_2, i_4, i_5 are in membrane 2, etc.), the only symbols moved are one copy of $d'_{c,q}$ which is now in the environment and c''_q which is in membrane 4. It is clear that we can iterate the process described above for all the types of rules in the counter automaton, so we correctly simulate the automaton.

The third phase, the finishing one, will stop the simulation and move the relevant objects into the output membrane. Specifically, when we reach a state $p \in F$ we can use the following rules:

$$
\begin{aligned}
R'''_1 &= \{(p, out; h, in) \mid p \in F\},\\
R'''_2 &= \{(h, in)\},\\
R'''_3 &= \emptyset,\\
R'''_4 &= \emptyset,\\
R'''_5 &= \{(h', out; h, in), (h'', out; h, in), (h, in)\}\\
&\quad \cup \{(h, out; c_{out\alpha}, in) \mid \alpha \in Q\},\\
R'''_6 &= \{(c_{out\alpha}, in) \mid \alpha \in Q\}.
\end{aligned}
$$

We first use $(p, out; h, in) \in R'''_1$, then h enters membrane 2 by (h, in) and at the next step h stops the oscillation of h' and h'' by putting them together in membrane 2 by means of $(h', out; h, in) \in R'''_5$ or $(h'', out; h, in) \in R'''_5$. After this h begins moving the content of output counter c_{out} into membrane 5 by using $(h, out; c_\alpha, in) \in R'''_5$. When the last c_α enters membrane 6 by using $(c_\alpha, in) \in R'''_6$ the system will be in a halting state only if a correct simulation was done in phases one and two, so the counter automaton was correctly simulated. This completes the proof.

5 Final Remarks

One can notice that membrane 6 was used only to collect the output. The same system without membrane 6 will simulate in the same way the counter automaton, but, when reaching the halt state will also contain the symbol h in the output membrane 5. This suggests that it could be possible to use a similar construct to improve the result from Theorem 3 to a result of the form:

Conjecture: $NOP_5(sym_1, anti_1) = RE$.

Obviously, P systems with minimal symport/antiport rules and using only one membrane can compute at most finite sets of numbers, at most as large as the number of objects present in the system in the initial configuration: the antiport rules do not increase the number of objects present in the system, the same with the symport rules of the form (a, out), while a symport rule of the form (a, in) should have $a \in V - E$ (otherwise the computation never stops, because the environment is inexhaustible).

The family $NOP_2(sym_1, anti_1)$ contains infinite sets of numbers. Consider, for instance, the system

$$\begin{aligned}\Pi &= (\{a, b\}, [_1 [_2\ ]_2]_1, a, \lambda, \{b\}, R_1, R_2, 2),\\ R_1 &= \{(a, out; b, in),\ (a, in)\},\\ R_2 &= \{(a, in),\ (b, in)\}.\end{aligned}$$

After bringing an arbitrary number of copies of b from the environment, the object a gets "hidden" in membrane 2, the output one.

An estimation of the size of families $NOP_m(sym_1, anti_1)$ for $m = 2, 3, 4, 5$ remains to be found.

The P systems with symport and antiport rules are interesting from several points of view: they have a precise biological inspiration, are mathematically elegant, the computation is done only by communication, by moving objects through membranes (hence the conservation law is observed), they are computationally complete. Thus, they deserve further investigations, including from the points of view mentioned above.

References

1. B. Alberts, *Essential Cell Biology. An Introduction to the Molecular Biology of the Cell*, Garland Publ. Inc., New York, London, 1998.
2. F. Bernardini, M. Gheorghe, On the Power of Minimal Symport/Antiport, *Workshop on Membrane Computing*, Tarragona, 2003.
3. R. Freund, A. Păun, Membrane Systems with Symport/Antiport: Universality Results, in *Membrane Computing. Intern. Workshop WMC-CdeA2002, Revised Papers* (Gh. Păun, G. Rozenberg, A. Salomaa, C. Zandron, eds.), *Lecture Notes in Computer Science*, 2597, Springer-Verlag, Berlin, 2003, 270–287.
4. P. Frisco, J.H. Hogeboom, Simulating Counter Automata by P Systems with Symport/Antiport, in *Membrane Computing. Intern. Workshop WMC-CdeA2002, Revised Papers* (Gh. Păun, G. Rozenberg, A. Salomaa, C. Zandron, eds.), *Lecture Notes in Computer Science*, 2597, Springer-Verlag, Berlin, 2003, 288–301.
5. J. Hopcroft, J. Ulmann, *Introduction to Automata Theory, Languages, and Computation*, Addison-Wesley, 1979.
6. C. Martín-Vide, A. Păun, Gh. Păun, On the Power of P Systems with Symport and Antiport Rules, *J. of Universal Computer Sci.*, 8, 2 (2002) 317–331.
7. A. Păun, Gh. Păun, The Power of Communication: P Systems with Symport/Antiport, *New Generation Computing*, 20, 3 (2002) 295–306.
8. Gh. Păun, Computing with Membranes, *J. of Computer and System Sciences*, 61, 1 (2000), 108–143, and *Turku Center for Computer Science-TUCS Report* No 208, 1998 (www.tucs.fi).

9. Gh. Păun, *Membrane Computing. An Introduction*, Springer-Verlag, Berlin, 2002.
10. Gh. Păun, M. Perez-Jimenez, F. Sancho-Caparrini, On the Reachability Problem for P Systems with Symport/Antiport, *Proc. Automata and Formal Languages Conf.*, Debrecen, Hungary, 2002.
11. G. Rozenberg, A. Salomaa, eds., *Handbook of Formal Languages*, 3 volumes, Springer-Verlag, Berlin, 1997.

An Algorithm for Testing Structure Freeness of Biomolecular Sequences

Satoshi Kobayashi[1], Takashi Yokomori[2], and Yasubumi Sakakibara[3]

[1] Dept. of Computer Science, Univ. of Electro-Communications
Yokohama, Japan
satoshi@cs.uec.ac.jp
[2] Dept. of Mathematics, School of Education
Waseda University, Tokyo, Japan
CREST, JST (Japan Science and Technology Corporation)
yokomori@waseda.jp
[3] Dept. of Biosciences and Informatics, Keio University, Chofu, Japan
CREST, JST (Japan Science and Technology Corporation)
yasu@bio.keio.ac.jp

Abstract. We are concerned with a problem of checking the structure freeness of S^+ for a given set S of DNA sequences. It is still open whether or not there exists an efficient algorithm for this problem. In this paper, we will give an efficient algorithm to check the structure freeness of S^+ under the constraint that every sequence may form only *linear* secondary structures, which partially solves the open problem.

1 Introduction

In the Adleman's pioneering work on the biomolecular experimental solution to the directed Hamiltonian path problem ([1]) and in many other works involving wet lab experiments performed afterward, it has been recognized to be very important how to encode information on DNA sequences, in order to guarantee the reliability of those encoded DNA sequences to avoid mishybridization. The problem of finding a good set of DNA sequences to use for computing is called the *DNA sequence design problem.* In spite of the importance of this problem, it seems that only rather recently research efforts have been paid to develop systematic methods for solving this problem. An excellent survey on this topic of DNA sequence design issues can be found in [6].

Being currently engaged in a research activity called *molecular programming project* in Japan, we are aiming as a final goal at establishing a systematic methodology for embodying desired molecular solutions within the molecular programming paradigm in which designing not only DNA sequences but also molecular reaction sequences of molecular machines are targeted as major goals ([12]). Here, by designing DNA sequences we mean a broader goal than the one mentioned above, e.g., the DNA sequence design may generally deal with the inverse folding problem, too, the problem of designing sequences so as to fold themselves into intended structural molecules.

N. Jonoska et al. (Eds.): Molecular Computing (Head Festschrift), LNCS 2950, pp. 266–277, 2004.

Taking a backward glance at the research area of DNA sequence design, there are already several works that employ some variants of Hamming distance between sequences and propose methods to minimize the similarity between sequences based on that measure ([8], [11]). Recently, Arita proposed a new design technique, called *template method* ([3]), which enables us to obtain a very large pool of sequences with uniform rate of the GC content such that any pair of the sequences is guaranteed to have a fairly large amount of mismatched distance (in the presence of the shift operation). The success of the template method is due to the use of sophisticated theory of error correcting codes. However, design methods based on the variants of Hamming distance do not take into consideration the possibility of forming secondary structures such as internal or bulge loops. Therefore, it would be very useful if we could devise a method for extracting a structure free subset from a given set of sequences, where structure freeness of the sequence set is defined as the property that the sequences in that set does not form stable secondary structures. (The notion of structure freeness will be defined precisely in Section 2.3.) In order to solve this extraction problem, it is of a crucial importance to devise an efficient algorithm to test the structure freeness of a given set of sequences. This motivated us to focus on this structure freeness test problem in this paper. There are in fact some works that took up for investigation the structure formation in the context of sequence design problem ([10], [15], [16]). An essential principle (or feature) commonly used in these papers involves the statistical thermodynamic theory of the DNA molecules to compute a hybridization error rate.

The present paper focuses on giving a necessary and sufficient condition to guarantee the global structure freeness of the whole set of sequences. More formally, we have interests in the structure freeness of S^+, where S is the set of sequences to be designed, and S^+ is the set of sequences obtained by concatenating the elements of S finitely many times. Concerning the structure freeness of S^+, Andronescu et al. proposed a method for testing whether S^m is structure-free or not, where m is a positive integer or $+$ ([2]). They gave a polynomial time algorithm for the case that m is a positive integer, but the proposed algorithm for the case of $m = +$ runs in exponential time. This leaves it open whether or not there exists an efficient algorithm for testing the structure freeness of S^+. In this paper, based on the idea of reducing the test problem to a classical shortest-path problem on a directed graph, we present an efficient algorithm for testing the structure freeness of S^+ with the condition that every sequence in S^+ may form only *linear* secondary structures, which partially solves the open problem posed in [2].

2 Preliminaries

Let Σ be an alphabet $\{\mathrm{A, C, G, T}\}$ or $\{\mathrm{A, C, G, U}\}$. A letter in Σ is also called a *base* in this paper. Furthermore, a string is regarded as a base sequence ordered from $5'$-end to $3'$-end direction. Consider a string α over Σ. By $|\alpha|$ we denote the length of α. On the other hand, for a set X, by $|X|$ we denote the cardinality of the set X. For integers i, j such that $1 \leq i \leq j \leq |\alpha|$, by $\alpha[i,j]$ we denote the substring of α starting from the ith letter and ending at the jth letter of α. In case of $i = j$, we simply write $\alpha[i]$. In case of $i > j$, $\alpha[i,j]$ represents a null string.

2.1 Secondary Structure

We will mainly follow the terminologies and notations used in [17]. Let us introduce the Watson-Crick complementarity function $\theta : \Sigma \rightarrow \Sigma$ defined by $\theta(\mathrm{A}) = \mathrm{T}$, $\theta(\mathrm{T}) = \mathrm{A}$, $\theta(\mathrm{C}) = \mathrm{G}$, and $\theta(\mathrm{G}) = \mathrm{C}$ [4]. A hydrogen bond between the ith base and jth base of α is denoted by (i, j), and called a *base pair*. A hydrogen bond (i, j) can be formed only if $\theta(\alpha[i]) = \alpha[j]$. Without loss of generality, we may always assume that $i < j$ for a base pair (i, j). A secondary structure of α is a finite set of such base pairs of α. A string α together with its secondary structure T is called a *structured string*, and written $\alpha(T)$. For representing the ith base in $\alpha(T)$, we often use the integer i.

For three bases i, j, and r in $\alpha(T)$, we say that i and j *surround* r if $i < r < j$. In case of $(i, j) \in T$, we can also say that the base pair (i, j) *surrounds* r. For two base pairs (i, j) and (p, q), we say that (i, j) *surrounds* (p, q), written $(p, q) < (i, j)$ or $(i, j) > (p, q)$, if (i, j) surrounds both p and q. In this paper, we consider only secondary structures which do not contain pseudo-knots, multiple loops, or parallel concatenation of hairpin structures. More formally, we consider only secondary structures T such that the base pairs of T can be linearly ordered with respect to the relation $<$. Such secondary structures are said to be *linear*. In the sequel, we will assume that every secondary structure is linear.

A base pair (p, q) or an unpaired base r is said to be *accessible from* a base pair (i, j), if it is surrounded by (i, j) and is not surrounded by any base pair (k, l) such that $(k, l) < (i, j)$.

For each pair $bp = (i, j) \in T$, we define a *cycle* $c(bp)$ as a substructure consisting of the pair (i, j) together with any pairs (p_1, q_1), (p_2, q_2), ... accessible from (i, j) and any unpaired bases accessible from (i, j). If $c(bp)$ contains k pairs (including the pair (i, j)), it is called a k-cycle or a *cycle of order* k. Since we consider only linear secondary structures, every cycle contained in $\alpha(T)$ is either a 1-cycle or a 2-cycle.

A cycle of order k defined by (i, j) is classified as follows: (See also Figure 1.)

1. In case of $k = 1$, it is called a *hairpin*.
2. In case that $k = 2$ and the accessible pair is $(i+1, j-1)$, it is called a *stacked pair*.
3. In case that $k = 2$ and the accessible pair is either $(i+1, p)$ or $(p, j-1)$ for some $i < p < j$, it is called a *bulge loop*.
4. In case that $k = 2$ but any condition above is not satisfied, it is called an *internal loop*.

[4] For the case of RNA strings, replace the letter T by U.

The *loop length* of a hairpin c with a base pair (i, j) is defined as the number of unpaired bases $j - i - 1$. The *loop length* of a bulge loop or an internal loop with base pairs (i, j) and (p, q) $((p, q) < (i, j))$ is also defined as the number of unpaired bases $p - i + j - q - 2$. The *loop length mismatch* of an internal loop with base pairs (i, j) and (p, q) $((p, q) < (i, j))$ is defined as $|(p - i) - (j - q)|$.

Let (i, j) be the element in T such that for any $(p, q) \in T$ with $(p, q) \neq (i, j)$, $(p, q) < (i, j)$ holds. Since T is linear, such an element is determined uniquely. Then, the substructure of $\alpha(T)$ consisting of the pair (i, j) and unpaired bases which are not surrounded by (i, j) is called a *free end structure* of $\alpha(T)$.

By $|\alpha\rangle$ we denote a hairpin consisting of a sequence α with a base pair between $\alpha[1]$ and $\alpha[|\alpha|]$. By $\boxed{\begin{matrix}\overrightarrow{\alpha}\\ \underleftarrow{\beta}\end{matrix}}$ we denote a 2-cycle consisting of two sequences α and β with base pairs between $\alpha[1]$ and $\beta[|\beta|]$ and between $\alpha[|\alpha|]$ and $\beta[1]$. By $\left.\begin{matrix}\overrightarrow{\alpha}\\ \underleftarrow{\beta}\end{matrix}\right|$ we denote a free end structure consisting of two sequences α and β with a base pair between $\alpha[|\alpha|]$ and $\beta[1]$.

2.2 Free Energy Calculation

Free energy value is assigned to each cycle or free end structure. Experimental evidence is used to determine such free energy values. The method for assigning free energy values is given as follows [5]: (See Figure 1.)

1. The free energy $E(c)$ of a hairpin c with a base pair (i, j) is dependent on the base pair (i, j), the bases $i+1, j-1$ adjacent to (i, j), and its loop length l:
$$E(c) = h_1(\alpha[i], \alpha[j], \alpha[i+1], \alpha[j-1]) + h_2(l),$$
where h_1, h_2 are experimentally obtained functions. The function h_2 is positive and there exists a constant L such that for the range $l > L$, h_2 is weakly monotonically increasing.
2. The free energy $E(c)$ of a stacked pair c with base pairs (i, j) and $(i+1, j-1)$ is dependent on the base pairs (i, j), $(i+1, j-1)$:
$$E(c) = s_1(\alpha[i], \alpha[j], \alpha[i+1], \alpha[j-1]),$$
where s_1 is an experimentally obtained function.
3. The free energy $E(c)$ of a bulge loop c with base pairs (i, j) and (p, q) $((p, q) < (i, j))$ is dependent on the base pairs (i, j), (p, q), and its loop length l:
$$E(c) = b_1(\alpha[i], \alpha[j], \alpha[p], \alpha[q]) + b_2(l),$$
where b_1, b_2 are experimentally obtained functions. The function b_2 is positive and weakly monotonically increasing.

[5] The calculation method presented here is in a general form so that it can be specialized to be equivalent to the one used in standard RNA packages such as ViennaRNA ([13]), mfold ([18]), etc.

4. The free energy $E(c)$ of an internal loop c with base pairs (i,j) and (p,q) ($(p,q) < (i,j)$) is dependent on the base pairs (i,j), (p,q), the bases $i+1, j-1$ adjacent to (i,j), the bases $p-1, q+1$ adjacent to (p,q), its loop length l, and its loop length mismatch d:

$$E(c) = e_1(\alpha[i], \alpha[j], \alpha[p], \alpha[q], \alpha[i+1], \alpha[j-1], \alpha[p-1], \alpha[q+1]) + e_2(l) + e_3(d),$$

where e_1, e_2, e_3 are experimentally obtained functions. The functions e_2, e_3 are positive and weakly monotonically increasing.
5. The free energy $E(c)$ of a free end structure c with a base pair (i,j) is dependent on the base pair (i,j) and the bases $i-1, j+1$ adjacent to (i,j):

$$E(c) = d_1(\alpha[i], \alpha[j], \alpha[i-1], \alpha[j+1]),$$

where d_1 is an experimentally obtained function.

We assume that all functions $h_1, h_2, s_1, b_1, b_2, e_1, e_2, e_3, d_1$ are computable in constant time.

Let $c_1, ..., c_k$ be the cycles contained in $\alpha(T)$, and c_0 be a free end structure of $\alpha(T)$. Then, the free energy $E(\alpha(T))$ of $\alpha(T)$ is given by:

$$E(\alpha(T)) = \Sigma_{i=0}^{k} E(c_i).$$

There exist efficient algorithms for prediction of secondary structures ([17], [18]), which is due to the fact that the free energy of a structured string is defined as the sum of free energies of cycles and free end structures contained in it. This fact also plays important roles in constructing the proposed algorithm in this paper. Furthermore, the property of energy functions that h_2, b_2, e_2, and e_3 are weakly monotonically increasing is important for the correctness of the proposed algorithm. In particular, this property plays an important role in Theorem 1.

2.3 Structure Freeness Test Problem

In this paper, we will consider the problem of testing whether a given finite set of strings of the same length n is structure free or not. The problem is formally stated as follows:

Input: A finite set S of strings of the same length n
Output: Answer "yes" if for any structured string $\alpha(T)$ such that $\alpha \in S^+$ and T is linear, $E(\alpha(T)) \geq 0$ holds, otherwise, answer "no".

We will give an efficient algorithm for solving this problem.

3 Configuration of Structured Strings

Let S be a finite set of strings of length n over Σ. Let $\alpha(T)$ be a structured string such that $\alpha \in S^+$. Let us consider a base pair $(i,j) \in T$ of $\alpha(T)$. Let

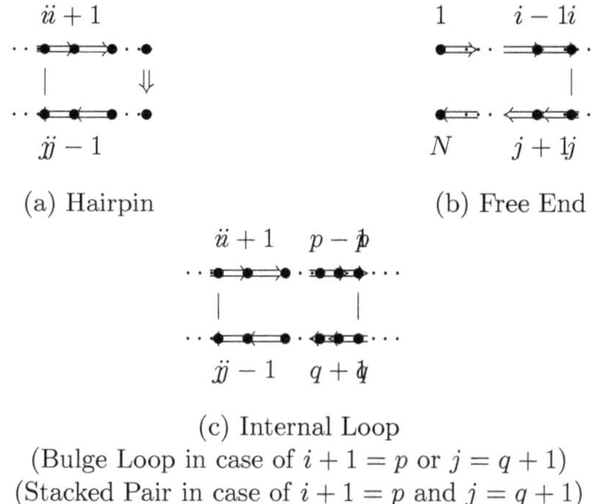

(a) Hairpin

(b) Free End

(c) Internal Loop
(Bulge Loop in case of $i+1=p$ or $j=q+1$)
(Stacked Pair in case of $i+1=p$ and $j=q+1$)

Fig. 1. Basic Secondary Structures

$\beta, \gamma \in S$. Then, (i, j) is said to have configuration (β, k, γ, l) in $\alpha(T)$ if the ith and jth bases of α correspond to the kth base of the segment β and the lth base of the segment γ, respectively. More formally, (i, j) is said to *have configuration* (β, k, γ, l) *in* $\alpha(T)$ if there exist $x, y, z, w \in S^*$ such that $\alpha = x\beta y = z\gamma w$, $|x| < i = |x| + k \leq |x\beta|$ and $|z| < j = |z| + l \leq |z\gamma|$. For a base pair bp in $\alpha(T)$, by $cf(bp)$ we denote the configuration of bp. For a configuration (β, k, γ, l), we prefer to use the following representation:

$$\begin{pmatrix} \overrightarrow{\beta} \ (k) \\ \cdot \\ \underleftarrow{\gamma} \ (l) \end{pmatrix}$$

that is graphically more appealing to the reader.

The *configuration* $cf(\alpha(T))$ *of* $\alpha(T)$, where $T = \{bp_1, ..., bp_m\}$ and $bp_1 > \cdots > bp_m$, is defined as a sequence $cf(bp_1), ..., cf(bp_m)$. For two structured strings $\alpha_1(T_1)$ and $\alpha_2(T_2)$, we write $\alpha_1(T_1) \equiv \alpha_2(T_2)$ if $cf(\alpha_1(T_1)) = cf(\alpha_2(T_2))$.

A structured string $\alpha(T)$ is said to be *E-minimal* if for any $\alpha'(T')$ such that $\alpha(T) \equiv \alpha'(T')$, $E(\alpha(T)) \leq E(\alpha'(T'))$ holds. The existence of such an E-minimal structured string is not clear at this point. But, we can show that for any configuration C, there always exists an E-minimal structured string $\alpha(T)$ such that $cf(\alpha(T)) = C$, which will be implicitly proved in the proof of (2)→(1) in Theorem 2.

A 2-cycle c with base pairs bp_1, bp_2 ($bp_2 < bp_1$) in $\alpha(T)$ is said to have boundary configuration (v_1, v_2) if $cf(bp_i) = v_i$ ($i = 1, 2$). A 1-cycle or a free end structure c with a base pair bp in $\alpha(T)$ is said to have the boundary configuration v if $cf(bp) = v$.

4 Minimum Free Energy Under Boundary Constraints

Let S be a finite set of strings of length n over Σ and consider the configurations,

$$v_1 = \begin{pmatrix} \overrightarrow{\alpha}\ (i) \\ \cdot \\ \underleftarrow{\beta}\ (j) \end{pmatrix} \quad \text{and} \quad v_2 = \begin{pmatrix} \overrightarrow{\gamma}\ (k) \\ \cdot \\ \underleftarrow{\delta}\ (l) \end{pmatrix},$$

where $\alpha, \beta, \gamma, \delta \in S$. We define $minI(v_1, v_2)$ as the minimum free energy of 2-cycles among all 2-cycles with the boundary configuration (v_1, v_2) in a structured string $x(T)$ with x ranging over S^+. By $minH(v_1)$ we denote the minimum free energy of 1-cycles among all 1-cycles with the boundary configuration v_1 in a structured string $x(T)$ with x ranging over S^+. The notation $minD(v_1)$ is defined to be the minimum free energy of free end structures among all free end structures with the boundary configuration v_1 in a structured string $x(T)$ with x ranging over S^+.

For a string $\alpha \in S$ and an integer i with $1 \leq i \leq n$, we introduce the notation $\alpha{:}i$ which indicates the ith position of the string α.

Then, for a set S, we define:

$$V(S) = \{\alpha{:}i \mid \alpha \in S,\ 1 \leq i \leq n\}.$$

Consider two elements $\alpha{:}i, \beta{:}j$ in $V(S)$.

(1) In case that either $\alpha \neq \beta$ or $i \not< j$ holds, we define:

$$W(\alpha{:}i, \beta{:}j) = \{\alpha[i,n]\gamma\beta[1,j] \mid \gamma \in S^*\}.$$

(2) In case that $\alpha = \beta$ and $i < j$ hold, we define:

$$W(\alpha{:}i, \beta{:}j) = \{\alpha[i,n]\gamma\beta[1,j] \mid \gamma \in S^*\} \cup \{\alpha[i,j]\}.$$

Theorem 1. For configurations,

$$v_1 = \begin{pmatrix} \overrightarrow{\alpha}\ (i) \\ \cdot \\ \underleftarrow{\beta}\ (j) \end{pmatrix} \text{ and } v_2 = \begin{pmatrix} \overrightarrow{\gamma}\ (k) \\ \cdot \\ \underleftarrow{\delta}\ (l) \end{pmatrix} \text{ with } \theta(\alpha[i]) = \beta[j],\ \theta(\gamma[k]) = \delta[l],$$

where $\alpha, \beta, \gamma, \delta \in S$, the following equations hold:

(1) $minH(v_1) = \min\{E(\ |x)\ ) \mid x \in W(\alpha{:}i, \beta{:}j)\}$,

(2) $minI(v_1, v_2) = \min\{E(\underleftarrow{\boxed{\begin{matrix} \overrightarrow{x} \\ y \end{matrix}}}) \mid x \in W(\alpha{:}i, \gamma{:}k),\ y \in W(\delta{:}l, \beta{:}j)\}$,

(3) $minD(v_1) = \min\{E(\begin{matrix} \overrightarrow{x\alpha[1,i]} \\ \underleftarrow{\beta[j,n]y} \end{matrix}) \mid x, y \in S^*\}$.

Furthermore, $minH(v_1)$, $minI(v_1, v_2)$, and $minD(v_1)$ are computable in $O(|S|)$ time.

Proof. (1) The set of hairpins $|x)$ such that $x \in W(\alpha\!:\!i, \beta\!:\!j)$ corresponds to the set of all hairpins which could have the boundary configuration v_1. However, since there exist infinitely many such x's, it is not clear whether the minimum of $E(\,|x)\,)$ exists or not.

Recall that n is the length of each string in S and L is the constant introduced in Section 2.2. Let $x \in W(\alpha\!:\!i, \beta\!:\!j)$ such that $|x| > L + 5n$. Since $|x| > 5n$, we can write $x = \alpha[i,n]\alpha'\gamma\beta'\beta[1,j]$ for some $\alpha', \beta' \in S^+$ and $\gamma \in S$. Then, we have that $x' = \alpha[i,n]\alpha'\beta'\beta[1,j]$ is also in $W(\alpha : i, \beta : j)$ and that $|x| > |x'| > L$. Since the values of h_1 for $|x)$ and $|x')$ are equal to each other and $h_2(|x|) \geq h_2(|x'|)$ holds (recall that h_2 is weakly monotonically increasing), we have $E(|x')) \leq E(|x))$. Therefore, we can conclude that $E(|x))$ takes the minimum value for some $x \in W(\alpha{:}i, \beta{:}j)$ with $|x| \leq L + 5n$. Thus, the first equation holds. Note that the value of $E(|x))$ depends only on the loop length and the four bases $x[1], x[2], x[|x|-1], x[|x|]$. Since, for each loop length, all possible combinations of those four bases can be computed in $O(|S|)$ time, $minH(v_1)$ can also be computed in $O(|S|)$ time.

(2) The set of 2-cycles $\left[\begin{smallmatrix}\overrightarrow{x}\\ \underleftarrow{y}\end{smallmatrix}\right]$ such that $x \in W(\alpha\!:\!i, \gamma\!:\!k)$ and $y \in W(\delta\!:\!l, \beta\!:\!j)$ corresponds to the set of all 2-cycles which could have boundary configuration (v_1, v_2). However, since there exist infinitely many such x's and y's, it is not clear whether the minimum of $E(\left[\begin{smallmatrix}\overrightarrow{x}\\ \underleftarrow{y}\end{smallmatrix}\right])$ exists or not.

Let $x \in W(\alpha\!:\!i, \gamma\!:\!k)$ and $y \in W(\delta\!:\!l, \beta\!:\!j)$. Suppose that $|x| > 6n$. Since $|x| > 6n$, we can write $x = \alpha[i,n]\alpha'\eta\gamma'\gamma[1,k]$ for some $\alpha', \gamma' \in S^+$ and $\eta \in S$. Then, we have that $x' = \alpha[i,n]\alpha'\gamma'\gamma[1,j]$ is also in $W(\alpha{:}i, \gamma{:}k)$ and that $|x| > |x'| > 5n$. If $|y| \leq 5n$, set $y' = y$. Otherwise, we can write $y = \delta[l,n]\delta'\rho\beta'\beta[1,j]$ for some $\delta', \beta' \in S^+$ and $\rho \in S$, and set $y' = \delta[l,n]\delta'\beta'\beta[1,j]$. Then, we have $y' \in W(\delta{:}l, \beta{:}j)$. Note that the loop length of $\left[\begin{smallmatrix}\overrightarrow{x}\\ \underleftarrow{y}\end{smallmatrix}\right]$ is greater than that of $\left[\begin{smallmatrix}\overrightarrow{x'}\\ \underleftarrow{y'}\end{smallmatrix}\right]$.

Further, note that the loop length mismatch of $\left[\begin{smallmatrix}\overrightarrow{x}\\ \underleftarrow{y}\end{smallmatrix}\right]$ is greater than or equal to that of $\left[\begin{smallmatrix}\overrightarrow{x'}\\ \underleftarrow{y'}\end{smallmatrix}\right]$. Since the values of s_1, b_1 and e_1 for $\left[\begin{smallmatrix}\overrightarrow{x}\\ \underleftarrow{y}\end{smallmatrix}\right]$ are equal to those for $\left[\begin{smallmatrix}\overrightarrow{x'}\\ \underleftarrow{y'}\end{smallmatrix}\right]$, respectively, we have $E(\left[\begin{smallmatrix}\overrightarrow{x'}\\ \underleftarrow{y'}\end{smallmatrix}\right]) \leq E(\left[\begin{smallmatrix}\overrightarrow{x}\\ \underleftarrow{y}\end{smallmatrix}\right])$ (recall that b_2, e_2 and e_3 are weakly monotonically increasing).

In case that $|y| > 6n$, in a similar manner, we can show the existence of $x' \in W(\alpha{:}i, \gamma{:}k)$ and $y' \in W(\delta{:}l, \beta{:}j)$ such that $E(\left[\begin{smallmatrix}\overrightarrow{x'}\\ \underleftarrow{y'}\end{smallmatrix}\right]) \leq E(\left[\begin{smallmatrix}\overrightarrow{x}\\ \underleftarrow{y}\end{smallmatrix}\right])$.

Therefore, we can conclude that $E(\left[\begin{smallmatrix}\overrightarrow{x}\\ \underleftarrow{y}\end{smallmatrix}\right])$ takes the minimum value for some $x \in W(\alpha : i, \gamma : k)$ and $y \in W(\delta : l, \beta : j)$ such that $|x| \leq 6n$ and $|y| \leq 6n$.

Thus, the second equation holds. Note that the value of $E(\begin{smallmatrix}\overrightarrow{x}\\\overleftarrow{y}\end{smallmatrix})$ depends only on the loop length, the loop length mismatch, and the eight bases $x[1], x[2], x[|x|-1], x[|x|], y[1], y[2], y[|y|-1], y[|y|]$. Since, for each loop length and loop length mismatch, all possible combinations of those eight bases can be computed in $O(|S|)$ time, $minI(v_1, v_2)$ can also be computed in $O(|S|)$ time.

(3) The set of free end structures $\begin{smallmatrix}\overrightarrow{x\alpha[1,i]}\\\overleftarrow{\beta[j,n]y}\end{smallmatrix}$ such that $x, y \in S^*$ corresponds to the set of all free end structures which could have boundary configuration v_1. It is clear that $E(\begin{smallmatrix}\overrightarrow{x\alpha[1,i]}\\\overleftarrow{\beta[j,n]y}\end{smallmatrix})$ takes the minimum value for some $x, y \in S^*$ such that $|x|, |y| \leq n$. Thus, the third equation holds. Let $s = x\alpha[1,i]$ and $t = \beta[j,n]y$. Then, note that the value of $E(\begin{smallmatrix}\overrightarrow{s}\\\overleftarrow{t}\end{smallmatrix})$ depends only on the four bases $s[|s|-1], s[|s|], t[1], t[2]$. Since for fixed lengths of s and t, all possible combinations of those four bases can be computed in $O(|S|)$ time, $minD(v_1)$ can also be computed in $O(|S|)$ time. □

5 Algorithm for Testing the Structure Freeness

For a set S of strings, we construct a weighted directed graph $G(S) = (V, E, w)$, called the *Hydrogen Bond Network graph* (HBN graph) of the set S, where V and E are defined as follows:

$$V = V' \cup \{d, h\},$$

$$V' = \{ \begin{pmatrix} \overrightarrow{\alpha}\ (i) \\ \cdot \\ \overleftarrow{\beta}\ (j) \end{pmatrix} \mid \alpha, \beta \in S,\ \theta(\alpha[i]) = \beta[j],\ 1 \leq i, j \leq n \},$$

$$E = (V' \times V') \cup (\{d\} \times V') \cup (V' \times \{h\}).$$

Furthermore, the weight function w is defined as follows:

(1) for $v_1, v_2 \in V'$, we define:

$$w((v_1, v_2)) = minI(v_1, v_2),$$

(2) for $v \in V'$, we define:

$$w((d, v)) = minD(v),$$

(3) for $v \in V'$, we define:

$$w((v, h)) = minH(v).$$

For a path p in G, by $w(p)$ we denote the sum of weights of the edges contained in p, i.e., the weight of the path p.

Theorem 2. For a finite set S of strings of the same length and a real value F, the following two statements are equivalent.

(1) There is an E-minimal structured string $\alpha(T)$ such that $\alpha \in S^+$, $T \neq \emptyset$ and $E(\alpha(T)) = F$,

(2) There is a path p from d to h of the graph $G(S)$ such that $w(p) = F$.

Proof. **(1)→(2) :** Let $\alpha(T)$ be an E-minimal structured string such that $\alpha \in S^+$, $T \neq \emptyset$. Let $cf(\alpha(T)) = v_0, ..., v_k$. Then, by c_i $(i = 1, ..., k)$ we denote a 2-cycle in $\alpha(T)$ with the boundary configuration (v_{i-1}, v_i), by c_0 we denote a free end structure in $\alpha(T)$ with the boundary configuration v_0, and by c_{k+1} we denote a hairpin in $\alpha(T)$ with the boundary configuration v_k. By the definition of the weight function w, for each $i = 1, ..., k$, we have $w((v_{i-1}, v_i)) = minI(v_{i-1}, v_i)$. By the definition of $minI(v_{i-1}, v_i)$, $E(c_i) \geq minI(v_{i-1}, v_i)$ holds for each $i = 1, ..., k$. Suppose $E(c_i) > minI(v_{i-1}, v_i)$ for some i, and let c' be a 2-cycle with the boundary configuration (v_{i-1}, v_i) such that $E(c') = minI(v_{i-1}, v_i)$. Then, by replacing c_i by c', we will obtain a new structured string $\alpha'(T')$ such that $\alpha(T) \equiv \alpha'(T')$ and $E(\alpha'(T')) < E(\alpha(T))$, which contradicts the fact that $\alpha(T)$ is E-minimal. Therefore, we have $E(c_i) = minI(v_{i-1}, v_i)$ for each $i = 1, ..., k$. Thus, for each $i = 1, ..., k$, $E(c_i) = w((v_{i-1}, v_i))$ holds. In similar ways, we have $E(c_0) = minD(v_0) = w((d, v_0))$ and $E(c_{k+1}) = minH(v_k) = w((v_k, h))$. Therefore, the weight of the path: $d \to v_0 \to \cdots \to v_k \to h$ is equivalent to $\Sigma_{i=0}^{k+1} E(c_i) = E(\alpha(T))$.

(2)→(1) : Let us consider a path $p : d \to v_0 \to \cdots \to v_k \to h$. For each $i = 1, ..., k$, let c_i be a 2-cycle with the boundary configuration (v_{i-1}, v_i) such that $E(c_i) = minI(v_{i-1}, v_i)$. Furthermore, let c_0 be a free end structure with the boundary configuration v_0 such that $E(c_0) = minD(v_0)$ and c_{k+1} be a hairpin with the boundary configuration v_k such that $E(c_{k+1}) = minH(v_k)$. Then, we can obtain a structured string $\alpha(T)$ by concatenating $c_0, c_1, ..., c_k, c_{k+1}$ in this order so that for each $i = 1, ..., k+1$, the two base pairs on the boundary between c_{i-1} and c_i could be a common *single* base pair of both c_{i-1} and c_i. Then, $E(\alpha(T))$ is equivalent to the weight $w(p)$ of the path p. Suppose that $\alpha(T)$ is not E-minimal. Then, there exists a structured string $\alpha'(T')$ such that $\alpha(T) \equiv \alpha'(T')$ and $E(\alpha'(T')) < E(\alpha(T))$. Let c'_i $(i = 1, ..., k)$ be a 2-cycle in $\alpha'(T')$ with the boundary configuration (v_{i-1}, v_i), c'_0 be a free end structure in $\alpha'(T')$ with the boundary configuration v_0, and c'_{k+1} be a hairpin in $\alpha'(T')$ with the boundary configuration v_k. By $E(\alpha'(T')) < E(\alpha(T))$, there exists some i $(0 \leq i \leq k+1)$ such that $E(c'_i) < E(c_i)$, which leads to a contradiction to the definition of either $minD$, $minI$, or $minH$. Thus, $\alpha(T)$ is E-minimal. □

By the above theorem, we can obtain the following simple algorithm to test the structure freeness of a given set S of strings of the same length:

Input: A finite set S of strings of the same length.

Output: Answer "yes" if for any structured string $\alpha(T)$ such that $\alpha \in S^+$ and T is linear, $E(\alpha(T)) \geq 0$ holds, otherwise, answer "no".

1. Construct the HBN graph $G(S)$.
2. Apply to $G(S)$ an algorithm for the single-source shortest-paths problem in which edge weights can be negative. If there exists no negative-weight path from d to h, answer "yes". Otherwise answer "no".

For the single-source shortest-paths problem in which edge weights can be negative, we can use the Bellman-Ford algorithm ([4], [9]). Given a weighted and directed graph $G = (V, E)$ with source s and weight function $w : E \to \mathbf{R}$, the Bellman-Ford algorithm returns a Boolean value indicating whether or not there exists a negative-weight cycle that is reachable from the source. If there is such a cycle, then the algorithm indicates that no shortest path exists. If there is no such cycle, then the algorithm produces the weights of the shortest paths.

Note that every vertex in $G(S)$ has a path to h. Thus, the existence of a negative-weight cycle reachable from d implies the existence of negative-weight path from d to h. So, we may answer "no" if there exists a negative-weight cycle that is reachable from the vertex d. Otherwise answer "no" if the computed weitht W of the path from d to h is negative. If W is not negative, answer "yes".

Let S be a given finite set of strings of length n. Let m be the number of strings in S. Then, the number of vertices of $G(S)$ is $O(m^2n^2)$. Therefore, the number of edges of $G(S)$ is $O(m^4n^4)$. As discussed in Theorem 1, since the time necessary for computing the weight of an edge is $O(m)$, the time necessary for the construction of $G(S)$ is $O(m^5n^4)$.

The time complexity of Bellman-Ford algorithm is $O(|V||E|)$. Consequently, the time complexity of the second step of the proposed algorithm is $O(m^6n^6)$. Therefore, the proposed algorithm runs in $O(m^6n^6)$ time in total.

6 Conclusions

In this paper, we focused on the problem of testing the structure freeness of S^+ for a given set S of sequences. We gave a partial answer to this problem, and proposed an efficient algorithm to test the structure freeness of S^+ under the constraint that every string may form only *linear* secondary structures. We are continuing our study in order to devise an efficient algorithm for solving the general problem in which sequences may form multiple loop structures.

Acknowledgements

This work is supported in part by Grant-in-Aid for Scientific Research on Priority Area No.14085205, Ministry of Education, Culture, Sports, Science and Technology, Japan. The first author is supported in part by Grant-in-Aid for Exploratory Research NO.13878054, Japan Society for the Promotion of Science. The first author is also under the Special Project Support from the President of the University of Electro-Communications.

References

1. L. Adleman, Molecular computation of solutions to combinatorial problems. *Science* **266**, pp.1021–1024, 1994.
2. M. Andronescu, D. Dees, L. Slaybaugh, Y. Zhao, A. Condon, B. Cohen, S. Skiena, Algorithms for testing that sets of DNA words concatenate without secondary structure, In *Proc. of 8th International Meeting on DNA Based Computers*, Lecture Notes in Computer Science, Vol.2568, Springer, pp.182–195, 2002.
3. M. Arita and S. Kobayashi, DNA sequence design using templates, *New Generation Computing*, **20**, pp.263–277, 2002.
4. R. Bellman, On a routing problem, *Quarterly of Applied Mathematics*, Vol.16, No.1, pp.87–90, 1958.
5. A. Ben-Dor, R. Karp, B. Schwikowski, and Z. Yakhini, Universal DNA tag systems: A combinatorial design scheme, *Proc. of the 4th Annual International Conference on Computational Molecular Biology* (RECOMB2000), pp.65–75, 2000.
6. A. Brenneman, A. E. Condon, Strand design for bio-molecular computation (Survey paper), *Theoretical Computer Science* **287** pp.39–58, 2002.
7. R. Deaton, M. Garzon, J.A. Rose, D.R. Franceschetti, R.C. Murphy and S.E. Jr. Stevens, Reliability and efficiency of a DNA based computation, *Physical Review Letter*, **80**, pp.417–420, 1998.
8. A. G. Frutos, Q. Liu, A. J. Thiel, A. M. W. Sanner, A. E. Condon, L. M. Smith and R. M. Corn, Demonstration of a word design strategy for DNA computing on surfaces, *Nucleic Acids Research*, Vol.25, No.23, pp.4748–4757, 1997.
9. L.R. Ford, Jr., and D.R. Fulkerson, *Flows in Networks*, Princeton University Press, 1962.
10. M. Garzon, R. Deaton, J.A. Rose, D.R. Franceschetti, Soft molecular computing, In *Proc. of Fifth International Meeting on DNA Based Computers*, June 14-15, MIT, pp.89–98, 1999.
11. M. Garzon, P. Neathery, R. Deaton, R.C. Murphy, D.R. Franceschetti, and S.E. Jr. Stevens, A new metric for DNA computing, In *Proc. of 2nd Annual Genetic Programming Conference*, Morgan Kaufmann, pp.472–478, 1997.
12. M. Hagiya, Towards molecular programming, in *Modeling in Molecular Biology* (ed. by G.Ciobanu), Natural Computing Series, Springer, 2003 (to appear).
13. I.L. Hofacker, W. Fontana, P.F. Stadler, L.S. Bonhoeffer, M. Tacker, P. Schuster, Fast folding and comparison of RNA secondary structures (The Vienna RNA package), *Monatshefte für Chemie*, **125**, pp.167–188, 1994.
14. S. Kobayashi, T. Kondo, M. Arita, On template method for DNA sequence design, In *Proc. of 8th International Meeting on DNA Based Computers*, Lecture Notes in Computer Science, Vol.2568, Springer, pp.205–214, 2002.
15. J.A. Rose, R. Deaton, M. Garzon, D.R. Franceschetti, S.E. Jr. Stevens, A statistical mechanical treatment of error in the annealing biostep of DNA computation, In *Proc. of GECCO-99 conference*, pp.1829–1834, 1999.
16. J.A. Rose, R. Deaton, The fidelity of annealing-ligation: a theoretical analysis, In *Proc. of 6th International Meeting on DNA Based Computers*, pp.207–221, 2000.
17. D. Sankoff, J.B. Kruskal, S. Mainville, R.J. Cedergren, Fast algorithms to determine RNA secondary structures containing multiple loops, in *Time Warps, String Edits, and Macromolecules: The Theory and Practice of Sequence Comparison*, D. Sankoff and J. Kruskal, Editors, Chapter 3, pp.93–120, 1983.
18. M. Zuker, P. Steigler, Optimal computer folding of large RNA sequences using thermodynamics and auxiliary information, *Nucleic Acids Research*, **9**, pp.133–148, 1981.

On Languages of Cyclic Words

Manfred Kudlek

Fachbereich Informatik, Universität Hamburg
Vogt-Kölln-Str. 30, D-22527 Hamburg, Germany
kudlek@informatik.uni-hamburg.de

Abstract. Languages of cyclic words and their relation to classical word languages are considered. To this aim, an associative and commutative operation on cyclic words is introduced.

1 Introduction

Related to DNA and splicing operations (see [1,2,4,11], etc.) it is of interest to investigate characterizations of languages of cyclic words.

Languages of *cyclic* words can be generated in various ways. One possibility is to consider languages of classical type, like *regular*, *linear*, or *context-free* languages, and then take the collection of all equivalence classes of words with respect to *cyclic permutation* from such languages. Another possibility is to consider languages of cyclic words defined by *rational*, *linear*, and *algebraic* systems of equations via least fixed point, with respect to an underlying associative operation on the monoid of equivalence classes. If the operation is also commutative, the classes of *rational*, *linear*, and *algebraic* languages coincide [10]. In the case of catenation as underlying operation for words such least fixed points give *regular*, *linear*, and *context-free* languages, respectively. A third way is to define languages of cyclic words by the algebraic closure under some (not necessarily associative) operation. A fourth way to generate languages of cyclic words is given by rewriting systems analogous to classical grammars for words, as *right linear*, *linear*, *context-free*, *monotone*, and *arbitrary* grammars [9].

For all notions not defined here we refer to [5,12].

An associative and commutative operation on cyclic words is introduced below. It is shown that the first two ways of defining languages of cyclic words do not coincide. It is also shown that the classical classes of *regular* and *context-free* languages are closed under *cyclic permutation*, but the class of *linear* languages is not.

2 Definitions

Let V be an alphabet. λ denotes the *empty* word, $|x|$ the length of a word, and $|x|_a$ the number of symbols $a \in V$ in x.

Furthermore, let **REG**, **LIN**, **CF**, **CS**, and **RE** denote the classes of *regular*, *linear*, *context-free*, *context-sensitive*, and *recursively enumerable* languages, respectively.

N. Jonoska et al. (Eds.): Molecular Computing (Head Festschrift), LNCS 2950, pp. 278–288, 2004.

Definition 1. *Define the relation* $\sim \subseteq V^* \times V^*$ *by*

$$x \sim y \Leftrightarrow x = \alpha\beta \wedge y = \beta\alpha \textit{ for some } \alpha, \beta \in V^*$$

Lemma 1. *The relation* $\sim$ *is an equivalence.*

Proof. Trivially, $\sim$ is reflexive since $x = \lambda x = x\lambda = x$, and symmetric by definition.

$\sim$ is also transitive since $x \sim y$, $y \sim z$ implies $x = \alpha\beta$, $y = \beta\alpha = \gamma\delta$, $z = \delta\gamma$.

If $|\beta| \leq |\gamma|$ then $\gamma = \beta\rho$, $\alpha = \rho\delta$, and therefore $x = \rho\delta\beta$, $z = \delta\beta\rho$. Hence $x \sim z$

If $|\beta| > |\gamma|$ then $\beta = \gamma\sigma$, $\delta = \sigma\alpha$, and therefore $x = \alpha\gamma\sigma$, $z = \sigma\alpha\gamma$. Hence $z \sim x$.

Thus $\sim$ is an equivalence relation. □

Definition 2. *Denote by* $[x]$ *the equivalence class of* x *consisting of all cyclic permutations of* x*, and by* $\mathcal{C}_V = V^*/\sim$ *the set of all equivalence classes of* $\sim$.

Definition 3. *For each cyclic word* $[x]$ *a norm may be defined in a natural way by* $||[x]|| = |x|$*. Clearly,* $||[x]||$ *is well defined since* $|\xi| = |x|$ *for all* $\xi \in [x]$*. The norm may be extended to sets of cyclic words by*

$$||A|| = max\{||[x]|| \mid [x] \in A\}.$$

It is obvious from the definition that $||\{[x]\} \circ \{[y]\}|| = ||\{[x]\}|| + ||\{[y]\}||$, and therefore $||A \circ B|| \leq ||A|| + ||B||$.

The next aim is to define an associative operation on $2^{\mathcal{C}_V}$.

Definition 4. *Define an operation* $\odot$ *on* $2^{\mathcal{C}_V}$ *as follows:*

$$\{[x]\} \odot \{[y]\} = \{[\xi\eta] \mid \xi \in [x], \eta \in [y]\}.$$

Note that $\{[\lambda]\} \odot \{[x]\} = \{[x]\} \odot \{[\lambda]\} = \{[x]\}$.

Unfortunately, we see that $\odot$ is only commutative but not associative.

Lemma 2. *The operation* $\odot$ *is a commutative but not an associative.*

Proof. $\{[x]\} \odot \{[y]\} = \{[\xi\eta] \mid \xi \in [x], \eta \in [y]\}$
$= \{[\eta\xi] \mid \xi \in [x], \eta \in [y]\} = \{[y]\} \odot \{[x]\}$.

Thus $\odot$ is commutative.

$(\{[ab]\} \odot \{[c]\}) \odot \{[d]\} = \{[abc], [acb]\} \odot \{[d]\}$
$= \{[abcd], [adbc], [abdc], [acbd], [adcb], [acdb]\}$,

$\{[ab]\} \odot (\{[c]\} \odot \{[d]\}) = \{[ab]\} \odot \{[cd]\} = \{[abcd], [acdb], [abdc], [adcb]\}$.

Thus $\odot$ is not associative. □

Another operation $\otimes$ can be defined as follows.

Definition 5. *Consider*

$$\{[x]\} \otimes \{[y]\} = \bigcup_{a \in V} \{[a\xi a\eta] \mid a\xi \in [x], a\eta \in [y]\}.$$

Note that $\{[x]\} \otimes \emptyset = \emptyset \otimes \{[x]\} = \emptyset \otimes \emptyset = \emptyset$ and
$\{[x]\} \otimes \{[\lambda]\} = \{[\lambda]\} \otimes \{[x]\} = \emptyset$.
Also this operation is commutative but not associative.

Lemma 3. *The operation $\otimes$ is a commutative but not an associative.*

Proof. Commutatitivity is obvious from the definition. But
$(\{[ac]\} \otimes \{[ab]\}) \otimes \{[bc]\} = \{[acab]\} \otimes \{[bc]\} = \{[bacabc], [cabacb]\}$,
$\{[ac]\} \otimes (\{[ab]\} \otimes \{[bc]\}) = \{[ac]\} \otimes \{[babc]\} = \{[acabcb], [cacbab]\}$,
showing that $\otimes$ is not associative. □

Thus, we define another operation $\circ$, using the shuffle operation $\sqcup\!\sqcup$, as follows.

Definition 6.

$$\{[x]\} \circ \{[y]\} = \{[\tau] \mid \tau \in \{\xi\} \sqcup\!\sqcup \{\eta\}, \xi \in [x], \eta \in [y]\} = \{[\tau] \mid \tau \in [x] \sqcup\!\sqcup [y]\}.$$

This operation may be called the *shuffle* of cyclic words.

Lemma 4. *The operation $\circ$ is a commutative and associative.*

Proof. Commutativity is obvious since $\sqcup\!\sqcup$ is a commutative operation.

$$\begin{aligned}
[\sigma] \in (\{[x]\} \circ \{[y]\}) \circ \{[z]\} &\Leftrightarrow \exists\, \tau \in [x] \sqcup\!\sqcup [y] \;\; [\sigma] \in \{[\tau]\} \circ \{[z]\} \\
&\Leftrightarrow \exists\, \tau \in [x] \sqcup\!\sqcup [y] \;\; \sigma \in [\tau] \sqcup\!\sqcup [z] \\
&\Leftrightarrow \sigma \in ([x] \sqcup\!\sqcup [y]) \sqcup\!\sqcup [z] = [x] \sqcup\!\sqcup ([y] \sqcup\!\sqcup [z]) \\
&\Leftrightarrow \exists\, \rho \in [y] \sqcup\!\sqcup [z] \;\; \sigma \in [x] \sqcup\!\sqcup [\rho] \\
&\Leftrightarrow \exists\, \rho \in [y] \sqcup\!\sqcup [z] \;\; [\sigma] \in \{[x]\} \circ \{[\rho]\} \\
&\Leftrightarrow [\sigma] \in \{[x]\} \circ (\{[y]\} \circ \{[z]\}).
\end{aligned}$$

Therefore, $\circ$ is associative. □

Consequently, the structure $\mathcal{M}_C = (2^{\mathcal{C}_V}, \circ, \{[\lambda]\})$ is a monoid, and the structure $\mathcal{S}_C = (2^{\mathcal{C}_V}, \cup, \circ, \emptyset, \{[\lambda]\})$ is an ω-complete semiring.

Thus, systems of equations may be defined [10]. Since $\circ$ is commutative, the classes of rational, linear and algebraic sets coincide. The class of such sets will be denoted by $\mathbf{RAT}(\circ)$.

Proposition 1. *Any $M \in \mathbf{RAT}(\circ)$ can be generated by a right-linear $\circ$-grammar $G = (\Delta, \Sigma, S, P)$ where Δ is a finite set of variables, $\Sigma \subset \mathcal{C}_V$ a finite set of constants, $S \in \Delta$, and P a finite set of productions of the forms $X \to Y \circ \{[x]\}$, $X \to \{[x]\}$ with $[x] \in \Sigma$, and $[x] = [\lambda]$ only for $S \to \{[\lambda]\}$ and S does not appear on any other right hand side of a production.*

Proposition 2. *(Iteration Lemma) For any $M \in \mathbf{RAT}(\circ)$ there exists a number $N \in \mathbb{N}$ such that for all $[x] \in M$ with $|||[x]||| > N$ there exist $[u], [v], [w] \in \mathcal{C}_V$ such that the following holds:*
$[x] \in \{[u]\} \circ \{[v]\} \circ \{[w]\}$, $\;|||[u] \circ [v]||| \leq N$, $\;|||[v]||| > 0$,
$\forall k \geq 0 \;\; \{[u]\} \circ \{[v]\}^k \circ \{[w]\} \subseteq M$.

Definition 7. *For any word language $L \subseteq V^*$ define the set of all equivalence classes with respect to $\sim$ by $\kappa(L) = \{[x] \mid x \in L\}$. κ is chosen for the Greek word κύκλος meaning circle.*

For any set $M \in \mathcal{C}_V$ let

$$\gamma(M) = \bigcup_{[x] \in M} [x] \;\subseteq\; V^*.$$

γ is chosen for γραμμή meaning line in Greek.

Trivially, $\kappa\gamma(M) = M$. But in general, $\gamma\kappa(L) \neq L$.
$\gamma\kappa(L)$ represents the closure of L under cyclic permutation.

Example 1. Let $V = \{a, b, c, d\}$. Then:

$$\begin{aligned}
&\{[a]\} \circ \{[b]\} = \{[ab]\},\\
&\{[a]\} \circ \{[a]\} \circ \{[b]\} = \{[aa]\} \circ \{[b]\} = \{[aab]\},\\
&\{[a]\} \circ \{[b]\} \circ \{[c]\} = \{[ab]\} \circ \{[c]\} = \{[a]\} \circ \{[bc]\} = \{[abc], [acb]\},\\
&\{[a]\} \circ \{[b]\} \circ \{[c]\} \circ \{[d]\} = \{[a]\} \circ \{[bc]\} \circ \{[d]\} = \{[ab]\} \circ \{[cd]\}\\
&\qquad = \{[abcd], [abdc], [adbc], [acbd], [acdb], [adcb]\},\\
&\{[abc]\} \circ \{[d]\} = \{[abcd], [adbc], [abdc]\},\\
&\{[ab]\} \circ \{[ab]\} = \{[aabb], [abab]\}.
\end{aligned}$$

Definition 8. *(Primitive Cyclic Words) A cyclic word x is called primitive with respect to $\circ$ iff it does not fulfill $\{[x]\} \subseteq \{[y]\}^k$ for some $y \in V^*$ and some $k > 1$, where the power is meant with respect to $\circ$.*

Note that if x is primitive with respect to catenation $\cdot$, then all $\xi \in [x]$ are primitive with respect to $\cdot$, too.

Definition 9. *Let $A \subseteq \mathcal{C}_V$. Then the $\oplus$-algebraic closure $L_\oplus(A)$ of the set A, with $\oplus \in \{\odot, \otimes, \circ\}$, is defined by*

$$\begin{aligned}
A_0 &= A,\\
A_{j+1} &= A_j \cup \bigcup_{[x],[y] \in A_j} \{[x]\} \oplus \{[y]\},\\
L_\oplus(A) &= \bigcup_{j=0}^{\infty} A_j.
\end{aligned}$$

Furthermore, if $\mathbf{X}$ is any class of sets (languages), let $\mathbf{ACL}_\oplus(\mathbf{X})$ denote the class of all $\oplus$-algebraic closures of sets $A \in \mathbf{X}$.

Lemma 5. *$\{[x]\} \circ \{[y]\} = \kappa(\gamma([x]) \sqcup\!\sqcup ([y]))$, $A \circ B = \kappa(\gamma(A) \sqcup\!\sqcup \gamma(B))$, and $\{[x_1]\} \circ \cdots \circ \{[x_k]\} = \kappa(\gamma([x_1]) \sqcup\!\sqcup \cdots \sqcup\!\sqcup \gamma([x_k]))$ for $k \geq 1$.*

Proof. The first statement is just the definition of $\circ$.

$$A \circ B = \bigcup_{[x]\in A,[y]\in B} \{[x]\} \circ \{[y]\} = \bigcup_{[x]\in A,[y]\in B} \kappa(\gamma([x]) \sqcup\!\sqcup \gamma([y]))$$
$$= \kappa(\bigcup_{[x]\in A,[y]\in B} \gamma([x]) \sqcup\!\sqcup \gamma([y])) = \kappa(\gamma(A) \sqcup\!\sqcup \gamma(B)).$$

The last statement follows by induction on k, using the second statement. □

3 Results

The first theorem presents a proof of Exercises 3.4 c in [6] or 4.2.11 in [5], respectively. (We give a proof because of technical reasons.)

Theorem 1. **REG** *is closed under cyclic permutation:* $\gamma\kappa(\mathbf{REG}) \subset \mathbf{REG}$.

Proof. Let $G = (V_N, V_T, S, P)$ be a type-3 grammar generating the language $L = L(G)$. Assume that G is in normal form, i.e. the productions are of the form $A \to aB$ or $A \to a$, and S does not occur on the right-hand side of any production. Assume also that each $x \in V_N \cup V_T$ is reachable from S.

From G construct a new type-3 grammar $G_1 = (V_{1N}, V_{1T}, S_1, P_1)$ with

$$V_{1N} = V_N \times V_N \cup \overline{V_N} \times V_N \cup \{S_1\},$$
$$V_{1T} = V_T, \text{ and productions}$$
$$P_1 = \{S_1 \to \langle B, A\rangle \mid \exists A \to aB \in P\}$$
$$\cup \{\langle B, A\rangle \to a\langle C, A\rangle \mid B \to aC \in P\} \cup \{\langle B, A\rangle \to a\langle \bar{S}, A\rangle \mid B \to a \in P\}$$
$$\cup \{\langle \bar{B}, A\rangle \to a\langle \bar{C}, A\rangle \mid B \to aC \in P\} \cup \{\langle \bar{A}, A\rangle \to a \mid A \to a \in P\}.$$

Then a derivation

$$S \Rightarrow a_0A_1 \Rightarrow \cdots \Rightarrow a_0 \cdots a_{k-1}A_k \Rightarrow a_0 \cdots a_kA_{k+1} \Rightarrow \cdots$$
$$\Rightarrow a_0 \cdots a_ka_{k+1} \cdots a_mA_{m+1} \Rightarrow a_0 \cdots a_ka_{k+1} \cdots a_ma_{m+1}$$

implies a derivation

$$S_1 \Rightarrow \langle A_{k+1}, A_k\rangle \Rightarrow a_{k+1}\langle A_{k+2}, A_k\rangle \Rightarrow a_{k+1} \cdots a_m\langle A_{m+1}, A_k\rangle$$
$$\Rightarrow a_{k+1} \cdots a_ma_{m+1}\langle \bar{S}, A_k\rangle \Rightarrow a_{k+1} \cdots a_{m+1}a_0\langle \bar{A}_1, A_k\rangle$$
$$\Rightarrow \cdots a_{k+1} \cdots a_{m+1}a_0 \cdots a_{k-1}\langle \bar{A}_k, A_k\rangle \Rightarrow a_{k+1} \cdots a_{m+1}a_0 \cdots a_k,$$

and therefore

$$S_1 \overset{*}{\Rightarrow} a_{k+1} \cdots a_{m+1}a_0 \cdots a_m.$$

On the other hand, a derivation

$$S_1 \Rightarrow \langle A_{k+1}, A_k\rangle \Rightarrow a_{k+1}\langle A_{k+2}, A_k\rangle \Rightarrow a_{k+1} \cdots a_m\langle A_{m+1}, A_k\rangle$$
$$\Rightarrow a_{k+1} \cdots a_ma_{m+1}\langle \bar{S}, A_k\rangle \Rightarrow a_{k+1} \cdots a_{m+1}a_0\langle \bar{A}_1, A_k\rangle$$
$$\Rightarrow \cdots a_{k+1} \cdots a_{m+1}a_0 \cdots a_{k-1}\langle \bar{A}_k, A_k\rangle \Rightarrow a_{k+1} \cdots a_{m+1}a_0 \cdots a_k$$

implies derivations

$$A_{k+1} \Rightarrow \cdots \Rightarrow a_{k+1} \cdots a_mA_{m+1} \Rightarrow a_{k+1} \cdots a_ma_{m+1} \text{ and}$$
$$S \Rightarrow a_0A_1 \Rightarrow \cdots a_0 \cdots a_{k-1}A_k \Rightarrow a_0 \cdots a_{k-1}a_kA_{k+1},$$

and therefore

$$S \overset{*}{\Rightarrow} a_0 \cdots a_ka_{k+1} \cdots a_{m+1}.$$

Thus, $L(G_1) = \gamma\kappa(L)$, the set of all cyclic permutations of L.

The proper inclusion follows from the fact that there are regular languages not being closed under cyclic permutation, like $L = \{a^m b^n \mid m, n \geq 0\}$. □

Theorem 2. **LIN** *is not closed under cyclic permutation, i.e.,* $\gamma\kappa(\mathbf{LIN}) \not\subset \mathbf{LIN}$.

Proof. Consider the language $L = \{a^n b^n \mid n \geq 0\} \in \mathbf{LIN}$. Assume $\gamma\kappa(L) \in \mathbf{LIN}$. Thus, $a^k b^{2k} a^k \in [a^{2k} b^{2k}]$. By the iteration lemma for linear context-free languages there exists $N = N(L) \in I\!N$ such that for $k \geq N$ there are words u, v, w, x, y such that $z = a^k b^{2k} a^k = uvwxy$ with $|uv| \leq N$, $|xy| \leq N$, $|vx| > 0$, and $\forall j \geq 0 : uv^j wx^j y \in L$. But then $uv \in \{a\}^*$, $xy \in \{a\}^*$, and only a's can be iterated, contradicting the fact that $|z|_a = |z|_b$ for all $z \in \gamma\kappa(L)$. □

The next Theorem is from [6].

Theorem 3. **CF** *is closed under cyclic permutation:* $\gamma\kappa(\mathbf{CF}) \subset \mathbf{CF}$.

Proof. $\gamma\kappa(\mathbf{CF}) \subset \mathbf{CF}$ is Exercise 6.4 c and its solution in [6], pp. 142-144. The problem is also stated as Exercise 7.3.1 c in [5].

The proper inclusion follows again from the fact that there are context-free (even regular) languages not being closed under cyclic permutation, like the example in Theorem 3.1. □

Theorem 4. $\kappa\mathbf{REG} \not\subset \mathbf{RAT}(\circ)$.

Proof. Consider $L = \{(ab)^n \mid n \geq 0\} \in \mathbf{REG}$. From this follows that $\kappa(L) = \{[(ab)^n] \mid n \geq 0\}$ with $[(ab)^n] = \{(ab)^n, (ba)^n\}$.

Now assume $\kappa(L) \in \mathbf{RAT}(\circ)$. Then, by the iteration lemma for $\mathbf{RAT}(\circ)$ there exists a $N \in I\!N$ such that for $k > N$ there exist u, v, w such that $[(ab)^k] \in \{[u]\} \circ \{[v]\} \circ \{[w]\}$ with $||\{[u]\} \circ \{[v]\}|| \leq N$, $||[v]|| > 0$, and with $\forall j \geq 0 :$ $\{[u]\} \circ \{[v]\}^j \circ \{[w]\} \subseteq \kappa(L)$.

Now $[uvw] \in \kappa(L)$ and $[uvvw] \in \kappa(L)$. Because of $||[v]|| > 0$ it follows that $v = av'$ or $v = bv'$. From $[v'a] = [av']$ and $[v'b] = [bv']$ follows that $[uv'aav'w] \in \kappa(L)$ or $[uv'bbv'w] \in \kappa(L)$, a contradiction to the structure of $\kappa(L)$. □

Theorem 5. $\gamma\mathbf{RAT}(\circ) \not\subset \mathbf{CF}$.

Proof. Consider the language M of cyclic words defined by $M = \{[a]\} \circ M \circ \{[bc]\}$. Clearly,

$$M = \bigcup_{k \geq 0} \{[a]\}^k \circ \{[bc]\}^k.$$

Now, $\{[a]\}^k = \{[a^k]\}$, and $[b^k c^k] \in \{[bc]\}^k$ for $k \geq 0$ which can be shown by induction on k, namely $[bc] \in \{[bc]\}$, and assuming $[b^k c^k] \in \{[bc]\}^k$ follows $[b^{k+1} c^{k+1}] = [b^k c^k cb] \in \{[b^k c^k]\} \circ \{[bc]\} \subseteq \{[bc]\}^{k+1}$.

From this follows that $[a^k b^k c^k] \in \{[a^k]\} \circ \{[b^k c^k]\} = \{[a]\}^k \circ \{[b^k c^k]\} \subseteq \{[a]\}^k \circ \{[bc]\}^k \subseteq M$.

Therefore, $\gamma(M) \cap \{a\}^* \{b\}^* \{c\}^* = \{a^k b^k c^k \mid k \geq 0\} \notin \mathbf{CF}$.

Thus, $\gamma(M) \notin \mathbf{CF}$. □

Theorem 6. $\mathbf{RAT}(\circ) \not\subseteq \kappa\mathbf{CF}$.

Proof. Assume the contrary, $\mathbf{RAT}(\circ) \subseteq \kappa\mathbf{CF}$. From this follows $\gamma\mathbf{RAT}(\circ) \subseteq \gamma\kappa\mathbf{CF} \subset \mathbf{CF}$, a contradiction to Theorem 5. □

Theorem 7. $\gamma\kappa\mathbf{REG} \not\subseteq \gamma\mathbf{RAT}(\circ)$.

Proof. Assume the contrary, $\gamma\kappa\mathbf{REG} \subseteq \gamma\mathbf{RAT}(\circ)$. From this follows $\kappa\mathbf{REG} = \kappa\gamma\kappa\mathbf{REG} \subseteq \kappa\gamma\mathbf{RAT}(\circ) = \mathbf{RAT}(\circ)$, a contradiction to Theorem 4 □

Theorem 8. $\mathbf{REG} \not\subseteq \gamma\kappa\mathbf{CF}$.

Proof. Consider the language $L = \{a\}\{b\}^* \in \mathbf{REG}$. $L \notin \gamma\kappa\mathbf{CF}$ is obvious since L is not closed under cyclic permutation. □

Theorem 9. $\gamma\kappa\mathbf{LIN} \subset \gamma\kappa\mathbf{CF}$.

Proof. Consider the language $L = \{a^k b^k c^m d^m e^n f^n \mid k, m, n \geq 0\} \in \mathbf{CF}$ and assume the contrary, $\gamma\kappa(L) \in \gamma\kappa\mathbf{LIN}$. Then there exists a language $L' \in \mathbf{LIN}$ such that $\gamma\kappa(L') = \gamma\kappa(L)$, and L' contains at least one cyclic permutation of $a^k b^k c^m d^m e^n f^n$ for any triple $(k, m, n) \in (I\!N \setminus \{0\})^3$. Thus $|L'| = \infty$.

Define the languages

$$L_1 = L' \cap \{a\}^* \cdot \{b\}^* \cdot \{c\}^* \cdot \{d\}^* \cdot \{e\}^* \cdot \{f\}^* \{a\}^*$$
$$\cup\, L' \cap \{b\}^* \cdot \{c\}^* \cdot \{d\}^* \cdot \{e\}^* \cdot \{f\}^* \cdot \{a\}^* \cdot \{b\}^*,$$
$$L_2 = L' \cap \{c\}^* \cdot \{d\}^* \cdot \{e\}^* \cdot \{f\}^* \cdot \{a\}^* \cdot \{b\}^* \{c\}^*$$
$$\cup\, L' \cap \{d\}^* \cdot \{e\}^* \cdot \{f\}^* \cdot \{a\}^* \cdot \{b\}^* \cdot \{c\}^* \cdot \{d\}^*,$$
$$L_3 = L' \cap \{e\}^* \cdot \{f\}^* \cdot \{a\}^* \cdot \{b\}^* \cdot \{c\}^* \cdot \{d\}^* \{e\}^*$$
$$\cup\, L' \cap \{f\}^* \cdot \{a\}^* \cdot \{b\}^* \cdot \{c\}^* \cdot \{d\}^* \cdot \{e\}^* \cdot \{f\}^*.$$

Because **LIN** is closed under intersection with regular languages, $L_1, L_2, L_3 \in \mathbf{LIN}$, and it is obvious that $L' = L_1 \cup L_2 \cup L_3$.

Now define the homomorphisms h_1, h_2, h_3 by

$h_1(a) = h_1(b) = \lambda$, $h_1(c) = c$, $h_1(d) = d$, $h_1(e) = e$, $h_1(f) = f$,
$h_2(a) = a$, $h_2(b) = b$, $h_2(c) = h_2(d) = \lambda$, $h_2(e) = e$, $h_2(f) = f$,
$h_3(a) = a$, $h_3(b) = b$, $h_3(c) = c$, $h_3(d) = d$, $h_3(e) = h_3(f) = \lambda$.

Because **LIN** is closed under morphisms, $h_1(L_1), h_2(L_2), h_3(L_3) \in \mathbf{LIN}$. Furthermore, $h_1(L_1)$ contains only words of the form $c^m d^m e^n f^n$, $h_2(L_2)$ only such of the form $e^n f^n a^k b^k$, and $h_3(L_3)$ only such of the form $a^k b^k c^m d^m$.

At least one of the languages $h_j(L_j)$, $j = 1, 2, 3$, has the property that both indices ((m, n), (k, n), or (k, m), respectively) are unbounded.

Assume the contrary, namely that one index is bounded. This implies for the languages L_j, $j = 1, 2, 3$, that at most two indices are unbounded (k, m), (k, n), (k, m), (m, n), (k, n), or (m, n), respectively). But from this follows that there is no triple (k, m, n) with all three indices unbounded, contradicting the property of L'.

Without loss of generality assume that $h_1(L_1)$ has the above property, implying that $h_1(L_1)$ contains words $c^m d^m e^n d^n$ with both indices (m, n) unbounded.

Now for $h_1(L_1)$ the iteration lemma for linear languages holds, with a constant $N > 0$. Thus there exist $m, n > N$ such that $c^m d^m e^n f^n \in h_1(L_1)$. But the iteration lemma implies $c^{m+j} d^m e^n f^{n+j} \in h_1(L_1)$ for $j \geq 0$, a contradiction to the structure of $h_1(L_1)$.

The two other cases are shown in the same way. Therefore, $\gamma\kappa(L) \notin \gamma\kappa\mathbf{LIN}$. □

Note that $\gamma\kappa(L) \in \gamma\kappa\mathbf{LIN}$ for $L = \{a^i b^i c^j d^j \mid i, j \geq 0\} \notin \mathbf{LIN}$. This is seen by considering $L' = \{b^i c^j d^j a^i \mid i, j \geq 0\}$.

Theorem 10. $\kappa\mathbf{LIN} \subset \kappa\mathbf{CF}$.

Proof. Assume $\kappa\mathbf{LIN} = \kappa\mathbf{CF}$ which implies $\gamma\kappa\mathbf{LIN} = \gamma\kappa\mathbf{CF}$, a contradiction to Theorem 9. □

The next two theorems state the closure of *context-sensitive* and *recursively enumerable* languages under *cyclic permutation*, also cited in [5].

Theorem 11. $\gamma\kappa\mathbf{CS} \subset \mathbf{CS}$.

Proof. This is obvious by adding productions to a length-increasing grammar G with $L(G) = L$, achieving all cyclic permutations of words $w \in L$. The proper inclusion follows from the fact that there exist context-sensitive languages not being closed under cyclic permutation. □

Theorem 12. $\gamma\kappa\mathbf{RE} \subset \mathbf{RE}$.

Proof. This is shown in a way analogous to Theorem 11. □

For the following note that $\sqcup\!\sqcup$ is an associative and commutative operation on words. For it, propositions analogous to Propositions 1 and 2 hold.

Lemma 6. $\gamma(\kappa(\{x\} \sqcup\!\sqcup \{y\})) = \gamma(\kappa(\{x\})) \sqcup\!\sqcup \gamma(\kappa(\{y\}))$ *for any* $x, y \in V^*$.

Proof. Any $w \in \{x\} \sqcup\!\sqcup \{y\}$ has the form $w = u_1 v_1 \cdots u_n v_n$ with $x = u_1 \cdots u_n$ and $y = v_1 \cdots v_n$ where it is also possible that $u_i = \lambda$ and $v_j = \lambda$ for some i, j. Then $z \in \gamma(\kappa(\{x\} \sqcup\!\sqcup \{y\}))$ has the form either $z = u_{i2} v_i \cdots u_n v_n u_1 v_1 \cdots u_{i1}$ or $z = v_{j2} u_{j+1} \cdots u_n v_n u_1 v_1 \cdots u_j v_{j1}$ with $u_i = u_{i1} u_{i2}$ and $v_j = v_{j1} v_{j2}$.

Then either $z = u_{i2} v_{i2} \cdots u_n v_n u_1 v_1 \cdots u_{i1} v_{i1}$ with $v_{12} = v_i$, $v_{i1} = \lambda$, and $u_{i2} \cdots u_n u_1 \cdots u_{i1} \in [x]$, $v_{i2} \cdots v_n v_1 \cdots v_{i1} \in [y]$,

or $z = u_{j2} v_{j2} u_{j+1} \cdots u_n v_n u_1 v_1 \cdots u_{j1} v_{j1}$ with $u_{j2} = \lambda$, $u_{j1} = u_j$, and $u_{j2} \cdots u_n u_1 \cdots u_{j1} \in [x]$, $v_{j2} \cdots v_n v_1 \cdots v_{j1} \in [y]$.

Therefore $z \in \gamma(\kappa(\{x\})) \sqcup\!\sqcup \gamma(\{y\}))$.

On the other hand, if $z \in \gamma(\kappa(\{x\})) \sqcup\!\sqcup \gamma(\{y\}))$, then $x = u_1 \cdots u_{i1} u_{i2} \cdots u_n$, $y = v_1 \cdots v_{i1} v_{i2} \cdots v_n$, and $z = u_{i2} v_{i2} \cdots u_n v_n u_1 v_1 \cdots u_{i1} v_{i1}$, where possibly $u_i = \lambda$ and $v_j = \lambda$ for some i, j.

But $w = u_1 v_1 \cdots u_{i1} v_{i1} u_{i2} v_{i2} \cdots u_n v_n \in \{x\} \sqcup\!\sqcup \{y\}$, and therefore it follows that $z = u_{i2} v_{i2} \cdots u_n v_n u_1 v_1 \cdots u_{i1} v_{i1} \in \gamma(\kappa(\{x\} \sqcup\!\sqcup \{y\}))$.

Hence $\gamma(\kappa(\{x\} \sqcup\!\sqcup \{y\})) = \gamma(\kappa(\{x\})) \sqcup\!\sqcup \gamma(\kappa(\{y\}))$. □

Theorem 13. $\gamma(\kappa(A \sqcup\!\sqcup B)) = \gamma(\kappa(A)) \sqcup\!\sqcup \gamma(\kappa(B))$ *for any* $A, B \subseteq V^*$.

Proof.

$$\gamma(\kappa(A \sqcup\!\sqcup B)) = \gamma(\kappa(\bigcup_{x\in A, y\in B} \{x\} \sqcup\!\sqcup \{y\})) = \bigcup_{x\in A, y\in B} \gamma(\kappa(\{x\} \sqcup\!\sqcup \{y\}))$$
$$= \bigcup_{x\in A, y\in B} (\gamma(\kappa(\{x\}) \sqcup\!\sqcup \gamma(\kappa(\{y\})) = \gamma(\kappa(A)) \sqcup\!\sqcup \gamma(\kappa(B)).$$

□

Lemma 7. $\gamma(\kappa(\{x_1\} \sqcup\!\sqcup \cdots \sqcup\!\sqcup \{x_k\})) = \gamma(\kappa(\{x_1\})) \sqcup\!\sqcup \cdots \gamma(\kappa(\{x_k\}))$ *for* $x_i \in V^*$ *and* $k \in \mathbb{N}$.

Proof. This is shown by induction on k. This is obviuos for $k = 1$. For $k = 2$ this is Lemma 6. Now, using Theorem 13,

$$\gamma(\kappa(\{x_1\} \sqcup\!\sqcup \cdots \sqcup\!\sqcup \{x_k\} \sqcup\!\sqcup \{x_{k+1}))$$
$$= \gamma(\kappa(\{x_1\} \sqcup\!\sqcup \cdots \sqcup\!\sqcup \{x_k\})) \sqcup\!\sqcup \gamma(\kappa(\{x_{k+1}))$$
$$= \gamma(\kappa(\{x_1\})) \sqcup\!\sqcup \cdots \sqcup\!\sqcup \gamma(\kappa(\{x_k\})) \sqcup\!\sqcup \gamma(\kappa(\{x_{k+1})).$$

□

Lemma 8. $\gamma\kappa\mathbf{RAT}(\sqcup\!\sqcup) \subset \mathbf{RAT}(\sqcup\!\sqcup)$.

Proof. Consider $L \in \mathbf{RAT}(\sqcup\!\sqcup)$. Then L can be generated by a right-linear $\sqcup\!\sqcup$-grammar $G = (\Delta, \Sigma, S, P)$ with productions of the form $X \rightarrow Y \sqcup\!\sqcup \{x\}$ or $X \rightarrow \{x\}$ where $x \in \Sigma$. $S \in \Delta$, and all sets Δ, $\Sigma \subseteq V^*$, and P are finite.

Therefore, for all $x \in L$ there exist $n \in \mathbb{N}$ and $x_i \in \Sigma$ such that $x \in \{x_1\} \sqcup\!\sqcup \cdots \{x_n\}$ with $x_i \in \Sigma$.

Now construct a new right-linear $\sqcup\!\sqcup$-grammar $G' = (\Delta, \Sigma', S, P')$ with
$\Sigma' = \bigcup_{x\in\Sigma}[x]$ and productions
$P' = \{X \rightarrow Y \sqcup\!\sqcup \{y\} \mid X \rightarrow Y\{x\} \in P, y \in [x]\}$
$\cup \{X \rightarrow \{y\} \mid X \rightarrow \{x\} \in P, y \in [x]\}$.

Now $x \in L$ implies $S \overset{*}{\Rightarrow} \{x_1\} \sqcup\!\sqcup \cdots \sqcup\!\sqcup \{x_n\}$ in G for some $n \in \mathbb{N}$ and $x_i \in \Sigma$. By the construction of P' follows that $S \overset{*}{\Rightarrow} \{y_1\} \sqcup\!\sqcup \cdots \sqcup\!\sqcup \{y_n\}$ in G' for all $y_i \in [x_i]$. By Lemma 7, this implies
$\gamma(\kappa(\{x_1\})) \sqcup\!\sqcup \cdots \sqcup\!\sqcup \gamma(\kappa(\{x_n\})) = \gamma(\kappa(\{x_1\} \sqcup\!\sqcup \cdots \sqcup\!\sqcup \{x_n\})) \subseteq L(G')$. Therefore $\gamma(\kappa\{x\})) \in L(G')$ for all $x \in L$, hence $\gamma(\kappa(L)) \subseteq L(G')$.

On the other hand, $z \in L(G')$ implies $z \in \{y_1\} \sqcup\!\sqcup \cdots \sqcup\!\sqcup \{y_n\}$ for some $n \in \mathbb{N}$ and $y_i \in [x_i]$, $x_i \in \Sigma$. From this follows that
$z \in \gamma(\kappa(\{x_1\})) \sqcup\!\sqcup \cdots \sqcup\!\sqcup \gamma(\kappa(\{x_n\})) = \gamma(\kappa(\{x_1\} \sqcup\!\sqcup \cdots \sqcup\!\sqcup \{x_n\}))$, hence $z \in \gamma(\kappa(L))$, implying $L(G') \subseteq \gamma(\kappa(L))$.

Therefore $L(G') = \gamma(\kappa(L))$.

Consider $L = \{ab\}^{\sqcup\!\sqcup} \in \mathbf{RAT}(\sqcup\!\sqcup)$. Trivially, $L \notin \gamma\kappa\mathbf{RAT}(\sqcup\!\sqcup)$ since L is not closed under cyclic permutation. E.g., $ab, abab \in L$, but $ba, baba \notin L$. □

Theorem 14. $\mathbf{RAT}(\sqcup\!\sqcup) \subset \mathbf{CS}$, $\kappa\mathbf{RAT}(\sqcup\!\sqcup) \subset \kappa\mathbf{CS}$.

Proof. $\mathbf{RAT}(\sqcup\!\sqcup) \subseteq \mathbf{CS}$ has been shown in [7,8], from which follows $\kappa\mathbf{RAT}(\sqcup\!\sqcup) \subseteq \kappa\mathbf{CS}$.

The proper inclusions follow from the fact that all languages in $\mathbf{RAT}(\sqcup\!\sqcup)$ are semilinear, but $\mathbf{CS}$ contains non-semilinear languages. $\square$

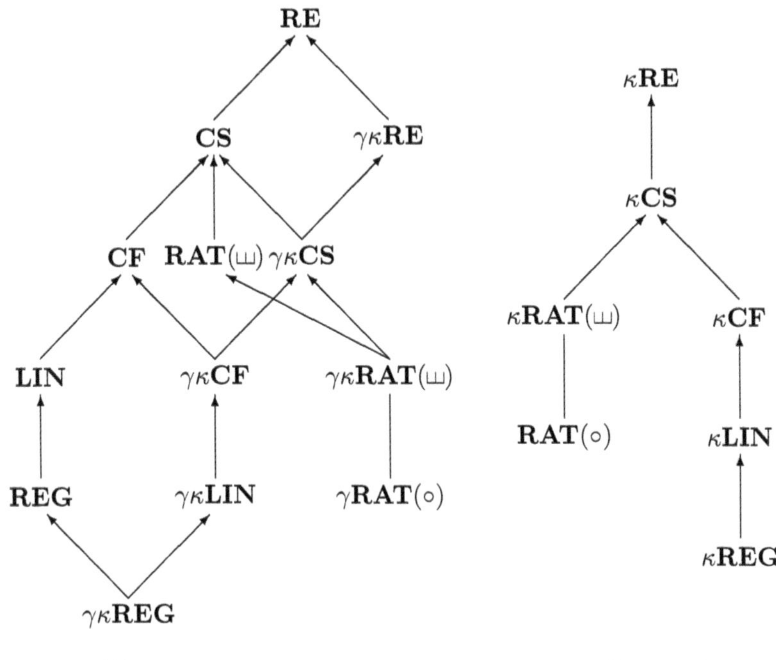

Figure 1 **Figure 2**

Theorem 15. $\mathbf{RAT}(\circ) \subsetneq \kappa\mathbf{RAT}(\sqcup\!\sqcup)$.

Proof. Consider $M \in \mathbf{RAT}(\circ)$. Then there exists a right-linear $\circ$-grammar $G = (\Delta, \Sigma, S, P)$, generating $M = L(G)$, with $\Sigma \subseteq V^*$ and productions of the form $X \rightarrow Y \circ \{[x]\}$ or $X \rightarrow \{[x]\}$, $x \in \Sigma$. Let $[z] \in M$. Then $z \in \{[x_1]\} \circ \cdots \circ \{[x_n]\}$ for some $n \in \mathbb{N}$, $x_i \in \Sigma$, and $S \overset{*}{\Rightarrow} \{[x_1]\} \circ \cdots \circ \{[x_n]\}$ in G.

Now $\{[x_1]\} \circ \cdots \circ \{[x_n]\} = \{[\tau] \mid \tau \in \{\xi_1\} \sqcup\!\sqcup \cdots \sqcup\!\sqcup \{\xi_n\}, \xi_i \in [x_i]\}$.

Construct a right-linear $\sqcup\!\sqcup$ -grammar $G' = (\Delta, \Sigma', S, P')$ with
$\Sigma' = \bigcup_{x \in \Sigma} [x]$ and productions
$P' = \{X \rightarrow Y \sqcup\!\sqcup \{\xi\} \mid \xi \in [x], X \rightarrow Y \circ \{[x]\} \in P\}$
$\quad \cup \{X \rightarrow \{\xi\} \mid \xi \in [x], X \rightarrow \{[x]\} \in P\}$.

Then $S \overset{*}{\Rightarrow} \{\xi_1\} \sqcup\!\sqcup \cdots \sqcup\!\sqcup \{\xi_n\}$ in G' iff $\xi_i \in [x_i]$ and $S \overset{*}{\Rightarrow} \{[x_1]\} \circ \cdots \circ \{[x_n]\}$ in G. Since

$\kappa(\{\xi_1\} \sqcup\!\sqcup \cdots \sqcup\!\sqcup \{\xi_n\}) = \{[\tau] \mid \tau \in \{\xi_1\} \sqcup\!\sqcup \cdots \sqcup\!\sqcup \{\xi_n\}\} \subseteq \{[x_1]\} \circ \cdots \circ \{[x_n]\}$ it follows that $\kappa(L(G')) \subseteq M$.

On the other hand, this is true for all $\xi_i \in [x_i]$. Therefore it follows that $\{[x_1]\} \circ \cdots \circ \{[x_n]\} \subseteq \kappa(L(G'))$, hence $M \subseteq \kappa(L(G'))$.

Thus $M = \kappa(L(G'))$. $\square$

Theorem 16. $\gamma\mathbf{RAT}(\circ) \subseteq \gamma\kappa\mathbf{RAT}(\sqcup\!\sqcup)$.

Proof. This follows immediately from Theorem 15. □

The relations between the language classes considered so far are shown in the diagrams of Figures 1 and 2.

4 Conclusion

An open problem is whether the inclusions $\mathbf{RAT}(\circ) \subseteq \kappa\mathbf{RAT}(\sqcup\!\sqcup)$ and $\gamma\mathbf{RAT}(\circ) \subseteq \gamma\kappa\mathbf{RAT}(\sqcup\!\sqcup)$ are proper or not. Other problems to be solved are the relation of language classes defined by grammars on cyclic words to those language classes considered above, and closure properties under certain language operations. Also the relations of language classes defined as algebraic closure under the operations $\odot$ and $\otimes$ have to be investigated. These are related to simple circular splicing operations [3]. Another open problem is to find alternative associative operations on cyclic words to generate classes of languages of cyclic words. Of interest is also the investigation of the language of primitive cyclic words and its relation to the set of primitive words.

References

1. P. Bonizzoni, C. de Felice, G. Mauri, R. Zizza: DNA and Circular Splicing. *Proc. DNA Computing (DNA'2000)*, eds. A. Condon, G. Rozenberg, LNCS 2054, pp 117–129, 2001.
2. P. Bonizzoni, C. de Felice, G. Mauri, R. Zizza: Decision Problems for Linear and Circular Splicing Systems. *Preproc. DLT'2002*, eds. M. Ito, M. Toyama, pp 7–27, Kyoto, 2002.
3. R. Ceterchi, K.G. Krithivasan: Simple Circular Splicing Systems. *Romanian J. Information Sci. and Technology*, 6, 1-2 (2003), 121–134.
4. T. Head: Splicing Schemes and DNA. In: *Lindenmayer Systems; Impacts on Theoretical Computer Science and Developmental Biology*, eds. G. Rozenberg, A. Salomaa, Springer, 1992, pp 371–383.
5. J.E. Hopcroft, R. Motwani, J.D. Ullman: *Introduction to Automata Theory, Languages, and Computation.* Addison-Wesley, 2001.
6. J.E. Hopcroft, J.D. Ullman: *Introduction to Automata Theory, Languages, and Computation.* Addison-Wesley, 1979.
7. M. Jantzen: Extending Regular Expressions with Iterated Shuffle. *Theoretical Computer Sci.*, 38, pp 223–247, 1985.
8. M. Kudlek, A. Mateescu: Mix Operation with Catenation and Shuffle. *Proceedings of the* 3rd *International Conference Developments in Language Theory*, ed. S. Bozapalidis, pp 387–398, Aristotle University Thessaloniki, 1998.
9. M. Kudlek, C. Martín-Vide, Gh. Păun: Toward FMT (Formal Macroset Theory). In: *Multiset Processing*, eds. C. Calude, Gh. Păun, G. Rozenberg, A. Salomaa, LNCS 2235, pp 123–133, 2001.
10. W. Kuich, A. Salomaa: *Semirings, Automata, Languages.* EATCS Monographs on TCS, Springer, 1986.
11. Gh. Păun, G. Rozenberg, A. Salomaa: *DNA Computing. New Computing Paradigms.* Springer, 1998.
12. A. Salomaa: *Formal Languages.* Academic Press, 1973.

A DNA Algorithm for the Hamiltonian Path Problem Using Microfluidic Systems

Lucas Ledesma*, Juan Pazos, and Alfonso Rodríguez-Patón**

Universidad Politécnica de Madrid
Facultad de Informática, Campus de Montegancedo s/n
Boadilla del Monte – 28660 Madrid, Spain
arpaton@fi.upm.es

Abstract. This paper describes the design of a linear time DNA algorithm for the Hamiltonian Path Problem (HPP) suited for parallel implementation using a microfluidic system. This bioalgorithm was inspired by the algorithm contained in [16] within the tissue P systems model. The algorithm is not based on the usual brute force generate/test technique, but builds the space solution gradually. The possible solutions/paths are built step by step by exploring the graph according to a breadth-first search so that only the paths that represent permutations of the set of vertices, and which, therefore, do not have repeated vertices (a vertex is only added to a path if this vertex is not already present) are extended. This simple distributed DNA algorithm has only two operations: concatenation (append) and sequence separation (filter). The HPP is resolved autonomously by the system, without the need for external control or manipulation. In this paper, we also note other possible bioalgorithms and the relationship of the implicit model used to solve the HPP to other abstract theoretical distributed DNA computing models (test tube distributed systems, grammar systems, parallel automata).

1 Introduction

A microfluidic device, microflow reactor or, alternatively, "lab-on-a-chip" (LOC) are different names of micro devices composed basically of microchannels and microchambers. These are passive fluidic elements, formed in the planar layer of a chip substrate, which serve only to confine liquids to small cavities. Interconnection of channels allows the formation of networks along which liquids can be transported from one location to another of the device, where controlled biochemical reactions take place in a shorter time and more accurately than in conventional laboratories. There are also 3D microfluidic systems with a limited number of layers. See [20,21] for an in-depth study of microflow devices.

* Research supported by a grant of AECI (Agencia Española de Cooperación Internacional).

** Research partially supported by Ministerio de Ciencia y Tecnología under project TIC2002-04220-C03-03, cofinanced by FEDER funds.

N. Jonoska et al. (Eds.): Molecular Computing (Head Festschrift), LNCS 2950, pp. 289–296, 2004.

The process of miniaturization and automation in analytical chemistry was an issue first addressed in the early 1980s, but the design (year 1992) of the device called "Capillary electrophoresis on a chip" [15] was the first important miniaturized chemical analysis system. Though still in its infancy, the interest in this technology has grown explosively over the last decade. The MIT Technology Review has rated to microreactors as one of the ten most promising technologies.

Microelectromechnical (MEMS) technologies are now widely applied to microflow devices for fabrication of microchannels in different substrate materials, the integration of electrical function into microfluidic devices, and the development of valves, pumps, and sensors. A thorough review of a large number of different microfluidic devices for nucleic acid analysis (for example, chemical amplification, hybridization, separation, and detection) is presented in [20].

These microfluidic devices can implement a dataflow-like architecture for the processing of DNA (see [11] and [17]) and could be a good support for the distributed biomolecular computing model called tissue P systems [16]. The underlying computational structure of tissue P systems are graphs or networks of connected processors that could be easily translated to microchambers (cells or processors) connected with microchannels.

There are several previous works on DNA computing with microfluidic systems. In one of them, Gehani and Reif [11] study the potential of microflow biomolecular computation, describe methods to efficiently route strands between chambers, and determine theoretical lower bounds on the quantities of DNA and the time needed to solve a problem in the microflow biomolecular model. Other two works [5,17] solve the Maximum Clique Problem with microfluidic devices. This is an NP-complete problem. McCaskill [17] takes a brute-force approach codifying every possible subgraph in a DNA strand. The algorithm uses the so-called Selection Transfer Modules (STM) to retain all possible cliques of the graph. The second step of McCaskill's algorithm is a sorting procedure to determine the maximum clique. By contrast, Whitesides group [5] describes a novel approach that uses neither DNA strands nor selection procedures. Subgraphs and edges of the graph are hard codified with wells and reservoirs, respectively, connected by channels and containing fluorescent beads. The readout is a measure of the fluorescence intensities associated with each subgraph. The weakness of this approach is an exponential increase in the hardware needed with the number of vertices of the graph.

2 A DNA Algorithm for the HPP

2.1 A Tissue P System for the HPP

We briefly and informally review the bioinspired computational model called tissue P systems (tP systems, for short). A detailed description can be found in [16].

A tissue P system is a network of finite-automata-like processors, dealing with multisets of symbols, according to local states (available in a finite number for each "cell"), communicating through these symbols, along channels ("axons") specified in advance. Each cell has a state from a given finite set and can process multisets of *objects*, represented by symbols from a given alphabet. The standard evolution rules are of the form $sM \rightarrow s'M'$, where s, s' are states and M, M' are multisets of symbols. We can apply such a rule only to one occurrence of M (that is, in a sequential, *minimal* way), or to all possible occurrences of M (a *parallel* way), or, moreover, we can apply a maximal package of rules of the form $sM_i \rightarrow s'M_i', 1 \leq i \leq k$, that is, involving the same states s, s', which can be applied to the current multiset (the *maximal* mode). Some of the elements of M' may be marked with the indication "go", and this means that they have to immediately leave the cell and pass to the cells to which there are direct links through synapses. This communication (transfer of symbol-objects) can be done in a *replicative* manner (the same symbol is sent to all adjacent cells), or in a *non-replicative* manner; in the second case, we can send all the symbols to only one adjacent cell, or we can distribute them non-deterministically.

One way to use such a computing device is to start from a given initial configuration and to let the system proceed until it reaches a halting configuration and to associate a result with this configuration. In these *generative* tP systems the output will be defined by sending symbols out of the system. To this end, one cell will be designated as the output cell.

Within this model, authors in [16] present a tP system (working in the maximal mode and using the replicative communication mode) to solve the Hamiltonian Path Problem (HPP). This problem involves determining whether or not, in a given directed graph $G = (V, A)$ (where $V = \{v_1, \ldots, v_n\}$ is the set of vertices, and $A \subseteq V \times V$ is the set of arcs) there is a path starting at some vertex v_{in}, ending at some vertex v_{out}, and visiting all vertices exactly once. It is known that the HPP is an NP-complete problem.

This tP system has one cell, σ_i, associated with each vertex, V_i, of the graph and the cells are communicating following the edges of the graph. The output cell is σ_n. The system works as follows (we refer to [16] for a detailed explanation). In all cells in parallel, the paths $z \in V^*$ in G grow simultaneously with the following restriction: each cell σ_i appends the vertex v_i to the path z if and only if $v_i \notin z$. The cell σ_n can work only after $n-1$ steps and a symbol is sent out of the net at step n. Thus, it is enough to watch the tP system at step n and if any symbol is sent out, then HPP has a solution; otherwise, we know that no such solution exists. (Note that the symbol z sent out describes a Hamiltonian path in G.)

2.2 The DNA Algorithm

The tP system for the HPP described above can be easily translated to a parallel and distributed DNA algorithm to be implemented in a microfluidic system. Remember we have a directed graph $G = (V, A)$ with n vertices. Our algorithm checks for the presence of Hamiltonian paths in the graph in linear time; moreover, it not only checks for a unique pair of special vertices v_{in} and v_{out}, but for all possible choices of the pair (v_{in}, v_{out}) from $V \times V$. Firstly, we describe the overall operations of the algorithm:

- *Coding*: Each vertex v_i has a short single strand associated. Each of these single strands must be carefully selected to avoid mismatch hybridizations along the run of the algorithm (see [3] for DNA words design criteria).
- In parallel, in $(n-1)$ steps and starting from each vertex of the graph the algorithm grows paths with non-repeated vertices (searching the graph breadth-first and starting from each vertex v_i).
- Time $t = n - 1$. Read the outputs of the system (empty or paths with n vertices non-repeated). If there are no output strands, then the graph has no Hamiltonian paths; otherwise, there are Hamiltonian paths (for some pair of vertices v_{in} and v_{out}).

The *hardware* of our system for a graph $G = (V, A)$ with n vertices is composed of one planar layer with $2n+1$ chambers: n filtering chambers F_i, n append/concatenation chambers C_i, and one DNA sequencing microchip for reading the Hamiltonian paths.

Chamber F_i. Inputs: DNA strands $z \in V^*$ from chambers C_j such that $(j, i) \in A$. Function: Retains the strands that contain the substrand associated with vertex v_i. Output: Strands that do not contain the substrand associated with v_i.

Chamber C_i. Inputs: DNA strands $z \in V^*$ from the output of F_i. Function: Appends to every strand z in its input the substrand associated to vertex v_i. Output: zv_i.

Pattern of connectivity (layout). The output of each chamber F_i is connected to the input of the associated chamber C_i. For all $(i, j) \in A$, there is a channel from chamber C_i to chamber F_j and there is a pump that forces the flow of liquid from C_i to F_j.

In this way, the chambers and the channels of the microfluidic system codify the vertices and edges of the graph G. The orientation of the edges is codified with the direction of the flow in the associated channel. The graph is hard codified as opposed to Adleman's soft/DNA codification. One microfluidic system of this type with $2n$ chambers can be reprogrammed (open/close microchannels) to codify the edges of any other graph with n vertices.

Implementation. It is beyond the scope of this paper to give details of the possible implementation of these microsystems. We merely indicate that filtering and append operations are widely used in DNA computing and a detailed description of microfluidic devices is given in [11,20].

Working (dynamics) of the system. We assume that filtering and append operations take one time unit and that in each time unit, the output of each F_i is available as input to each C_i.

- Step 0: ($t = 0$) (*pre-loading*). Put into each chamber F_j, enough copies of the strands associated with vertices v_i for which there exists an edge $(i, j) \in A$.
- Steps 1 to $(n-1)$: (from $t = 1$ to $t = (n-1)$).
  - *Computations:* In each time unit, for all $1 \leq i \leq n$, in parallel a filtering in all F_i and an append operation in all C_i is performed.
  - *Movement* (pumping) of strands from step t to step $t+1$. $\text{Input}(F_{j_{(t+1)}}) = \bigcup \text{Output}(C_{i_{(t)}})$ for all $(i, j) \in A$. The inlet of each chamber F_j in time $t+1$ is the union of the outlets of chambers C_i in time t such that $(i, j) \in A$.
- *Readout:* After step $(n-1)$ collects the strands in the output of chambers C_i. If there are no output strands, the graph has no Hamiltonian paths. If there are output strands in chamber C_i, there are Hamiltonian paths ending with vertex v_i; we need yet to determine the initial vertex of the strands in the output of each chamber C_i. The initial vertex of each Hamiltonian path and the exact sequence of vertices could be determined with a DNA sequencing microchip.

2.3 Example

We show an example of the execution of the algorithm for a graph with four vertices. This graph has 3 Hamiltonian paths: 2143, 2134 and 1234. Figure 1 shows the graph and the associated microfluidic device. Table 1 shows the contents of each chamber in each step. Remember that $\text{Input}(F_{j_{(t+1)}}) = \bigcup \text{Output}(C_{i_{(t)}})$ for all $(i, j) \in A$. In step 3 (time $t = 3 = n-1$), we must check if there is any output in any chamber C_i. In this case, for this graph, there are outputs in chambers C_3 and C_4 so we know there are Hamiltonian paths ending at vertices v_3 and v_4 respectively. To find out the first vertex and the exact order of the other vertices in each path, we need to sequence these strands: paths 2143, 2134, and 1234 are found.

$t = 1$	$t = 2$	$t = 3$	$t = 4$
$F_1 = \{2\}$	$F_1 = \{12\}$	$F_1 = \{-\}$	$F_1 = \{-\}$
$C_1 = \{21\}$	$C_1 = \{-\}$	$C_1 = \{-\}$	$C_1 = \{-\}$
$F_2 = \{1\}$	$F_2 = \{21\}$	$F_2 = \{-\}$	$F_2 = \{-\}$
$C_2 = \{12\}$	$C_2 = \{-\}$	$C_2 = \{-\}$	$C_2 = \{-\}$
$F_3 = \{1, 2, 4\}$	$F_3 = \{21, 12\}$	$F_3 = \{214, 134, 234\}$	$F_3 = \{2134, 1234\}$
$C_3 = \{13, 23, 43\}$	$C_3 = \{213, 123\}$	$C_3 = \{\mathbf{2143}\}$	$C_3 = \{-\}$
$F_4 = \{1, 3\}$	$F_4 = \{21, 13, 23, 43\}$	$F_4 = \{213, 123, 143\}$	$F_4 = \{2143\}$
$C_4 = \{14, 34\}$	$C_4 = \{214, 134, 234\}$	$C_4 = \{\mathbf{2134, 1234}\}$	$C_4 = \{-\}$

Table 1. Running of the algorithm. In time $t = 3$ we must readout the output of the chambers C_i. In bold, strands codifying Hamiltonian paths.

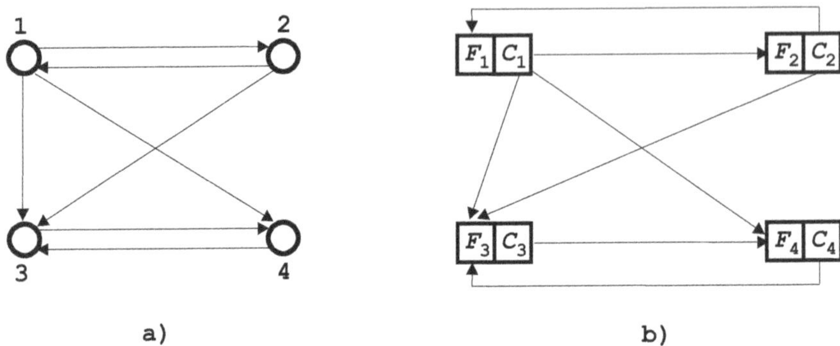

Fig. 1. a) Graph with 4 vertices and b) microfluidic system associated with this graph.

3 Final Remarks

This article describes a preliminary work on a DNA algorithm for HPP, which, like many others in the field of DNA computing (see [1,14]) has a linear execution time but an exponential consumption of DNA molecules. However, this algorithm runs autonomously in parallel without manual external intervention and does not include complex operations with high error rates (PCR, electrophoresis, enzyme restrictions, etc.). Additionally, we believe that it could be easily implemented in microfluidic systems bearing in mind the actual state of this technology. Another notable feature of the proposed bioalgorithm is that it follows a constructive problem-solving strategy (like [18] but without complex operations), pruning the unfeasible solutions (it only generates paths that do not have repeated vertices). The details of a possible implementation of this algorithm will be examined in a future paper.

However, we believe that other problem types (like sequence problems or so-called DNA2DNA computation) are more appropriate and of greater biological interest. With this in mind, we are working on designing an autonomous and distributed DNA algorithm for solving the Shortest Common Superstring (known NP-complete problem only solved in the DNA computing field in [12] with a brute-force approach) using a conditional version of the parallel overlapping assembly operation.

Another variant of the algorithm proposed here for the HPP can be used to generate in linear time all the permutations of the n integers $1, 2, \ldots, n$. Each number i is codified with a different single strand and has two associated chambers F_i and C_i. The pattern of connectivity is a complete graph, so, from every chamber C_i, there is a channel to every chamber F_j, for all $1 \leq i, j \leq n$. In $n-1$ steps, the microsystem generates DNA strands codifying all permutations. Again, our system generates only the correct strings because the algorithm fails to extend strings with the same integer in at least two locations. This problem is presented in [2] as a useful subprocedure for the DNA solution of several NP-complete problems.

This paper also poses theoretical questions, as the parallel and distributed computing model on which it relies, as well as the operations it uses—conditional concatenation or filtering plus concatenation—are already present in theoretical distributed DNA computing models, like test tubes distributed systems, grammar systems theory, parallel finite automata systems and also tissue P systems.

In particular, the redistribution of the contents of the test tubes according to specific filtering conditions (presence or absence of given subsequences) is present in the Test Tube Distributed Systems model [7,8] and, similarly, in Parallel Communicating (PC) Automata Systems [4] or [19]. However, our algorithm defines no master chamber. It also differs from PC grammar systems [6] because communication is not performed on request but is enacted automatically.

The generative power of this new system with distributed conditional concatenation remains to be examined. Only a few preliminary comments. In the HPP algorithm we use a very simple conditional concatenation: a symbol a is concatenated to a sequence w if and only if $a \notin w$. In a similar way, we could extend the restrictions to check the presence of specific suffixes, prefixes or subwords as in [9]. This distributed conditional concatenation is a combination of restricted evolution and communication (append restricted by filtering) that implicitly assumes the existence of regular/finite filters between the concatenation chambers. A similar study [10] was conducted for the test tube systems model but with context-free productions and regular filters. Looking at these results, we think our model probably needs a more powerful operation than conditional concatenation to be able to achieve computational completeness (like splicing [13] in the grammar systems area).

Microfluidic systems seem to be an interesting and promising future support for many distributed DNA computing models (shift from Adleman/Lipton manual wet test tubes to DNA computing on surfaces to microflow DNA computing), and its full potential (the underlying computational paths could be a graph with cycles that will allow, for example, the iterative construction and selection of the solutions) needs to be thoroughly examined.

References

1. L.M. Adleman. Molecular computation of solutions to combinatorial problems. *Science*, 266 (1994), 1021–1024.
2. M. Amos, A. Gibbons, and D. Hodgson. Error-resistant implementation of DNA computation. In *Proceedings of the Second Annual Meeting on DNA Based Computers*, held at Princeton University, June 10-12, 1996.
3. M. Andronescu, D. Dees, L. Slaybaugh, Y Zhao, A. Condon, B. Cohen, and S. Skiena. Algorithms for testing that DNA word designs avoid unwanted secondary structure. In *DNA Computing, 8th international Workshop on DNA-Based Computers*, DNA 2002, Sapporo, Japan, 10-13 June 2002. Hokkaido University, 2002. Also in LNCS 2568. M. Hagiya and A. Ohuchi, editors, pages 92–104.

4. J. Castellanos, C. Martín-Vide, V. Mitrana, and J. Sempere. Solving NP-complete problems with networks of evolutionary processors. In *Proc. of the 6th International Work-Conference on Artificial and Natural Neural Networks*, IWANN (2001), LNCS 2048, pages 621–628.
5. D.T. Chiu, E. Pezzoli, H. Wu, A.D. Stroock, and G. M. Whitesides. Using three-dimensional microfluidic networks for solving computationally hard problems. *PNAS*, 98, 6 (March 13, 2001), 2961–2966.
6. E. Csuhaj-Varju, J. Dassow, J. Kelemen, and Gh. Păun. *Grammar Systems. A Grammatical Approach to Distribution and Cooperation.* Gordon and Breach, London, 1994.
7. E. Csuhaj-Varju, L. Kari, and Gh. Păun. Test tube distributed systems based on splicing. *Computers and AI*, 15, 2-3 (1996), 211–232.
8. E. Csuhaj-Varju, R. Freund, L. Kari, and Gh. Păun. DNA computing based on splicing: universality results. *Proc. First Annual Pacific Symp. on Biocomputing*, Hawaii, (1996), 179–190.
9. J. Dassow, C. Martín-Vide, Gh. Păun, and A. Rodríguez-Patón. Conditional concatenation, *Fundamenta Informaticae*, **44**(4), (2000), 353–372.
10. R. Freund and F. Freund. Test tube systems or how to bake a DNA cake. *Acta Cybernetica*, 12, 4 (1996), 445–459.
11. A. Gehani and J.H. Reif. Microflow bio-molecular computation. *Biosystems*, 52, 1-3 (1999), 197–216.
12. G. Gloor, L. Kari, M. Gaasenbeek, and S. Yu. Towards a DNA solution to the shortest common superstring problem. In *4th Int. Meeting on DNA-Based Computing*, Baltimore, Penns., June, 1998.
13. T. Head. Formal language theory and DNA: an analysis of the generative capacity of specific recombinant behaviors. *Bull. Math. Biology*, 49 (1987), 737–759.
14. T. Head. Hamiltonian Paths and Double Stranded DNA. In *Computing with Bio-Molecules. Theory and Experiments* (Gh. Păun, ed.), World Scientific, (1998), 80–92.
15. A. Manz, D. J. Harrison, E.M.J. Verpoorte, J.C. Fettinger, A. Paulus, H. Ludi, and H. M. Widmer. Planar chips technology for miniaturization and integration of separation techniques into monitoring systems: "Capillary electrophoresis on a chip". *J. Chromatogr.*, 593 (1992), 253–258.
16. C. Martín-Vide, Gh. Păun, J. Pazos, and A. Rodríguez-Patón. Tissue P systems. *Theoretical Computer Science*, 296, 2 (2003), 295–326.
17. J.S. McCaskill. Optically programming DNA computing in microflow reactors. *Biosystems*, 59, 2 (2001), 125–138.
18. N. Morimoto, M. Arita, and A. Suyama. Solid phase DNA solution to the Hamiltonian path problem. In *Proceedings of the 3rd DIMACS Workshop on DNA Based Computers*, held at the University of Pennsylvania, June (1997), 83–92.
19. Gh. Păun and G. Thierrin. Multiset processing by means of systems of finite state transducers. In *Proc. of Workshop on implementing automata WIA99*, Lecture Notes in Computer Science 2213, Springer-Verlag, (2001), 140–157.
20. P.R. Selvaganapathy, E.T. Carlen, and C.H. Mastrangelo. Recent Progress in Microfluidic Devices for Nucleic Acid and Antibody Assays. *Proceedings of the IEEE*, 91, 6 (2003), 954–973.
21. E. Verpoorte and N.F. De Rooij. Microfluidics Meets MEMS. *Proceedings of the IEEE*, 91, 6 (2003), 930–953.

Formal Languages Arising from Gene Repeated Duplication

Peter Leupold[1], Victor Mitrana[2], and José M. Sempere[3]

[1] Research Group on Mathematical Linguistics
Rovira i Virgili University
Pça. Imperial Tàrraco 1, 43005 Tarragona, Spain
pl.doc@estudiants.urv.es

[2] Faculty of Mathematics and Computer Science, Bucharest University
Str. Academiei 14, 70109 Bucureşti, Romania
and
Research Group on Mathematical Linguistics
Rovira i Virgili University
Pça. Imperial Tàrraco 1, 43005 Tarragona, Spain
vmi@fll.urv.es

[3] Departamento de Sistemas Informáticos y Computación
Universidad Politécnica de Valencia
Camino de Vera s/n, 46022 Valencia, Spain
jsempere@dsic.upv.es

Abstract. We consider two types of languages defined by a string through iterative factor duplications, inspired by the process of tandem repeats production in the evolution of DNA. We investigate some decidability matters concerning the unbounded duplication languages and then fix the place of bounded duplication languages in the Chomsky hierarchy by showing that all these languages are context-free. We give some conditions for the non-regularity of these languages. Finally, we discuss some open problems and directions for further research.

1 Introduction

In the last years there have been introduced some operations and generating devices based on duplication operations, motivated by considerations from molecular genetics. It is widely accepted that DNA and RNA structures may be viewed to a certain extent as strings; for instance, a DNA strand can be presented as a string over the alphabet of the complementary pairs of symbols $(A,T), (T,A), (C,G), (G,C)$. Consequently, point mutations as well as large scale rearrangements occurring in the evolution of genomes may be modeled as operations on strings.

One of the most frequent and less well understood mutations among the genome rearrangements is the gene duplication or the duplication of a segment of a chromosome. Chromosomal rearrangements include pericentric and paracentric inversions, intrachromosomal as well as interchromosomal transpositions,

N. Jonoska et al. (Eds.): Molecular Computing (Head Festschrift), LNCS 2950, pp. 297–308, 2004.

translocations, etc. Crossover results in recombination of genes in a pair of homologous chromosomes by exchanging segments between parental chromatides. We refer to [3], [4], [5] and [19] for discussions on different formal operations on strings related to the language of nucleic acids. This feature is also known in natural languages. For motivations coming from linguistics, we refer to [11] and [17].

In the process of duplication, a stretch of DNA is duplicated to produce two or more adjacent copies, resulting in a tandem repeat. An interesting property of tandem repeats is that they make it possible to do "phylogenetic analysis" on a single sequence which might be useful to determine a minimal or most likely duplication history.

Several mathematical models have been proposed for the production of tandem repeats including replication, slippage and unequal crossing over [10,24,18]. These models have been supported by biological studies [20,?].

The so-called crossing over between "sister" chromatides is considered to be the main way of producing tandem repeats or block deletions in chromosomes. In [2], modeling and simulation suggests that very low recombination rates (unequal crossing over) can result in very large copy number and higher order repeats. A model of this type of crossing over has been considered in [6]. It is assumed that every initial string is replicated so that two identical copies of every initial string are available. The first copy is cut somewhere within it, say between the segments α and β, and the other one is cut between γ and δ (see Figure 1). Now, the last segment of the second string gets attached to the first segment of the first string, and a new string is obtained. More generally, another string is also generated, by linking the first segment of the second string to the last segment of the first string.

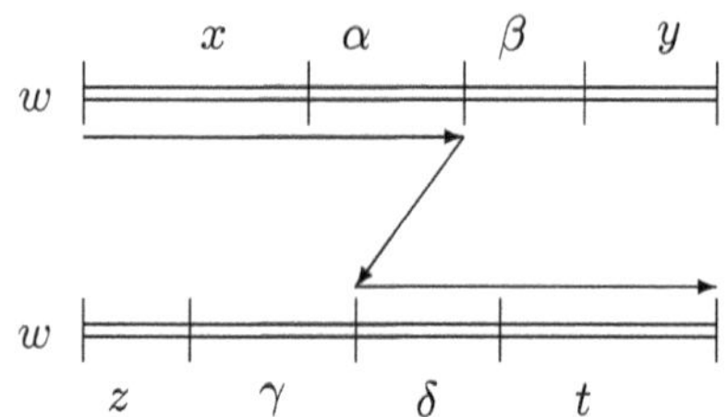

Figure 1: A scheme for gene duplication

It is easily seen that one obtains the insertion of a substring of w in w; this has the potential for inducing duplications of genes within a chromosome. We note here that, despite this model is inspired from recombination in vivo, it actually makes use of splicing rules in the sense of [8], where a computational model based on the DNA recombination under the influence of restriction enzymes and ligases essentially in vitro is introduced. This model turned out to be very attractive for computer scientists, see, e.g., the chapter [9] in [16] and [15].

Based on [3], Martín-Vide and Păun introduced in [12] a generative mechanism (similar to the one considered in [4]) based only on duplication: one starts with a given finite set of strings and produces new strings by copying specified substrings to certain places in a string, according to a finite set of duplication rules. This mechanism is studied in [12] from the generative power point of view. In [14] one considers the context-free versions of duplication grammars, solves some problems left open in [12], proves new results concerning the generative power of context-sensitive and context-free duplication grammars, and compares the two classes of grammars. Context-free duplication grammars formalize the hypothesis that duplications appear more or less at random within the genome in the course of its evolution.

In [7] one considers a string and constructs the language obtained by iteratively duplicating any of its substrings. One proves that when starting from strings over two-letter alphabets, the obtained languages are regular; an answer for the case of arbitrary alphabets is given in [13], where it is proved that each string over a three-letter alphabet generates a non-regular language by duplication.

This paper continues this line of investigation. Many questions are still unsolved; we list some of them, which appear more attractive to us – some of then will be investigated in this work:

- Is the boundary of the duplication unique, is it confined to a few locations or it is seemingly unrestricted?

- Is the duplication unit size unique, does it vary in a small range or is it unrestricted?

- Does pattern size affect the variability of duplication unit size?

- Does duplication occur preferentially at certain sites?

In [7] the duplication unit size is considered to be unrestricted. We continue here with a few properties of the languages defined by unbounded duplication unit size and then investigate the effect of restricting this size within a given range.

The paper is organized as follows: in the next section we give the basic definitions and notations used throughout the paper. Then we present some properties of the unbounded duplication languages based essentially on [7,13]. The fourth section is dedicated to bounded duplication languages. The main results of this section are: 1. Each bounded duplication language is context-free. 2. Any square-free word over an at least three-letter alphabet defines a k-bounded duplication languages which is not regular for any $k \geq 4$. The papers ends with a discussion on several open problems and directions for further research.

2 Preliminaries

Now, we give the basic notions and notations needed in the sequel. For basic formal language theory we refer to [15] or [16]. We use the following basic notation. For sets X and Y, $X \setminus Y$ denotes the set-theoretic difference of X and Y. If X is finite, then $card(X)$ denotes its cardinality; $\emptyset$ denotes the empty set. The set of all strings (words) over an alphabet V is denoted by V^* and $V^+ = V^* \setminus \{\varepsilon\}$, where ε denotes the empty string. The length of a string x is denoted by $|x|$, hence $|\varepsilon| = 0$, while the number of all occurrences of a letter a in x is denoted by $|x|_a$. For an alphabet $V = \{a_1, a_2, \ldots, a_k\}$ (we consider an ordering on V), the Parikh mapping associated with V is a homomorphism Ψ_V from V^* into the monoid of vector addition on $\mathbb{N}^k$, defined by $\Psi_V(s) = (|s|_{a_1}, |s|_{a_2}, \ldots, |s|_{a_k})$; moreover, given a language L over V, we define its image through the Parikh mapping as the set $\Psi_V(L) = \{\Psi_V(x) \mid x \in L\}$. A subset X of $\mathbb{N}^k$ is said to be *linear* if there are the vectors $c_0, c_1, c_2, \ldots, c_n \in \mathbb{N}^k$, for some $n \geq 0$ such that $X = \{c_0 + \sum_{i=1}^{n} x_i c_i \mid x_i \in \mathbb{N}, 1 \leq i \leq n\}$. A finite union of linear sets is called *semilinear.* For any positive integer n we write $[n]$ for the set $\{1, 2, \ldots, n\}$.

Let V be an alphabet and $X \in \{\mathbb{N}\} \cup \{[k] \mid k \geq 1\}$. For a string $w \in V^+$, we set

$$D_X(w) = \{uxxv \mid w = uxv, u, v \in V^*, x \in V^+, |x| \in X\}.$$

We now define recursively the languages:

$$D_X^0(w) = \{w\}, \qquad D_X^i(w) = \bigcup_{x \in D_X^{i-1}(w)} D_X(x),\ i \geq 1,$$

$$D_X^*(w) = \bigcup_{i \geq 0} D_X^i(w).$$

The languages $D_{\mathbb{N}}^*(w)$ and $D_{[k]}^*(w)$, $k \geq 1$, are called the *unbounded duplication language* and the *k-bounded duplication language*, respectively, defined by w. In other words, for any $X \in \{\mathbb{N}\} \cup \{[k] \mid k \geq 1\}$, $D_X^*(w)$ is the smallest language $L' \subseteq V^*$ such that $w \in L'$ and whenever $uxv \in L'$, $uxxv \in L'$ holds for all $u, v \in V^*$, $x \in V^+$, and $|x| \in X$.

A natural question concerns the place of unbounded duplication languages in the Chomsky hierarchy. In [7] it is shown that the unbounded duplication language defined by any word over a two-letter alphabet is regular, while [13] shows that these are the only cases when the unbounded language defined by a word is regular. By combining these results we have:

Theorem 1. [7,13] *The unbounded duplication language defined by a word w is regular if and only if w contains at most two different letters.*

3 Unbounded Duplication Languages

We do not know whether or not all unbounded duplication languages are context-free. A straightforward observation leads to the fact that all these languages are linear sets, that is, the image of each unbounded duplication language through the Parikh mapping is linear. Indeed, if $w \in V^+$, $V = \{a_1, a_2, \ldots, a_n\}$, then one can easily infer that

$$\Psi_V(D_{\mathbb{N}}^*(w)) = \{\Psi_V(w) + \sum_{i=1}^{n} x_i e_i^{(n)} \mid x_i \in \mathbb{N}, 1 \leq i \leq n\},$$

where $e_i^{(n)}$ is the vector of size n having the ith entry equal to 1 and all the other entries equal to 0.

Theorem 2. *Given a regular language L one can algorithmically decide whether or not L is an unbounded duplication language.*

Proof. We denote by $alph(x)$ the smallest alphabet such that $x \in (alph(x))^*$. The algorithm works as follows:

(i) We find the shortest string $z \in L$ (this can be done algorithmically). If there are more strings in L of the same length as z, then L is not an unbounded duplication language.
(ii) We now compute the cardinality of $alph(z)$.
(iii) If $card(alph(z)) \geq 3$, then there is no x such that $L = D^*_{\mathbb{N}}(x)$.
(iv) If $card(alph(z)) = 1$, then L is an unbounded duplication language if and only if $L = \{a^{|z|+m} \mid m \geq 0\}$, where $alph(z) = a$.
(v) If $k = 2$, $z = z_1z_2 \dots z_n$, $z_i \in alph(z)$, $1 \leq i \leq n$, then L is an unbounded duplication language if and only if

$$L = z_1^+ e_1 z_2 e_2 \dots e_{n-1} z_n^+, \tag{1}$$

where

$$e_i = \begin{cases} z_{i+1}^*, & \text{if } z_i = z_{i+1} \\ \{z_i + z_{i+1}\}^*, & \text{if } z_i \neq z_{i+1} \end{cases}$$

for all $1 \leq i \leq n-1$. Note that one can easily construct a deterministic finite automaton recognizing the language in the right-hand side of equation (1).

□

Theorem 3. *1. The following problems are algorithmically decidable for unbounded duplication languages:*

Membership: Given x and y, is x in $D^*_{\mathbb{N}}(y)$?
Inclusion: Given x and y, does $D^*_{\mathbb{N}}(x) \subseteq D^*_{\mathbb{N}}(y)$ hold?

2. The following problems are algorithmically decidable in linear time:

Equivalence: Given x and y, does $D^*_{\mathbb{N}}(x) = D^*_{\mathbb{N}}(y)$ hold?
Regularity: Given x, is $D^*_{\mathbb{N}}(x)$ a regular language?

Proof. Clearly, the membership problem is decidable and

$$D^*_{\mathbb{N}}(x) \subseteq D^*_{\mathbb{N}}(y) \text{ iff } x \in D^*_{\mathbb{N}}(y).$$

For $D^*_{\mathbb{N}}(x) = D^*_{\mathbb{N}}(y)$, it follows that $|x| = |y|$, hence $x = y$. In conclusion, $x = y$ iff $D^*_{\mathbb{N}}(x) = D^*_{\mathbb{N}}(y)$. This implies that the equivalence problem is decidable in linear time.

The regularity can be decided in linear time by Theorem 1. □

4 Bounded Duplication Languages

Unlike the case of unbounded duplication languages, we are able to determine the place of bounded duplication languages in the Chomsky hierarchy. This is the main result of this section.

Theorem 4. *For any word r and any integer $n \geq 1$, the n-bounded duplication language defined by r is context-free.*

Proof. For our alphabet $V = alph(r)$ we define the extended alphabet V_e by $V_e := V \cup \{\langle a \rangle \mid a \in V\}$. Further we define $L^{\leq l} := \{w \in L \mid |w| \leq l\}$ for any language L and integer l. We now define the pushdown automaton

$$A_n^r = \left(Q, V, \Gamma, \delta, \begin{bmatrix} \varepsilon \\ \varepsilon \\ r \end{bmatrix}, \perp, \left\{ \begin{bmatrix} \varepsilon \\ \varepsilon \\ \varepsilon \end{bmatrix} \right\}\right),$$

where $Q = \left\{ \begin{bmatrix} \mu \\ v \\ w \end{bmatrix} \mid \mu \in (V_e^* \cdot V^* \cup V^* \cdot V_e^*)^{\leq n}, v \in (V^*)^{\leq n}, w \in (V^*)^{\leq |r|} \right\}$,

and $\Gamma = \{\perp\} \cup \left\{ \overline{\mu} \mid \begin{bmatrix} \mu \\ v \\ w \end{bmatrix} \in Q, v \in (V^*)^{\leq n}, w \in (V^*)^{\leq |r|} \right\}$.

Here we call the three strings occurring in a state from bottom to top *pattern*, *memory*, and *guess*, respectively. Now we proceed to define an intermediate deterministic transition function δ'. In this definition the following variables are always quantified universally over the following domains: $u, v \in (V^*)^{\leq n}$, $w \in (V^*)^{\leq |r|}$, $\mu \in (V_e^*)^{\leq n}$, $\eta \in (V_e^* \cdot V^* \cup V^* \cdot V_e^*)^{\leq n}$, $\gamma \in \Gamma$, $x \in V$ and $Y \in V_e$.

(i) $\delta'\left(\begin{bmatrix} \varepsilon \\ \varepsilon \\ xw \end{bmatrix}, x, \perp\right) = \left(\begin{bmatrix} \varepsilon \\ \varepsilon \\ w \end{bmatrix}, \perp\right)$ and $\delta'\left(\begin{bmatrix} \varepsilon \\ xu \\ xw \end{bmatrix}, \varepsilon, \perp\right) = \left(\begin{bmatrix} \varepsilon \\ u \\ w \end{bmatrix}, \perp\right)$

(ii) $\delta'\left(\begin{bmatrix} \mu x u \\ \varepsilon \\ w \end{bmatrix}, x, \gamma\right) = \left(\begin{bmatrix} \mu\langle x\rangle u \\ \varepsilon \\ w \end{bmatrix}, \gamma\right)$ and $\delta'\left(\begin{bmatrix} \mu x u \\ xv \\ w \end{bmatrix}, \varepsilon, \gamma\right) = \left(\begin{bmatrix} \mu\langle x\rangle u \\ v \\ w \end{bmatrix}, \gamma\right)$

(iii) $\delta'\left(\begin{bmatrix} u\langle x\rangle Y\mu \\ \varepsilon \\ w \end{bmatrix}, x, \gamma\right) = \left(\begin{bmatrix} uxY\mu \\ \varepsilon \\ w \end{bmatrix}, \gamma\right)$ and

$\delta'\left(\begin{bmatrix} u\langle x\rangle Y\mu \\ xv \\ w \end{bmatrix}, \varepsilon, \gamma\right) = \left(\begin{bmatrix} uxY\mu \\ v \\ w \end{bmatrix}, \gamma\right)$

(iv) $\delta'\left(\begin{bmatrix} u\langle x\rangle \\ \varepsilon \\ w \end{bmatrix}, x, \overline{\eta}\right) = \left(\begin{bmatrix} \eta \\ ux \\ w \end{bmatrix}, \varepsilon\right)$, $\delta'\left(\begin{bmatrix} u\langle x\rangle \\ xv \\ w \end{bmatrix}, \varepsilon, \overline{\eta}\right) = \left(\begin{bmatrix} \eta \\ uxv \\ w \end{bmatrix}, \varepsilon\right)$.

For all triples $(q, x, \gamma) \in Q \times (V \cup \{\varepsilon\}) \times \Gamma$ not listed above, we put $\delta'(q, x, \gamma) = \emptyset$.

To ensure that our finite state set suffices, we take a closer look at the memory of the states – since after every reduction the reduced word is put there (see transition set (iv)), there is a danger of this being unbounded. However, during any reduction of a duplication, which in the end puts $k \leq n$ letters into the memory, either $2k$ letters from the memory or all letters of the memory (provided the memory is shorter than $2k$) have been read, since reading from memory has priority (note that the tape is read in states with empty memory only). This gives us a bound on the length of words in the memory which is n. It is worth noting that the transitions of δ' actually match either the original word r or the guess against the input or memory. To obtain the transition function δ of our automaton, we add the possibility to interrupt in any point the computation of δ' and change into a state that starts the reduction of another duplication. To this end we define for all $\begin{bmatrix} \eta \\ v \\ w \end{bmatrix} \in Q \setminus F$, and $\gamma \in \Gamma$

$$\delta\left(\begin{bmatrix} \eta \\ v \\ w \end{bmatrix}, \varepsilon, \gamma\right) = \delta'\left(\begin{bmatrix} \eta \\ v \\ w \end{bmatrix}, \varepsilon, \gamma\right) \cup \left\{\left(\begin{bmatrix} z \\ v \\ w \end{bmatrix}, \overline{\eta}\gamma\right) \mid z \in (\Sigma^+)^{\leq n}\right\}.$$

Such a transition guesses that at the current position a duplication of the form zz can be reduced (note that also here the bound of the length of the memory is not violated). For the sake of understandability we first describe also the function of the sets of transitions of δ'. Transitions (i) match the input word and r; this is only done on empty guess and stack, which ensures that every letter is matched only after all duplications affecting it have been reduced. Sets (ii) and (iii) check whether the guess, which is the segment of a guessed duplication, does indeed occur twice adjacently in the memory followed by the input. This is done by converting first guess letters from letters in V to the corresponding letters in V_e (set (ii)) exactly if the respective letter is read from either memory or input. Then set (iii) recovers the original letters. Finally the transitions in (iv) check the last letter; if it also matches, then the duplication is reduced by putting only one copy (two have been read) of the guess in the memory, and the computation is continued by putting the string encoded in the topmost stack symbol back into the guess. Now we can state an important property of $\mathcal{A}_n^r$ which allows us to conclude that it accepts exactly the language $D_{[n]}^*(r)$.

Property. *If there is an accepting computation in $\mathcal{A}_n^r$ starting from the configuration $\left(\begin{bmatrix} \eta \\ \varepsilon \\ w \end{bmatrix}, v, \alpha\right)$, then there is also an accepting computation starting from any configuration $\left(\begin{bmatrix} \eta \\ v_1 \\ w \end{bmatrix}, v_2, \alpha\right)$ where $v = v_1v_2$, $|v_1| \leq n$.*

Proof of the property. The statement is trivially true for $v_1 = \varepsilon$, therefore in the rest of the proof we consider $v_1 \neq \varepsilon$. We prove the statement by induction on the length of the computation. First we note that there exists an unique accepting configuration that is $\left(\begin{bmatrix} \varepsilon \\ \varepsilon \\ \varepsilon \end{bmatrix}, \varepsilon, \bot\right)$. Let Π be an accepting computation in $\mathcal{A}_n^r$ starting from the configuration $\left(\begin{bmatrix} \eta \\ \varepsilon \\ w \end{bmatrix}, v, \alpha\right)$.

If the length of Π is one, then $\eta = \varepsilon$, $\alpha = \bot$, and $v = w \in V$ which makes the statement obviously true. Let us assume that the statement is true for any computation of length at most p and consider a computation Π of length $p+1$ where we emphasize the first step. We distinguish three cases:

Case 1.

$$\left(\begin{bmatrix} \eta \\ \varepsilon \\ w \end{bmatrix}, v, \alpha\right) \vdash \left(\begin{bmatrix} z \\ \varepsilon \\ w \end{bmatrix}, v, \overline{\eta}\alpha\right) \vdash^* \left(\begin{bmatrix} \varepsilon \\ \varepsilon \\ \varepsilon \end{bmatrix}, \varepsilon, \bot\right).$$

Clearly, we have also $\left(\begin{bmatrix}\eta\\ v_1\\ w\end{bmatrix}, v_2, \alpha\right) \vdash \left(\begin{bmatrix}z\\ v_1\\ w\end{bmatrix}, v_2, \overline{\eta}\alpha\right)$ for any $v_1v_2 = v$, $|v_1| \leq n$. By the induction hypothesis, $\left(\begin{bmatrix}z\\ v_1\\ w\end{bmatrix}, v_2, \overline{\eta}\alpha\right) \vdash^* \left(\begin{bmatrix}\varepsilon\\ \varepsilon\\ \varepsilon\end{bmatrix}, \varepsilon, \perp\right)$ holds as well.

Case 2.

$$\left(\begin{bmatrix}\eta\\ \varepsilon\\ w\end{bmatrix}, xv', \alpha\right) \vdash \left(\begin{bmatrix}\eta'\\ \varepsilon\\ w\end{bmatrix}, v', \alpha\right) \vdash^* \left(\begin{bmatrix}\varepsilon\\ \varepsilon\\ \varepsilon\end{bmatrix}, \varepsilon, \perp\right),$$

where $v = xv'$ and the first step is based on a transition in the first part of the sets (i–iii). We have also $\left(\begin{bmatrix}\eta\\ xv_1'\\ w\end{bmatrix}, v_2, \alpha\right) \vdash \left(\begin{bmatrix}\eta'\\ v_1'\\ w\end{bmatrix}, v_2, \alpha\right)$ based on one of the corresponding transitions in the second part of the sets (i–iii). Since $v_1 = xv_1'$, by the induction hypothesis we are done in the second case.

Case 3.

$$\left(\begin{bmatrix}\eta\\ \varepsilon\\ w\end{bmatrix}, xv', \overline{\sigma}\alpha'\right) \vdash \left(\begin{bmatrix}\sigma\\ y\\ w\end{bmatrix}, v', \alpha'\right) \vdash^* \left(\begin{bmatrix}\varepsilon\\ \varepsilon\\ \varepsilon\end{bmatrix}, \varepsilon, \perp\right),$$

where the first step is based on a transition in the first part of the set (iv). Since reading from memory has priority, there exist w', τ, and β such that

$$\left(\begin{bmatrix}\sigma\\ y\\ w\end{bmatrix}, v', \alpha'\right) \vdash^* \left(\begin{bmatrix}\tau\\ \varepsilon\\ w'\end{bmatrix}, v', \beta\right) \vdash^* \left(\begin{bmatrix}\varepsilon\\ \varepsilon\\ \varepsilon\end{bmatrix}, \varepsilon, \perp\right).$$

We have $\left(\begin{bmatrix}\eta\\ xv_1'\\ w\end{bmatrix}, v_2, \overline{\sigma}\alpha'\right) \vdash \left(\begin{bmatrix}\sigma\\ yv_1'\\ w\end{bmatrix}, v_2, \alpha'\right)$ for any $v_1v_2 = v$, $|v_1| \leq n$. Further, following the same transitions as above, we get $\left(\begin{bmatrix}\sigma\\ yv_1'\\ w\end{bmatrix}, v_2, \alpha'\right) \vdash^*$ $\left(\begin{bmatrix}\tau\\ v_1'\\ w'\end{bmatrix}, v_2, \beta\right)$, and by the induction hypothesis we conclude the proof of the third case, and thus the proof of the property.

After the elaborate definition of our pushdown automaton, the proof for $L(\mathcal{A}_n^r) = D_{[n]}^*(r)$ by induction on the number of duplications used to create a word in $D_{[n]}^*(r)$ is now rather straightforward. We start by noting that the original word r is obviously accepted. Now we suppose that all words reached from r by $m-1$ duplications are also accepted and look at a word s reached by m duplications. Clearly there is a word s' reached from r by $m-1$ duplications such that s is the result of duplicating one part of s'. By the induction hypothesis, s' is accepted by $\mathcal{A}_n^r$, which means that there is a computation, we call it Ξ, which accepts s'. Let $s'[l \dots k]$ (the subword of s' which starts at the position l and ends at the position k in s', $l \leq k$) be the segment of s' which is duplicated at the final stage of producing s. Therefore

$$s = s'[1 \dots k]s'[l \dots |s'|] = s'[1 \dots l-1]s'[l \dots k]s'[l \dots k]s'[k+1 \dots |s'|].$$

Because $\mathcal{A}_n^k$ reads in an accepting computation every input letter exactly once, there is exactly one step in Ξ, where $s'[l]$ is read. Let this happen in a state $\begin{bmatrix}\mu\\ \varepsilon\\ w\end{bmatrix}$

with $s'[l \dots |s'|] = s[k+1 \dots |s|]$ left on the input tape and α the stack contents. Now we go without reading the input tape to state $\begin{bmatrix} s'[l \dots k] \\ \varepsilon \\ w \end{bmatrix}$ and push $\overline{\mu}$ onto the stack. Then we reduce the duplication of the subword $s'[l \dots k]$ of s, which is the guess of the current state, by matching it twice against the letters read on the input tape. After reducing this duplication, we arrive at a configuration having a state containing μ in the guess, $s'[l \dots k]$ in the memory, and w in the pattern, $s[k+1 \dots |s|$ left on the tape, and the stack contents as before. By the property above, there exists an accepting computation starting with this configuration, hence s is accepted. Since all words accepted by $\mathcal{A}_n^r$ are clearly in $D^*_{[n]}(r)$, and we are done. □

Clearly, any language $D^*_{[1]}(w)$ is regular. The same is true for any language $D^*_{[2]}(w)$. Indeed, it is an easy exercise (we leave it to the reader) to check that

$$D^*_{[2]}(w) = (w[1]^+ w[2]^+)^* (w[2]^+ w[3]^+)^* \dots (w[|w-1|]^+ w[|w|]^+)^*.$$

The following question appears in a natural way: Which are the minimal k and n such that there are words w over an n-letter alphabet such that $D^*_{[k]}(w)$ is not regular?

Theorem 5. *1. $D^*_{[k]}(w)$ is always regular for any word w over a two-letter alphabet and any $k \geq 1$.*

2. For any word w of the form $w = xabcy$ with $a \neq b \neq c \neq a$, $D^*_{[k]}(w)$ is not regular for any $k \geq 4$.

Proof. 1. The equality $D^*_{\mathbb{N}}(w) = D^*_{[k]}(w)$ holds for any $k \geq 2$ and any word w over a two-letter alphabet, hence the first item is proved.

2. Our reasoning for proving the second item, restricted without loss of generality to the word $w = abc$, is based on a similar idea to that used in [13]. So, let $w = abc$, $V = \{a, b, c\}$, and $k \geq 4$. First we prove that for any $u \in V^+$ such that wu is square-free, there exists $v \in V^*$ such that $wuv \in D^*_{[k]}(w)$. We give a recursive method for constructing wuv starting from w; at any moment we have a string $wu'v'$ where u' is a prefix of u and v' is a suffix of v. Initially, the current string is w which satisfies these conditions. Let us assume that we reached $x = wu[1]u[2] \dots u[i-1]v'$ and we want to get $y = wu[1]u[2] \dots u[i]v''$. To this end, we duplicate the shortest factor of x which begins with $u[i]$ and ends on the position $2+i$ in x. Because $wu[1]u[2] \dots u[i-1]$ is square-free and any factor of a square-free word is square-free as well, the length of this factor which is to be duplicated is at most 4. Now we note that, given u such that wu is square free and v is the shortest word such that $wuv \in D^*_{[k]}(w)$, we have on the one hand $k|u| + 3 \geq |wuv|$ (each duplication produces at least one symbol of u), and on the other hand $(k-1)|v| \geq |u|$ (each duplication produces at least one symbol of v since wu is always square-free). Therefore,

$$(k-1)|u| \geq |v| \geq \frac{|u|}{k-1}. \tag{2}$$

We are ready now to prove that $D^*_{[k]}(w)$ is not regular using the Myhill-Nerode characterization of regular languages. We construct an infinite sequence of square-free words $w_1, w_2, \ldots,$, each of them being in a different equivalence class:

$$w_1 = w \text{ and } w_{i+1} = wu \text{ such that } wu \text{ is square-free and } (k-1)|w_i| < \frac{|u|}{k-1}.$$

Clearly, we can construct such an infinite sequence of square-free words since there are arbitrarily long square-free words having the prefix abc [21,22,1]. For instance, all words $h^n(a)$, $n \geq 1$, are square-free and begin with abc, where h is an endomorphism on V^* defined by $h(a) = abcab$, $h(b) = acacb$, $h(c) = acbcacb$. Let v_i be the shortest word such that $w_i v_i \in D^*_{[k]}(w)$, $i \geq 1$. By relation (2), $w_{i+1} v_j \notin D^*_{[k]}(w)$ for any $1 \leq j \leq i$. Consequently, $D^*_{[k]}(w)$ is not regular. □

Since each square-free word over an at least three-letter alphabet has the form required by the previous theorem, the next corollary directly follows.

Corollary 1. *$D^*_{[k]}(w)$ is not regular for any square-free word w over an alphabet of at least three letters and any $k \geq 4$.*

5 Open Problems and Further Work

We list here some open problems which will be in our focus of interest in the near future:

1. Is any unbounded duplication language context-free? A way for attacking this question, in the aim of an affirmative answer, could be to prove that for any word w there exists a natural k_w such that $D^*_{\mathbb{N}}(w) = D^*_{[k_w]}(w)$. Note that a similar result holds for words w over alphabets with at most two letters.

2. What is the complexity of the membership problem for unbounded duplication languages? Does the particular case when y is square-free make any difference?

3. We define the *X-duplication distance* between two strings x and y, denoted by $Dupd_X(x,y)$, as follows:

$$Dupd_X(x,y) = \min\{k \mid x \in D^k_X(y) \text{ or } y \in D^k_X(x)\}, X \in \{\mathbb{N}\} \cup \{[n] \mid n \geq 2\}.$$

Clearly, *Dupd* is a distance. Are there polynomial algorithms for computing this distance? What about the particular case when one of the input strings is square-free?

4. An *X-duplication root*, $X \in \{\mathbb{N}\} \cup \{[n] \mid n \geq 2\}$, of a string x is an X-square-free string y such that $x \in D^*_X(y)$. A string y is X-square-free if it does not contain any factor of the form zz with $|z| \in X$. It is known that there are words having more than one $\mathbb{N}$-duplication root. Then the following question is natural: If x and y have a common duplication root and X is as above, then $D^*_X(x) \cap D^*_X(y) \neq \emptyset$? We strongly suspect an affirmative answer. Again, the case of at most two-letter alphabets is quite simple: Assume that x and y are two strings over the alphabet $V = \{a,b\}$ which have the same duplication root,

say aba (the other cases are identical). Then $x = a^{k_1}b^{p_1}a^{k_2}\dots a^{k_n}b^{p_n}a^{k_{n+1}}$ and $y = a^{j_1}b^{q_1}a^{j_2}\dots a^{j_m}b^{q_m}a^{j_{m+1}}$ for some $n, m \geq 1$. It is an easy exercise to get by duplication starting from x and y, respectively, the string $(a^ib^i)^sa^i$, where $s = \max(n, m)$ and $i = \max(A)$, with

$$A = \{k_t \mid 1 \leq t \leq n+1\} \cup \{p_t \mid 1 \leq t \leq n\}$$
$$\cup \{j_t \mid 1 \leq t \leq m+1\} \cup \{q_t \mid 1 \leq t \leq m\}.$$

What is the maximum number of X-duplication roots of a string? How hard is it to compute this number? Given n and $X \in \{\mathbb{N}\} \cup \{[n] \mid n \geq 2\}$, are there words having more than n X-duplication roots? (the *non-triviality* property) Given n, are there words having exactly n X-duplication roots? (the *connectivity* property) Going further, one can define the X-duplication root of a given language. What kind of language is the X-duplication root of a regular language? Clearly, if it is infinite, then it is not context-free.

References

1. Bean, D.R., Ehrenfeucht, A., Mc Nulty, G.F. (1979) Avoidable patterns in strings of symbols, Pacific J. of Math. 85:261–294.
2. Charlesworth, B., Sniegowski, P., Stephan, W. (1994) The evolutionary dynamics of repetitive DNA in eukaryotes, Nature 371:215–220.
3. Dassow, J., Mitrana, V. (1997) On some operations suggested by the genome evolution. In: Altman, R., Dunker, K., Hunter, L., Klein, T. (eds.) Pacific Symposium on Biocomputing'97, 97–108.
4. Dassow, J., Mitrana, V. (1997) Evolutionary grammars: a grammatical model for genome evolution. In: Hofestädt, R., Lengauer, T., Löffler, M., Schomburg, D. (eds.) Proceedings of the German Conference in Bioinformatics GCB'96, LNCS 1278, Springer, Berlin, 199–209.
5. Dassow, J., Mitrana, V., Salomaa, A. (1997) Context-free evolutionary grammars and the language of nucleic acids. BioSystems 4:169–177.
6. Dassow, J., Mitrana, V. (1998) Self cross-over systems. In: Păun, G. (ed.) Computing with Bio-Molecules, Springer, Singapore, 283–294.
7. Dassow, J., Mitrana, V., Păun, G. (1999) On the regularity of duplication closure, Bull. EATCS, 69:133–136.
8. Head, T. (1987) Formal language theory and DNA: an analysis of the generative capacity of specific recombinant behaviours, Bull. Math. Biology 49:737–759.
9. Head, T., Păun, G., Pixton, D. (1997) Language theory and molecular genetics. Generative mechanisms suggested by DNA recombination. In: [16]
10. Levinson, G., Gutman, G. (1987) Slipped-strand mispairing: a major mechanism for DNA sequence evolution, Molec. Biol. Evol. 4:203–221.
11. Manaster Ramer, A. (1999) Some uses and misuses of mathematics in linguistics. In: Martín-Vide, C. (ed.) Issues from Mathematical Linguistics: A Workshop, John Benjamins, Amsterdam, 70–130.
12. Martín-Vide, C., Păun, G. (1999) Duplication grammars, Acta Cybernetica 14:101–113.
13. Ming-wei, W. (2000) On the irregularity of the duplication closure, Bull. EATCS, 70:162–163.

14. Mitrana, V., Rozenberg, G. (1999) Some properties of duplication grammars, Acta Cybernetica, 14:165–177.
15. Păun, G., Rozenberg, G., Salomaa, A. (1998) DNA Computing. New Computing Paradigms, Springer, Berlin.
16. Rozenberg, G., Salomaa, A. (eds.) (1997) Handbook of Formal Languages, vol. I-III, Springer, Berlin.
17. Rounds, W.C., Manaster Ramer, A., Friedman, J. (1987) Finding natural languages a home in formal language theory. In: Manaster Ramer, A. (ed.) Mathematics of Language, John Benjamins, Amsterdam, 349–360.
18. Schlotterer, C., Tautz, D. (1992) Slippage synthesis of simple sequence DNA, Nucleic Acids Res. 20:211-215.
19. Searls, D.B. (1993) The computational linguistics of biological sequences. In: Hunter, L. (ed.) Artificial Intelligence and Molecular Biology, AAAI Press/MIT Press, Menlo Park, CA/Cambridge, MA, 47–120.
20. Strand, M., Prolla, T., Liskay, R., Petes, T. (1993) Destabilization of tracts of simple repetitive DNA in yeast by mutations affecting DNA mismatch repair, Nature 365:274–276.
21. Thue, A. (1906) Uber unendliche Zeichenreihen, Norske Videnskabers Selskabs Skrifter Mat.-Nat. Kl. (Kristiania), 7:1–22.
22. Thue, A (1912) Uber die gegenseitige Lage gleicher Teile gewiisser Zeichenreihen, Norske Videnskabers Selskabs Skrifter Mat.-Nat. Kl. (Kristiania), 1:1–67.
23. Weitzmann, M., Woodford, K., Usdin, K. (1997) DNA secondary structures and the evolution of hyper-variable tandem arrays, J. of Biological Chemistry 272:9517–9523.
24. Wells, R. (1996) Molecular basis of genetic instability of triplet repeats, J. of Biological Chemistry 271:2875–2878.

A Proof of Regularity for Finite Splicing

Vincenzo Manca

Università di Verona
Dipartimento di Informatica
Ca' Vignal 2 - Strada Le Grazie 15, 37134 Verona, Italy
vincenzo.manca@univr.it

Abstract. We present a new proof that languages generated by (non extended) H systems with finite sets of axioms and rules are regular.

1 Introduction

Splicing is the basic combinatorial operation which DNA Computing [8] is based on. It was introduced in [4] as a formal representation of DNA recombinant behavior and opened new perspectives in the combinatorial analysis of strings, languages, grammars, and automata. Indeed, biochemical interpretations were found for concepts and results in formal language theory [11,8] and Molecular Computing emerged as a new field covering these subjects, where synergy between Mathematics, Computer Science and Biology yields an exceptional stimulus for developing new theories and applications based on discrete formalizations of biological processes. In this paper we express the combinatorial form of splicing by four cooperating combinatorial rules: two cut rules (suffix and prefix deletion), one paste rule, and one (internal) deletion rule. This natural translation of splicing allows us to prove in a new manner the regularity of (non extended) splicing with finite sets of axioms and of rules.

2 Preliminaries

Consider an alphabet V and two symbols $\#, \$$ not in V. A *splicing rule* over V is a string $r = u_1\#u_2\$u_3\#u_4$, where $u_1, u_2, u_3, u_4 \in V^*$. For a rule r and for any $x_1, x_2, y_1, y_2 \in V^*$ we define the (ternary) splicing relation $\Longrightarrow_r$ such that:

$$x_1u_1u_2x_2 \ , \ \ y_1u_3u_4y_2 \Longrightarrow_r x_1u_1u_4y_2.$$

In this case we say that $x_1u_1u_4y_2$ is obtained by a *splicing step*, according to the rule r, from the *left argument* $x_1u_1u_2x_2$ and the *right argument* $y_1u_3u_4y_2$. The strings u_1u_2 and u_3u_4 are respectively the left and right *splicing points* or *splicing sites* of the rule r. An *H system*, according to [8], can be defined by a structure $\Gamma = (V, A, R)$ where V is an *alphabet*, that is, a finite set of elements called *symbols*, A is a set of strings over this alphabet, called *axioms* of the system, and R is a set of splicing rules over this alphabet.

N. Jonoska et al. (Eds.): Molecular Computing (Head Festschrift), LNCS 2950, pp. 309–317, 2004.

$L_0(\Gamma)$ consists of the axioms of Γ. For $n \geq 0$, $L_{n+1}(\Gamma)$ consists of $L_n(\Gamma)$ as well as the strings generated by one splicing step from strings of $L_n(\Gamma)$, by applying all the rules in R that can be applied. The language $L(\Gamma)$ *generated* by Γ is the union of all languages $L_n(\Gamma)$ for $n \geq 0$. If a terminal alphabet $T \subset V$ is considered, and $L(\Gamma)$ consists of the strings over T^* generated by Γ, then we obtain an *extended* H system $\Gamma = (V, T, A, R)$. H systems are usually classified by means of two classes of languages FL_1, FL_2: a H system is of type $H(FL_1, FL_2)$ when its axioms form a language in the class FL_1 and its rules, which are strings of $(V \cup \{\#, \$\})^*$, form a language in the class FL_2; $EH(FL_1, FL_2)$ is the subtype of *extended* H systems of type $H(FL_1, FL_2)$. We identify a type $C = H(FL_1, FL_2)$ of H systems with the class of languages generated by C. Let FIN, REG, RE indicate the classes of finite, regular, and recursively enumerable languages respectively. It is known that: $H(FIN, FIN) \subset REG$, $H(REG, FIN) = REG$, $EH(FIN, FIN) = REG$, $EH(FIN, REG) = RE$. Comprehensive details can be found in [8]. We refer to [11] and [8] for definitions and notations in formal language theory.

3 Cut-and-Paste Splicing

A most important mathematical property of splicing is that the class of languages generated by a *finite splicing*, that is, by a finite number of splicing rules, from a finite initial set of strings, is a subclass of regular languages: $H(FIN, FIN) \subset REG$ and, more generally, $H(REG, FIN) = REG$. The proof of this result has a long history. It originates in [1,2] and was developed in [9], in terms of a complex inductive construction of a finite automaton. In [8] Pixton's proof is presented (referred to as *Regularity preserving Lemma*). More general proofs, in terms of closure properties of abstract families of languages, are given in [5,10]. In [6] a direct proof was obtained by using ω-splicing. In this section we give another and more direct proof of this lemma, as a natural consequence of a representation of splicing rules.

Let Γ be a H system of alphabet V, with a finite number of splicing rules. The language $L(\Gamma)$ generated by a H system Γ can be obtained in the following way. Replace every splicing rule

$$r_i = u_1\#u_2\$u_3\#u_4$$

of Γ by the following four rules (two cut rules, a paste rule, a deletion rule) where $\bullet_i$ and $\odot_i$ are symbols that do not belong to V, and variables $x_1, x_2, y_1, y_2, x, y, w, z$ range over the strings on V.

1. $x_1 u_1 u_2 x_2 \Longrightarrow_{r_i} x_1 u_1 \bullet_i$	right cut rule
2. $y_1 u_3 u_4 y_2 \Longrightarrow_{r_i} \odot_i u_4 y_2$	left cut rule
3. $x u_1 \bullet_i,\ \odot_i u_4 y \Longrightarrow_{r_i} x u_1 \bullet_i \odot_i u_4 y$	paste rule
4. $w \bullet_i \odot_i z \Longrightarrow_{r_i} wz$	deletion rule

If we apply these rules in all possible ways, starting from the axioms of Γ, then the set of strings so generated that also belong to V^* and coincides with the language $L(\Gamma)$.

This representation of splicing has a very natural biochemical reading [3]. Actually, the first two rules are an abstract formulation of the action of restriction enzymes, where the symbols $\bullet_i$ and $\odot_i$ correspond to the sticky ends (here complementarity is between $\bullet_i$ and $\odot_i$). The third rule is essentially the annealing process that joins strands with matching sticky ends but leaves a hole (represented by the string $\bullet_i\odot_i$) in the fosphodiesteric bond between the 5' and 3' loci. The final rules express the hole repair performed by a ligase enzyme. The proof that will follow is inspired by this analysis of the splicing mechanism and develops an informal idea already considered in [7].

Theorem 1. *If Γ is an H system with a finite number of axioms and rules, then $L(\Gamma)$ is regular.*

Proof. Let $r_1, r_2, \ldots, r_n$ be the rules of Γ. For each rule $r_i = u_1\#u_2\$u_3\#u_4$, introduce two new symbols $\bullet_i$ and $\odot_i$, for $i = 1, \ldots n$, which we call *bullet* and *antibullet* of the rule r_i. If u_1u_2 and u_3u_4 are the left and the right splicing sites of r_i, then symbols $\bullet_i$ and $\odot_i$ can be used in order to highlight the left and right splicing sites that occur in a string. More formally, let h be the morphism

$$h : (V \cup \{\bullet_i \mid i = 1, \ldots n\} \cup \{\odot_i \mid i = 1, \ldots n\})^* \rightarrow V^*$$

that coincides with the identity on V and erases all the symbols that do not belong to V, that is, associates the empty string λ to them. In the following V' will abbreviate $(V \cup \{\bullet_i \mid i = 1, \ldots n\} \cup \{\odot_i \mid i = 1, \ldots n\})^*$. For any rule $r_i = u_1\#u_2\$u_3\#u_4$, with $i = 1, \ldots, n$, we say that a string of V' is $\bullet_i$-factorizable if it includes a substring $u_1'u_2' \in V'$ such that $h(u_1') = u_1$, $h(u_2') = u_2$. In this case the relative $\bullet_i$-factorization of the string is obtained by replacing $u_1'u_2'$ with $u_1' \bullet_i u_2'$. Analogously, a string of V' is $\odot_i$-factorizable if it includes a substring $u_3'u_4' \in V'$ such that $h(u_3') = u_3$, $h(u_4') = u_4$. In this case the relative $\odot_i$-factorization of the string is obtained by replacing $u_3'u_4'$ with $u_3' \odot_i u_4'$. A string α is a maximal factorization of a string η if α is a factorization such that $h(\alpha) = \eta$, α contains no two consecutive occurrences of the same bullet or antibullet, while any further factorization of α contains two consecutive occurrences of the same bullet or antibullet. It is easy to verify that the maximal factorization of a string is unique.

Now, given an H system, we factorize its axioms in a maximal way. Let $\alpha_1, \alpha_2, \ldots, \alpha_m$ be these factorizations and let $\triangleright\alpha_1\triangleright, \triangleright\alpha_2\triangleright, \ldots, \triangleright\alpha_m\triangleright$ be their extensions with a symbol marking the start and the end of these factorizations. From the set Φ of these *factorization strings* we construct the following labeled directed graph G_0, which we call *axiom factorization graph* of the system, where:

1. A node is associated to each occurrence of a bullet or antibullet that occurs in a strings of Φ, while a unique *entering node* $\triangleright\!\bullet$ is associated to all the occurrences of symbol $\triangleright$ at beginning of the strings of Φ, and a unique *exiting*

node $\bullet\!\!\triangleright$ is associated to all the occurrences of symbol $\triangleright$ at end of the strings of Φ.
2. From any node n there is an arc to the node m if the occurrence to which m is associated is immediately after the occurrence to which n is associated; this arc is labeled with the string between these two occurrences.
3. Each bullet is linked by an arc with the empty label to the antibullet of the same index.

As an example, let us apply this procedure to the following H system specified by:

Alphabet: $\{a, b, c, d\}$,
Axioms: $\{dbab\ ,\ cc\}$,
Rules: $\{r_1 :\ a\#b\$\lambda\#ab\ , r_2 :\ baa\#aaa\$\lambda\#c\}$.

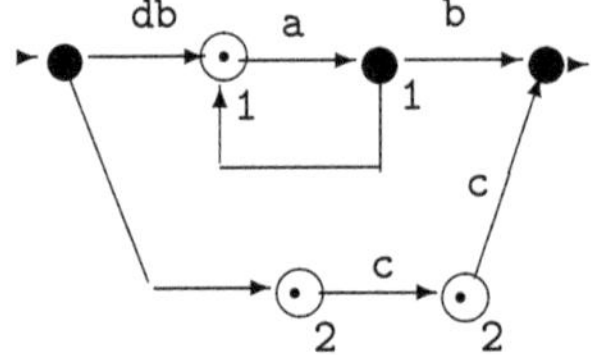

Fig. 1. An Axiom Factorization Graph

In Figure 1 the graph G_0 of our example is depicted. Hereafter, unless it is differently specified, by a path we understand a path going from the entering node to the exiting node. If one concatenates the labels along a path, one gets a string generated by the given H system. However, there are strings generated by the system that are not generated as (concatenation of labels of) paths of this graph. For example, the strings *dbaac*, *dbaacc* do not correspond to any path of the graph above. In order to overcome this inadequacy, we extend the graph that factorizes the axioms by adding new possibilities of factorizations, where the strings u_1 and u_4 of the rules are included and other paths with these labels are possible.

Before going on, let us sketch the main intuition underlying the proof, that should appear completely clear when the formal details are developed. The axiom factorization graph suggests that all the strings generated by a given H system are combinations of: i) strings that are labels of the axiom factorization graph, and ii) strings u_1 and u_4 of splicing rules of the system. Some combinations of these pieces can be iterated, but, although paths may include cycles, there is only a finite number of ways to combine these pieces in paths going from the entering node to the exiting node. An upper bound on this number is determined by: i) the number of substrings of the axioms and of the rules, and ii) the number of factorization nodes that can be inserted in a factorization graph when extending the axiom factorization with the rules, as it will be shown in the proof.

The detailed construction of the proof is based on two procedures that expand the axiom factorization graph G_0. We call the first procedure **Rule expansion** and the second one **Cross bullet expansion**. In the following, for the sake of brevity, we identify paths with factorization strings.

3.1 Rule Expansion

Let G_0 be the graph of the maximal factorizations of the axioms of Γ. Starting from G_0, we generate a new graph G_e which we call *rule expansion* of G_0. Consider a symbol $\otimes$ which we call *cross bullet*. For this symbol h is assumed to be a deleting function, that is, $h(\otimes) = \lambda$. For every rule $r_i = u_1\#u_2\$u_3\#u_4$ of Γ, add to G_0 two rule components: the u_1 component that consists of a pair of nodes $\otimes$ and $\bullet_i$, with an arc from $\otimes$ to $\bullet_i$ labeled by the string u_1, and the u_4 component that consists of a pair of nodes $\otimes, \odot_i$ with an arc from $\odot_i$ to $\otimes$ labeled by the string u_4.

Then, add arcs with the empty label from the new node $\bullet_i$ to the corresponding antibullet nodes $\odot_i$ that were already in the graph. Analogously, add arcs with the empty label from the nodes $\bullet_i$ that were already in the graph to the new antibullet node $\odot_i$.

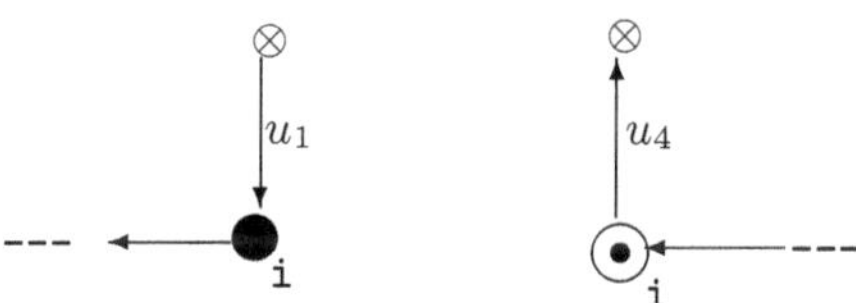

Fig. 2. Rule i Expansion Components

In the case of the graph in Figure 1, if we add the rule expansions of the two rules of the system, then we get the graph of Figure 2.

3.2 Cross Bullet Expansion

Now we define the cross bullet expansion procedure. Consider a symbol $\circ$, which we call *empty bullet*. For this symbol, h is assumed to be a deleting function, that is, $h(\circ) = \lambda$. Suppose that in G_e there is a cycle $\odot_j - \bullet_j$. A cycle introduces new factorization possibilities that are not explicitly present in the graph, but that appear when we go around the cycle a certain number of times. Let us assume that, by iterating this cycle, a path θ is obtained that generates a new splicing site for some rule. We distinguish two cases. In the first case, which

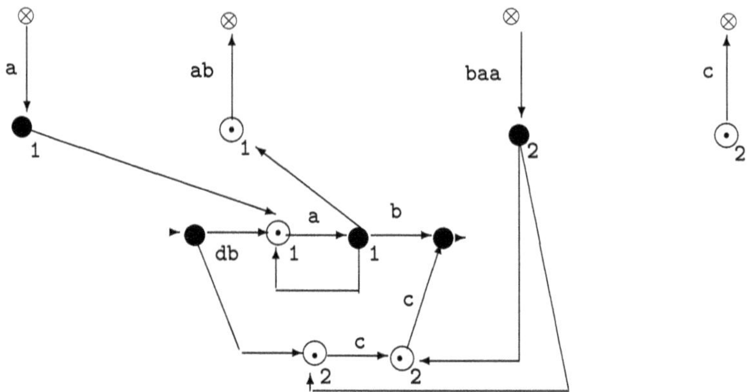

Fig. 3. The Rule Expansion of the Graph of Figure 1

we call *1-cross bullet expansion*, the path θ, considered as a string, includes the splicing site u_1u_2 of a rule

$$\theta = \eta\xi\beta$$

for some ξ beginning with a symbol of V such that $h(\xi) = u_1u_2$. In the second case, which we call *4-cross bullet expansion*, the path θ, considered as a string, includes the splicing site u_3u_4 of a rule

$$\theta = \eta\xi\beta$$

for some ξ ending with a symbol of V such that $h(\xi) = u_3u_4$. As it is illustrated in the following pictures, the positions where the beginning of ξ or the end of ξ are respectively located could be either external to the cycle or internal to it. In both cases we insert an empty bullet in the path θ.

- In the case of a 1-cross bullet expansion, we insert an empty bullet $\circ$, exactly before ξ, with an arrow going from this empty bullet to the cross bullet of the u_1 expansion component of r_i. Let $1/i$ be the type of this empty bullet. This expansion is performed unless η does not already include an empty bullet of type $1/i$ after its last symbol of V.
- In the cases of a 4-cross bullet expansion, we insert an empty bullet $\circ$, exactly after ξ, with an arrow, entering this empty bullet and coming from the cross bullet of the u_4 expansion component of r_i. Let $4/i$ be the type of this empty bullet. This expansion is performed unless β does not already include an empty bullet of type $4/i$ before its first symbol of V.

The two cases are illustrated in the following pictures (in 1-cross bullet expansion the beginning of the splicing site is external to the cycle, in the 4-cross bullet expansion the end of the splicing point is internal to the cycle).

The general situation of cross bullet expansion is illustrated in Figure 3.

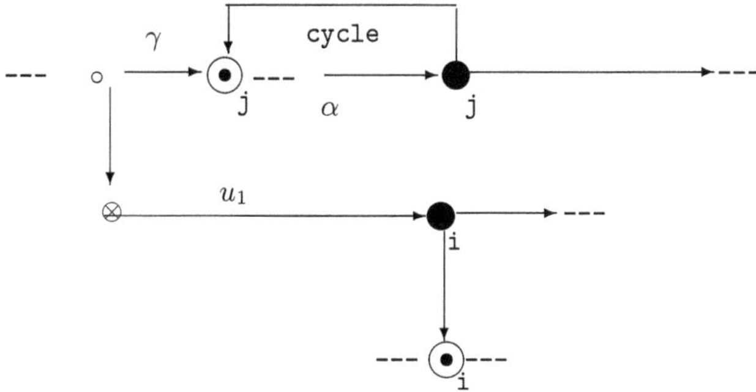

Fig. 4. 1-Cross bullet Expansion: $\gamma\alpha^n$ includes a string of $h^{-1}(u_1u_2)$ as its prefix

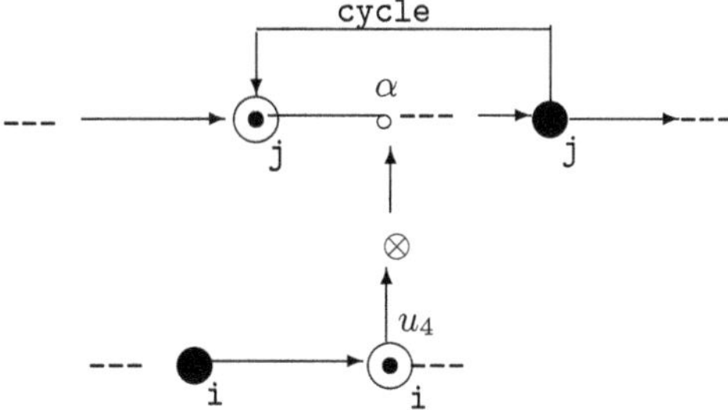

Fig. 5. 4-Cross bullet Expansion: α^n includes a string of $h^{-1}(u_3u_4)$ as its suffix

If we apply cross bullet expansion to the graph of Figure 3 we get the graph of Figure 7 where only one 1-cross bullet expansion was applied. The following lemma establishes the termination of the cross bullet expansion procedure.

Lemma 1. *If, starting from G_e, we apply again and again the cross bullet expansion procedure, then the resulting process eventually terminates, that is, after a finite number of steps we get a final graph where no new cycles can be introduced.*

Proof. This holds because in any cross bullet expansion we insert an empty bullet $\circ$ and an arc with the empty label connecting it to a cross node, but empty bullets, at most one for each type, are always inserted between two symbols of V starting from the graph G_e, which is fixed at beginning of the cross bullet expansion process. Therefore, only a finite number of empty bullets can be inserted. This ensures that the expansion process starting from G_e will eventually stop.

Let G be the *completely expanded graph*. The cross bullet expansion procedure is performed by a (finite) sequence of steps. At each step an empty bullet is inserted and an arc is added that connects the empty bullet with a cross bullet.

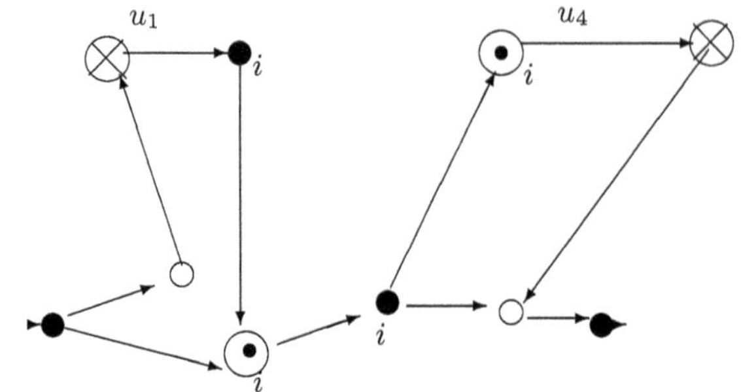

Fig. 6. Cross bullet Expansions of rule i

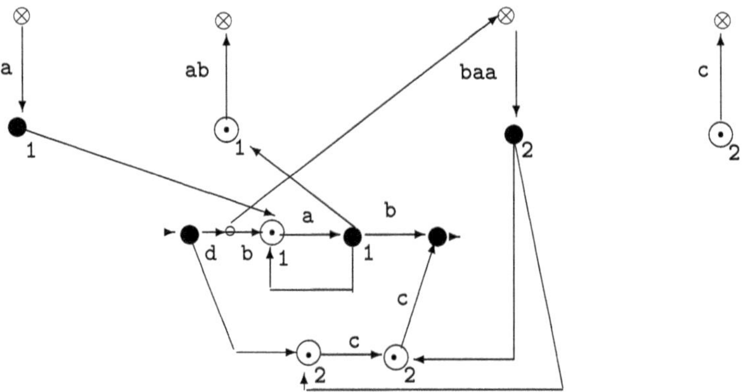

Fig. 7. Cross bullet Expansion of the Graph of Figure 3

Let $L(G)$ be the language of the strings generated by paths of G. The inclusion $L(G) \subseteq L(\Gamma)$ can be easily shown by induction on the number of cross bullet expansion steps: obviously $L(G_0) \subseteq L(\Gamma)$, thus, assume that all the paths of the graph at step i generate strings of $L(\Gamma)$. Then the paths in the expanded graph at step $i+1$ generate strings spliced from paths which are present at step i. Therefore, the inclusion holds for the completely expanded graph.

For the inverse inclusion we need the following lemma which follows directly from the method of the cross bullet expansion procedure.

Lemma 2. *In the completely expanded graph G, when there is a path $\eta\theta\sigma\rho$ where $h(\theta) = u_1$, $h(\sigma) = u_2$ and u_1u_2 is the splicing site of the rule r_i, then also a path $\eta\theta'\bullet_i$ occurs in G with $h(\theta) = h(\theta')$. Analogously, if in G there is a path $\eta\theta\sigma\rho$ where $h(\theta) = u_3$, $h(\sigma) = u_4$ and u_3u_4 is the splicing site of the rule r_i, then also a path $\odot_i\sigma'\rho$ occurs in G with $h(\sigma) = h(\sigma')$.*

The inclusion $L(\Gamma) \subseteq L(G)$ can be shown by induction on the number of splicing steps. If η is an axiom, the condition trivially holds. Assume that η derives, by means of a rule r_i, from two strings. This means that these strings can be factored as:

$$\alpha \bullet_i \beta$$

$$\gamma \odot_i \delta$$

and, by the induction hypothesis, in G there are two paths θ, ρ generating $\alpha\beta$ and $\gamma\delta$, respectively. These paths include as sub-paths $h^{-1}(u_1u_2)$, $h^{-1}(u_3u_4)$, respectively, therefore, according to the previous lemma, a path $\sigma\bullet_i$ is in G where $h(\sigma) = \alpha$ and a path $\odot_i\pi$ is in G where $h(\pi) = \delta$. This means that the path $\sigma \bullet_i \odot_i\pi$ is in G, but $h(\sigma \bullet_i \odot_i\pi) = \eta$, therefore η is generated by a path of the completely expanded graph. In conclusion, $L(\Gamma) = L(G)$. The language $L(G)$ is regular because it is easy to define it by means of a regular expression deduced from G.

Acknowledgments. I want to express my gratitude to Giuditta Franco, Tom Head, Victor Mitrana, Gheorghe Păun, and Giuseppe Scollo for their suggestions that were essential in improving some previous versions of this paper.

References

1. Culik II K., Harju T., The regularity of splicing systems and DNA, in Proc. ICALP 1989, LNCS 372, pp. 222-233, 1989.
2. Culik II K., Harju T., Splicing semigroups of dominoes and DNA, Discrete Applied Mathematics, 31, pp. 261-277, 1991.
3. Garret R.H., Grisham C.M., Biochemistry, Saunders College Publishing, Harcourt, 1997.
4. Head T., Formal language theory and DNA: an analysis of the generative capacity of recombinant behaviors, Bulletin of Mathematical Biology, 49, pp. 737-759, 1987.
5. Head T., Păun G., Pixton D., Language theory and molecular genetics, Chapter 7 in: [11], pp. 295-360, 1997.
6. Manca V., Splicing normalization and regularity, in: C. Calude, Gh. Păun, eds., Finite versus Infinite. Contributions to an Eternal Dilemma, Springer-Verlag, London, 2000.
7. Manca V., On some forms of splicing, in: C. Martin-Vide, V. Mitrana eds., Words, Sequences, Grammars, Languages: Where Biology, Computer Science, Linguistics and Mathematics Meet, Kluwer, Dordrecht, 2000.
8. Păun G., Rozenberg G., Salomaa A., DNA Computing. New Computing Paradigms, Springer-Verlag, Berlin, 1998.
9. Pixton D., Regularity of splicing languages, Discrete Applied Mathematics (69)1-2, pp. 101-124, 1996.
10. Pixton D., Splicing in abstract families of languages, Theoretical Computer Science, 234:135–166, 2000.
11. Rozenberg G., Salomaa A. (eds.), Handbook of Formal Languages, 3 Vols., Springer-Verlag, Berlin, Heidelberg, 1997.

The Duality of Patterning in Molecular Genetics

Solomon Marcus

Romanian Academy, Mathematics
Calea Victoriei 125
Bucuresti, Romania
solomon.marcus@imar.ro

The first time I met Professor Tom Head, just in the year 1971, was when I became interested in the triangle *linguistics-molecular genetics-mathematics*. Tom Head is a pioneer in the theory of splicing, which became an important component of the new domain of DNA computing ([5,14,7]). It is nice for me to remember that in December 1971 I had the chance to be invited by Professor Head at the University of Alaska, where I gave a series of lectures. A few months earlier, in July-August 1971, I organized within the framework of the Linguistic Institute of America (at State University of New York at Buffalo, under the direction of Professor David Hays) a research seminar on the above mentioned triangle. My starting point for the seminar was a book by Professor Z. Pawlak, a famous Polish computer scientist, and the writings of Roman Jakobson on the link between linguistics and molecular genetics. One of the participants at this Seminar was Bernard Vauquois, one of the initiators of ALGOL 60. As a result of this seminar, I published the article [10] and much later [11], where the interplay *nucleotide bases-codons-amino acids-proteins* is analyzed in the perspective of structural linguistics and of formal language theory.

In a retrospect, the notion of splicing was in the immediate neighbourhood of the ideas investigated there, but we were not able to invent it and to give it an explicit status. It was the privilege of Tom Head [5,6] to identify it and to show its generative capacity. In a further steps, [14,7,16,4] developed a comprehensive theory of DNA computing, based on the splicing operation. However, concomitantly there was another line of development where splicing was not involved, although potentially it is implied and its relevance can be shown: the Human Genome Project (HGP). It started around 1990 and was directed towards sequencing the DNA, identifying the genes and establishing the gene-protein function correlation. This line of research brings to the center of attention the analytic aspects related to DNA, genes and codons. Obviously, in the context of Watson-Crick double strand organization, splicing, as a purely combinatorial-sequential operation related to DNA, finds its natural place and should be involved in the gene-protein interaction. But the aim of this note is more modest and of a very preliminary nature in respect to the problem we just mentioned. We have in view the so-called duality of patterning principle and its possible relevance for molecular genetics. This aspect is not without relation with the gene-protein interaction, because the duality is in strong relation with the arbitrariness of the genetic sign, a genetic equivalent of the problem of arbitrariness of the linguistic sign, discussed by Ferdinand de Saussure.

N. Jonoska et al. (Eds.): Molecular Computing (Head Festschrift), LNCS 2950, pp. 318–321, 2004.

The duality of patterning principle (shortly: duality), discussed for the first time by Martinet [12,13] and called in French 'principe de la double articulation' asserts the existence in language of two levels of structural organization, phonological and grammatical, such that the segments of the latter level are composed of segments of the former level, called phonemes (see also [9] pages 71-76). There is here no reference to meaning and no reference to the quantitative aspect of the units involved in the respective levels. However, phonemes are meaningless units while forms of the higher level are meaningful units (the meaning may be lexical or grammatical). Moreover, there is a discrepancy between the relatively small number of units involved in the lower level and the relatively large number of units involved in the higher level. As in many other situations, the principle of least effort seems to be involved here too: try to obtain as many as possible lexical and grammatical forms by using as few as possible elements of the lower level (the higher level corresponds to the first articulation, while the lower level corresponds to the second articulation). Indeed, the number of phonemes is usually between ten and hundred, while the number of words (or of morphemes) is usually larger than ten thousands.

When Martinet [12,13] claims that duality is a specific feature of natural languages, things become very provocative. In other words, no other semiotic system, particularly no artificial language is endowed with the duality of patterning. In contrast with this claim, researchers in molecular genetics who trusted the language status of DNA, RNA and proteins claimed just the opposite: duality is an important aspect of molecular genetics or, how Ji calls it [8], of the cell language.

In [10,11], I gave a detailed presentation of various proposals of interpretation of duality in molecular genetics and also I gave my interpretation, arguing in its favor and against other interpretations. In the meantime, other proposals were published, for instance, those of Collado-Vides [2,3], of Bel-Enguix [1] and of Ji [8]. Our interpretation was to assimilate the first articulation with the level of codons and the second articulation with the level of nucleotide bases. This means that the former are the morphemes of the genetic language, while the latter are its phonemes. The genetic phonemes are endowed with chemical significance, while the genetic morphemes are endowed with biological significance (because they encode, via the genetic dictionary, some amino acids or some 'ortographic' indication, such as 'start' or 'stop'). Ji ([8] page 155) proposes a unique framework for what he calls "human and cell languages" [1] under the form of a 6-tuple $L = (A, W, S, G, P, M)$, where A is the alphabet, W is the vocabulary or lexicon (i.e., a set of words), S is an arbitrary set of sentences, G is a set of rules governing the formation of sentences from words (the first articulation, in Ji's interpretation), as well as the formation of words from letters (the second articulation), P is a set of mechanisms realizing and implementing a language, and finally M is a set of objects (both mental and material) or processes referred to by words and sentences. For Ji, the first articulation of the cell language is

[1] As a matter of fact, for natural and genetic languages; there are many human languages other than the natural ones, for instance all artificial languages.

identified with the spatio-temporal organization of gene expressions through the control of DNA folding patterns via conformational (or noncovalent) interactions, while the second articulation is identified with the linear arrangement of nucleotides to form structural genes through covalent (or configurational) interactions. Ji rediscovers another duality at the level of proteins, where the second articulation is identified with covalent structures, while the first articulation is identified with the three-dimensional structures of polypeptides formed through conformational (or noncovalent) interactions. As it can be seen, for Ji there are a total of four articulations of the genetic language and the condition of linearity and sequentiality of a language structure is no longer satisfied. The concept of a language as a special sign system is in question, and it is no longer clear where is the border, if there exists such a border, between an arbitrary sign-system and a language-like sign system. Both Martinet [13] and Lyons ([9] page 74-76) stress the link between duality and arbitrariness. Duality points out to some limits of the arbitrariness of the linguistic sign. Can we speak, equivalently, of the arbitrariness of the "genetic sign"? To what extent is heredity the result of the combinatorial game expressed by the first articulation, under the limits and the freedom imposed by the second articulation?

References

1. Bel-Enguix, G. *Molecular Computing Methods for Natural Language Syntax*, PhD Thesis, 2000, Univ. of Tarragona, Spain.
2. Collado-Vides, J. A syntactic representation of units of genetic information. *J. Theor. Biol.* **148** (1991) 401–429.
3. Collado-Vides, J. The elements for a classification of units of genetic information with a combinatorial component. *J. Theor. Biol.*, **163** (1993) 527–548.
4. Freund, R., Kari, L., Păun, Gh. DNA computing based on splicing: the existence of universal computers. *Theory of Computing Systems*, **32** (1999) 69–112.
5. Head, T. Formal language theory and DNA: an analysis of the generative capacity of specific recombinant behaviors. *Bull. Math. Biol.* **49** (1987) 737–739.
6. Head, T. Splicing schemes and DNA, in *Lindenmayer Systems: Impact on Theoretical Computer Science and Developmental Biology* (G. Rozenberg, A. Salomaa, eds.) Springer-Verlag, Berlin, (1992) 371–383.
7. Head, T., Păun, Gh., Pixton, D. Language theory and molecular genetics, Chapter 7 in *Handbook of Formal Languages* (G. Rozenberg, A. Salomaa, eds.), vol. 2 (1997) 295–360.
8. Ji, S. The Bhopalator: an information/energy dual model of the living cell (II). *Fumndamenta Informaticae*, **49** 1/3 (2002) 147–165.
9. Lyons, J. *Semantics I.* Cambridge Univ. Press, 1997.
10. Marcus, S. Linguistic structures and generative devices in molecular genetics. *Cahiers de Ling. Theor. et Appl.*, **11** 1 (1974) 77–104.
11. Marcus, S. Language at the crossroad of computation and biology. In *Computing with Bio-molecules: Theory and Experiments* (Păun, Gh., ed.), Springer-Verlag, Singapore, (1998) 1–35.
12. Martinet, A. La double articulation linguistique. *Travaux du Cercle Linguistique de Copenhague*, **5** (1949) 30–37.

13. Martinet, A. Arbitraire linguistique et double articulation. *Cahiers Ferdinand de Saussure*, **15** (1957)105–116.
14. Păun, Gh. Splicing: a challenge to formal language theorists. *Bull. EATCS* **57** (1995) 183–194.
15. Păun, Gh. On the splicing operation. *Discrete Applied Mathematics*, **70** (1996) 57–79.
16. Păun, Gh., Rozenberg, G., Salomaa, A. *DNA Computing. New Computing Paradigms*, Springer-Verlag, Berlin, 1998.

Membrane Computing: Some Non-standard Ideas

Gheorghe Păun[1,2]

[1] Institute of Mathematics of the Romanian Academy
PO Box 1-764, 70700 Bucureşti, Romania,
george.paun@imar.ro

[2] Research Group on Mathematical Linguistics
Rovira i Virgili University
Pl. Imperial Tàrraco 1, 43005 Tarragona, Spain
gp@astor.urv.es

Abstract. We introduce four new variants of P systems, which we call non-standard because they look rather "exotic" in comparison with systems investigated so far in the membrane computing area: (1) systems where the rules are moved across membranes rather than the objects processed by these rules, (2) systems with reversed division rules (hence entailing the elimination of a membrane when a membrane with an identical contents is present nearby), (3) systems with accelerated rules (or components), where any step except the first one takes half of the time needed by the previous step, and (4) *reliable* systems, where, roughly speaking, all possible events actually happen, providing that "enough" resources exist. We only briefly investigate these types of P systems, the main goal of this note being to formulate several related open problems and research topics.

1 Introduction

This paper is addressed to readers who are already familiar with membrane computing – or who are determined to become familiar from sources other than a standard prerequisites section – which will not be present in this paper. Still, we repeat here a many times used phrase: "A friendly introduction to membrane computing can be found in [5], while comprehensive details are provided by [4], or can be found at the web address http://psystems.disco.unimib.it."

Of course, calling "exotic" the classes of P systems we are considering here is a matter of taste. This is especially true for the first type, where the objects never change the regions, but, instead, the rules have associated target indications and can leave the region where they are used. Considering migratory rules can remind of the fact that in a cell the reactions are catalyzed/driven/controlled by chemicals which can move through membranes like any other chemicals – the difference is that here the "other chemicals" are not moving at all. Actually, P systems with moving rules were already investigated, by Rudi Freund and his co-workers, in a much more general and flexible setup, where both the objects

N. Jonoska et al. (Eds.): Molecular Computing (Head Festschrift), LNCS 2950, pp. 322–337, 2004.

and the rules can move (and where universality results are obtained – references can be found in [4]).

We find the functioning of such systems pretty tricky (not to say difficult), and only a way to generate the Parikh sets of matrix languages (generated by grammars without appearance checking) is provided here. Maybe the reader will prove – or, hopefully, disprove – that these systems are universal.

The second type of "exotic" systems starts from the many-times-formulated suggestion to consider P systems where for each rule $u \rightarrow v$ one also uses the rule $v \rightarrow u$. This looks strange in the case of membrane dividing rules, $[_i a]_i \rightarrow [_i b]_i [_i c]_i$, where we have to use the reversed rule $[_i b]_i [_i c]_i \rightarrow [_i a]_i$, with the meaning that if the two copies of membrane i differ only in the objects b, c, then we can remove one of them, replacing the distinguished object of the remaining membrane by a. This is a very powerful operation – actually, we have no precise estimation of "very powerful", but only the observation that in this way we can compute in an easy manner the intersection of two sets of numbers/vectors which are computable by means of P systems with active membranes.

The third idea is to imagine P systems where the rules "learn from experience", so that the application of any given rule takes one time unit at the first application, half of this time at the second application, and so on – the $(i+1)$th application of the rule takes half of the time requested by the ith application. In this way, arbitrarily many applications of a rule – even infinitely many! – take at most two time units... Even a system which never stops (that is, is able to continuously apply rules) provides a result after a known number of *external* time units. Note this important difference, between the *external time*, measured in "physical" units of equal length, and the *computational time*, dealing with the number of rules used. When the rules are to be used in parallel, delicate synchronization problems can appear, because of the different time interval each rule can request, depending on how many times each rule has already been used in the computation.

However, characterizations of Turing computable numbers are found by means of both cooperative multiset rewriting-like rules and by antiport rules of weight 2 where the rules are accelerated. Clearly, the computation of infinite sets is done in a finite time. Universality reasons show that *all* Turing computable sets of numbers can be computed in a time bounded in advance. A concrete value of this bound remains to be found, and this seems to be an interesting (mathematical) question.

Of course, the acceleration can be considered not only at the level of rules, but also at the level of separate membranes (the first transition in such a membrane takes one time unit, the second one takes half of this time, and so on), or directly at the level of the whole system. These cases are only mentioned and left as a possible research topic (for instance, as a possible way to compute "beyond Turing" – as it happens wuth accelerated Turing machines, see [1], [2], [3]).

The same happens with the idea to play a similar game with the space instead of the time. What about space compression (whatever this could mean) or expansion (idem)? The universe itself is expanding, so let us imagine that in some regions of a P system the objects (and membranes) are doubled from time to time, just like that, without applying individual rules – still, some rules for controlling this global doubling should be considered. Any quantum computing links? Especially: any application in "solving" hard problems in a fast way?

Finally, the fourth idea originates in the observation that there is a striking difference between classic computer science (complexity theory included), based on (assumed) deterministic functioning of computers, and bio-computations, as carried in DNA computing and as imagined in membrane computing: if we have "enough" molecules in a test tube, then all possible reactions will actually happen. This is far from a mathematical assertion, but many DNA computing experiments, including the history-making one by Adleman, are based on such assumptions (and on the mathematical fact that a positive solution can be effectively checked in a feasible time). The belief is that reality arranges things around the average, it does not deal with worst cases; probabilistically stated, "the frequency tends to the probability". If something has a non-zero probability, then eventually it will happen. We just have to provide "enough" possibilities to happen (attempts, resources, etc).

Here, we formulate this optimistic principle in the following way: if we have "enough" objects in a membrane, then *all* rules are applied in each transition. The problem is to define "enough"... and we take again an optimistic decision: if each combination of rules has polynomially many chances with respect to the number of combinations itself, then each combination is applied. This is especially suitable (and clear) for the case of string-objects: if we have n combinations of rules which can be applied to an existing string, then each combination will be applied to at least one string, providing that we have at least $n \cdot p(n)$ copies of the string available, for some polynomial $p(x)$ specific to the system we deal with. A P system having this property is said to be *reliable*.

Clearly, a deterministic system is reliable, and the polynomial $p(x)$ is the constant, $p(n) = 1$, so of more interest is to assume this property for nondeterministic P systems. Such systems are able to solve **NP**-complete problems in polynomial time. We illustrate this possibility for SAT, which is solved in linear time by means of reliable systems with string-objects able to replicate strings of length one (in some sense, we start with symbol-objects processed by rules of the form $a \to bb$, we pass to strings by rules $b \to w$, and then we process the string-objects w and the strings derived from them, as usual in P systems, with string-objects).

2 Moving Rules, Not Objects

We consider here only the symbol-object case, with multiset rewriting-like rules. After being applied, the rules are supposed to move through membranes; this migration is governed by a *metarules* of the form (r, tar), existing in all regions, with the meaning that the multiset-processing rule r, if present in the region, after being used has to go to the region indicated by tar; as usual, tar can be one of *here, out, in*.

More formally, *a P system with migrating rules* is a construct

$$\Pi = (O, C, T, R, \mu, w_1, \ldots, w_m, R_1, \ldots, R_m, D_1, \ldots, D_m, i_o),$$

where:

1. O is the alphabet of *objects*;
2. $C \subseteq O$ is the set of *catalysts*;
3. $T \subseteq (O - C)$ is the set of *terminal* objects;
4. R is a finite set of *rules* of the form $u \to v$, where $u \in O^+, v \in O^*$; the catalytic rules are of the form $ca \to cu$ for $c \in C, a \in O - C$, and $u \in (O - C)^*$;
5. μ is a membrane structure of degree m, with the membranes labeled in a one to one manner with $1, 2, \ldots, m$;
6. $w_1, \ldots, w_m$ are strings over O specifying the multisets of objects present in the m regions of μ at the beginning of the computation;
7. $R_1, \ldots, R_m$ are subsets of R, specifying the rules available at the beginning of the computation in the m regions of μ;
8. $D_1, \ldots, D_m$ are sets of pairs – we call them *metarules* – of the form (r, tar), with $r \in R$ and $tar \in \{here, out, in\}$ associated with the m regions of μ;
9. $1 \leq i_o \leq m$ is the *output* membrane of the system, an elementary one in μ.

Note that the objects from regions are present in the multiset sense (their multiplicity matters), but the rules are present in the set sense – if present, a rule is present *in principle*, as a possibility to be used. For instance, if at some step of a computation we have a rule r in a region i and the same rule r is sent in i from an inner or outer region to i, then we will continue to have *the rule* r present, not two or more copies of it.

However, if a rule r is present in a region i, then it is applied in the nondeterministic maximally parallel manner as usual in P systems: we nondeterministically assign the available objects to the available rules, such that the assignment is exhaustive, no further object can evolve by means of the existing rules. Objects have no target indication in the rules of R, hence all objects obtained by applying rules remain in the same region. However, the applied rules move according to the pairs (r, tar) from D_i.

Specifically, we assume that we apply pairs (r, tar), not simply rules r. For instance, if in a region i we have two copies of a and two metarules $(r, here), (r, out)$ for some $r : a \to b$ which is present in region i, then the two copies of a can evolve both by means of $(r, here)$, or one by $(r, here)$ and one by (r, out), or both by means of (r, out). In the first case, the rule r remains in the same region, in the second case the rule will be available both in the same region and outside it, while in the third case the rule r will be available only in the region outside membrane i (if this is the environment, then the rule might be "lost", because we cannot take it back from the environment). Anyway, both copies of a should evolve (the maximality of the parallelism). If r is not present, then a remains unchanged; if a is not present, then the rule is not used, hence it remains available for the next step.

Thus, a rule can be used an arbitrarily large number of times in a region, and for each different target *tar* from pairs (r, tar) used, a different region of the system will have the rule available in the next step.

The sets $D_i, 1 \leq i \leq m$, are not necessarily "R-complete", that is, not each rule $r \in R$ has a metarule (r, tar) in each region; if a rule r arrives in a region i for which no metarule (r, tar) is provided by D_i, then the rule cannot be used, it is ignored from that step on (in region i). Thus, in each set D_i we can have none, one, two, or three different pairs (r, tar) for each r, with different $tar \in \{here, out, in\}$.

The computation starts from the initial configuration of the system, that described by $w_1, \ldots, w_m, R_1, \ldots, R_m$, and evolves according to the metarules from $D_1, \ldots, D_m$. With a halting computation we associate a result, in the form of the vector $\Psi_T(w)$ describing the multiplicity of elements from T present in the output membrane i_o in the halting configuration (thus, $w \in T^*$, the objects from $O - T$ are ignored). We denote by $Ps(\Pi)$ the set of vectors of natural numbers computed in this way by Π.

As usual, the rules from R can be cooperative, catalytic, or noncooperative. We denote by $PsR_MP_m(\alpha)$ the family of sets of vectors $Ps(\Pi)$ computed as above by P systems with rule migration using at most $m \geq 1$ membranes, with rules of type $\alpha \in \{coo, cat, ncoo\}$.

The systems from the proof of the next theorem illustrate the previous definition – and also shed some light on the power of P systems with migrating rules. We *conjecture* (actually, we mainly hope) that these systems are not universal. (In the theorem below, $PsMAT$ is the family of Parikh images of languages generated by matrix grammars without appearance checking.)

Theorem 1. *$PsMAT \subset PsR_MP_2(cat)$.*

Proof. Let us consider a matrix grammar without appearance checking $G = (N, T, S, M)$ in the binary normal form, that is, with $N = N_1 \cup N_2 \cup \{S\}$ and the matrices from M of the forms (1) $(S \to XA)$, $X \in N_1, A \in N_2$, (2) $(X \to Y, A \to x)$, $X, Y \in N_1, A \in N_2, x \in (N_2 \cup T)^*$, and (3) $(X \to \lambda, A \to x)$, $X \in N_1, A \in N_2, x \in T^*$; the sets $N_1, N_2, \{S\}$ are mutually disjoint and there is only one matrix of type (1). We assume all matrices of types (2) and (3) from M labeled in a one to one manner with elements of a set H. Without any loss of generality we may assume that for each symbol $X \in N_1$ there is at least one matrix of the form $(X \to \alpha, A \to x)$ in M (if a symbol X does not appear in such a matrix, then it can be removed, together with all matrices which introduce it by means of rules of the form $Z \to X$).

We construct the P system with migrating rules

$$\Pi = (O, \{c\}, T, R, [_1[_2\ ]_2]_1, w_1, w_2, R_1, R_2, D_1, D_2, 1),$$

where:

$$
\begin{aligned}
O = {} & N_1 \cup N_2 \cup T \cup \{A' \mid A \in N_2\} \cup \{h, h', h'', h''' \mid h \in H\} \cup \{c, \#\}, \\
R = {} & \{r_{1,h} : X \to h, \\
& \quad r_{2,h} : h \to h', \\
& \quad r_{3,h} : h' \to h'', \\
& \quad r_{4,h} : h'' \to h''', \\
& \quad r_{5,h} : ch''' \to c\#, \\
& \quad r_{6,h} : h''' \to \alpha, \\
& \quad r_{7,h} : h \to Ah'', \\
& \quad r_{8,h} : cA \to cx' \mid \text{for all } h : (X \to \alpha, A \to x) \in M, \\
& \quad X \in N_1, \alpha \in N_1 \cup \{\lambda\}, A \in N_2, x \in (N_2 \cup T)^*\} \\
& \cup \{r_B : B' \to B, \\
& \quad r'_B : B \to B \mid \text{for all } B \in N_2\} \\
& \cup \{r_X : X \to XX, \\
& \quad r'_X : X \to \lambda \mid \text{for all } X \in N_1\} \\
& \cup \{r_\infty : \# \to \#\}, \\
& \quad \text{where } x' \text{ is the string obtained by priming all symbols} \\
& \quad \text{from } N_2 \text{ which appear in } x \in (N_2 \cup T)^* \text{ and} \\
& \quad \text{leaving unchanged the symbols from } T, \\
w_1 = {} & cX_1 \ldots X_s, \text{ for } N_1 = \{X_1, X_2, \ldots, X_s\}, s \geq 1, \\
w_2 = {} & cXA, \text{ for } (X \to XA) \text{ being the initial matrix of } G, \\
R_1 = {} & \{r_{4,h}, \ r_{6,h}, \ r_{7,h}, \ r_{8,h} \mid h \in H\} \\
& \cup \{r_X, \ r'_X \mid X \in N_1\}, \\
R_2 = {} & \{r_{1,h}, \ r_{2,h}, \ r_{3,h}, \ r_{4,h}, \ r_{5,h} \mid h \in H\} \\
& \cup \{r_B, \ r'_B \mid B \in N_2\} \\
& \cup \{r_\infty\},
\end{aligned}
$$

and with the following sets of metarules:

D_1 contains the pair $(r, here)$ for all rules $r \in R$ with the exception of the rules from the next pairs
$(r_{1,h}, out)$, $(r_{6,h}, out)$, $(r_{8,h}, out)$, for all $h \in H$, which are in D_1, too;
D_2 contains the pair $(r, here)$ for all rules $r \in R$ with the exception of the rules from the next pairs
$(r_{1,h}, in)$, $(r_{6,h}, in)$, $(r_{8,h}, in)$, for all $h \in H$, which are in D_2, too.

Note that the metarules are "deterministic", in the sense that for each $r \in R$ there is only one pair (r, tar) in each of D_1 and D_2, and that the only migrating rules are $r_{1,h}, r_{6,h}, r_{8,h}$, for each $h \in H$. Initially, $r_{1,h}$ is in region 2 and $r_{6,h}, r_{8,h}$ are in region 1.

For the reader convenience, the initial configuration of the system, including the metarules different from $(r, here)$ from the two regions, is pictorially represented in Figure 1.

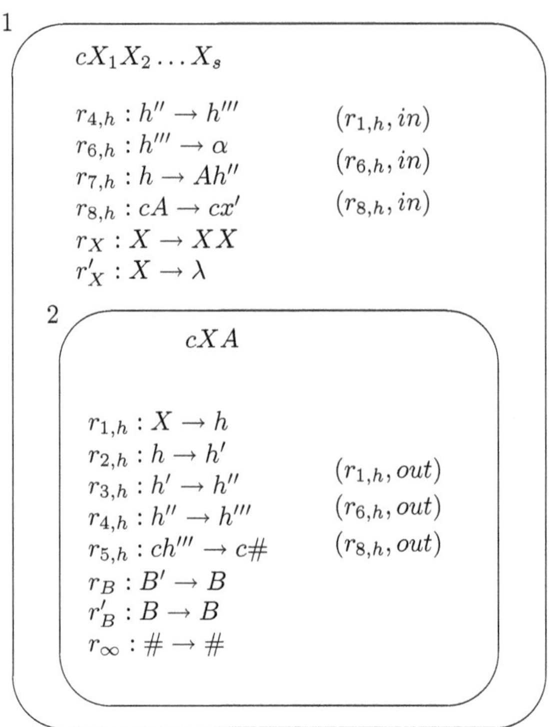

Figure 1: The P systems with migrating rules from Theorem 1.

The skin region contains the rules $X \to XX$ and $X \to \lambda$ for each $X \in N_1$; they can produce any necessary copy of symbols from N_1 and can also remove all such symbols in the end of a computation. The inner region contains a rule $B \to B$ for each $B \in N_2$, hence the computation will continue as long as any symbol from N_2 is present.

Assume that we start with a multiset cXw, $X \in N_1, w \in (N_2 \cup T)^*$ in region 2 – initially, we have here cXA, for XA introduced by the initial matrix of G. The only possibility (besides rules $r'_B : B \to B$ which we will ignore from now on) is to use a rule of the form $r_{1,h}$ for some matrix $h : (X \to \alpha, A \to x)$. We obtain the object h in region 2, and the rule $r_{1,h}$ goes to region 1.

Assume that in region 1 we have available a copy of X, moreover, we use it for applying the rule $r_{1,h}$; thus, the rule $r_{1,h}$ returns to the inner region. At the same time, in region 2 we use the rule $r_{2,h} : h \to h'$. Now we have h in region 1 and h' in region 2; no nonterminal from N_1 is available in region 2, so no rule of type $r_{1,g}, g \in H$, can be used. Simultaneously, we use now $r_{3,h} : h' \to h''$ in region 2 and $r_{7,h} : h \to Ah''$ in region 1. Both rules remain in the respective regions.

We continue with $r_{4,h} : h'' \to h'''$ in region 2, and with both $r_{4,h} : h'' \to h'''$ and $r_{8,h} : cA \to cx'$ in region 1. The first two rules remain in the respective regions, the last one goes to region 2. In region 2 we have available both rules $r_{5,h} : ch''' \to c\#$ and $r_{8,h} : cA \to cx'$ (but not also $r_{6,h} : h''' \to \alpha$, which is still in region 1). If the rule $r_{r,5} : ch''' \to c\#$ is used, then the computation will never finish. Thus, we have to use the rule $r_{8,h} : cA \to cx'$, which simulates the second rule of the matrix h, with the nonterminals of x primed. In parallel, in region 1 we have to use the rule $r_{6,h} : h''' \to \alpha$, which has then to enter region 2. In this way, at the next step in region 2 we can avoid using the rule $r_{5,h} : ch''' \to c\#$ and use the rule $r_{6,h} : h''' \to \alpha$, which completes the simulation of the matrix h (in parallel, the rules r_B – always available – return x' to x).

If X is not present in region 1, or only one copy of X is present and we do not apply to it the rule $X \to h$, then the rule $r_{7,h}$ is not used in the next step, hence neither $r_{6,h}$ and $r_{8,h}$ after that, and then the use of the rule $r_{5,h} : ch''' \to c\#$ cannot be avoided, and # is intoduced in membrane 2, making the computation never halt.

In the case when $\alpha \in N_1$, the process can be iterated: after simulating any matrix, the rules $r_{1,h}, r_{6,h}, r_{8,h}$ are back in the regions where they were placed in the beginning; note that the symbols from N_2 present in region 1 are primed, hence the rules $r_{8,h}$ cannot use them, while the possible symbol Y introduced in region 1 is just one further copy of such symbols already present here.

In the case when $\alpha = \lambda$, hence h was a terminal matrix, the computation stops if no symbol $B \in N_2$ is present in region 2 (hence the derivation in G is terminal), or continue forever otherwise. Consequently, $Ps(N) = \Psi_T(L(G))$, which shows that we have the inclusion $PsMAT \subseteq PsR_MP_2(cat)$.

This is a proper inclusion. In fact, we have the stronger assertion $PsR_MP_1(ncoo) - PsMAT \neq \emptyset$, which is proved by the fact that one-membrane P systems can compute one-letter non-semilinear sets of numbers (such sets cannot be Parikh sets of one-letter matrix languages). This is the case with the following system

$$\Pi = (\{a\}, \emptyset, \{a\}, \{h : a \to aa\}, [_1\ ]_1, a, \{h\}, \{(h, here), (h, out)\}, 1).$$

In each step we double the number of copies of a existing in the system; if we use at least once the metarule $(h, here)$, then the computation continues, otherwise the rule is "lost" in the environment and the computation halts. Clearly, $Ps(\Pi) = \{(2^n) \mid n \geq 1\}$, which is not in $PsMAT$. □

We do not know whether the result in Theorem 1 can be improved, by computing all Parikh images of recursively enumerable languages, or – especially interesting – by using only non-cooperative rules.

The system constructed at the beginning of the previous proof has never replicated a rule, because in each region each rule had only one target associated by the local metarules. In the case when the metarules specify different targets for the same rule, we can either proceed as in the definition given above, or we can restrict the use of metarules so that only one target is used. On the other hand, we can complicate the matter by adding further features, such as the possibility to dissolve a membrane (then both the objects and the rules should remain free in the immediately upper membrane), or the possibility to also move objects through membranes. Furthermore, we can take as the result of a computation the trace of a designated rule in its passage through membranes, in the same way as the trace of a traveller-object was considered in "standard" (symport/antiport) P systems. Then, we can pass to systems with symport/antiport rules; in such a case, *out* will mean that the rule moves up and gets associated with the membrane immediately above the membrane with which the rule was associated (and applied), while *in* will mean going one step down in the membrane structure. Otherwise stated, the targets are, in fact, *here, up, down.*

Plenty of problems, but we switch to another "exotic" idea.

3 Counter-Dividing Membranes

The problem to consider reversible P systems was formulated several times – with "reversible" meaning both the possibility to reverse computations, like in dynamic systems, and local/strong reversibility, at the level of rules: considering $v \to u$ at the same time with $u \to v$. For multiset rewriting-like rules this does not seems very spectacular (although most proofs from the membrane computing literature, if not all proofs, will get ruined by imposing this restriction to the existing sets of rules, at least because the synchronization is lost). The case of membrane division rules looks different/strange. A rule $[_i a]_i \to [_i b]_i [_i c]_i$ produces two identical copies of the membrane i, replicating the contents of the divided membrane, the only difference between the two copies being that objects b and c replace the former object a. Moreover, generalizations can be considered, with division into more than two new membranes, with new membranes having (the same contents but) different labels. Also, we can divide both elementary and non-elementary membranes.

Let us stay at the simplest level, of rules $[_i a]_i \to [_i b]_i [_i c]_i$ which deal with elementary membranes only, 2 division, and the same label for the new membranes. Reversing such a rule, hence considering a rule $[_i b]_i [_i c]_i \to [_i a]_i$, will mean to pass from two membranes with the label i and identical contents up to the two objects b and c present in them to a single membrane, with the same contents and with the objects b, c replaced by a new object, a. This operation looks rather powerful, as in only one step it compares the contents of two membranes, irrespective how large they are, and removes one of them, also changing one object.

How to use this presumed power, for instance, for obtaining solutions to computationally hard problems, remains to be investigated. Here we are only mentioning the fact that this counter-division operation can be used in order to build a system which computes the intersection of the sets of numbers/vectors computed by two given systems. The idea is suggested in Figure 2. Consider two

P systems, Π_1 and Π_2. Associate with each of them one membrane with label i. Embed one of the systems, say Π_2, together with the associated membrane i, in a further membrane, with label 2, and then all these membranes into a skin membrane (1, in Figure 2). Let the systems Π_1, Π_2 work and halt; the objects defining the results are moved out of the two systems and then into the corresponding membranes with label i. At some (suitable) moment, dissolve membrane 2. Now, by a rule $[_i b]_i [_i c]_i \rightarrow [_i a]_i$ we check whether or not the two systems have produced the same number (or vector of numbers), and, in the affirmative case, we stop. (For instance, each membrane i can evolve forever by means of suitable rules applied to b and c, respectively; after removing one of the membranes, the object a newly introduced will not evolve, hence the computation can stop.) Of course, several details are to be completed, depending on the type of the systems we work with.

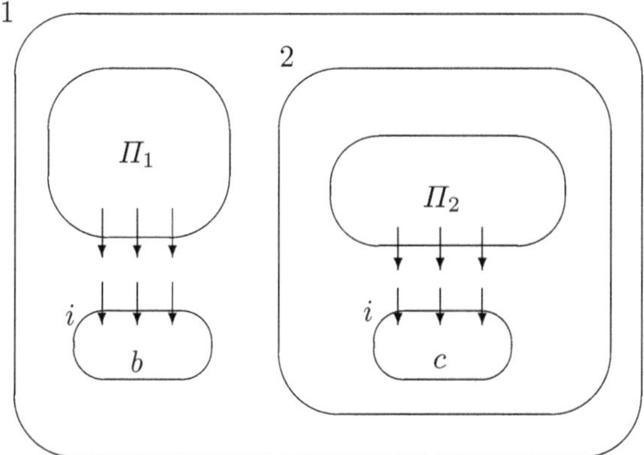

Figure 2: Computing the intersection.

Now, again the question arises: is this way to compute the intersection of any usefulness? We avoid addressing this issue, and pass to the third new type of P systems.

4 Rules (and Membranes) Acceleration

A new type of P systems, but not a new idea: accelerated Turing machines were considered since many years, see [1], [2], [3] and their references. Assume that our machine is so clever that it can learn from its own functioning, and this happens at the spectacular level of always halving the time needed to perform a step; the first step takes, as usual, one time unit. Then, in $1 + \frac{1}{2} + \frac{1}{4} + \ldots + \frac{1}{2^n} + \ldots = 2$ time units the machine performs an infinity of steps, hence finishing the computation... Beautiful, but exotic...

Let us consider P systems using similarly clever rules, each one learning from its own previous applications and halving the time needed at the next application. Because at the same time we can have several rules used in parallel, in the same region or in separate regions, a problem appears with the cooperation of rules, with the objects some rules are producing for other rules to use. As it is natural to assume, such objects are available only after completing the use of a rule. Because the rules have now different "speeds" of application, we have to take care of the times when they can take objects produced by other rules.

How long – in terms of time units, not of computational steps (of used rules) – lasts a computation? For Turing machines, two time units are sufficient for carrying out any computation. In our case, each rule can take two time units, so a system with n rules, not necessarily used in paralel, will compute for at most $2n$ time units. Thus, if we were able to find accelerated P systems able to compute all Turing computable sets of numbers/vectors, then an upper bound can be obtained on the time needed for all these computations, by considering a universal P system (hence with a given number of rules). The only problem is to find P systems (of various types) able to compute all computable sets of numbers/vectors in the accelerated mode.

It is relatively easy to find such systems – the trick is to ensure that no two rules are used in parallel, and then no synchronization problem appears.

Consider first the case of symbol-object P systems with multiset rewriting-like rules of a cooperative type. It is easy to see that the P system constructed in the proof of Theorem 3.3.3 from [4] computes the same set of numbers with or without having accelerated rules (with the exception of rules dealing with the trap symbol #, which anyway make the computation to never stop, all other rules are not used in parallel).

The same assertion is obtained also for P systems with symport/antiport rules – specifically, for systems with antiport rules of weight 2 and no symport rule. Because of the technical interest in the proof of this assertion, we give it in some details. Let $N(\Pi)$ be the set of numbers computed by a P system Π and let $NOP_m(acc, sym_r, anti_s)$ be the family of sets $N(\Pi)$ computed by systems with at most m membranes, using accelerated symport rules of weight at most r and antiport rules of weight at most s. By NRE we denote the family of Turing computable sets of natural numbers (the length sets of recursively enumerable languages).

Theorem 2. $NRE \subseteq NOP_1(acc, sym_0, anti_2)$.

Proof. Consider a matrix grammar with appearance checking $G = (N, T, S, M, F)$ in the Z-binary normal form. Therefore, we have $N = N_1 \cup N_2 \cup \{S, Z, \#\}$, with these three sets mutually disjoint, and the matrices in M are in one of the following forms:

1. $(S \to XA)$, with $X \in N_1, A \in N_2$,
2. $(X \to Y, A \to x)$, with $X, Y \in N_1, A \in N_2, x \in (N_2 \cup T)^*, |x| \leq 2$,
3. $(X \to Y, A \to \#)$, with $X \in N_1, Y \in N_1 \cup \{Z\}, A \in N_2$,
4. $(Z \to \lambda)$.

Moreover, there is only one matrix of type 1, F consists exactly of all rules $A \to \#$ appearing in matrices of type 3, and if a sentential form generated by

G contains the object Z, then it is of the form Zw, for some $w \in (T \cup \{\#\})^*$ (that is, the appearance of Z makes sure that, except for Z and, possibly, #, all objects which are present in the system are terminal); # is a trap-object, and the (unique) matrix of type 4 is used only once, in the last step of a derivation.

We construct the following P system with accelerated antiport rules:

$$\begin{aligned}
\Pi &= (O, T, [_1\ ]_1, XA, O, R_1, 1),\\
O &= N_1 \cup N_2 \cup T \cup \{A', Y_A, Y'_A \mid Y \in N_1, A \in N_2 \cup \{Z\}\}\\
&\quad \cup \{\langle Y\alpha\rangle \mid Y \in N_1, \alpha \in N_2 \cup T \cup \{\lambda\}\} \cup \{Z, \#\},\\
R_1 &= \{(XA, out; \langle Y\alpha_1\rangle\alpha_2, in),\\
&\qquad (\langle Y\alpha_1\rangle, out; Y\alpha_1, in) \mid \text{for } (X \to Y, A \to \alpha_1\alpha_2) \in M,\\
&\qquad \text{with } X, Y \in N_1, A \in N_2, \alpha_1, \alpha_2 \in N_2 \cup T \cup \{\lambda\}\}\\
&\quad \cup \{(X, out; Y_A A', in),\\
&\qquad (A'A, out; \#, in),\\
&\qquad (Y_A, out; Y'_A, in),\\
&\qquad (Y'_A A', out; Y, in) \mid \text{for } (X \to Y, A \to \#) \in M,\\
&\qquad \text{with } X \in N_1, Y \in N_2 \cup \{Z\}, A \in N_2\}\\
&\quad \cup \{(\#, out; \#, in)\}\\
&\quad \cup \{(X, out; X, in) \mid X \in N_1\}.
\end{aligned}$$

The equality $N(\Pi) = \{|w| \mid w \in L(G)\}$ can easily be checked, even for the accelerated way of working. Indeed, never two rules are used in parallel, hence the acceleration makes no trouble. Now, the matrices $(X \to Y, A \to \alpha_1\alpha_2)$ without appearance checking rules can be directly simulated by means of rules $(XA, out; \langle Y\alpha_1\rangle\alpha_2, in), (\langle Y\alpha_1\rangle, out; Y\alpha_1, in)$. A matrix $(X \to Y, A \to \#)$ is simulated as follows. First, we use $(X, out; Y_A A', in)$, hence X is sent to the environment and Y_A, A' enter the system (note that all objects are available in the environment in arbitrarily many copies). If any copy of A is present in the system, then the rule $(A'A, out; \#, in)$ must be used and the computation will never halt. If A is not present, then the object A' waits in the system for the object Y'_A, which is introduced by the rule $(Y_A, out; Y'_A, in)$. By the rule $(Y'_A A', out; Y, in)$ we then send out the auxiliary objects Y'_A, A' and bring in the object Y, thus completing the simulation of the matrix. As long as any symbol $X \in N_1$ is present, the computation continues, at least by the rule $(X, out; X, in)$. Because always we have at most one object $X \in N_1$ in the system, no synchronization problem appears. When the symbol Z is introduced, this means that the derivation in G is concluded. No rule processes the object Z in the system Π. If the object # is present, then the computation will continue forever, otherwise also the computation in Π stops. Thus, we stops if and only if the obtained multiset – minus the object Z – corresponds to a terminal string generated by G, and this concludes the proof. □

Several research topics remain to be investigated also in this case. One of them was already mentioned: find a constant U such that each set $H \in NRE$

can be computed in at most U time units by a P system of a given type (for instance, corresponding to the family $NOP_1(acc, sym_0, anti_2)$). To this aim, it is first necessary to find a universal matrix grammar with appearance checking in the Z-normal form (or to start the proof from other universal devices equivalent with Turing machines).

Then, it is natural to consider P systems where certain components are accelerated. In two time units, such a component finishes its work, possibly producing an element from a set from NRE. Can this be used by the other membranes of the system in order to speed-up computations or even to go beyond Turing, computing sets of numbers which are not Turing computable? This speculation has been made from time to time with respect to bio-computing models, membrane systems included, but no biologically inspired idea was reported able to reach such a goal. The acceleration of rules and/or of membranes can be a solution – though not necessarily biologically inspired.

Indeed, accelerated Turing machines can solve Turing undecidable problems, for instance, the halting problem, in the following sense. Consider a Turing machine M and an input w for it. Construct an accelerated Turing machine M', obtained by accelerating M and also providing M' with a way to output a signal s if and only if M halts when starting with w on its tape. Letting M' work, in two time units we have the answer to the question whether or not M halts on w: if M halts, then also M' halts and provides the sugnal s; if M does not halt, then neither M' halts, but its computation, though infinite, lasts only two time units. So, M halts on w if and only if M' outputs s after two time units. Again, it is important to note the difference between the external time (measured in "time units"), and the internal duration of a computation, measured by the number of rules used by the machine. Also important is the fact that M' is not a proper Turing machine, because "providing a signal s" is not an instruction in a Turing machine. (This way to get an information about a computation reminds of the way of solving decision problems by P systems, by sending into the environment a special object *yes*. The difference is that sending objects outside the system is a "standard instruction" in P systems.) Further discussions about accelerated Turing machines and their relation with "so-called Turing-Church thesis" can be found, e.g., in [2].

It is now natural to use acceleration for constructing P systems which can compute Turing non-computable sets of numbers (or functions, or languages, etc). In the string-objects case with cooperative rules this task is trivial: just take a Turing machine and simulate it by a P system; an accelerated Turing machine able of non-Turing computations will lead to a P system with the same property.

The symbol-object case does not look similarly simple. On the one hand, we do not know such a thing like an accelerated deterministic register machine (or matrix grammar with appearance checking). On the other hand, we have to make sure that the simulation of such a machine by means of a P system of a given type faithfully corresponds to the functioning of the machine: the system has to stop if and only if the machine stops. This is not the case, for instance, with the system from the proof of Theorem 2, because of the intrinsically nondeterministic behavior of Π: the trap symbol # can be introduced because of the attempt to simulate a "wrong" matrix, the derivation in G could correctly continue and eventually correctly halt, but the system fails to simulate it. It seems to be a challenging task to find an accelerated symport/antiport system which can avoid this difficulty (and hence can compute Turing non-computable sets of numbers).

Actually, in [3] one discusses ten possibilities of changing a Turing machine in such a way to obtain "hypercomputations" (computations which cannot be carried out by usual Turing machines); we do not recall these possibilities here, but we only point out them to the reader interested in "going beyond Turing".

5 Reliable P Systems

As suggested in the Introduction, the intention behind the definition of reliable P systems is to make use of the following observation. Assume that we have some copies of an object a and two rules, $a \to b$ and $a \to c$, in the same region. Then, some copies of a will evolve by means of the first rule and some others by means of the second rule. All combinations are possible, from "all copies of a go to b" to "all copies of a go to c". In biology and chemistry the range of possibilities is not so large: if we have "enough" copies of a, then "for sure" part of them will become b and "for sure" the other part will become c. What "enough" can mean depends on the circumstances. At our symbolic level, if we have two copies of a we cannot expect that "for sure" one will become b and one c, but starting from, say, some dozens of copies can ensure that both rules are applied.

We formulate this observation for string rewriting. Assume that a string w can be rewritten by n rules (each of them can be applied to w). The system we work with is *reliable* if all the n rules are used as soon as we have at least $n \cdot n^k$ copies of w, for a given constant k depending on the system. If we work with parallel rewriting and n possible derivations $w \Longrightarrow w_i$, $1 \leq i \leq n$, are possible, then each one happens as soon as we have at least $n \cdot n^k$ copies of w.

We consider here polynomially many opportunities for each of the n alternatives, but other functions than polynomials can be considered; constant functions would correspond to a "very optimistic" approach, while exponential functions would indicate a much less optimistic approach.

Instead of elaborating more at the general level (although many topics arise here: give a fuzzy sets, rough sets, or probabilistic definition of reliability; provide some sufficient conditions for it, a sort of axioms entailing reliability; investigate its usefulness at the theoretical level and its adequacy/limits in practical applications/experiments), we pass directly to show a way to use the idea of reliability in solving SAT in linear time.

Consider a propositional formula C in the conjunctive normal form, consisting of m clauses C_j, $1 \leq j \leq m$, involving n variables x_i, $1 \leq i \leq n$. We construct the following P system (with string-objects, and both replicated and parallel rewriting – but the rules which replicate strings always applicable to strings of length one only) of degree m:

$$
\begin{aligned}
\Pi &= (V, \mu, b_0, \lambda, \ldots, \lambda, R_1, \ldots, R_m), \\
V &= \{b_i \mid 0 \leq i \leq r\} \cup \{a_i, t_i, f_i \mid 1 \leq i \leq n\}, \\
\mu &= [_1[_2 \cdots [_{m-1}[_m\ ]_m]_{m-1} \cdots]_2]_1, \\
R_m &= \{b_i \to b_{i+1}||b_{i+1} \mid 0 \leq i \leq r-1\} \\
&\quad \cup \{b_r \to a_1a_2\ldots a_n\} \\
&\quad \cup \{a_i \to t_i,\ a_i \to f_i \mid 1 \leq i \leq n\} \\
&\quad \cup \{(t_i \to t_i, out) \mid \text{if } x_i \text{ appears in } C_m, 1 \leq i \leq n\} \\
&\quad \cup \{(f_i \to f_i, out) \mid \text{if } \sim x_i \text{ appears in } C_m, 1 \leq i \leq n\}, \\
R_j &= \{(t_i \to t_i, out) \mid \text{if } x_i \text{ appears in } C_j, 1 \leq i \leq n\} \\
&\quad \cup \{(f_i \to f_i, out) \mid \text{if } \sim x_i \text{ appears in } C_j, 1 \leq i \leq n\}, \\
&\quad \text{for all } j = 1, 2, \ldots, m-1.
\end{aligned}
$$

The parameter r from this construction depends on the polynomial which ensures the reliability of the system – see immediately bellow what this means.

The work of the system starts in the central membrane, the one with the label m. In the first r steps, by means of replicating rules of the form $b_i \to b_{i+1}||b_{i+1}$, we generate 2^r strings b_r (of length one; this phase can be considered as using symbol objects, and then the rules are usual multiset rewriting rules, $b_i \to b_{i+1}^2$). In the next step, each b_r is replaced by the string $a_1a_2\ldots, a_n$. No parallel rewriting was used up to now, but in the next step each a_i is replaced either by t_i or by f_i. We have 2^n possibilities of combining the rules $a_i \to t_i,\ a_i \to f_i,\ 1 \leq i \leq n$. *If 2^r is "large enough" with respect to 2^n, then we may assume that all the 2^n combinations really happen.* This is the crucial reliability-based step. For each of the 2^n possibilities we need $(2^n)^k$ opportunities to happen. This means that we need to have a large enough r such that $2^r \geq 2^n \cdot (2^n)^k$, for a constant k (depending on the form of rules, hence on SAT). This means $r \geq n(k+1)$ – hence this is the number of steps we need to perform before introducing the strings $a_1a_2\ldots a_n$ in order to ensure that we have "enough" copies of these strings.

After having all truth-assignments generated in membrane m, we check whether or not there is at least one truth-assignment which satisfies clause C_m; all truth-assignments for which C_m is true exit membrane m. In membrane $m-1$ we check the satisfiability of clause C_{m-1}, and again we pass one level up all truth-assignments which satisfy C_{m-1}. We continue in this way until reaching the skin region, where we check the satisfiability of C_1. This means that after the $r+3$ steps in membrane m we perform further $m-1$ steps; in total, this means $r+m+2$ steps.

The formula C is satisfiable if and only if at least one string is sent out of the system in step $r+m+2$. Because r is linearly bounded with respect to n, the problem was solved in a linear time (with respect to both n and m).

Note that the system is of a polynomial size with respect to n and m, hence it can be constructed in polynomial time by a Turing machine starting from C. If we start from a formula in the 3-normal form (at most three variables in each clause), then the system will be of a linear size in terms of n and m.

Reliability seems to be both a well motivated property, closely related to the biochemical reality, and a very useful one from a theoretical point of view; further investigations in this area are worth pursuing.

References

1. B.J. Copeland, Even Turing machines can compute uncomputable functions, in *Unconventional Models of Computation* (C.S. Calude, J. Casti, M.J. Dinneen, eds.), Springer-Verlag, Singapore, 1998, 150–164.
2. B.J. Copeland, R. Sylvan, Beyond the universal Turing machine, *Australasian Journal of Phylosophy*, **77** (1999), 46–66.
3. T. Ord, *Hypercomputation: compute more than the Turing machine*, Honorary Thesis, Department of Computer Science, Univ. of Melbourne, Australia, 2002 (available at `http://arxiv.org/ftp/math/papers/0209/0209332.pdf`).
4. Gh. Păun, *Membrane Computing. An Introduction*, Springer-Verlag, Berlin, 2002.
5. Gh. Păun, G. Rozenberg, A guide to membrane computing, *Theoretical Computer Science*, **287** (2002), 73–100.

The P Versus NP Problem Through Cellular Computing with Membranes

Mario J. Pérez-Jiménez, Alvaro Romero-Jiménez, and Fernando Sancho-Caparrini

Dpt. Computer Science and Artificial Intelligence
E.T.S. Ingeniería Informática. University of Seville
Avda. Reina Mercedes s/n, 41012 Sevilla, Spain
{Mario.Perez, Alvaro.Romero, Fernando.Sancho}@cs.us.es

Abstract. We study the **P** *versus* **NP** problem through membrane systems. Language accepting P systems are introduced as a framework allowing us to obtain a characterization of the $\mathbf{P} \neq \mathbf{NP}$ relation by the polynomial time unsolvability of an **NP**–complete problem by means of a P system.

1 Introduction

The **P** *versus* **NP** problem [2] is the problem of determining whether every language accepted by some non-deterministic algorithm in polynomial time is also accepted by some deterministic algorithm in polynomial time. To define the above problem precisely we must have a formal definition for the concept of an algorithm. The theoretical model to be used as a computing machine in this work is the Turing machine, introduced by Alan Turing in 1936 [10], several years before the invention of modern computers.

A deterministic Turing machine has a transition function providing a functional relation between configurations; so, for every input there exists only one computation (finite or infinite), allowing us to define in a natural way when an input is *accepted* (through an accepting computation).

In a non-deterministic Turing machine, for a given configuration several successor configurations can exist. Therefore, it could happen that for a given input different computations exist. In these machines, an input is accepted if there exists at least one finite accepting computation associated with it.

The class **P** is the class of languages accepted by some deterministic Turing machine in a time bounded by a polynomial on the length (size) of the input. From an informal point of view, the languages in the class **P** are identified with the problems having an efficient algorithm that gives an answer in a feasible time; the problems in **P** are also known as *tractable* problems.

The class **NP** is the class of languages accepted by some non-deterministic Turing machine where for every accepted input there exists at least one accepting computation taking an amount of steps bounded by a polynomial on the length of the input.

N. Jonoska et al. (Eds.): Molecular Computing (Head Festschrift), LNCS 2950, pp. 338–352, 2004.

Every deterministic Turing machine can be considered as a non-deterministic one, so we have $\mathbf{P} \subseteq \mathbf{NP}$. In terms of the previously defined classes, the **P** *versus* **NP** problem can be expressed as follows: is it verified the relation $\mathbf{NP} \subseteq \mathbf{P}$?

The $\mathbf{P} \stackrel{?}{=} \mathbf{NP}$ question is one of the outstanding open problems in theoretical computer science. The relevance of this question does not lie only in the inherent pleasure of solving a mathematical problem, but in this case an answer to it could provide an information of a high practical interest. For instance, a negative answer to this question would confirm that the majority of current cryptographic systems are secure from a practical point of view. On the other hand, a positive answer could not only entail the vulnerability of cryptographic systems, but this kind of answer is expected to come together with a general procedure which will provide a deterministic algorithm solving any **NP**-complete problem in polynomial time.

Moreover, the problems known to be in the class **NP** but not known to be in **P** are varied and of highest practical interest. An **NP**–complete problem is a *hardest* (in certain sense) problem in **NP**; that is, any problem in **NP** could be efficiently solved using an efficient algorithm which solves a fixed **NP**–complete problem. These problems are the suitable candidates to attack the **P** *versus* **NP** problem.

In the last years several computing models using powerful and inherent tools inspired from nature have been developed (because of this reason, they are known as *bio-inspired* models) and several solutions in polynomial time to problems from the class **NP** have been presented, making use of non-determinism or of an exponential amount of space. This is the reason why a practical implementation of such models (in biological, electronic, or other media) could provide a quantitative improvement for the resolution of **NP**-complete problems.

In this work we focus on one of these models, the *cellular computing model with membranes*, specifically, on one of its variants, the language accepting P systems, in order to develop a computational complexity theory allowing us to attack the **P** *versus* **NP** problem from other point of view than the classical one.

The paper is structured as follows. The next section is devoted to the definition of language accepting P systems. In section 3 a polynomial complexity class for the above model is introduced. Sections 4 and 5 provides simulations of deterministic Turing machines by P systems and language accepting P systems by deterministic Turing machines. Finally, in section 6 we establish a characterization of the **P** *versus* **NP** problem through P systems.

2 Language Accepting P Systems

Until the end of 90's decade several natural computing models have been introduced simulating the way nature *computes* at the genetic level (genetic algorithms and DNA based molecular computing) and at the neural level (neural networks). In 1998, Gh. Păun [5] suggests a new level of computation: *the cellular level.*

Cells can be considered as machines performing certain computing processes; in the distributed framework of the hierarchical arrangement of internal vesicles, the communication and alteration of the chemical components of the cell are carried out. Of course, the processes taking place in the cell are complex enough for not attempting to completely model them. The goal is to create an abstract cell-like computing model allowing to obtain alternative solutions to problems which are intractable from a classical point of view.

The first characteristic to point out from the internal structure of the cell is the fact that the different units composing the cell are delimited by several types of *membranes* (in a broad sense): from the membrane that separates the cell from the environment into which the cell is placed, to those delimiting the inner vesicles. Also, with regard to the functionality of these membranes in nature, it has to be emphasized the fact that they do not generate isolated compartments, but they allow the chemical compounds to flow between them, sometimes in selective forms and even in only one direction. Similar ideas were previously considered, for instance, in [1] and [3].

P systems are described in [4] as follows: a *membrane structure* consists of several membranes arranged in a hierarchical structure inside a main membrane (called the *skin*) and delimiting *regions* (each region is bounded by a membrane and the immediately lower membranes, if there are any). Regions contain *multisets* of *objects*, that is, sets of objects with multiplicities associated with the elements. The objects are represented by symbols from a given alphabet. They evolve according to given *evolution rules*, which are also associated with the regions. The rules are applied non-deterministically, in a maximally parallel manner (in each step, all objects which *can* evolve *must* do so). The objects can also be moved (*communicated*) between regions. In this way, we get *transitions* from one *configuration* of the system to the next one. This process is synchronized: a global clock is assumed, marking the time units common to all compartments of the system. A sequence (finite or infinite) of transitions between configurations constitutes a *computation*; a computation which reaches a configuration where no rule is applicable to the existing objects is a *halting computation*. With each halting computation we associate a *result*, by taking into consideration the objects collected in a specified *output membrane* or in the environment.

For an exhaustive overview of transition P systems and of their variants and properties, see [4].

Throughout this paper, we will study the capacity of cellular systems with membranes to attack the efficient solvability of presumably intractable decision problems. We will focus on a specific variant of transition P systems: *language accepting P systems*. These systems have an *input membrane*, and work in such a way that when introducing in the input membrane a properly encoded string, a "message" is sent to the environment, encoding whether this string belongs or not to a specified language.

Definition 1. *A* membrane structure *is a rooted tree, where the nodes are called* membranes, *the root is called* skin, *and the leaves are called* elementary membranes.

Definition 2. *Let* $\mu = (V(\mu), E(\mu))$ *be a membrane structure. The* membrane structure with external environment *associated with* μ *is the rooted tree such that: (a) the root of the tree is a new node that we denote by env; (b) the set of nodes is* $V(\mu) \cup \{env\}$*; and (c) the set of edges is* $E(\mu) \cup \{\{env, skin\}\}$.

The node *env* is called *environment* of the structure μ. So, every membrane structure has associated in a natural way an environment.

Definition 3. *A* language accepting P system *(with input membrane and external output) is a tuple*

$$\Pi = (\Sigma, \Gamma, \Lambda, \#, \mu_\Pi, \mathcal{M}_1, ..., \mathcal{M}_p, (R_1, \rho_1), ..., (R_p, \rho_p), i_\Pi)$$

verifying the following properties:

- *The input alphabet of* Π *is* Σ.
- *The working alphabet of* Π *is* Γ, *with* $\Sigma \subsetneq \Gamma$ *and* $\# \in \Gamma - \Sigma$.
- μ_Π *is a membrane structure consisting of* p *membranes, with the membranes (and hence the regions) injectively labelled with* $1, 2, \dots, p$.
- i_Π *is the label of the input membrane.*
- *The output alphabet of* Π *is* $\Lambda = \{Yes, No\}$.
- $\mathcal{M}_1, ..., \mathcal{M}_p$ *are multisets over* $\Gamma - \Sigma$, *representing the initial contents of the regions of* $1, 2, \dots, p$ *of* μ_Π.
- $R_1, ..., R_p$ *are finite sets of evolution rules over* Γ *associated with the regions* $1, 2, \dots, p$ *of* μ_Π.
- ρ_i, $1 \leq i \leq p$, *are partial order relations over* R_i *specifying a priority relation among rules of* R_i.

An *evolution rule* is a pair (u, v), usually represented $u \to v$, where u is a string over Γ and $v = v'$ or $v = v'\delta$, with v' a string over

$$\Gamma \times (\{here, out\} \cup \{in_i \mid i = 1, \dots, p\}).$$

Consider a rule $u \to v$ from a set R_i. To apply this rule in membrane i means to remove the multiset of objects specified by u from membrane i (the latter must contain, therefore, sufficient objects so that the rule can be applied), and to introduce the objects specified by v, in the membranes indicated by the *target commands* associated with the objects from v.

Specifically, for each $(a, out) \in v$ an object a will exit the membrane i and will become an element of the membrane immediately outside it (that is, the father membrane of membrane i), or will leave the system and will go to the environment if the membrane i is the skin membrane. If v contains a pair $(a, here)$, then the object a will remain in the same membrane i where the rule is applied (when specifying rules, pairs $(a, here)$ are simply written a, the indication *here* is omitted). For each $(a, in_j) \in v$ an object a should be moved in the membrane with label j, providing that this membrane is immediately inside membrane i (that is, membrane i is the father of membrane j); if membrane j is not directly accesible from membrane i (that is, if membrane j is not a child membrane of membrane i), then the rule cannot be applied. Finally, if δ appears in v, then membrane i is dissolved; that is, membrane i is removed from the membrane structure, and all objects and membranes previously present in it become elements of the immediately upper membrane (the father membrane) while the evolution rules and the priority relations of the dissolved membrane are removed. The skin membrane is never dissolved; that is, no rule of the form $u \to v'\delta$ is applicable in the skin membrane.

All these operations are done in parallel, for all possible applicable rules $u \to v$, for all occurrences of multisets u in the membrane associated with the rules, and for all membranes at the same time.

The rules from the set R_i, $1 \leq i \leq p$, are applied to objects from membrane i synchronously, in a non-deterministic maximally parallel manner; that is, we assign objects to rules, non-deterministically choosing the rules and the objects assigned to each rule, but in such a way that after this assignation no further rule can be applied to the remaining objects. Therefore, a rule can be applied in the same step as many times as the number of copies of objects allows it.

On the other hand, we interpret the priority relations between the rules in a *strong sense*: a rule $u \to v$ in a set R_i can be used only if no rule of a higher priority exists in R_i and can be applied at the same time with $u \to v$.

A *configuration* of Π is a tuple $(\mu, M_E, M_{i_1}, \ldots, M_{i_q})$, where μ is a membrane structure obtained by removing from μ_Π all membranes different from $i_1, \ldots, i_q$ (of course, the skin membrane cannot be removed), M_E is the multiset of objects contained in the environment of μ, and M_{i_j} is the multiset of objects contained in the region i_j.

For every multiset m over Σ (the input alphabet of the P system), the *initial configuration of Π with input m* is the tuple $(\mu_\Pi, \emptyset, \mathcal{M}_1, ..., \mathcal{M}_{i_\Pi} \cup m, ..., \mathcal{M}_p)$. That is, in any initial configuration of Π the environment is empty. We will denote by I_Π the collection of possible inputs for the system Π.

Given a configuration C of a P system Π, applying properly the evolution rules as described above, we obtain, in a non-deterministic way, a new configuration C'. We denote by $C \Rightarrow_\Pi C'$, and we say that we have a *transition* from C to C'. A *halting configuration* is a configuration in which no evolution rule can be applied.

A *computation* $\mathcal{C}$ of a P system is a sequence of configurations, $\{C^i\}_{i<r}$, where: C^0 is an initial configuration of the system; $C^i \Rightarrow_\Pi C^{i+1}$, for every $i < r$; and, either $r \in \mathbb{N}^+$ (that is, it is a non-zero natural number) and C^{r-1} is a *halting configuration*, or $r = \infty$, in which case it is said that $\mathcal{C}$ is not halting.

For a computation $\mathcal{C} = \{C^i\}_{i<r}$ we will denote by M_E^j the content of the environment in the configuration C^j. Next we define the output of the P system.

Definition 4. *The output of a computation* $\mathcal{C} = \{C^i\}_{i<r}$ *is:*

$$Output(\mathcal{C}) = \begin{cases} Yes, & \text{if } \mathcal{C} \text{ is halting, } Yes \in M_E^{r-1} \text{ and } No \notin M_E^{r-1}, \\ No, & \text{if } \mathcal{C} \text{ is halting, } No \in M_E^{r-1} \text{ and } Yes \notin M_E^{r-1}, \\ \text{not defined}, & \text{otherwise.} \end{cases}$$

If $\mathcal{C}$ *satisfies any of the two first conditions, then we say that it is a* successful computation.

Definition 5. *A language accepting P system is said to be* valid *if every halting computation is a successful computation and every halting computation, and only them, sends out the symbol # (and only in the last step).*

We denote by $\mathcal{LA}$ the class of valid language accepting P systems.

Next we define what it means that such P systems *accept* or *decide* a language.

Definition 6. *Let* L *be a language over an alphabet* Ω. *We say that the system* $\Pi \in \mathcal{LA}$ accepts *the language* L *if the following properties are verified:*

- *There exists a total function,* $cod : \Omega^* \to I_\Pi$, *computable and injective, encoding strings over* Ω *by means of multisets over the input alphabet of* Π.
- *For every string* $w \in \Omega^*$ *it is verified that:*
  - *If* $w \in L$, *then there exists* a *computation* $\mathcal{C}$ *of* Π *with input* $cod(w)$ *such that* $\mathcal{C}$ *is halting and* $Output(\mathcal{C}) = Yes$.
  - *If there exists* a *computation* $\mathcal{C}$ *of* Π *with input* $cod(w)$ *such that* $\mathcal{C}$ *is halting and* $Output(\mathcal{C}) = Yes$, *then* $w \in L$.

Definition 7. *Let* L *be a language over an alphabet* Ω. *We say that the system* $\Pi \in \mathcal{LA}$ decides *the language* L *if the following properties are verified:*

- *Every computation of* Π *is halting.*
- *There exists a total function,* $cod : \Omega^* \to I_\Pi$, *computable and injective, encoding strings over* Ω *by means of multisets over the input alphabet of* Π.
- *For every string* $w \in \Omega^*$ *it is verified that:*
  - *If* $w \in L$, *then for* every *computation* $\mathcal{C}$ *of* Π *with input* $cod(w)$ *it is verified that* $Output(\mathcal{C}) = Yes$.
  - *If* $w \notin L$, *then for* every *computation* $\mathcal{C}$ *of* Π *with input* $cod(w)$ *it is verified that* $Output(\mathcal{C}) = No$.

3 A Polynomial Complexity Class in Cellular Systems

In order to give a formal definition of computational complexity classes in this model, we have to first specify what we mean by a decision problem.

Definition 8. *A decision problem, X, is a pair (I_X, θ_X) such that I_X is a language (over a finite alphabet) whose elements are called* instances *of the problem and θ_X is a total Boolean function over I_X.*

A decision problem X is solvable by a Turing machine TM if I_X is the set of inputs of TM, for any $w \in I_X$ the Turing machine halts over w, and w is accepted if and only if $\theta_X(w) = 1$.

To solve a problem by means of P systems, we usually construct a family of such devices so that each element decides the instances of *equivalent size*, in a certain sense which will be specified below.

Definition 9. *Let $g : \mathbf{N}^+ \to \mathbf{N}^+$ be a total computable function. We say that a decision problem X is solvable by a family of valid language accepting P systems, in a time bounded by g, and we denote this by $X \in MC_{\mathcal{LA}}(g)$, if there exists a family of P systems, $\mathbf{\Pi} = \left(\Pi(n)\right)_{n \in \mathrm{N}^+}$, with the following properties:*

1. *For every $n \in \mathbb{N}$ it is verified that $\Pi(n) \in \mathcal{LA}$.*
2. *There exists a Turing machine constructing $\Pi(n)$ from n in polynomial time (we say that Π is polynomially uniform by Turing machines).*
3. *There exist two functions, $cod : I_X \to \bigcup_{n \in \mathrm{N}^+} I_{\Pi(n)}$ and $s : I_X \to \mathrm{N}^+$, computable in polynomial time, such that:*
   - *For every $w \in I_X$, $cod(w) \in I_{\Pi(s(w))}$.*
   - *The family $\mathbf{\Pi}$ is bounded, with regard to (X, cod, s, g); that is, for each $w \in I_X$ every computation of the system $\Pi(s(w))$ with input $cod(w)$ is halting and, moreover, it performs at most $g(|w|)$ steps.*
   - *The family $\mathbf{\Pi}$ is sound, with regard to (X, cod, s); that is, for each $w \in I_X$ if there exists an accepting computation of the system $\Pi(s(w))$ with input $cod(w)$, then $\theta_X(w) = 1$.*
   - *The family $\mathbf{\Pi}$ is complete, with regard to (X, cod, s); that is, for each $w \in I_X$ if $\theta_X(w) = 1$, then every computation of the system $\Pi(s(w))$ with input $cod(w)$ is an accepting computation.*

Note that we impose a certain kind of *confluence* of the systems, in the sense that every computation with the same input must return the same output.

As usual, the polynomial complexity class is obtained using as bounds the polynomial functions.

Definition 10. *The class of decision problems solvable in polynomial time by a family of cellular computing systems belonging to the class $\mathcal{LA}$, is*

$$\mathbf{PMC}_{\mathcal{LA}} = \bigcup_{g \ poly.} MC_{\mathcal{LA}}(g).$$

This complexity class is closed under polynomial-time reducibility.

Proposition 1. *Let X and Y be two decision problems such that X is polynomial-time reducible to Y. If $Y \in \mathbf{PMC}_{\mathcal{LA}}$, then $X \in \mathbf{PMC}_{\mathcal{LA}}$.*

4 Simulating Deterministic Turing Machines by P Systems

In this section we consider deterministic Turing machines as language decision devices. That is, the machines halt over any string on the input alphabet, with the halting state equal to the accepting state, in the case that the string belongs to the decided language, and with the halting state equal to the rejecting state in the case that the string does not belong to the language.

It is possible to associate with a Turing machine a decision problem, and this will permit us to define what means that such a machine is simulated by a family of P systems.

Definition 11. *Let TM be a Turing machine with input alphabet Σ_{TM}. The* decision problem associated with TM *is the problem $X_{TM} = (I, \theta)$, where $I = \Sigma_{TM}^*$, and for every $w \in \Sigma_{TM}^*$, $\theta(w) = 1$ if and only if TM accepts w.*

Obviously, the decision problem X_{TM} is solvable by the Turing machine TM.

Definition 12. *We say that a Turing machine TM is simulated in polynomial time by a family of systems of the class $\mathcal{LA}$, if $X_{TM} \in \mathbf{PMC}_{\mathcal{LA}}$.*

Next we state that every deterministic Turing machine can be simulated in polynomial time by a family of systems of the class $\mathcal{LA}$.

Proposition 2. *Let TM be a deterministic Turing machine working in polynomial time. Then $X_{TM} \in \mathbf{PMC}_{\mathcal{LA}}$.*

See chapter 9 of [8], which follows ideas from [9], for details of the proof.

5 Simulating Language Accepting P Systems by Deterministic Turing Machines

In this section we are going to prove that if a decision problem can be solved in polynomial time by a family of language accepting P systems, then it can also be solved in polynomial time by a deterministic Turing machine.

For the design of the Turing machine we were inspired by the work of C. Zandron, C. Ferretti and G. Mauri [11], with the difference that the mentioned paper deals with P systems with active membranes.

Proposition 3. *For every decision problem solvable in polynomial time by a family of valid language accepting P systems, there exists a Turing machine solving the problem in polynomial time.*

Proof. Let X be a decision problem such that $X \in \mathbf{PMC}_{\mathcal{LA}}$. Then, there exists a family of valid language accepting P systems $\mathbf{\Pi} = \big(\Pi(n)\big)_{n \in \mathbb{N}^+}$ such that:

1. The family $\mathbf{\Pi}$ is polynomially uniform by Turing machines.
2. There exist two functions $cod : I_X \to \bigcup_{n \in \mathbb{N}^+} I_{\Pi(n)}$ and $s : I_X \to \mathbb{N}^+$, computable in polynomial time, such that:
   - For every $w \in I_X$, $cod(w) \in I_{\Pi(s(w))}$.
   - The family $\mathbf{\Pi}$ is polynomially bounded, with regard to (X, cod, s).
   - The family $\mathbf{\Pi}$ is sound and complete, with regard to (X, cod, s).

Given $n \in \mathbb{N}^+$, let A_n be the number of symbols in the input alphabet of $\Pi(n)$, B_n the number of symbols in the working alphabet, C_n the number of symbols in the output alphabet, D_n the number of membranes, E_n the maximum size of the multisets initially associated with them, F_n the total number of rules of the system, and G_n the maximum length of them. Since the family $\mathbf{\Pi}$ is polynomially uniform by Turing machines, these numbers are polynomial with respect to n.

Let m be an input multiset of the system $\Pi(n)$. Given a computation $\mathcal{C}$ of $\Pi(n)$ with input m, we denote by $H_n(m)$ the maximum number of digits, in base 2, of the multiplicities of the objects contained in the multisets associated with the membranes of the systems and with the environment, in any step of $\mathcal{C}$. Naturally, this number depends on $\mathcal{C}$, but what we are interested in, and we will prove at the end of the proof, is that *any* computation of the system $\Pi(s(w))$ with input $cod(w)$ verifies that $H_{s(w)}(cod(w))$ is polynomial in the size of the string w.

Next, we associate with the system $\Pi(n)$ a deterministic Turing machine, $TM(n)$, *with multiple tapes*, such that, given an input multiset m of $\Pi(n)$, the machine reproduces *a* specific computation of $\Pi(n)$ over m.

The input alphabet of the machine $TM(n)$ coincides with that of the system $\Pi(n)$. On the other hand, the working alphabet contains, besides the symbols of the input alphabet of $\Pi(n)$ the following symbols: a symbol for each label assigned to the membranes of $\Pi(n)$; the symbols 0 and 1, that will allow to operate with numbers represented in base 2; three symbols indicating if a membrane has not been dissolved, has to be dissolved or has been dissolved; and three symbols that will indicate if a rule is awaiting, is applicable or is not applicable.

Subsequently, we specify the tapes of this machine.

- We have one *input tape*, that keeps a string representing the input multiset received.
- For each membrane of the system we have:
  - One *structure tape*, that keeps in the second cell the label of the father membrane, and in the third cell one of the three symbols that indicate if the membrane has not been dissolved, if the membrane has to dissolve, or if the membrane has been dissolved.
  - For each object of the working alphabet of the system:
    - One *main tape*, that keeps the multiplicity of the object, in base 2, in the multiset contained in the membrane.
    - One *auxiliary tape*, that keeps temporary results, also in base 2, of applying the rules associated with the membrane.

- One *rules tape*, in which each cell starting with the second one corresponds to a rule associated with the membrane (we suppose that the set of those rules is ordered), and keeps one of the three symbols that indicate whether the rule is awaiting, it is applicable, or it is not applicable.

– For each object of the output alphabet we have:
  - One *environment tape*, that keeps the multiplicity of the object, in base 2, in the multiset associated with the environment.

Next we describe the steps performed by the Turing machine in order to simulate the P system. Take into account that, making a breadth first search traversal (with the skin as source) on the *initial* membrane structure of the system $\Pi(n)$, we obtain a natural order between the membranes of $\Pi(n)$. In the algorithms that we specify below we consider that they always traverse *all* the membranes of the *original* membrane structure and that, moreover, they do it in the order induced by the breadth traversal of that structure.

I. Initialization of the system. In the first phase of the simulation process followed by the Turing machine the symbols needed to reflect the initial configuration of the computation with input m that is going to be simulated are included in the corresponding tapes.

for all membrane *mb* of the system **do**
  if *mb* is not the skin membrane **then**
    – Write in the second cell of the structure tape of *mb* the label corresponding to the father of *mb*
  end if
  – Mark *mb* as non-dissolved membrane in the third cell of the structure tape of *mb*
  for all symbol *ob* of the working alphabet **do**
    – Write in the main tape of *mb* for *ob* the multiplicity, in base 2, of *ob* in the multiset initially associated with *mb*
  end for
end for
for all symbol *ob* of the input alphabet **do**
  – Read the multiplicity, in base 2, of *ob* in the input tape
  – Add that multiplicity to the main tape of the input membrane for *ob*
end for

II. Determine the applicable rules. To simulate a step of the cellular computing system, what the machine has to do first is to determine the set of rules that are applicable (each of them independently) to the configuration considered in the membranes they are associated with.

for all membrane *mb* of the system **do**
  if *mb* has not been dissolved **then**
    for all rule r associated with *mb* **do**
      – Mark r as awaiting rule
    end for

for all rule r associated with mb **do**
 if – r is awaiting and
  – mb contains the antecedent of r and
  – r only sends objects to child membranes of mb that have not been dissolved and
  – r does not try to dissolve the skin membrane
 then
  – Mark r as applicable rule
  for all rule r' associated with mb of lower priority than r **do**
   – Mark r' as non-applicable rule
  end for
 else
  – Mark r as non-applicable rule
 end if
end for
end if
end for

III. Apply the rules. Once the applicable rules are determined, they are applied in a maximal manner to the membranes they are associated with. The fact that the rules are considered in a certain order (using local maximality for each rule, according to that order) determines a specific applicable multiset of rules, thus fixing the computation of the system that the Turing machine simulates. However, from Definition 9 of complexity class it will follow that the chosen computation is not relevant for the proof, due to the *confluence* of the system.

for all membrane mb of the system **do**
 if mb has not been dissolved **then**
  for all rule r associated with mb that is applicable **do**
   for all object ob in the antecedent of r **do**
    –Compute the integer quotient that results from dividing the multiplicity of ob in the main tape of mb by the multiplicity of ob in the antecedent of r
   end for
   – Compute the minimum of the values obtained in the previous loop (that minimum is the maximum number of times that the rule r can be applied to membrane mb). Let us call it *index* of the rule r.
   for all object ob in the antecedent of r **do**
    – Multiply the multiplicity of ob in the antecedent of r by the index of r
    – Erase the result obtained from the main tape of mb for the object ob
   end for
   for all object ob in the consequent of r **do**
    – Multiply the multiplicity of ob in the consequent of r by the index of r

```
        – Add the result obtained to the auxiliary tape for ob in the corre-
          sponding membrane
      end for
      if r dissolves mb then
        – Mark mb as membrane to dissolve in the third cell of the structure
          tape of mb
      end if
    end for
  end if
end for
```

IV. Update the multisets. After applying the rules, the auxiliary tapes keep the results obtained, and then these results have to be moved to the corresponding main tapes.

```
for all membrane mb of the system do
  if mb has not been dissolved then
    – Copy the content of the auxiliary tapes of mb into the corresponding
      main tapes
  end if
end for
```

V. Dissolve the membranes. To finish the simulation of one step of the computation of the P system it is necessary to dissolve the membranes according to the rules that have been applied in the previous phase and to rearrange accordingly the structure of membranes.

```
for all membrane mb of the system do
  if – mb has not been dissolved and
     – the father of mb is marked as membrane to dissolve
  then
    – Make the father of mb equal to the father of the father of mb
  end if
end for
for all membrane mb of the system do
  if mb is marked as membrane to dissolve then
    – Copy the contents of the main tapes of mb into the main tapes of
      the (possibly new) father of mb
    – Mark mb as dissolved membrane in the third cell of the structure
      tape of mb
  end if
end for
```

VI. Check if the simulation has ended. Finally, after finishing the simulation of one transition step of the computation of $\Pi(n)$, the Turing machine has to check if a halting configuration has been reached and, in that case, if the computation is an accepting or a rejecting one.

if the environment tape contains the symbol # **then**
 if the environment tape contains the symbol Yes **then**
  – Halt and accept the multiset m
 else
  – Halt and reject the multiset m
 end if
else
 – Simulate again a step of the computation of the P system
end if

It is easy to check that the family $(TM(n))_{n \in \mathbb{N}^+}$ can be constructed in an uniform way and in polynomial time from $n \in \mathbb{N}^+$.

Let us finally consider the deterministic Turing machine $TM_{\mathbf{\Pi}}$ that works as follows:

Input: $w \in I_X$

- Compute $s(w)$
- Construct $TM(s(w))$
- Compute $cod(w)$
- Simulate the functioning of $TM(s(w))$ with input $cod(w)$

Then, the following assertions are verified:

1. The machine $TM_{\mathbf{\Pi}}$ works in polynomial time over $|w|$.
   Since the functions *cod* and s are polynomial in $|w|$, the numbers $A_{s(w)}$, $B_{s(w)}$, $C_{s(w)}$, $D_{s(w)}$, $E_{s(w)}$, $F_{s(w)}$, and $G_{s(w)}$ are polynomial in $|w|$.
   On the other hand, the family $\mathbf{\Pi}$ is polynomially bounded, with regard to (X, cod, s). Therefore, every computation of the system $\Pi(s(w))$ with input $cod(w)$ performs a polynomial number of steps on $|w|$. Consequently, the number of steps, P_w, performed by the computation simulated by the machine $TM(s(w))$ over $cod(w)$ is polynomial in $|w|$.
   Hence, the maximal multiplicity of the objects contained in the multisets associated with the membranes is in the order of $O(E_{s(w)} \cdot G_{s(w)}^{P_w})$. This implies that $H_{s(w)}(cod(w))$ is in the order of $O(P_w \cdot log_2(E_{s(w)} \cdot G_{s(w)}))$; that is, polynomial in $|w|$.
   It follows that the total time spent by $TM_{\mathbf{\Pi}}$ when receiving w as input is polynomial in $|w|$.
2. Let us suppose that $TM_{\mathbf{\Pi}}$ accepts the string w. Then the computation of $\Pi(s(w))$ with input $cod(w)$ simulated by $TM(s(w))$ is an accepting computation. Therefore $\theta_X(w) = 1$.
3. Let us suppose that $\theta_X(w) = 1$. Then every computation of $\Pi(s(w))$ with input $cod(w)$ is an accepting computation. Therefore, it is also the computation simulated by $TM(s(w))$. Hence $TM_{\mathbf{\Pi}}$ accepts the string w.

Consequently, we have proved that $TM_{\mathbf{\Pi}}$ solves X in polynomial time.

6 Characterizing the P ≠ NP Relation Through P Systems

Next, we establish characterizations of the **P** ≠ **NP** relation by means of the polynomial time unsolvability of **NP**–complete problems by families of language accepting P systems.

Theorem 1. *The following propositions are equivalent:*

1. $\mathbf{P} \neq \mathbf{NP}$.
2. $\exists X \big(X$ *is an* **NP**–*complete decision problem* $\wedge X \notin \mathbf{PMC}_{\mathcal{LA}} \big)$.
3. $\forall X \big(X$ *is an* **NP**–*complete decision problem* $\rightarrow X \notin \mathbf{PMC}_{\mathcal{LA}} \big)$.

Proof. To prove the implication $1 \Rightarrow 3$, let us suppose that there exists an **NP**–complete problem X such that $X \in \mathbf{PMC}_{\mathcal{LA}}$. Then, from Proposition 3 there exists a deterministic Turing machine solving the problem X in polynomial time. Hence, $X \in \mathbf{P}$. Therefore, $\mathbf{P} = \mathbf{NP}$, which leads to a contradiction.

The implication $3 \Rightarrow 2$ is trivial, because the class of **NP**–complete problems is non empty.

Finally, to prove the implication $2 \Rightarrow 1$, let X be an **NP**–complete problem such that $X \notin \mathbf{PMC}_{\mathcal{LA}}$. Let us suppose that $\mathbf{P} = \mathbf{NP}$. Then $X \in \mathbf{P}$. Therefore there exists a deterministic Turing machine TM that solves the problem X in polynomial time.

By Proposition 2, the problem X_{TM} is in $\mathbf{PMC}_{\mathcal{LA}}$. Then there exists a family $\mathbf{\Pi}_{TM} = \big(\Pi_{TM}(k)\big)_{k \in \mathbb{N}^+}$ of valid language accepting P systems simulating TM in polynomial time (with associated functions cod_{TM} and s_{TM}).

We consider the function $cod_X : I_X \rightarrow \bigcup_{k \in \mathbb{N}^+} I_{\Pi_{TM}(k)}$, given by $cod_X(w) = cod_{TM}(w)$, and the function $s_X : I_X \rightarrow \mathbb{N}^+$, given by $s_X(w) = |w|$. Then:

- The family $\mathbf{\Pi}_{TM}$ is polynomially uniform by Turing machine, and polynomially bounded, with regard to (X, cod_X, s_X).
- The family $\mathbf{\Pi}_{TM}$ is sound, with regard to (X, cod_X, s_X). Indeed, let $w \in I_X$ be such that there exists a computation of the system $\Pi_{TM}(s_X(w)) = \Pi_{TM}(s_{TM}(w))$ with input $cod_X(w) = cod_{TM}(w)$ that is an accepting computation. Then $\theta_{TM}(w) = 1$. Therefore $\theta_X(w) = 1$.
- The family $\mathbf{\Pi}_{TM}$ is complete, with regard to (X, cod_X, s_X). Indeed, let $w \in I_X$ be such that $\theta_X(w) = 1$. Then TM accepts the string w. Therefore $\theta_{TM}(w) = 1$. Hence, every computation of the system $\Pi_{TM}(s_{TM}(w)) = \Pi_{TM}(s_X(w))$ with input $cod_{TM}(w) = cod_X(w)$ is an accepting computation.

Consequently, $X \in \mathbf{PMC}_{\mathcal{LA}}$, and this leads to a contradiction.

Acknowledgement. The authors wish to acknowledge the support of the project TIC2002-04220-C03-01 of the Ministerio de Ciencia y Tecnología of Spain, cofinanced by FEDER funds.

References

1. Berry, G.; Boudol, G. The chemical abstract machine. *Theoretical Computer Science*, 96, 1992, pp. 217–248.
2. Cook, S. The P versus NP problem. Manuscript prepared for the Clay Mathematics Institute for the Millennium Prize Problems (revised November, 2000).
3. Manca, V. String rewriting and metabolism: A logical perspective. In *Computing with Bio-Molecules. Theory and Experiments* (Gh. Păun, ed.), Springer-Verlag, Singapore, 1998, pp. 36–60.
4. Păun, G. *Membrane Computing. An Introduction*, Springer-Verlag, Berlin, 2002.
5. Păun G. Computing with membranes, *Journal of Computer and System Sciences*, 61 (1), 2000, pp. 108–143, and *Turku Center for Computer Science-TUCS Report* Nr. 208, 1998.
6. Păun, G.; Rozenberg, G. A guide to membrane computing, *Theoretical Computer Science*, 287, 2002, pp. 73–100.
7. Pérez–Jiménez, M.J.; Romero-Jiménez, A.; Sancho-Caparrini, F. *Teoría de la Complejidad en Modelos de Computación con Membranas*, Ed. Kronos, Sevilla, 2002.
8. Romero-Jiménez, A. *Complexity and Universality in Cellular Computing Models*, PhD. Thesis, University of Seville, Spain, 2003.
9. Romero-Jiménez, A.; Pérez Jiménez, M.J. Simulating Turing machines by P systems with external output. *Fundamenta Informaticae*, vol. 49 (1-3), 2002, pp. 273–287.
10. Turing, A. On computable numbers with an application to the Entscheidnungsproblem. *Proceeding London Mathematical Society*, serie 2, 42, 1936-7, pp. 230–265.
11. Zandron, C.; Ferreti, C.; Mauri, G. Solving NP-complete problems using P systems with active membranes. In *Unconventional Models of Computation, UMC'2K* (I. Antoniou; C. Calude; M.J. Dinneen, eds.), Springer-Verlag, Berlin, 2000, pp. 289–301.

Realizing Switching Functions Using Peptide-Antibody Interactions*

M. Sakthi Balan and Kamala Krithivasan

Department of Computer Science and Engineering,
Indian Institute of Technology, Madras
Chennai - 600036, India
sakthi@cs.iitm.ernet.in
kamala@iitm.ernet.in

Abstract. We propose a theoretical model for performing gate operations – OR, AND, and NOT – using peptide-antibody interactions. The idea is extended to further gates, such as XOR, $NAND$, and NOR.

1 Introduction

With unconventional models of computing holding a central stage in this modern computing world, a lot of work is going on to define new bio-computing models to perform computations efficiently.

Peptide-antibody interactions are very natural and systematic and they can be used as a basis for a computational model. This way of computing using antibodies which specifically recognize peptide sequences was introduced by H. Hug et al in [3], where one solves the well-known NP Complete problem SAT. In [2], it the computational completeness of peptide computing was proven and it was shown how to solve other two well-known NP-complete problems, namely the Hamiltonian path problem and the exact cover by 3-sets problem (a variation of the set cover problem) using the interactions between peptides and antibodies.

A peptide is a sequence of aminoacids attached by covalent bonds called peptide bonds. A peptide consists of recognition sites called *epitopes* for the antibodies. A peptide can contain more than one epitope for the same or different antibodies. For each antibody which attaches to a specific epitope there is a binding power associated with it called *affinity*. If more than one antibody participate in recognition of its sites which overlap in the given peptide, then the antibody with a greater affinity has a higher priority.

In the model proposed in [2,3] the peptides represent the sample space of a given problem and antibodies are used to select certain subsets of this sample space, which will eventually give the solution for the given problem. Similar to DNA-computing, parallel interactions between the peptide sequences and the antibodies should make it possible to solve NP-complete problems in polynomial time.

* Financial support from Infosys Technologies Limited, India, is acknowledged

N. Jonoska et al. (Eds.): Molecular Computing (Head Festschrift), LNCS 2950, pp. 353–360, 2004.

The specificity of an epitope recognition by an antibody need not be absolute. Sometimes the antibody may recognize a different site which is very similar to its own binding site. This issue is called *cross-reactivity*. This paper does not address the issue of cross-reactivity between antibodies and peptides.

Using DNA strands H. Hug et al [4] propose a model to solve simple binary addition in parallel. In [1] we propose a model to do simple arithmetic (addition and subtraction) operations using peptides and antibodies.

The organization of the paper is as follows. In the next section we define our proposed model of performing gate operations OR, AND and NOT. We explain with some examples how the operations are carried out. In section 4 our method is extended to do XOR, $NAND$ and NOR gates. Our paper concludes with some discussion on the bio-chemical implementation of our proposed model.

2 Proposed Model

The proposed model uses a peptide sequence and a set of antibodies. The peptide sequence consists of five epitopes; if P denotes the peptide sequence, then P can be represented as follows:

$$P = yx_1xx_2z,$$

where each y, x_1, x, x_2 and z are epitopes. We take six antibodies denoted by $A_1, A_2, B_1, B_2, C_{init}$ and C_f. Basically,

1. the antibodies A_1, A_2, B_1 and B_2 denote the inputs,
2. C_{init} denote the initial value of the result of the operation,
3. C_{init} and C_f denote the output of the operation,
4. the epitopes x and y are the binding places for the antibodies denoting the input,
5. the epitope x_1xx_2 is the place where the antibody representing the initial output binds,
6. the epitopes x_1x, xx_2 and x are the binding places for the antibodies denoting the output.

The peptide sequence P is given in Fig. 1. The peptide sequence with possible antibodies binding on it is represented in Fig. 2. Note that at a time only one antibody can be there in the overlapping epitopes. The figure is meant only to show the binding sites of the antibodies pictorically.

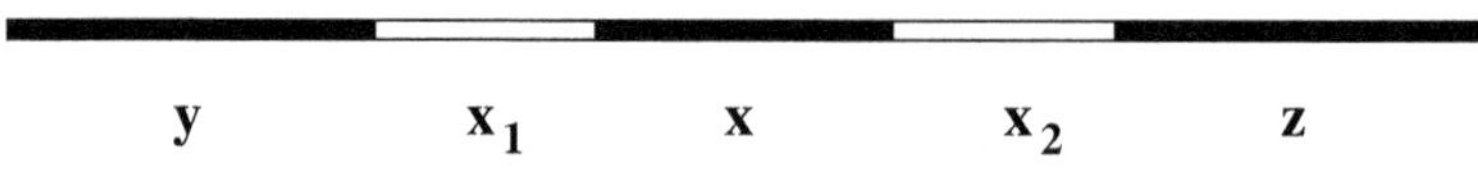

Fig. 1. Peptide sequence

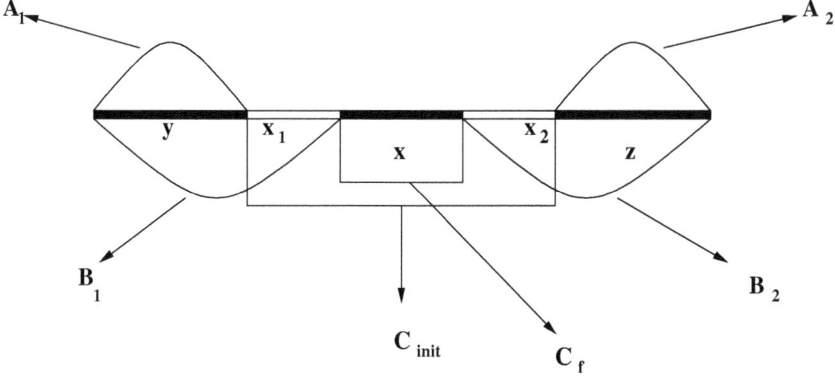

Fig. 2. Peptide sequence with possible antibodies

3 Implementation of OR, AND, and NOT Gates

First we will discuss OR gate implementation. AND gate is very similar to the OR gate implementation.

For the OR gate we follow the representation given in the sequel:

1. Input bits 0 and 1 are represented by the antibodies A_i and B_i respectively, where $1 \leq i \leq 2$.
2. The antibody C_{init} denotes the bit 0.
3. The antibody C_f (labeled antibody) denotes the bit 1.
4. $epitope(A_1) = \{y\}$, $epitope(A_2) = \{z\}$,
5. $epitope(B_1) = \{yx_1\}$, $epitope(B_2) = \{x_2z\}$,
6. $epitope(C_{init}) = \{x_1xx_2\}$, $epitope(C_f) = \{x\}$.
7. $aff(B_i) > aff(C_{init}) > aff(C_f)$, $1 \leq i \leq 2$.

The simple idea used in the implementation of OR gate follows from the fact that the output 1 occurs even if only one of the inputs is 1. So if we start with an initial output of 0, which is done by getting the peptide sequence with the antibody C_{init} binding on it, then it should be toggled even if only one 1 comes as an input. For this to be carried out we have made the epitopes for the antibody C_{init} and the antibody $B_i, 1 \leq i \leq 2$, overlapping. This facilitates toggle of the output bit to 1. The algorithm is as follows:

1. Take the peptide sequence P in an aqueous solution.
2. Add the antibody C_{init}.
3. Add antibodies corresponding to the input bits. For example, if the first bit is 1 and the second bit is 0, then add antibodies B_1 and A_2.
4. Add antibody C_f.

In the above algorithm only if 1 occurs as an input bit the initial antibody C_{init} is removed (since $aff(B_i) > aff(C_{init})$), which facilitates the binding of the antibody C_f. If both bits are 0, then the antibody C_{init} is not removed. If the output has to be seen, then the antibody C_f can be given some color so that at the end of the algorithm if fluorescence is detected, then the output will be 1 or else it will be 0.

The working of the *OR* gate model is depicted in Fig. 3.

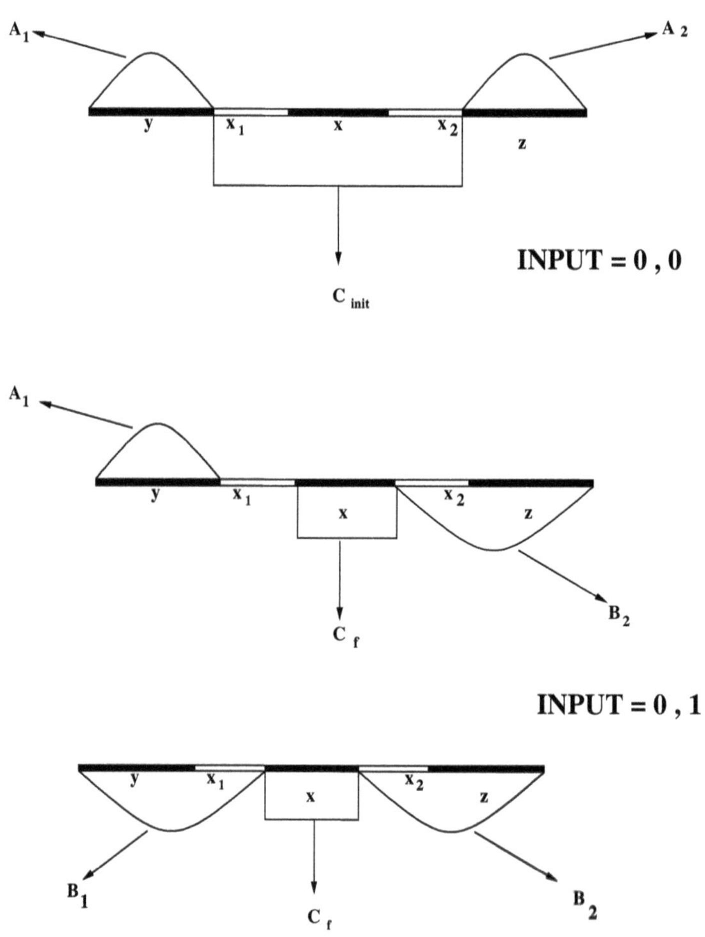

Fig. 3. OR gate

AND Gate

In the same manner as above we can implement the AND gate. The details are the following.

1. The antibody C_{init} denotes the bit 1.
2. The antibody C_f (labeled antibody) denotes the bit 0.
3. $epitope(B_1) = \{y\}$, $epitope(B_2) = \{z\}$,
4. $epitope(A_1) = \{yx_1\}$, $epitope(A_2) = \{x_2z\}$,
5. $epitope(C_{init}) = \{x_1xx_2\}$, $epitope(C_f) = \{x\}$.
6. $aff(A_i) > aff(C_{init}) > aff(C_f)$, $1 \leq i \leq 2$.

The simple idea used in the implementation of the AND gate follows from the fact that the output 0 occurs even if only one of the inputs is 0. The algorithm for the AND gate is the same as for the OR gate. The working of the AND gate should be easy to understand as it is very similar to the working of the OR gate.

NOT Gate

Since the NOT gate requires only one input, we take the peptide sequence as

$$P = xx_2z$$

and take only the antibodies A_1 and B_1. The model is as follows:

1. The antibody C_{init} denotes the bit 0.
2. The antibody C_f (labeled antibody) denotes the bit 1.
3. $epitope(B_2) = \{z\}$,
4. $epitope(A_2) = \{x_2z\}$,
5. $epitope(C_{init}) = \{xx_2\}$, $epitope(C_f) = \{x\}$.
6. $aff(A_i) > aff(C_{init}) > aff(C_f)$, $1 \leq i \leq 2$.

The algorithm for NOT gate is as follows:

1. Take the peptide sequence P in an aqueous solution.
2. Add the antibody C_{init}.
3. Add antibody corresponding to the input bit.
4. Add antibody C_f.

It can be noted that the initial bit denoting 0 is toggled only if the input bit is 0.

4 Extension to XOR, NOR, and $NAND$ Gates

The same idea used in the implementation of the OR and AND gates is extended to XOR, NOR and $NAND$. First we take the XOR gate; the other two gates follow from OR and AND gates very easily.

The model for the *XOR* gate requires little change in the proposed model defined for other gates in the previous section. The peptide sequence is the same but this gate requires one more antibody and the binding sites for the output antibodies are different. It should be easy to follow from the picture in Fig. 4.

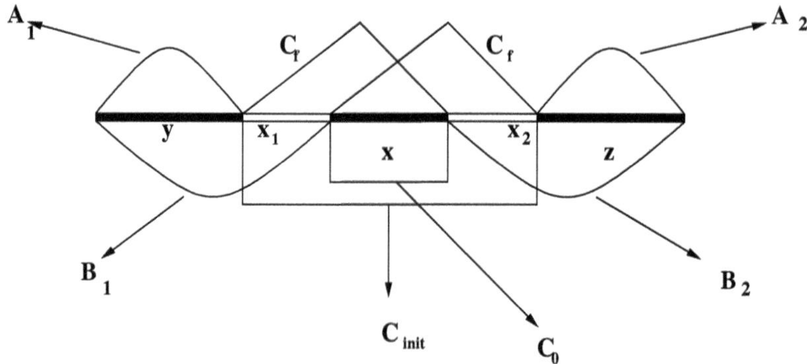

Fig. 4. Peptide sequence with possible antibodies

The construction for the *XOR* gate is as follows:

1. Input bits 0 and 1 are represented by the antibodies A_i and B_i respectively where $1 \leq i \leq 2$.
2. The antibodies C_{init} and C_0 denotes the bit 0.
3. The antibody C_f (labeled antibody) denotes the bit 1.
4. $epitope(A_1) = \{y\}$, $epitope(A_2) = \{z\}$,
5. $epitope(B_1) = \{yx_1\}$, $epitope(B_2) = \{x_2z\}$,
6. $epitope(C_{init}) = \{x_1xx_2\}$, $epitope(C_f) = \{x_1x, xx_2\}$ and $epitope(C_0) = \{x\}$,
7. $aff(B_i) > aff(C_{init}) > aff(C_f) > aff(C_0)$, $1 \leq i \leq 2$.

The simple idea used in the implementation of the *XOR* gate is that whenever the inputs are the same, either the epitope x_1xx_2 is bounded by the antibody C_{init} or the epitope x will be free for the antibody C_0 to come and bind to it. This makes sure that the output is 0 whenever the inputs are not the same. When the inputs are different either the epitope x_1x or xx_2 will be free. So that the antibody C_f can bind to either of them. This will guarantee that the output will be 1 when the inputs are different. The algorithm is given below:

1. Take the peptide sequence P in an aqueous solution.
2. Add the antibody C_{init}.
3. Add antibodies corresponding to the input bits.
4. Add antibody C_f.
5. Add antibody C_0.

NOR Gate

If we perform the following changes in the OR gate construction

1. C_{init} denotes the bit 1 and
2. C_f denotes 0,

then it should be easy to note that the switching operation represents the NOR function. The idea is that even if one 1 comes as an input, the output is 0. It should be easy to check that the above construction works exactly as the NOR gate.

NAND Gate

In the AND gate construction we perform the following changes:

1. The antibody C_{init} denotes the bit 0.
2. The antibody C_f (labeled antibody) denotes the bit 0.

The idea is that even if one 0 comes as an input, the output is 1. It should be easy to check that the above construction works exactly as $NAND$ gate.

5 Remarks

In this work we have proposed a model to do simple switching operations. Basically there are two ways of getting the output. One way is just detection, and the other is to decipher bio-chemically the output. For detection, we can give a label to either of the output antibody, so that at the end of the process, depending on whether the fluorescence is detected or not we can detect the answer.

There are many biochemical methods to decipher the interaction of peptides with antibodies [5]. To extract the numbers from the peptide-antibody interacting system, Nuclear Magnetic Resonance Spectroscopy (NMR) is one of the two powerful tools to map the binding regions of peptide and/or proteins, at atomic level, during their interactions; the other tool is X-ray crystallography. Depending upon the binding affinity of the peptide with antibody, there are several NMR techniques which can be used for this purpose [7]. If the antibody is spin labeled, then mapping of the peptide binding domain in antibody can also be accomplished by Surface Activity relationship (SAR) technique [6].

It may be interesting to see whether this model can be implemented in a laboratory. It may also be worthwhile to determine how this model can be extended to switching operations on strings of bits with several strings.

References

1. M.S. Balan and K. Krithivasan, Parallel computation of simple arithmetic using peptide-antibody interactions, *Proceedings of International Workshop on Information Processing in Cells and Tissues*, Lausanne, Switzerland, 2003.

2. M.S. Balan, K. Krithivasan and Y. Sivasubramanyam, Peptide computing – Universality and complexity, *Proceedings of Seventh International Conference on DNA Based Computers - DNA7, LNCS 2340* (N. Jonoska, N, Seeman, eds.), Springer-Verlag, 2002, 290–299.
3. H. Hug and R. Schuler, Strategies for the developement of a peptide computer, *Bioinformatics*, 17 (2001), 364–368.
4. H. Hug and R. Schuler, DNA-based parallel computation of simple arithmetic, *Proceedings of Seventh International Conference on DNA Based Computers - DNA7, LNCS 2340*, (N. Jonoska, N, Seeman, eds.), Springer-Verlag, 2002, 321–328.
5. E.M. Phizicky and S. Fields, Protein-protein interactions: methods for detection and analysis, *Microbiol. Rev.*,59 (1995), 94–123.
6. S.B. Shuker, P.J. Hajduk, R.P. Meadows, and S.W. Fesik, SAR by NMR: A method for discovering high affinity ligands for proteins, *Science*, 274 (1996), 1531–1534.
7. B.J. Stockman, NMR spectroscopy as a tool for structure-based drug design, *Progress in Nuclear Magnetic Resonance Spectroscopy*, 33 (1998), 109–151.

Plasmids to Solve #3SAT

Rani Siromoney[1] and Bireswar Das[2]

[1] Madras Christian College
Chennai 600 059, India, and
Chennai Mathematical Institute
Chennai 600 017, India
ranisiro@sify.com

[2] Institute of Mathematical Sciences
Chennai 600 113, India
bireswar@imsc.res.in

Abstract. Tom Head, [1] has given a simple and elegant method of aqueous computing to solve 3SAT. The procedure makes use of Divide-Delete-Drop operations performed on plasmids. In [4], a different set of operations, Cut-Expand-Ligate, are used to solve several **NP**-Complete problems. In this paper, we combine the features in the two procedures and define Cut-Delete-Expand-Ligate which is powerful enough to solve #3SAT, which is a counting version of 3SAT known to be in **IP**. The solution obtained is advantageous to break the propositional logic based cryptosystem introduced by J. Kari [5].

1 Plasmids for Aqueous Computing

DNA molecules occur in nature in both linear and circular form. Plasmids are small circular double-stranded DNA molecules. They carry adequate information encoded in their sequences necessary for their replication and this is used in genetic engineering. The technique used is cut and paste – cutting by restriction enzymes and pasting by a ligase. The sequential application of a set of restriction enzymes acting at distinct non-overlapping, different sites in circular DNA molecules is fundamental to the procedure suggested below.

2 Divide-Delete-Drop (D-D-D)

Tom Head [1] has given the following procedure which provides the correct YES/NO answer for instances of 3-SAT in a number of steps linearly bounded by the sum of the number of atomic propositional variables and the number of triples that are disjoined.

The fundamental data structure for the computational work involved in D-D-D is an *artificial plasmid* constructed as follows.

For a specified set of restriction enzymes $\{RE_1, RE_2, \ldots, RE_n\}$, the plasmid contains a segment of the form $c_1 s_1 c_1 c_2 s_2 c_2 \ldots c_i s_i c_i \ldots c_{n-1} s_{n-1} c_{n-1} c_n s_n c_n$; the subsegments $c_1, c_2, \ldots, c_n$ are sites at which the enzymes $RE_1, \ldots, RE_n$ can

N. Jonoska et al. (Eds.): Molecular Computing (Head Festschrift), LNCS 2950, pp. 361–366, 2004.

cut the plasmid. These are the only places where the enzymes can cut the plasmid. The subsegments $s_1, s_2, \ldots, s_n$ are of *fixed length* and are called *stations*. The sequence for each station is to be chosen so that no other subsegment of the plasmid has the same base pair sequence.

The key computational operation used in plasmids is the compound operation consisting of the removal of one station from a (circular) plasmid followed by recircularizing the linear molecule. This is a cut and paste process. The net effect of the compound operation is the removing of the station. The compound biochemical deletion process is Delete(Station-Name). Two further operations used are Drop and Divide.

3 Procedure Cut-Expand-Light (C-E-L) to Solve Satisfiability of Sets of Boolean Clauses

In Binghamton, Tom Head, S. Gal and M. Yamamura provided a prototype solution of a three variables, four clauses satisfiability (SAT) problem as an aqueous computation.

The memory register molecule was chosen to be a commercially available plasmid. Restriction enzyme sites were chosen in the Multiple Cloning Site (MCS) of the plasmid, each of which serving as a station for the computation.

The computation begins with a test-tube of water (for appropriate buffer) that contains a vast number of identical, n-station plasmids. During a computation, the plasmids are modified so as to be readable later. Modifications take place only at the stations. Each station s, at any time during the computation, is in one of the two stations representing one of the bits 1 or 0. Thus, each plasmid plays a role corresponding to that of an n-bit data register in a conventional computer.

The initial condition of a station (restriction enzyme site) represents the bit 1. A zero is written at a station by altering the site making use of the following steps:

1. Linearize the plasmid by cutting it at the station (site), with the restriction enzyme associated with it.
2. Using a DNA polymerase, extend the $3'$ ends of the strands lying under the $5'$ overhangs to produce a linear molecule having blunt ends.
3. When a station is altered to represent the bit zero, the length of the station is increased and the station no longer encodes a site for the originally associated enzyme.

3.1 Satisfiability of Sets of Boolean Clauses

The SAT instance solved in the web lab was to find a truth assignment for the variables p, q, r for which each of the four clauses

$$p \wedge q, \ \neg p \wedge q \wedge \neg r, \ \neg q \wedge \neg r, \ p \wedge r$$

evaluates to true.

A commercially available cloning plasmid was used as memory register molecule. Six restriction enzyme sites in the multiple cloning site (MCS) of the plasmid were chosen to serve as stations for the computation. The first step is to remove contradictions. At the end, there are plasmids that represent only logically consistent truth assignments for the variables and their negations.

The next step involves elimination. Molecules that do not encode truth assignments to each one of the clauses are eliminated, one clause after the other. After the elimination phase, the remaining plasmids are consistent with truth settings for which all the given clauses have the value true.

4 Cut-Delete-Expand-Ligate (C-D-E-L)

The features in Divide-Delete-Drop (D-D-D, [1]) and Cut-Expand-Ligate (C-E-L, [2,3,4]) Tom Head procedures are combined to form Cut-Delete-Expand-Ligate (C-D-E-L). This enables us to get an aqueous solution to #3SAT which is a counting problem and known to in **IP**.

An algorithm for a counting problem takes as input an instance of a decision problem and produces as output a non-negative integer that is the number of solutions for that instance.

In the 3SAT problem, we are required to determine whether a given 3CNF is satisfiable. Here we are interested in a counting versions of this problem called #3SAT. Given a 3CNF formula F and an integer s, we need to verify that the number of distinct satisfying truth assignments for F is s. Echivalently, we have to find the number of truth assignments that satisfy F.

4.1 #3SAT

Instance: A propositional formula of the form $F = C_1 \wedge C_2 \wedge \ldots \wedge C_m$, where $C_i, i = 1, 2, \ldots, m$ are clauses. Each C_i is of the form $(l_{i_1} \vee l_{i_2} \vee l_{i_3})$, where $l_{i_j}, j = 1, 2, 3$, are literals for the set of variables $\{x_1, x_2, \ldots, x_n\}$.

Question: What is the number of truth assignments that satisfy F?

It is known that 3SAT is in IP and **IP** = **PSPACE**.

Computation: As in C-E-L, only one variety of plasmid is used, but trillions of artificial plasmids containing multiple cloning sites (MCS) are provided for the initial solution. MCS consists of a number of sites identified by different restriction enzymes and determined by the number of variables. All bio-molecular operations are done on MCS. As in D-D-D, the segment of plasmid used is of the form

$$c_1 s_1 c_1 \ldots c_2 s_2 c_2 \ldots c_n s_n c_n,$$

where $c_i, i = 1, \ldots, n$, are the sites such that no other subsequence of the plasmid matches with this sequence, $s_i, i = 1, \ldots, n$, are the stations.

In D-D-D, the lengths of the stations were required to be the same, while one of the main differences in C-D-E-L is that the lengths of the stations are all required to be different. This is fundamental to solving #3SAT. Biomolecular operations used in our C-D-E-L procedure are similar to those used in C-E-L.

Design. Let $x_1, \ldots, x_n$ be the variables in F, $\neg x_1, \ldots, \neg x_n$ their negations, c_i a site associated with station s_i, $\neg c_i$ a site associated with station $\neg s_i$, r_i the length of the station associated with $x_i, i = 1, \ldots, n$, and r_{n+j} the length of the station associated with literal $\neg x_j, j = 1, \ldots, n$.

We chose stations in such a way that the sequence $[r_1, \ldots, r_{2n}]$ satisfies the property $\sum_{i=1}^{k} r_i < r_{k+1}$, $k = 1, \ldots, 2n - 1$, i.e., a super-increasing Easy Knapsack sequence.

From the sum, the subsequence can be efficiently recovered.

4.2 Operations in C-D-E-L to Solve #3SAT

The initial plasmid is represented by a binary string of length $2n$, with all ones. Writing a zero instead of a one is done by deleting a station from the plasmid; in this way, the site is disabled so that the restriction enzyme cannot cut anymore.

This is done in 4 steps as follows:

1. Apply restriction enzyme R which corresponds to site c that flanks s. This produces two linear DNA strands, a short one containing station s and a portion of site c, and the other long, having sticky ends at both ends (the sticky ends are portions of c).
2. Remove the short DNA strand containing the station s.
3. Using DNA polymerase, extend the DNA below the sticky ends of the long strand to make blunt ends.
4. Apply ligase to recircularise the linear molecues ligating the blunt ends.

Step 1 of Procedure: Remove all contradictions If the initial plasmid contains stations associated with both x_i and $\neg x_i, i = 1, \ldots, n$, then these are contradictions. To remove them for a variable x_i, we proceed as follows. Divide the solution into 2 test tubes T_1 and T_2. In T_1, write zero for station s_i (corresponding to x_i). In T_2, write zero for station $\neg s_i$ (corresponding to $\neg x_i$). T_1 and T_2 are mixed together. Do this for all the variables in F, to remove all contradictions.

Step 2 of Procedure: Suppose $C_i = l_{i_1} \vee l_{i_2} \vee l_{i_3}$ is the ith clause in F; c_{ij} is the restriction enzyme site corresponding to the literal $l_{i_j}, j = 1, 2, 3$. Divide the solution into three test tubes L, M, R; in L apply restriction enzyme for $\neg c_{l i_1}$ to linearise all strands that has 0 for l_{i_1}, i.e., 1 for $\neg l_{i_1}$. Then, remove all linear strands from L. In M apply restriction enzyme $\neg c_{l_{i_2}}$ and remove all linear strands to eliminate all strands in M that do not have 1 for literal l_{i_2}. In R do this for l_{i_3}. Pour L, M, and R into a new test tube, to get the resultant solution. Starting with clause c_1, do this for all clauses c_i by taking the resultant from clause $c_{i-1}, i = 2, \ldots, n$. The resultant solution in test tube for clause c_n encodes all the satisfying assignments of F, if any. The resultant solution is analysed by gel separation. If more than one satisfying assignment is present in the final solution, the plasmids encoding the different assignments have different lengths. Since they contain different sets of sequences they form distinct bands in the gel platform. Any subsequence of an Easy Knapsack sequence has different sum from the sums of other subsequences. Since the sequences are chosen from Easy Knapsack, the r_i can be retrieved from the sum. From the sum, the subsequence can be efficiently recovered.

5 Breaking the Propositional Logic-Based Cryptosystem

J. Kari [5] introduced a cryptosystem based on the propositional logic. He proved that the system was optimal in the sense that any cryptanalytic attack against this can be used to break any other system as well. He also showed that the problem of cryptanalytic attack encountered by the eavesdropper is **NP**hard.

In [6] we had adapted the D-D-D procedure of T. Head [1] to give a method of cryptanalytic attack against this cryptosystem. Now using the C-D-E-L procedure and the method explained above to solve #3SAT, we note that a more efficient and better method of cryptanalytic attack against the propositional logic-based cryptosystem of Kari can be given. This is done by applying #SAT to $F = \neg p_0 \wedge p_1$, where p_0 and p_1 are the public keys.

The final solution wil give all those truth assignment that make p_0 false and p_1 true. There are several advantages over the earlier method. In the cryptanalytic attack proposed in [6] modifying D-D-D, it was required to execute the DNA algorithm for each bit in the cryptotext, while in the present C-D-E-L method (combining features of D-D-D and C-D-E-L) the final solution gives all the truth assignments.

6 Conclusion

In Adleman's and Lipton's methods, the construction of the exponential number of all potential solutions is considered at the initial step. In Tom Head's procedure, this is replaced by initialization with one single circular molecular variety, which is seen to be capable of providing representation of all solutions as computation progresses. However, an exponential number of plasmids is necessary, and this limits the size of problems which can be handled in the laboratory. We hope to work on procedures which will overcome this difficulty.

References

1. T. Head, Circular Suggestions for DNA Computing, in*Pattern Formation in Biology, Vision and Dynamics* (A. Carbone, M Gromov, and P. Prusinkeiwicz, Eds.), World Scientific, Singapore, 2000, 325–335.
2. T. Head, Splicing Systems, Aqueous Computing, and Beyond, in *Unconventional Models of Computation, UMC 2K* (I. Antoniou, C.S. Calude, and M.J. Dinneen, Eds.), Springer, Berlin, 2001, 68–84.
3. T. Head, G. Rozenberg, R.S. Bladergreen, C.K.D. Breek, P.H.M. Lommerse, and H.P.S. Spaink, Computing with DNA by Operating on Plasmids, *BioSystems*, 57 (2000), 87–93.

4. T. Head, M. Yamamura, and S.Gal, Aqueous Computing: Writing on Molecules, in *Proc. Congress on Evolutionary Computation* 1999, IEEE Service Center, Piscataway, NJ, 1006–1010.
5. J. Kari, A Cryptosystem based on Propositional Logic, in *Machines, Languages and Complexity, 5th International Meeting of Young Computer Scientists*, Czeckoslovakia, Nov. 14-18, 1989 (J. Dassow and J. Kelemen, Eds.), LNCS 381, Springer, 1989, 210–219.
6. R. Siromoney and D. Bireswar, DNA Algorithm for Breaking a Propositional Logic Based Cryptosystem, *Bulletin of the EATCS*, 79 (February 2003), 170–176

Communicating Distributed H Systems with Alternating Filters

Sergey Verlan

Laboratoire d'Informatique Théorique et Appliquée
Université de Metz, France
verlan@sciences.univ-metz.fr

Abstract. We present a variant of communicating distributed H systems where each filter is substituted by a tuple of filters. Such systems behave like original ones with the difference that at each moment we use one element of the tuple for the filtering process and this element is replaced after each use, periodically. We show that it suffices to use tuples of two filters in order to generate any recursively enumerable language, with two tubes only. We also show that it is possible to obtain the same result having no rules in the second tube which acts as a garbage collector. Moreover, the two filters in a tuple differ only in one letter. We also present different improvements and open questions.

1 Introduction

Splicing systems (H systems) were the first theoretical model of biomolecular computing and they were introduced by T. Head. [5,6]. The molecules from biology are replaced by words over a finite alphabet and the chemical reactions are replaced by a *splicing* operation. An H system specifies a set of rules used to perform a splicing and a set of initial words or axioms. The computation is done by applying iteratively the rules to the set of words until no more new words can be generated. This corresponds to a bio-chemical experiment where we have enzymes (splicing rules) and initial molecules (axioms) which are put together in a tube and we wait until the reaction stops.

Unfortunately, H systems are not very powerful and a lot of other models introducing additional control elements were proposed (see [14] for an overview). One of these well-known models are *communicating distributed H systems* (CDH systems) or *test tube systems* (TT systems) introduced in [1]. This model introduces *tubes* (or *components*). Each tube contains a set of strings and a set of splicing rules and evolves as an H system. The result of its work can be redistributed to other tubes according to a certain filter associated with each tube. Each filter is an alphabet and a string will enter a tube only if it is composed only by symbols present in the filter associated with the tube. After that these strings form the new starting molecules for the corresponding tube.

In the same paper it was shown that the model can generate any recursively enumerable language using a finite set of strings and a finite set of splicing rules.

N. Jonoska et al. (Eds.): Molecular Computing (Head Festschrift), LNCS 2950, pp. 367–384, 2004.

The number of test tubes used in [1] to generate a language was dependent on the language itself. It was also shown that only one tube generates the class of regular languages.

A series of papers then showed how to create test tube systems able to generate any recursively enumearble language using fewer and fewer test tubes. In [2] this result is obtained with 9 test tubes, in [11] with 6, and in [15] with 3 tubes.

In [1], it is shown that test tube systems with two test tubes can generate non regular languages, but the problem whether or not we can generate all recursively enumerable languages with two test tubes is still open. Nevertheless, in [3] R. and F. Freund showed that two test tubes are able to generate all recursively enumerable languages if the filtering process is changed. In the variant they propose, a filter is composed by a finite union of sets depending upon the simulated system. In [4] P. Frisco and C. Zandron propose a variant which uses two symbols filter, which is a set of single symbols and couples of symbols. A string can pass the filter if it is made from single symbols or it contains both elements from a couple. In the case of this variant two test tubes suffice to generate any recursively enumerable language.

In this paper we propose a new variant of test tube systems which differs from the original definition by the filtering process. Each filter is replaced by a tuple of filters, each of them being an alphabet (*i.e.*, like in the original definition), and the current filter i (the one which shall be used) is replaced by the next element of the tuple ($i+1$) after its usage. When the last filter in the tuple is reached the first filter is taken again and so on.

We show that systems with two tubes are enough to generate any recursively enumerable language. We present different solutions depending on the number of elements in the filter tuple. In Section 3 we describe a system having tuples of two filters. Additionally both filters of the first tube coincide. In Section 4 we present a system having tuples of four filters, but which contains no rules in the second tube. In this case, all the work is done in the first tube and the second one acts like a garbage collector. In the same section it is shown how to reduce the number of filters to three. Finally, in Section 5 we present a system with two filters which contains no rules in the second tube and for which the two filters in a tuple differ only in one letter.

2 Basic Definitions

2.1 Grammars and Turing Machines

For definitions on Chomsky grammars and Turing machines we shall refer to [7].

In what follows we shall fix the notations that we use. We consider non-stationary deterministic Turing machines, *i.e.*, at each step the head shall move to the left or right. We denote such machines by $M = (Q, T, a_0, q_0, F, \delta)$, where Q is the set of states and T is the tape alphabet, $a_0 \in T$ is the blank symbol, $q_0 \in Q$ is the initial state, $F \subseteq Q$ is the set of final (halting) states. By δ we denote the set of instructions of machine M. Each instruction is represented in the following way: $q_i a_k D a_l q_j$. The meaning is the following: if the head of machine M being in state q_i is scanning a cell which contains a_k then the contents of the scanned cell is replaced by a_l, the head moves to the left (if $D = L$) or to the right (if $D = R$) and its state changes to q_j.

By a *configuration* of a Turing machine we shall understand the string $w_1 q w_2$, where $w_1, w_2 \in T^*$ ($w_2 \notin \varepsilon$) and $q \in Q$. A configuration represents the contents of non-empty cells of the working tape of the machine (*i.e.*, all other cells to the left and to the right are blank), its state and the position of the head on the tape. The machine head is assumed to read the leftmost letter of w_2. Initially all cells on the tape are blank except finitely many cells.

2.2 The Splicing Operation

An *(abstract) molecule* is simply a word over some alphabet. A *splicing rule* (over alphabet V), is a quadruple (u_1, u_2, u_1', u_2') of words $u_1, u_2, u_1', u_2' \in V^*$, which is often written in a two dimensional way as follows: $\frac{u_1 | u_2}{u_1' | u_2'}$. We shall denote the empty word by ε.

A splicing rule $r = (u_1, u_2, u_1', u_2')$ is said to be applicable to two molecules m_1, m_2 if there are words $w_1, w_2, w_1', w_2' \in V^*$ with $m_1 = w_1 u_1 u_2 w_2$ and $m_2 = w_1' u_1' u_2' w_2'$, and produces two new molecules $m_1' = w_1 u_1 u_2' w_2'$ and $m_2' = w_1' u_1' u_2 w_2$. In this case, we also write $(m_1, m_2) \vdash_r (m_1', m_2')$.

A pair $h = (V, R)$, where V is an alphabet and R is a finite set of splicing rules, is called a *splicing scheme* or an H *scheme.*

For an H scheme $h = (V, R)$ and a language $L \subseteq V^*$ we define:

$$\sigma_h(L) \stackrel{\text{def}}{=} \{w, w' \in V^* \mid (w_1, w_2) \vdash_r (w, w') \text{ for some } w_1, w_2 \in L,\ r \in R\},$$
$$\sigma_h^0(L) = L,$$
$$\sigma_h^{i+1}(L) = \sigma_h^i(L) \cup \sigma_h(\sigma_h^i(L)),\ i \geq 0,$$
$$\sigma_h^*(L) = \cup_{i \geq 0} \sigma_h^i(L).$$

A *Head splicing system* [5,6], or H *system*, is a construct $H = (h, A) = ((V, R), A)$ of an alphabet V, a set $A \subseteq V^*$ of initial molecules over V, the *axioms*, and a set $R \subseteq V^* \times V^* \times V^* \times V^*$ of splicing rules. H is called finite if A and R are finite sets.

The language generated by H system H is:

$$L(H) \stackrel{\text{def}}{=} \sigma_h^*(A).$$

Thus, the language generated by H system H is the set of all molecules that can be generated in H starting with A as initial molecules by iteratively applying splicing rules to copies of the molecules already generated.

2.3 Distributed Splicing Systems

A *communicating distributed H system* (CDH system) or a *test tube system* (TT system) [1] is a construct

$$\Delta = (V, T, (A_1, R_1, F_1), \ldots, (A_n, R_n, F_n)),$$

where V is an alphabet, $T \subseteq V$ (terminal alphabet), A_i are finite languages over V (axioms), R_i are finite sets of splicing rules over V, and $F_i \subseteq V$ are finite sets called *filters* or *selectors*, $1 \leq i \leq n$.

Each triple (A_i, R_i, F_i), $1 \leq i \leq n$, is called a *component* (or *tube*) of Δ.

An n-tuple $(L_1, \ldots, L_n)$, $L_i \subseteq V^*$, $1 \leq i \leq n$, is called a *configuration* of the system; L_i is also called the *contents* of ith component. For two configurations $(L_1, \ldots, L_n)$ and $(L'_1, \ldots, L'_n)$ we define:

$(L_1, \ldots, L_n) \Rightarrow (L'_1, \ldots, L'_n)$ iff

$$L'_i = \left(\bigcup_{j=1}^{n} \sigma_j^*(L_j) \cap F_i^* \right) \cup \left(\sigma_i^*(L_i) \cap \left(V^* - \bigcup_{k=1}^{n} F_k^* \right) \right),$$

where $\sigma_i = (V, R_i)$ is the H scheme associated with the component i of the system.

In words, the contents of each component is spliced according to the associated set of rules like in the case of an H system (splicing step) and the result is redistributed among the n components according to filters F_i; the part which cannot be redistributed remains in the component (communication step). If a string can be distributed over several components then each of them receives a copy of the string.

We shall consider all computations between two communications as a macro-step, so macro-steps are separated by communication.

The language generated by Δ is:

$$L(\Delta) = \{w \in T^* \mid w \in L_1, \exists L_1, \ldots, L_n \subseteq V^* : (A_1, \ldots, A_n) \Rightarrow^* (L_1, \ldots, L_n)\}.$$

A *communicating distributed H system with m alternating filters* (CDHF system or TTF system) is a construct

$$\Gamma = (V, T, (A_1, R_1, F_1^{(1)}, \ldots, F_1^{(m)}), \ldots, (A_n, R_n, F_n^{(1)}, \ldots, F_n^{(m)})),$$

where V, T, A_i and R_i are defined as in a communicating distributed H system. Now instead of one filter F_i for each component i we have an m-tuple of filters $F_i^{(r)}$, $1 \leq r \leq m$, where each filter $F_i^{(r)}$ is defined as in the above case. At each macro-step $k \geq 1$ only filter $F_i^{(r)}$, $r = (k-1) \pmod{m} + 1$, is active for component i.

We define:

$$(L_1^{(1)}, \ldots, L_n^{(1)}) = (A_1, \ldots, A_n),$$

$$L_i^{(k+1)} = \left(\bigcup_{j=1}^{n} \sigma_j^*(L_j^{(k)}) \cap F_i^{(r)*} \right) \cup \left(\sigma_i^*(L_i^{(k)}) \cap \left(V^* - \bigcup_{k=1}^{n} F_k^{(r)*} \right) \right), k \geq 1,$$

where $\sigma_i = (V, R_i)$ is the H scheme associated with component i of the system.

This is very similar to the previous system. The only difference is that instead of one stationary filter F_i there is a tuple of filters $F_i^{(1)}, \dots, F_i^{(m)}$ and the filter to be used depends on the number of communications steps being done.

The language generated by Γ is:

$$L(\Gamma) = \{w \in T^* \mid w \in L_1^{(k)} \text{ for some } k \geq 1\}.$$

We denote by CDH_n (or TT_n) the family of communicating distributed H systems having at most n tubes and by $\text{CDHF}_{n,m}$ (or $\text{TTF}_{n,m}$) the family of communicating distributed H systems with alternating filters having at most n tubes and m filters.

3 $\text{TTF}_{2,2}$ Are Computationally Universal

In this section we present a CDHF system with two components and two filters which simulates a type-0 grammar. Moreover, both filters in the first component are identical.

Theorem 1. *For any type-0 grammar $G = (N, T, P, S)$ there is a communicating distributed H system with alternating filters having two components and two filters $\Gamma = (V, T, (A_1, R_1, F_1^{(1)}, F_1^{(2)}), (A_2, R_2, F_2^{(1)}, F_2^{(2)}))$ which simulates G and $L(\Gamma) = L(G)$. Additionally $F_1^{(1)} = F_1^{(2)}$.*

We construct Γ as follows. Let

$$N \cup T \cup \{B\} = \{a_1, a_2, \dots, a_n\}$$

(where we assume $B = a_n$).

In what follows we will assume the following:

$$\begin{aligned} &1 \leq i \leq n, \ \mathbf{a} \in N \cup T \cup \{B\}, \ \gamma \in \{\alpha, \beta\}, \\ &\mathbf{b} \in N \cup T \cup \{B\} \cup \{\diamond\}, \\ V = {} &N \cup T \cup \{B\} \cup \{\alpha, \beta\} \cup \{\diamond\} \\ &\cup \{X, Y, X_\alpha, X_\beta, X'_\alpha, Y_\alpha, Y'_\alpha, Y_\beta, Z, Z'\}. \end{aligned}$$

The terminal alphabet T is the same as for the grammar G.

Test tube I:

Rules of R_1:

$$1.1: \frac{\varepsilon \mid uY}{Z \mid vY}; \qquad 1.2: \frac{\mathbf{a} \mid a_i Y}{Z \mid \beta\alpha^{i-1}Y'_\alpha}; \qquad 1.3: \frac{\mathbf{a} \mid BY}{Z \mid \diamond Y_\beta};$$

$$1.4: \frac{X \mid \mathbf{a}}{X_\alpha\beta \mid Z}; \qquad 1.5: \frac{X \mid \mathbf{a}}{X_\beta\beta\beta \mid Z}; \qquad 1.6: \frac{\mathbf{b} \mid \beta Y_\alpha}{Z \mid Y_\beta};$$

$$1.7: \frac{X'_\alpha \mid \alpha}{X_\alpha \mid Z}; \qquad 1.8: \frac{X'_\alpha \mid \alpha}{X_\beta\beta \mid Z}; \qquad 1.9: \frac{\gamma \mid \alpha Y_\alpha}{Z \mid Y'_\alpha};$$

Filter:
$F_1 = F_1^{(1)} = F_1^{(2)} = N \cup T \cup \{B\} \cup \{\alpha, \beta\} \cup \{\diamond\} \cup \{X, Y, X'_\alpha, Y_\alpha\}$.
Axioms (A_1):
$\{XBSY, Z\beta\alpha^i Y'_\alpha, Z \diamond Y_\beta, X_\alpha \beta Z, X_\beta \beta\beta Z, ZY_\beta, X_\alpha Z, X_\beta \beta Z, ZY'_\alpha\}$
$\cup \{ZvY \mid u \to v \in P\}$.

Test tube II:
Rules of R_2:

$$2.1 : \frac{\gamma \mid Y'_\alpha}{Z \mid Y_\alpha}; \qquad 2.2 : \frac{X_\alpha \mid \gamma}{X'_\alpha \alpha \mid Z}; \qquad 2.3 : \frac{a_i \mid Y_\beta}{Z \mid Y};$$

$$2.4 : \frac{X_\beta \beta\alpha^i \beta \mid \mathbf{a}}{X a_i \mid Z}; \qquad 2.5 : \frac{X_\beta \beta\beta \mid \mathbf{a}}{\varepsilon \mid Z'}; \qquad 2.6 : \frac{\mathbf{a} \mid \diamond Y_\beta}{Z' \mid \varepsilon};$$

Filter:
$F_2^{(1)} = N \cup T \cup \{B\} \cup \{\alpha, \beta\} \cup \{\diamond\} \cup \{X_\alpha, Y'_\alpha\}$,
$F_2^{(2)} = N \cup T \cup \{B\} \cup \{\alpha, \beta\} \cup \{\diamond\} \cup \{X_\beta, Y_\beta\}$.
Axioms (A_2):
$\{ZY_\alpha, X'_\alpha \alpha Z, ZY, X a_i Z, Z'\}$.

Ideas of the Proof

The "Rotate-and-Simulate" Method The system uses the method of words' rotation [12,13,14]. Let us recall it briefly. For any word $w = w'w'' \in (N \cup T)^*$ of the grammar G the word $Xw''Bw'Y$ ($X, Y, B \notin N \cup T$) of CDHF system Γ, is called a "rotational version" of the word w. The system Γ models the grammar G as follows. It rotates the word Xw_1uw_2BY into Xw_2Bw_1uY and applies a splicing rule $\frac{\varepsilon \mid uY}{Z \mid vY}$. So we model the application of a rule $u \to v$ of the grammar G by a single scheme-rule. Γ rotates the word Xwa_iY ($a_i \in (N \cup T \cup \{B\})$ "symbol by symbol", *i.e.*, the word Xa_iwY will be obtained after some steps of working of system Γ.

The rotation technique is implemented as follows [15,13,14]. Suppose that we have Xwa_iY and we rotate a_i, *i.e.*, we shall obtain Xa_iwY after some steps of computation. We perform this rotation in three stages. First we encode a_i by $\beta\alpha^i$ and we get $X_\alpha \beta w \beta \alpha^{i-1} Y'_\alpha$. After that we transfer α's from the right end of the molecule to the left end obtaining $X_\beta \beta\alpha^i \beta w Y_\beta$. Finally we decode $\beta\alpha^i\beta$ into a_i which gives us Xa_iwY. More precisely, we start with the word Xwa_iY in the first tube. The second tube receives the word $X_\alpha \beta w \beta \alpha^{i-1} Y'_\alpha$ from the first one. After that point the system works in a cycle where α's are inserted at the left end of the molecule and simultaneously deleted at the right end. The cycle stops when there is no α at the right end. After that we obtain the word $X_\beta \beta\alpha^i \beta w Y_\beta$ in the second tube and we decode a_i obtaining Xa_iwY.

If we have $XwBY$, then we take off X, Y and B obtaining w as a possible result.

Proof

Our proof is done in the following way. We present how we can simulate the behaviour of grammar G by Γ. In the same time we consider all other possible evolutions of molecules and we show that they do not lead to a valid result.

Checking the Simulation We note that the molecules from the bottom part of each rule are already present in the corresponding tube. Moreover, one of the results of the splicing will contain Z and will not contribute for obtaining the result. So we will omit these molecules and we will write:

$$Xwa_iY \underset{1.2}{\longrightarrow} Xw\beta\alpha^{i-1}Y'_\alpha, \text{ instead of}$$
$$(Xw|a_iY, Z|\beta\alpha^{i-1}Y'_\alpha) \vdash_{1.2} (Xw\beta\alpha^{i-1}Y'_\alpha, Za_iY),$$

where by | we highlighted the splicing sites. During the presentation we mark molecules which can evolve in the same tube with h and we mark molecules which must communicate to another tube with 1 or 2 (depending on tube). We also write $m \uparrow$ if molecule m cannot enter any rule and cannot be communicated to another tube.

We shall present the evolution of words of the form Xwa_iY. We shall describe splicing rules which are used and the resulting molecules. We note that, due to parallelism, at the same time in the system there can be several molecules of this form or intermediate forms which evolve in parallel. These molecules do not interact with each other, so we shall concentrate only on a single evolution.

Rotation We shall give the evolution of word Xwa_iY which is in the first tube.

Step 1.

Tube I.

$Xwa_iY\mathtt{h} \underset{1.2}{\longrightarrow} \mathtt{X}\mathtt{w}\beta\alpha^{\mathtt{i}-1}\mathtt{Y}'_\alpha\mathtt{h} \underset{1.4}{\longrightarrow} \mathtt{X}_\alpha\beta\mathtt{w}\beta\alpha^{\mathtt{i}-1}\mathtt{Y}'_\alpha 2.$

We can use these two rules in opposite order which gives the same result. We can also use rule 1.5 instead of the last application. This gives us:

$Xw\beta\alpha^{i-1}Y'_\alpha\mathtt{h} \underset{1.5}{\longrightarrow} \mathtt{X}_\beta\beta\beta\mathtt{w}\beta\alpha^{\mathtt{i}-1}\mathtt{Y}'_\alpha \uparrow.$

We can also have:

$Xwa_iY\mathtt{h} \underset{1.5}{\longrightarrow} \mathtt{X}_\beta\beta\beta\mathtt{wYh}.$

Molecule $X_\alpha\beta w\beta\alpha^{i-1}Y'_\alpha$ goes to the second tube because it can pass the filter $F_2^{(1)}$ and all other molecules cannot pass any of filters of the second tube.

As we noted before there are other applications of splicing rules on other molecules but they are not relevant for our evolution. We shall not mention this fact in the future.

Tube II.

As we noted before there are other applications of splicing rules on other molecules but they are not relevant for our evolution. We shall not mention this fact in the future.

Step 2.
Tube II.
$X_\alpha\beta w\beta\alpha^{i-1}Y'_\alpha h \underset{2.1}{\longrightarrow} X_\alpha\beta w\beta\alpha^{i-1}Y_\alpha h \underset{2.2}{\longrightarrow} X'_\alpha\alpha\beta w\beta\alpha^{i-1}Y_\alpha 1.$
Only molecule $X'_\alpha\alpha\beta w\beta\alpha^{i-1}Y_\alpha$ can pass filter F_1.
In this way we transferred one α from the right end to the left end.
Step 3.
Tube I.
$X'_\alpha\alpha\beta w\beta\alpha^{i-1}Y_\alpha h \underset{1.9}{\longrightarrow} X'_\alpha\alpha\beta w\beta\alpha^{i-2}Y'_\alpha h \underset{1.7}{\longrightarrow} X_\alpha\alpha\beta w\beta\alpha^{i-2}Y'_\alpha 2.$
We can also have:
$X'_\alpha\alpha\beta w\beta\alpha^{i-1}Y_\alpha h \underset{1.8}{\longrightarrow} X_\beta\beta\alpha\beta w\beta\alpha^{i-1}Y_\alpha h \underset{1.9}{\longrightarrow} X_\beta\beta\alpha\beta w\beta\alpha^{i-2}Y'_\alpha \uparrow.$
Only $X_\alpha\alpha\beta w\beta\alpha^{i-1}Y'_\alpha$ can pass filter $F_2^{(1)}$.
Step 4.
Tube II.
$X_\alpha\alpha\beta w\beta\alpha^{i-2}Y'_\alpha h \underset{2.1}{\longrightarrow} X_\alpha\alpha\beta w\beta\alpha^{i-2}Y_\alpha h \underset{2.2}{\longrightarrow} X'_\alpha\alpha\alpha\beta w\beta\alpha^{i-2}Y_\alpha 1.$
Only molecule $X'_\alpha\alpha\alpha\beta w\beta\alpha^{i-2}Y_\alpha$ can pass filter F_1.

In this way we transferred one more α from the right end to the left end. We can continue in this way until we obtain for the first time $X'_\alpha\alpha^i\beta w\beta Y_\alpha$ in tube 2 at some step p. This molecule is communicated to tube 1 where we can use the rule 1.6 and we complete the rotation of a_i.

Step p+1.
Tube I.
$X'_\alpha\alpha^i\beta w\beta Y_\alpha h \underset{1.6}{\longrightarrow} X'_\alpha\alpha^i\beta w Y_\beta h \underset{1.8}{\longrightarrow} X_\beta\beta\alpha^i\beta w Y_\beta 2.$
We can also have:
$X'_\alpha\alpha^i\beta w\beta Y_\alpha h \underset{1.7}{\longrightarrow} X_\alpha\alpha^i\beta w\beta Y'_\alpha h \underset{1.6}{\longrightarrow} X_\alpha\alpha^i\beta w Y_\beta \uparrow.$

Only $X_\beta\beta\alpha^i\beta w Y_\beta$ can pass filter $F_2^{(2)}$. We also add that p+1 is an odd number, so molecule $X_\beta\beta\alpha^i\beta w Y_\beta$ will remain for one step in the first component because the filter used in the second component during odd steps is $F_2^{(1)}$. During next step (p+2) this molecule will not change and it will be communicated to the second tube because the filter to be used will be $F_2^{(2)}$.

Step p+3.
Tube II.
$X_\beta\beta\alpha^i\beta w Y_\beta h \underset{2.4}{\longrightarrow} X a_i w Y_\beta h \underset{2.3}{\longrightarrow} X a_i w Y 1.$
$X a_i w Y$ goes to the first tube.
In this way we competed the rotation of a_i.

Simulation of Grammar Productions If we have a molecule $XwuY$ in the first tube and there is a rule $u \to v \in P$, then we can apply rule 1.1 to simulate the corresponding production of the grammar.

Obtaining Results Now we shall show how we can obtain a result. If we have a molecule of type $XwBY$, then we can perform a rotation of B as described above. We can also take off X and Y from the ends. In this way, if $w \in T^*$, then

w will be a result. So, now we shall show how to do this. Suppose that $XwBY$ appears for the first time in tube 1 at some step q (it is easy to check that q is even).

Step q.

Tube 1.

$XwBY\mathtt{h} \underset{1.3}{\longrightarrow} \mathtt{Xw} \diamond \mathtt{Y}_\beta\mathtt{h} \underset{1.5}{\longrightarrow} \mathtt{X}_\beta\beta\beta\mathtt{w} \diamond \mathtt{Y}_\beta 2.$

We can also have:

$Xw \diamond Y_\beta\mathtt{h} \underset{1.4}{\longrightarrow} \mathtt{X}_\alpha\beta\mathtt{w} \diamond \mathtt{Y}_\beta \uparrow.$

$X_\beta\beta\beta w \diamond Y_\beta$ can pass filter $F_2^{(2)}$.

Step q+1.

Tube II.

$X_\beta\beta\beta w \diamond Y_\beta\mathtt{h} \underset{2.5}{\longrightarrow} \mathtt{w} \diamond \mathtt{Y}_\beta\mathtt{h} \underset{2.6}{\longrightarrow} \mathtt{w1}.$

So, if $w \in T^*$ then w is a result.

Final Remarks We start with molecule $XBSY$ in the first tube. After that we are doing a rotation and simulation of productions of G. Finally we reach a configuration $XwBY$ where $w \in T^*$. At this moment we eliminate X, Y and B obtaining $w \in T^*$.

As all cases have been investigated, the proof is complete.

4 $\mathbf{TTF_{2,4}}$ with Garbage Collection

In this section we shall present a TTF system having two components and four filters. Additionally this system contains no rules in the second component, so it is used only as a garbage collector. Also, all filters in the second tube are the same.

Theorem 2. *For any type-0 grammar $G = (N, T, P, S)$ there is a communicating distributed H system with alternating filters having two components and four filters $\Gamma = (V, T, (A_1, R_1, F_1^{(1)}, \dots, F_1^{(4)}), (A_2, R_2, F_2^{(1)}, \dots, F_2^{(4)}))$ which simulates G and $L(\Gamma) = L(G)$. Additionally, the second component of Γ has no splicing rules, i.e., $R_2 = \emptyset$, and also $F_2^{(1)} = \dots = F_2^{(4)}$.*

We construct Γ as follows.

Let $N \cup T \cup \{B\} = \{a_1, a_2, \dots, a_n\}$ $(B = a_n)$.

In what follows we will assume the following:

$1 \le i \le n, 1 \le j \le 4, \mathbf{a} \in N \cup T \cup \{B\}$,

$\mathbf{b} \in N \cup T \cup \{B\} \cup \{\diamond\}, \gamma \in \{\alpha, \beta\}$.

Also let $\mathcal{V} = N \cup T \cup \{B\} \cup \{\alpha, \beta\} \cup \{\diamond\}$.

$V = \mathcal{V} \cup \{X, Y, X_\alpha, X_\beta, X'_\alpha, Y_\alpha, Y'_\alpha, Y_\beta, Z, Z', Z_j, R_j, L_j\}$.

The terminal alphabet T is the same as for grammar G.

Test tube I:

Rules of R_1:

$1.1.1 : \dfrac{\varepsilon \mid uY}{R_1 \mid vY};$ $\quad 1.1.2 : \dfrac{X \mid \mathbf{a}}{X_\alpha\beta \mid L_1};$ $\quad 1.1.3 : \dfrac{X \mid \mathbf{a}}{X_\beta\beta\beta \mid L_1};$

$1.1.4 : \dfrac{\mathbf{a} \mid a_iY}{R_1 \mid \beta\alpha^{i-1}Y'_\alpha};$ $\quad 1.1.5 : \dfrac{\mathbf{a} \mid BY}{R_1 \mid \diamond Y_\beta};$

$1.2.1 : \dfrac{\gamma \mid Y'_\alpha}{R_2 \mid Y_\alpha};$ $\quad 1.2.2 : \dfrac{X_\alpha \mid \gamma}{X'_\alpha\alpha \mid L_2};$

$1.3.1 : \dfrac{X'_\alpha \mid \alpha}{X_\alpha \mid L_3};$ $\quad 1.3.2 : \dfrac{X'_\alpha \mid \alpha}{X_\beta\beta \mid L_3};$ $\quad 1.3.3 : \dfrac{\gamma \mid \alpha Y_\alpha}{R_3 \mid Y'_\alpha};$

$1.3.4 : \dfrac{\mathbf{b} \mid \beta Y_a}{R_3 \mid Y_\beta};$

$1.4.1 : \dfrac{a_i \mid Y_\beta}{R_4 \mid Y};$ $\quad 1.4.2 : \dfrac{X_\beta\beta\alpha^i\beta \mid \mathbf{a}}{Xa_i \mid L_4};$ $\quad 1.4.3 : \dfrac{X_\beta\beta\beta \mid \mathbf{a}}{\varepsilon \mid Z'L_4};$

$1.4.4 : \dfrac{\mathbf{a} \mid \diamond Y_\beta}{Z'L_4 \mid \varepsilon};$

$1.1.1' : \dfrac{Z_1 \mid vY}{R_1 \mid L_1};$ $\quad 1.1.2' : \dfrac{X_\alpha\beta \mid Z_1}{R_1 \mid L_1};$ $\quad 1.1.3' : \dfrac{X_\beta\beta\beta \mid Z_1}{R_1 \mid L_1};$

$1.1.4' : \dfrac{Z_1 \mid \beta\alpha^iY'_\alpha}{R_1 \mid L_1};$ $\quad 1.1.5' : \dfrac{Z_1 \mid \diamond Y_\beta}{R_1 \mid L_1};$

$1.2.1' : \dfrac{Z_2 \mid Y_\alpha}{R_2 \mid L_2};$ $\quad 1.2.2' : \dfrac{X'_\alpha\alpha \mid Z_2}{R_2 \mid L_2};$

$1.3.1' : \dfrac{X_\alpha \mid Z_3}{R_3 \mid L_3};$ $\quad 1.3.2' : \dfrac{X_\beta\beta \mid Z_3}{R_3 \mid L_3};$ $\quad 1.3.3' : \dfrac{Z_3 \mid Y'_\alpha}{R_3 \mid L_3};$

$1.3.4' : \dfrac{Z_3 \mid Y_\beta}{R_3 \mid L_3};$

$1.4.1' : \dfrac{Z_4 \mid Y}{R_4 \mid L_4};$ $\quad 1.4.2' : \dfrac{Xa_i \mid Z_4}{R_4 \mid L_4};$ $\quad 1.4.3' : \dfrac{Z' \mid Z_4}{R_4 \mid L_4};$

Filter:

$F_1^{(1)} = \mathcal{V} \cup \{X_\alpha, Y'_\alpha, R_2, L_2\},$

$F_1^{(2)} = \mathcal{V} \cup \{X'_\alpha, Y_\alpha, R_3, L_3\},$

$F_1^{(3)} = \mathcal{V} \cup \{X_\beta, Y_\beta, R_4, L_4\},$

$F_1^{(4)} = \mathcal{V} \cup \{X, Y, R_1, L_1\}.$

Axioms (A_1):
$\{XBSY, X_\alpha\beta Z_1, X_\beta\beta\beta Z_1, Z_1\beta\alpha^i Y'_\alpha, Z \diamond Y_\beta, Z_2 Y_\alpha, X'_\alpha \alpha Z_2,$
$X_\alpha Z_3, X_\beta\beta Z_3, Z_3 Y'_\alpha, ZY_\beta, Z_4 Y, X a_i Z_4, Z' Z_4, R_1 L_1\}$
$\cup\{ZvY \mid u \to v \in P\}$.

Test tube II:

No rules ($R_2 = \emptyset$).
Filter:
$F_2^{(j)} = V \cup \{X, Y, X_\alpha, X_\beta, X'_\alpha, Y_\alpha, Y'_\alpha, Y_\beta, Z', R_j, L_j\}$.
Axioms (A_2):
$\{R_2 L_2, R_3 L_3, R_4 L_4\}$

Ideas of the Proof

The system is similar to the system from the previous section. It contains the same rules which are grouped in the first tube. The rules are split into four subsets (1.1.x–1.4.x) which cannot be used all at the same time. In order to achieve this we use the method described in [16,9]. We use molecules $R_j L_j$ which travel from one tube to another. When a molecule $R_j L_j$ appears in the first tube it triggers subset j of rules which can be applied (in [16] these molecules are called directing molecules). More precisely, the bottom part of each rule is made in a special way: the molecule which is present there can be produced only if we have the corresponding $R_j L_j$ in the first tube. For example, rule 1.1.3 can be used only if molecule $X_\beta\beta\beta L_1$ is present. But this molecule can be present only if $R_1 L_1$ is present in the first tube (it is produced by applying rule 1.1.3' on the axiom $X_\beta\beta\beta Z_1$ and $R_1 L_1$, and on the next step it is communicated to the second tube).

This mechanism can be realized by using alternating filters. The filters act also like selectors for correct molecules and the computation goes in the following way: molecules are sent into the second tube and after that the correct molecules are extracted from this tube into the first one where they are processed.

We also note that the operations performed in the first group and the third group are independent and the molecules are of different forms. So, we can join the first group with the third group (the corresponding filters 2 and 4 as well) and we obtain:

Theorem 3. *For any type-0 grammar $G = (N, T, P, S)$ there is a communicating distributed H system with alternating filters having two components and three filters $\Gamma = (V, T, (A_1, R_1, F_1^{(1)}, \ldots, F_1^{(3)}), (A_2, R_2, F_2^{(1)}, \ldots, F_2^{(3)}))$ which simulates G and $L(\Gamma) = L(G)$. Additionally, the second component of Γ has no splicing rules, i.e., $R_2 = \emptyset$, and also $F_2^{(1)} = \cdots = F_2^{(3)}$.*

5 $\mathbf{TTF_{2,2}}$ with Garbage Collection

In this section we shall present a TTF system with two components and two filters which has no rules in the second component. Moreover, the two filters for both components differ only in one letter. Also the proof is different: we shall use a simulation of a Turing machine instead of a Chomsky grammar.

We define a coding function ϕ as follows. For any configuration w_1qw_2 of a Turing machine M we define $\phi(w_1qw_2) = Xw_1Sqw_2Y$, where X, Y and S are three symbols which are not in T.

We also define the notion of an *input* for a TTF system. An input word for a system Γ is simply a word w over the non-terminal alphabet of Γ. The computation of Γ on input w is obtained by adding w to the axioms of the first tube and after that by making Γ to evolve as usual.

Lemma 1. *For any Turing machine $TM = (Q, T, a_0, s_0, F, \delta)$ and for any input word w there is a communicating distributed H system with alternating filters having two components and two filters, $\Gamma = (V, T_\Gamma, (A_1, R_1, F_1^{(1)}, F_1^{(2)}), (A_2, R_2, F_2^{(1)}, F_2^{(2)}))$ which given the input $\phi(w)$ simulates M on input w, i.e., such that:*

1. *for any word $w \in L(M)$ that reaches a halting configuration w_1qw_2, Γ will produce a unique result $\phi(w_1qw_2)$;*
2. *for any word $w \notin L(M)$, Γ will produce the empty language.*

Let $T = \{a_0, \dots, a_{m-1}\}$, $Q = \{q_0, \dots, q_{n-1}\}$, $\mathbf{a} \in T \cup \{X\}$, $\mathbf{b}, \mathbf{d}, \mathbf{e} \in T$, $\mathbf{c} \in T \cup \{Y\}$, $\mathbf{q} \in Q$, a_0 – blank symbol.

We construct Γ as follows:

$V = T \cup Q \cup \{X, Y, S, S', R_Y, R, L, R', L', R^R, Z_1^R, Z_X, R_1^L, Z_1^L, R_1', Z_1'\}$.

$T_\Gamma = T \cup \{X, Y, S\} \cup \{q \mid q \in F\}$.

Test tube I:

Rules of R_1:

For any rule $q_ia_kRa_lq_j \in \delta$ we have the following group of 4 rules:

1.1.1.1. $\begin{array}{c|c} \mathbf{a}Sq_ia_k & Y \\ \hline R_Y & a_0Y \end{array}$, 1.1.1.2. $\begin{array}{c|c} \mathbf{a} & Sq_ia_k\mathbf{b} \\ \hline R & L \end{array}$,

1.1.1.3. $\begin{array}{c|c} RSq_ia_k & \mathbf{b} \\ \hline R_1^Ra_lS'q_j & Z_1^R \end{array}$, 1.1.1.4. $\begin{array}{c|c} \mathbf{a} & L \\ \hline R_1^R & \mathbf{b}S'\mathbf{qd} \end{array}$,

For any rule $q_ia_kLa_lq_j \in \delta$ we have the following group of 5 rules:

1.1.2.1. $\begin{array}{c|c} X & Sq_ia_k \\ \hline Xa_0 & Z_X \end{array}$, 1.1.2.2. $\begin{array}{c|c} \mathbf{b} & \mathbf{d}Sq_ia_k \\ \hline R & L \end{array}$, 1.1.2.3. $\begin{array}{c|c} R\mathbf{b}Sq_ia_k & \mathbf{c} \\ \hline R_1^LS'q_j\mathbf{b}a_l & Z_1^L \end{array}$,

1.1.2.1′. $\begin{array}{c|c} X & \mathbf{b}Sq_ia_k \\ \hline Xa_0 & Z_X \end{array}$ 1.1.2.4. $\begin{array}{c|c} \mathbf{b} & L \\ \hline R_1^L & S'\mathbf{qdec} \end{array}$,

We have also the following group of 3 rules:

1.2.1. $\begin{array}{c|c} \mathbf{ab} & S'\mathbf{qd} \\ \hline R' & L' \end{array}$, 1.2.2. $\begin{array}{c|c} R'S' & \mathbf{qb} \\ \hline R_1'S & Z_1' \end{array}$, 1.2.3. $\begin{array}{c|c} \mathbf{b} & L' \\ \hline R_1' & S\mathbf{qd} \end{array}$

Filter:
$F_1^{(1)} = \{R', L, L'\}$,
$F_1^{(2)} = \{R, L, L'\} = (F_1^1 \setminus \{R'\}) \cup \{R\}$.
Axioms (A_1):
$XSq_0wY, R_1^R a_l S' q_j Z_1^R (\exists q_i a_k R a_l q_j \in \delta), R_1^L S' q_j \mathbf{b} a_l Z_1^L (\exists q_i a_k L a_l q_j \in \delta)$,
$R_Y a_0 Y, X a_0 Z_X, R_1' S Z_1', RL$.

Test tube II: No rules (R_2 is empty).
Filter:
$F_2^{(1)} = Q \cup T \cup \{X, Y, S, R, L, R', L', R_1^L, R_1^R, R_1'\}$,
$F_2^{(2)} = Q \cup T \cup \{X, Y, S', R, L, R', L', R_1^L, R_1^R, R_1'\} = (F_2^1 - \{S\}) \cup \{S'\}$.
Axioms (A_2): $R'L'$.

Ideas of the Proof

The tape of M is encoded in the following way: for any configuration $w_1 q w_2$ of M we have a configuration $X w_1 S q w_2 Y$ in the system Γ. So, the tape is enclosed by X and Y and we have a marker Sq_i which marks the head position and the current state on the tape.

Γ simulates each step of M in 2 stages. During the first stage it simulates a step of the Turing machine and it primes symbol S using rules 1.1.x.2–1.1.x.4. During the second stage it unprimes S' using rules 1.2.x. The tape can be extended if necessary by rules 1.1.x.1. The system produces a result only if M halts on the initial string w.

In order to achieve the separation in 2 stages we use two subsets of rules in the first tube and the directing molecules RL and $R'L'$ which appear alternatively in the first and the second test tube will trigger the first (1.1.x) or the second (1.2.x) subset (see also section 4 and [16]). All extra molecules are passed to the second test tube which is considered as a garbage collector because only RL and $R'L'$ may go from that tube to the first one.

Molecules $R_Y a_0 Y, R_1^R a_l S' q_j Z_1^R, X a_0 Z_X, R_1^L S' q_j \mathbf{b} a_l Z_1^L, R_1' S Z_1'$, are present in test tube 1 all the time and they do not leave it. Also all molecules having Z with indices at the end remain in the first test tube.

We start with $X w_1 S q_0 w_2 Y$ in the first test tube where $w_1 q_0 w_2$ is the initial configuration of M.

Checking the Simulation

We completely describe the simulation of the application of a rule with a move to right ($q_i a_k R a_l q_j \in \delta$) to configuration $X w_1 S q_i a_k w_2$Y. Now also we shall omit molecules $R_Y a_0 Y, R_1^R a_l S' q_j Z_1^R, X a_0 Z_X, R_1^L S' q_j \mathbf{b} a_l Z_1^L, R_1' S Z_1'$ which are always present in the first tube. Also the molecules having Z with indices at the end and which do not alter the computation will be omitted from the description after one step of computation.

We also note that the second tube may contain some other molecules (obtained during the previous steps) as it is a garbage collector. We shall omit these molecules as they do not alter our simulation.

Step 1.

Splicing:

Tube 1:

We have: $Xw_1Sq_ia_kw_2Y, RL$.

We apply the following rules which are all from group 1.1.1.x:

$(Xw_1|Sq_ia_kw_2Y, R|L) \vdash_{1.1.1.2} (Xw_1L, RSq_ia_kw_2Y)$.

$(RSq_ia_k|w_2Y, R_1^Ra_lS'q_j|Z_1^R) \vdash_{1.1.1.3} (R_1^Ra_lS'q_jw_2Y, RSq_ia_kZ_1^R)$.

$(Xw_1|L, R_1^R|a_lS'q_jw_2Y) \vdash_{1.1.1.4} (Xw_1a_lS'q_jw_2Y, R_1^RL)$.

Tube 2:

We have: $R'L'$. As it was said above we can also have a lot of other molecules here which do not alter the computation. We omit them and we do not mention this in the future.

No application of rules.

Communication:

The current filter is $F_2^{(1)} = Q \cup T \cup \{X, Y, S, R, L, R', L', R_1^L, R_1^R, R_1'\}$.

Molecules going from tube 1 to tube 2:

$RL, Xw_1Sq_ia_kw_2Y, Xw_1L, RSq_ia_kw_2Y, R_1^RL$.

Molecules remaining in tube 1:

$R_1^Ra_lS'q_jw_2Y, RSq_ia_kZ_1^R, Xw_1a_lS'q_jw_2Y$.

Molecules going from tube 2 to tube 1:

$R'L'$.

Molecules remaining in tube 2:

none.

Molecule $RSq_ia_kZ_1^R$ cannot evolve any further so we shall omit it from the checking. Also molecule $R_1^Ra_lS'q_jw_2Y$ will go from the first test tube to the second tube during the next iteration.

Step 2.

Splicing:

Tube 1:

We have: $R_1^Ra_lS'q_jw_2Y, Xw_1a_lS'q_jw_2Y, R'L'$.

We apply the following rules which are all from group 1.2.x:

$(Xw_1a_l|S'q_jw_2Y, R'|L') \vdash_{1.2.1} (Xw_1a_lL', R'S'q_jw_2Y)$.

$(R'S'|q_jw_2Y, R_1'S|Z_1') \vdash_{1.2.2} (R_1'Sq_jw_2Y, R'S'Z_1')$.

$(Xw_1a_l|L', R_1'|Sq_jw_2Y) \vdash_{1.2.3} (Xw_1a_lSq_jw_2Y, R_1'L')$.

Tube 2:

We have: $Xw_1Sq_ia_kw_2Y, RL, Xw_1L, RSq_ia_kw_2Y, R_1'L'$.

No application of rules.

Communication:

The current filter is $F_2^{(2)} = Q \cup T \cup \{X, Y, S', R, L, R', L', R_1^L, R_1^R, R_1'\}$.

Molecules going from tube 1 to tube 2:

$R'L', R_1^Ra_lS'q_jw_2Y, Xw_1a_lS'q_jw_2Y, Xw_1a_lL', R'S'q_jw_2Y, R_1'L'$.

Molecules remaining in tube 1:
$R'S'Z_1, R_1'Sq_jw_2Y, Xw_1a_lSq_jw_2Y$.
Molecules going from tube 2 to tube 1:
RL.
Molecules remaining in tube 2:
$R'L', R_1^Ra_lS'q_jw_2Y, Xw_1a_lS'q_jw_2Y, Xw_1a_lL', R'S'q_jw_2Y, R_1'L', Xw_1Sq_ia_kw_2Y, Xw_1L, RSq_ia_kw_2Y, R_1^RL$.

On the next iteration $R_1'Sq_jw_2Y$ will go from the first test tube to the second tube.

Thus we modelled the application of the rule (q_i, a_k, R, a_l, q_j) of the Turing machine M. It is easy to observe that if $w_2 = \varepsilon$ we first apply rule 1.1.1.1 in order to extend the tape to the right and after that $w_2 = a_0$. Also it is easy to check that additional molecules obtained (such as $Xw_1Sq_ia_kw_2Y$, Xw_1L, $RSq_ia_kw_2Y$, R_1^RL, $R_1^Ra_lS'q_jw_2Y, RSq_ia_kZ_1^R, Xw_1a_lS'q_jw_2Y$, $R'S'q_jw_2Y$, $R'S'Z_1$, $R_1'Sq_jw_2Y$, $Xw_1a_lS'q_jw_2Y, Xw_1a_lL'$) either go to the second test tube or do not alter the computation.

For the left shift rules the things are similar, except that we ensure that there are at least 2 symbols at the left of S.

We see that we model step by step the application of rules of M. It is easy to observe that this is the only possibility for the computations to go on as the molecule which encodes the configuration of M triggers the application of rules of Γ.

Theorem 4. *For any recursively enumerable language $\mathcal{L} \subseteq T^*$ there is a communicating distributed H system with alternating filters with two components and two filters $\Gamma = (V, T, (A_1, R_1, F_1^{(1)}, F_1^{(2)}), (A_2, R_2, F_2^{(1)}, F_2^{(2)}))$, which generates $\mathcal{L}$, i.e., $\mathcal{L} = L(\Gamma)$.*

Proof. For the proof we use ideas similar to ones used in [10].

Let $\mathcal{L} = \{w_0, w_1, \dots\}$. Then there is a Turing machine $T_{\mathcal{L}}$ which computes $w_i0\dots0$ starting from 01^{i+1} where 0 is the blank symbol.

It is possible to construct such a machine in the following way. Let G be a grammar which generates $\mathcal{L}$. Then we can construct a Turing machine M which will simulate derivations in G, *i.e.*, we shall obtain all sentential forms of G. We shall simulate bounded derivation in order to deal with a possible recursion in a derivation. After deriving a new word the machine checks if this word is a terminal word and, if this is the case, it checks if this word is the i-th terminal word which is obtained. If the last condition holds, then the machine erases everything except the word.

Moreover, this machine is not stationary (*i.e.*, the head has to move at each step) and it never visits cells to the left of the 0 from the initial configuration (*i.e.*, the tape is semi-infinite to the right). Ideas for doing this can be found in [8]. So, machine $T_{\mathcal{L}}$ transforms configuration q_001^{k+1} into $q_fw_k0\dots0$.

Now starting from $T_{\mathcal{L}}$ it is possible to construct a machine $T_{\mathcal{L}}'$ which computes $01^{k+2}Mq_f'w_k0\dots0$ starting from q_001^{k+1}. Now, using Γ we shall cut off w_k and pass to configuration q_001^{k+2}. In this way we will generate $\mathcal{L}$ word by word.

In order to simplify the proof we shall consider machine $T''_{\mathcal{L}}$ which will do the same thing as $T'_{\mathcal{L}}$ but at the end will move its head to the right until the first 0, *i.e.* will produce $01^{k+2}Mw_kq''_f0\ldots0$ from q_001^{k+1}. We use also two additional Turing machines T_1 and T_2 having the tape alphabet of $T''_{\mathcal{L}}$ extended by the set $\{X,Y,S\}$. The first machine, T_1, moves to the left until it reaches M and then stops in the state q^1_f. The starting state of T_1 is q^1_s. The second machine, T_2 moves to the left until it reaches 0 and then stops in the state q_0. The starting state of this machine is q^2_s.

Now we consider Γ which is very similar to the system constructed in the previous lemma. In fact, we take machine $T''_{\mathcal{L}}$ and we construct Γ for this machine as it is done before. After that we take the rules of T_1 and T_2 and we add the corresponding splicing rules and axioms to Γ. Finally we add the following splicing rules:

$$x.1.\ \frac{\mathbf{a} \,|\, \mathbf{bd}Sq''_f0}{Z_1 \,|\, \mathbf{a}Sq^1_s\mathbf{bd}}, \qquad x.2.\ \frac{\mathbf{e} \,|\, 11Sq^1_fM}{X_2 \,|\, \mathbf{e}Sq^2_s11}, \qquad x.2'.\ \frac{X_211Sq^1_fM \,|\, \mathbf{c}}{\varepsilon \,|\, Z'}$$

where $\mathbf{e} = \{0,1\}$.

We also add Xq_001Y to the axioms in the first tube and we consider T as a terminal alphabet.

We claim that Γ can generate $\mathcal{L}$. First of all it simulates machine $T''_{\mathcal{L}}$ so it permits to pass from configuration $XSq_001^{k+1}Y$ to $X01^{k+2}Mw_kSq'_f0\ldots0Y$. Now we can use rule $x.1$. As a result we have the part $0\ldots0Y$ which is cut off and sent to the second tube, and a new molecule $X01^{k+2}Mw'_kSq^1_sab$ ($w_k = w'_kab$) which is obtained. Now we can easily observe that we can simulate the work of T_1 because the only condition which must be satisfied is the presence of one symbol at the right from the head position. So at this moment Γ will simulate the work of T_1 and we will arrive to the following molecule: $X01^{k+2}Sq^1_fMw_k$. Now by rules $x.2$ and $x.2'$ we will cut off w_k which will represent a result and we will obtain the molecule $X01^kSq^2_s11Y$. In a similar way we can simulate the work of T_2 and we will arrive to configuration $XSq_001^{k+2}Y$. So, we showed that we succeeded to implement the mechanism of generation of words which we presented before.

6 Conclusions

We presented a new variant of communicating distributed H systems which uses a tuple of filters instead of one single filter. We showed that using two components it is possible to generate any recursively enumerable language. We presented four systems which differ in the number of filters being used. In the first system we use two filters and additionally both filters for the first tube coincide. The second system uses four filters, but do not contain any rules in the second component which serves for garbage collection. We showed that it is possible to reduce the number of filters down to three. The last system contains two filters, but do not contain any rule in the second component and the two filters in each tuple differ only in one letter. In the first three systems we simulate a Chomsky grammar and in the last system we simulate a Turing machine.

It is an open problem whether or not it is possible to have no rules in the second component and only one tuple containing different filters (as in the case of the first system). Also it is interesting to simulate Chomsky grammars instead of Turing machines in the case of the last system.

Acknowledgments The author wants to thank M. Margenstern and Yu. Rogozhin for their very helpful comments and for their attention for the present work. The author acknowledges also the "MolCoNet" IST-2001-32008 project and the *Laboratoire d'Informatique Théorique et Apliquée de Metz* which provided him the best conditions for producing the present result. And finally the author acknowledges the Ministry of Education of France for the financial support of his PhD.

References

1. Csuhaj-Varju E., Kari L., Păun Gh., Test tube distributed systems based on splicing. *Computers and Artificial Intelligence*, **15** 2 (1996) 211–232.
2. Ferretti C., Mauri G., Zandron C., Nine test tubes generate any RE language, *Theoretical Computer Science*, **231** N2 (2000) 171–180.
3. Freund R., Freund F., Test tube systems: when two tubes are enough, *Preproceedings of DLT99: Developments in Languages Theory,* Aachen, July 6-9, (1999) 275–286.
4. Frisco P., Zandron C., On variants of communicating distributed H systems, *Fundamenta Informaticae*, **48** 1(2001) 9–20.
5. Head T., Formal language theory and DNA: an analysis of the generative capacity of recombinant behaviors. *Bulletin of Mathematical Biology*, **49** (1987) 737–759.
6. Head T., Păun Gh., Pixton D., Language theory and molecular genetics. Generative mechanisms suggested by DNA recombination, ch. 7 in vol.2 of G. Rozenberg, A. Salomaa, eds., *Handbook of Formal Languages*, Springer-Verlag, Heidelberg, 1997.
7. Hopcroft J.E., Motwani R., Ullman J.D., *Introduction to Automata Theory, Languages, and Computation,* 2nd ed., Addison-Wesley, Reading, Mass., 2001.
8. Kleene S., *Introduction to Metamathematics.* Van Nostrand Comp. Inc., New-York, 1952.
9. Margenstern M., Rogozhin Yu., Verlan S., Time-varying distributed H systems with parallel computations: the problem is solved. *Preproceedings of DNA9*, Madison, USA, June 1-4, 2003, 47–51.
10. Margenstern M., Rogozhin Yu., Time-varying distributed H systems of degree 1 generate all recursively enumerable languages, in vol. *Words, Semigroups and Transductions* (M. Ito, G. Păun, S. Yu, eds.), World Scientific, Singapore, 2001, 329–340.
11. Păun G., DNA computing: distributed splicing systems. In *Structures in Logic and Computer Science. A Selection of Essays in honor of A. Ehrenfeucht*, *Lecture Notes in Computer Science*, **1261** (1997), 353–370.
12. Păun G., Regular extended H systems are computationally universal, *J. Automata, Languages, and Combinatorics*, **1** 1 (1996) 27–36.

13. Păun G., DNA computing based on splicing: universality results. *Theoretical Computer Science*, **231** 2 (2000) 275–296.
14. Păun G., Rozenberg G., Salomaa A., *DNA Computing: New Computing Paradigms.* Springer, Berlin, 1998.
15. Priese L., Rogozhin Yu., Margenstern M., Finite H-Systems with 3 Test Tubes are not Predictable. In *Proceedings of Pacific Symposium on Biocomputing,* Kapalua, Maui, January 1998 (R.B. Altman, A.K. Dunker, L. Hunter, T.E. Klein, eds.), World Sci. Publ., Singapore, 1998, 545–556.
16. Verlan S., A frontier result on enhanced time-varying distributed H systems with parallel computations. *Preproceeding of DCFS'03,* Budapest, Hungary, July 12–14, 2003, 221–232.

Publications by Thomas J. Head

Books:

1. *Modules: A Primer of Structure Theorems*, Brooks/Cole, 1974.
2. (with J. Golan) *Modules and the Structure of Rings*, Marcel Dekker, 1991.

Research Papers:

1. Dense submodules, *Proc. Amer. Math. Soc.*, **13** (1962), 197–199.
2. Remarks on a problem in primary abelian groups, *Bull. Soc. Math. France*, **91** (1963), 109–112.
3. An application of abelian groups to geometries with a transitive set of translations, in *Topics in Abelian Groups* (J.M. and E.A. Walker Scott, eds.), Foresman& Co., 1963, 349–355.
4. Note on the occurrence of direct factors in groups, *Proc. Amer. Math. Soc.*, **15** (1964), 193–195.
5. Note on groups and lattices, *Nieuw Archief Wiskunde*, **13** (1965), 110–112.
6. Normal subgroups and Cartesian powers of simple groups, *Nieuw Archief Wiskunde*, **13** (1965), 177–179.
7. Purity in compactly generated modular latices, *Acta Math. Acad. Sci. Hungaricae*, **17** (1966), 55–59.
8. A direct limit representation for abelian groups with an aplication to tensor sequences, *Acta Math. Acad. Sci. Hungaricae*, **18** (1967), 231–234.
9. Homomorphisms of commutative semigroups as tensor maps, *J. Nat. Sci. & Math.*, **7** (1967), 39–49.
10. A tensor product of a group with a semigroup; The tensor product of semigroups with minimal ideals; Functor properties of semigroup tensor products, *J. Nat. Sci. & Math.*, **7** (1967), 155–171.
11. The varieties of commutative monoids, *Nieuw Archief Wiskunde*, **16** (1968), 203–206.
12. Tensor products and maximal subgroups of commutative semigroups, *Pub. Math. Debrecen*, **16** (1969), 145–147.
13. Groups as sequences and direct products of countable groups, *J. Nat. Sci. & Math.*, **10** (1970), 45–47.
14. Commutative semigroups having greatest regular images, *Semigroup Forum*, **2** (1971), 130–137.
15. Embedded ordinals in modular algebraic lattices, *Algebra Universalis*, **1** (1971), 200–203.
16. Preservation of coproducts by Hom(,), *Rocky Mtn. J. Math.*, **2** (1972), 235–237.
17. (with N. Kuroki) Greatest regular images of tensor products of commutative semigroups, *Kodai Math. Sem. Rep.*, **26** (1975), 132–136.
18. An algebraic characterization of multiplicative semigroups of real valued functions (short note), *Semigroup Forum*, **14** (1977), 93–94.

N. Jonoska et al. (Eds.): Molecular Computing (Head Festschrift), LNCS 2950, pp. 385–389, 2004.

19. (with M. Blattner) Single valued a–transducers, *J. Computer & System Sci.*, **15** (1977), 310–327.
20. (with M. Blatner) Automata that recognize intersections of free monoids, *Information & Control*, **35** (1977), 173–176.
21. Quotenet monoids and greatest commutative monoid images of several types, *Semigroup Forum*, **17** (1979), 351–363.
22. (with M. Blattner) The decidability of equivalence for deterministic finite transducers, *J. Computer & System Sci.*, **19** (1979), 45–49.
23. Codes, languages, 0L schemes, and a–transducers, *Soochow J. Math.*, **5** (1979), 45–62.
24. Codeterministic Lindenmayer schemes and systems, *J. Computer & Systems Sci.*, **19** (1979), 203–210.
25. A–transducers and the monotonicity of IL schemes, *J. Computer & Systems Sci.*, **21** (1980), 87–91.
26. (with J. Anderson) On injective and flat commutative regular semigroups (short note), *Semigroup Forum*, **21** (1980), 283–284.
27. Unique decipherability relative to a language, *Tamkang J. Math.*, **11** (1980), 59–66.
28. (with G. Thierrin) Hypercodes in deterministic and slender 0L languages, *Information & control*, **45** (1980), 251–262.
29. Fixed languages and the adult languages of 0L schemes, *Intern. J. Computer Math.*, **10** (1981), 103–107.
30. Expanded subalphabets in the theories of languages and semigroups, *Intern. J. Computer. Math.*, **12** (1982), 113–123.
31. (with J. Wilkinson) Finite D0L languages and codes (note), *Theoretical Computer Science*, **21** (1982), 357–361.
32. (with Culik II) Transductions and the parallel generation of languages, *Intern. J. Computer. Math.*, **13** (1983), 3–15.
33. (with G. Therrin, J. Wilkinson) D0L schemes and the periodicity of string embeddings (note), *Theoretical Computer Science*, **23** (1983), 83–89.
34. (with G. Therrin) Polynomially bounded D0L systems yield codes, in *Combinatorics on Words: Progress and Perspectives* (L.J. Cummings, ed.), Academic Press, 1983, 167–174.
35. Adherences of D0L languages, *Theoretical Computer Science*, **31** (1984), 139–149.
36. Adherence equivalence is decidable for D0L languages, in *Lecture Notes in Computer Science* **166** (M. Fontet, K. Melhorn, eds.), Springer–Verlag, 1984, 241–249.
37. (with J. Wilkinson) Code properties and homomorphisms of D0L systems, *Theoretical Comp. Sci.*, **35** (1985), 295–312.
38. (with J. Wilkinson) Code properties and derivatives of D0L systems, *Combinatorial Algorithms on Words* (A. Apostolico & Z. Galil, eds.), Springer–Verlag, 1985, 315–327.
39. The topological structure of adherences of regular languages *R.A.I.R.O. Theor. Informatics & Applications*, **20** (1986), 31–41.

40. The adherences of languages as topological spaces, in *Automata on Infinite Words* (M. Nivat, D. Perrin, eds.), Lecture Notes in Computer Science **192**, Springer–Verlag, 1985, 147–163.
41. (with B. Lando) Fixed and stationary omega–words and omega–languages, in *The Book of L* (G. Rozenberg, A. Salomaa, eds.), Springer–Verlag, 1986, 147–156.
42. (with B. Lando) Periodic D0L languages, *Theoretical Computer Science*, **46** (1986), 83–89.
43. (with B. Lando) Regularity of sets of initial strings of periodic D0L systems, *Theoretical Computer Science*, **48** (1986), 101–108.
44. (with B. Lando) Bounded D0L languages, *Theoretical Computer Science*, **51** (1987), 255–264.
45. The topological structure of the space of unending paths of a graph, *Congressus Numerantium*, **60** (1987), 131–140.
46. Formal language theory and DNA: an analysis of generative capacity of specific recombinant behaviors, *Bull. Math. Biology*, **49** (1987), 737–759.
47. Deciding the immutability of regular codes and languages under finite transductions, *Information Processing Letters*, **31** (1989), 239–241.
48. One–dimensional cellular automata: injectivity from unambiguity, *Complex Systems*, **3** (1989), 343–348.
49. (with B. Lando) Eignewords and periodic behaviors, in *Sequences: Combinatorics, Compression, Security and Transmission* (R. Capocelli, ed.), Springer–Verlag, 1990, 244–253.
50. The set of strings mapped into a submonoid by iterates of a morphism, *Theoretical Computer Science*, **73** (1990), 329–333.
51. (with W. Forys) The poset of retracts of a free monoid, *International J. Computer Math.*, **37** (1990), 45–48.
52. (with W. Forys) Retracts of free monoids are nowhere dense with respect to finite–group topologies and p–adic topologies (short note), *Semigroup Forum*, **42** (1991), 117–119.
53. The topologies of sofic subshifts have comutable Pierce invariants, *R.A.I.R.O. Theor. Informatics & Applications*, **25** (1991), 247–254.
54. (with J. Anderson) The lattice of semiretracts of a free monoid, *Intern. J. Computer Math.*, **41**, ???
55. Splicing schemes and DNA, in *Lindenmayer Systems: Impact on Theoretical Computer Science and Developmental Biology* (G. Rozenberg, A. Salomaa, eds.) Springer–Verlag, 1992, 371–383.
56. (with A. Weber) Deciding code related properties by means of finite transducers, *Sequences, II (Positano, 1991)*, Springer–Verlag, 1993, 260–272.
57. (with A. Weber) The finest homophonic partition and related code concepts, *Proc. Mathematical Foundations of Computer Science (Košice, 1994)*, Lecture Notes in Computer Scvience **841**, Springer–Verlag, 1994, 618–628.
58. (with M. Ito) Power absorbing languages and semigroups, *Proc. Words, Languages and Combinatorics, II (Kyoto, 1992)*, World Scientific, Singapore, 1994, 179–191.

59. (with N. Jonoska) Images of cellular maps on sofic shifts, *Congr. Numer.*, **101** (1994), 109–115.
60. A metatheorem for deriving fuzzy theorems from crisp versions, *Fuzzy Sets and Systems*, **73** (1995), 349–358.
61. Cylindrical and eventual languages, *Mathematical Linguistics and Related Topics* (Gh. Păun, ed.), Ed. Acad. Române, Bucharest, 1995, 179–183.
62. (with A. Weber) Deciding multiset decipherability, *IEEE Trans. Inform. Theory*, **41** (1995), 291–297.
63. (with A. Weber) The finest homophonic partition and related code concepts, *IEEE Trans. Inform. Theory*, **42** (1996), 1569–1575.
64. The finest homophonic partition and related code concepts, *IEEE Trans. Inform. Theory* **42** 5 (1996), 1569–1575.
65. (with K. Satoshi, T. Yokomori) Locality, reversibility, and beyond: learning languages from positive data, *Proc. Algorithmic Learning Theory (Otzenhausen, 1998)*, Lecture Notes in Computer Science **1501**, Springer–Verlag, 1998, 191–204.
66. Splicing representations of strictly locally testable languages, *Discrete Appl. Math.*, **87** (1998), 139–147.
67. (with G. Rozenberg, R. Bladergroen, C.K.D. Breek, P.H.M. Lomerese, H. Spaink) Computing with DNA by operating on plasmids, *BioSystems*, **57** (2000), 87–93.
68. Writing by communication by documents in communities of organisms, *Millenium III*, Issue #4 (winter 1999/2000), 33–42.
69. Circular suggestions for DNA computing, in *Pattern Formation in Biology, Vision and Dynamics* (A. Carbone, M. Gromov, P. Prusinkiewicz, eds.), World Scientific, Singapore, 2000, 325–335.
70. Relativised code properties and multi–tube DNA dictionaries, in *Finite vs. Infinite* (C. Calude, Gh. Păun, eds.), Springer–Verlag, 2000, 175–186.
71. (with M. Yamamura, S. Gal) Aqueous computing – mathematical principles of molecular memory and its biomolecular implementation, Chap. 2 in *Genetic Algorithms* (H. Kitano, ed.), **4** (2000), 49–73 (in Japanese).
72. Writing by methylation proposed for aqueous computing, Chap. 31 in *Where Mathematics, Computer Science, Linguistics, and Biology Meet* (C. Martin–Vide, V. Mitrana, eds.), Kluwer, Dordrecht, 2001, 353–360.
73. Splicing systems, aqueous computing, and beyond, in *Unconventional Models of Computation UMC'2K* (I. Antoniou, C.S. Calude, M.J. Dineen, eds.), Springer–Verlag, 2001, 68–84.
74. Biomolecular realizations of a parallel architecture for solving combinatorial problems, *New Generation Computing*, **19** (2001), 301–312.
75. (with S. Gal) Aqueous computing: Writing into fluid memory, *Bull. Europ. Assoc. Theor. Comput. Sci.*, **75** (2001), 190–198.
76. Visualizing languages using primitive powers, in *Words, Semigroups, Transducers* (M. Ito, Gh. Păun, S. Yu, eds.), World Scientific, Singapore, 2001, 169–180.
77. Finitely generated languages and positive data, *Romanian J. Inform. Sci. & Tech.*, **5** (2002), 127–136.

78. Aqueous simulations of membrane computations, *Romanian J. Inform. Sci. & Tech.*, **5**, 4 (2002), 355–364.
79. An aqueous algorithm for finding the bijections in a binary relation, in *Formal and Natural Computing: Essays Dedicated to Grzegorz Rozenberg* (W. Brauer, H. Ehrig, J. Karhumaki, A. Salomaa, eds.), Lecture Notes in Computer Science **2300**, Springer–Verlag, 2002, 354–360.
80. (with X. Chen, M.J. Nichols, M. Yamamura, S. Gal) Aqueous solutions of algorithmic problems: emphasizing knights on a 3X3, *DNA Computing – 7th International Workshop on DNA-Based Computers June 2001* (N. Jonoska, N.C.Seeman, eds.), Lecture Notes in Computer Science **2340**, Springer–Verlag, 2002, 191–202.
81. (with X. Chen, M. Yamamura, S. Gal) Aqueous computing: a survey with an invitation to participate, *J. Computer Sci. Tech.*, **17** (2002), 672–681.
82. (with D. Pixton, E. Goode) Splicing systems: regularity and below, *DNA Computing – 8th International Workshop on DNA-Based Computers* (M. Hagiya, A. Ohuchi, eds.), Lecture Notes in Computer Science **2568**, Springer–Verlag, 2003, 262–268.

Author Index

Lecture Notes in Computer Science

For information about Vols. 1–2843
please contact your bookseller or Springer-Verlag

Vol. 2844: J.A. Jorge, N.J. Nunes, J.F. e Cunha (Eds.), Interactive Systems. Proceedings, 2003. XIII, 429 pages. 2003.

Vol. 2846: J. Zhou, M. Yung, Y. Han (Eds.), Applied Cryptography and Network Security. Proceedings, 2003. XI, 436 pages. 2003.

Vol. 2847: R. de Lemos, T.S. Weber, J.B. Camargo Jr. (Eds.), Dependable Computing. Proceedings, 2003. XIV, 371 pages. 2003.

Vol. 2848: F.E. Fich (Ed.), Distributed Computing. Proceedings, 2003. X, 367 pages. 2003.

Vol. 2849: N. García, J.M. Martínez, L. Salgado (Eds.), Visual Content Processing and Representation. Proceedings, 2003. XII, 352 pages. 2003.

Vol. 2850: M.Y. Vardi, A. Voronkov (Eds.), Logic for Programming, Artificial Intelligence, and Reasoning. Proceedings, 2003. XIII, 437 pages. 2003. (Subseries LNAI)

Vol. 2851: C. Boyd, W. Mao (Eds.), Information Security. Proceedings, 2003. XI, 443 pages. 2003.

Vol. 2852: F.S. de Boer, M.M. Bonsangue, S. Graf, W.-P. de Roever (Eds.), Formal Methods for Components and Objects. Proceedings, 2003. VIII, 509 pages. 2003.

Vol. 2853: M. Jeckle, L.-J. Zhang (Eds.), Web Services – ICWS-Europe 2003. Proceedings, 2003. VIII, 227 pages. 2003.

Vol. 2854: J. Hoffmann, Utilizing Problem Structure in Planning. XIII, 251 pages. 2003. (Subseries LNAI)

Vol. 2855: R. Alur, I. Lee (Eds.), Embedded Software. Proceedings, 2003. X, 373 pages. 2003.

Vol. 2856: M. Smirnov, E. Biersack, C. Blondia, O. Bonaventure, O. Casals, G. Karlsson, George Pavlou, B. Quoitin, J. Roberts, I. Stavrakakis, B. Stiller, P. Trimintzios, P. Van Mieghem (Eds.), Quality of Future Internet Services. IX, 293 pages. 2003.

Vol. 2857: M.A. Nascimento, E.S. de Moura, A.L. Oliveira (Eds.), String Processing and Information Retrieval. Proceedings, 2003. XI, 379 pages. 2003.

Vol. 2858: A. Veidenbaum, K. Joe, H. Amano, H. Aiso (Eds.), High Performance Computing. Proceedings, 2003. XV, 566 pages. 2003.

Vol. 2859: B. Apolloni, M. Marinaro, R. Tagliaferri (Eds.), Neural Nets. Proceedings, 2003. X, 376 pages. 2003.

Vol. 2860: D. Geist, E. Tronci (Eds.), Correct Hardware Design and Verification Methods. Proceedings, 2003. XII, 426 pages. 2003.

Vol. 2861: C. Bliek, C. Jermann, A. Neumaier (Eds.), Global Optimization and Constraint Satisfaction. Proceedings, 2002. XII, 239 pages. 2003.

Vol. 2862: D. Feitelson, L. Rudolph, U. Schwiegelshohn (Eds.), Job Scheduling Strategies for Parallel Processing. Proceedings, 2003. VII, 269 pages. 2003.

Vol. 2863: P. Stevens, J. Whittle, G. Booch (Eds.), «UML» 2003 – The Unified Modeling Language. Proceedings, 2003. XIV, 415 pages. 2003.

Vol. 2864: A.K. Dey, A. Schmidt, J.F. McCarthy (Eds.), UbiComp 2003: Ubiquitous Computing. Proceedings, 2003. XVII, 368 pages. 2003.

Vol. 2865: S. Pierre, M. Barbeau, E. Kranakis (Eds.), Ad-Hoc, Mobile, and Wireless Networks. Proceedings, 2003. X, 293 pages. 2003.

Vol. 2866: J. Akiyama, M. Kano (Eds.), Discrete and Computational Geometry. Proceedings, 2002. VIII, 285 pages. 2003.

Vol. 2867: M. Brunner, A. Keller (Eds.), Self-Managing Distributed Systems. Proceedings, 2003. XIII, 274 pages. 2003.

Vol. 2868: P. Perner, R. Brause, H.-G. Holzhütter (Eds.), Medical Data Analysis. Proceedings, 2003. VIII, 127 pages. 2003.

Vol. 2869: A. Yazici, C. Şener (Eds.), Computer and Information Sciences – ISCIS 2003. Proceedings, 2003. XIX, 1110 pages. 2003.

Vol. 2870: D. Fensel, K. Sycara, J. Mylopoulos (Eds.), The Semantic Web - ISWC 2003. Proceedings, 2003. XV, 931 pages. 2003.

Vol. 2871: N. Zhong, Z.W. Raś, S. Tsumoto, E. Suzuki (Eds.), Foundations of Intelligent Systems. Proceedings, 2003. XV, 697 pages. 2003. (Subseries LNAI)

Vol. 2873: J. Lawry, J. Shanahan, A. Ralescu (Eds.), Modelling with Words. XIII, 229 pages. 2003. (Subseries LNAI)

Vol. 2874: C. Priami (Ed.), Global Computing. Proceedings, 2003. XIX, 255 pages. 2003.

Vol. 2875: E. Aarts, R. Collier, E. van Loenen, B. de Ruyter (Eds.), Ambient Intelligence. Proceedings, 2003. XI, 432 pages. 2003.

Vol. 2876: M. Schroeder, G. Wagner (Eds.), Rules and Rule Markup Languages for the Semantic Web. Proceedings, 2003. VII, 173 pages. 2003.

Vol. 2877: T. Böhme, G. Heyer, H. Unger (Eds.), Innovative Internet Community Systems. Proceedings, 2003. VIII, 263 pages. 2003.

Vol. 2878: R.E. Ellis, T.M. Peters (Eds.), Medical Image Computing and Computer-Assisted Intervention - MICCAI 2003. Part I. Proceedings, 2003. XXXIII, 819 pages. 2003.

Vol. 2879: R.E. Ellis, T.M. Peters (Eds.), Medical Image Computing and Computer-Assisted Intervention - MICCAI 2003. Part II. Proceedings, 2003. XXXIV, 1003 pages. 2003.

Vol. 2880: H.L. Bodlaender (Ed.), Graph-Theoretic Concepts in Computer Science. Proceedings, 2003. XI, 386 pages. 2003.

Vol. 2881: E. Horlait, T. Magedanz, R.H. Glitho (Eds.), Mobile Agents for Telecommunication Applications. Proceedings, 2003. IX, 297 pages. 2003.

Vol. 2882: D. Veit, Matchmaking in Electronic Markets. XV, 180 pages. 2003. (Subseries LNAI)

Vol. 2883: J. Schaeffer, M. Müller, Y. Björnsson (Eds.), Computers and Games. Proceedings, 2002. XI, 431 pages. 2003.

Vol. 2884: E. Najm, U. Nestmann, P. Stevens (Eds.), Formal Methods for Open Object-Based Distributed Systems. Proceedings, 2003. X, 293 pages. 2003.

Vol. 2885: J.S. Dong, J. Woodcock (Eds.), Formal Methods and Software Engineering. Proceedings, 2003. XI, 683 pages. 2003.

Vol. 2886: I. Nyström, G. Sanniti di Baja, S. Svensson (Eds.), Discrete Geometry for Computer Imagery. Proceedings, 2003. XII, 556 pages. 2003.

Vol. 2887: T. Johansson (Ed.), Fast Software Encryption. Proceedings, 2003. IX, 397 pages. 2003.

Vol. 2888: R. Meersman, Zahir Tari, D.C. Schmidt et al. (Eds.), On The Move to Meaningful Internet Systems 2003: CoopIS, DOA, and ODBASE. Proceedings, 2003. XXI, 1546 pages. 2003.

Vol. 2889: Robert Meersman, Zahir Tari et al. (Eds.), On The Move to Meaningful Internet Systems 2003: OTM 2003 Workshops. Proceedings, 2003. XXI, 1096 pages. 2003.

Vol. 2890: M. Broy, A.V. Zamulin (Eds.), Perspectives of System Informatics. Proceedings, 2003. XV, 572 pages. 2003.

Vol. 2891: J. Lee, M. Barley (Eds.), Intelligent Agents and Multi-Agent Systems. Proceedings, 2003. X, 215 pages. 2003. (Subseries LNAI)

Vol. 2892: F. Dau, The Logic System of Concept Graphs with Negation. XI, 213 pages. 2003. (Subseries LNAI)

Vol. 2893: J.-B. Stefani, I. Demeure, D. Hagimont (Eds.), Distributed Applications and Interoperable Systems. Proceedings, 2003. XIII, 311 pages. 2003.

Vol. 2894: C.S. Laih (Ed.), Advances in Cryptology - ASIACRYPT 2003. Proceedings, 2003. XIII, 543 pages. 2003.

Vol. 2895: A. Ohori (Ed.), Programming Languages and Systems. Proceedings, 2003. XIII, 427 pages. 2003.

Vol. 2896: V.A. Saraswat (Ed.), Advances in Computing Science – ASIAN 2003. Proceedings, 2003. VIII, 305 pages. 2003.

Vol. 2897: O. Balet, G. Subsol, P. Torguet (Eds.), Virtual Storytelling. Proceedings, 2003. XI, 240 pages. 2003.

Vol. 2898: K.G. Paterson (Ed.), Cryptography and Coding. Proceedings, 2003. IX, 385 pages. 2003.

Vol. 2899: G. Ventre, R. Canonico (Eds.), Interactive Multimedia on Next Generation Networks. Proceedings, 2003. XIV, 420 pages. 2003.

Vol. 2901: F. Bry, N. Henze, J. Maluszyński (Eds.), Principles and Practice of Semantic Web Reasoning. Proceedings, 2003. X, 209 pages. 2003.

Vol. 2902: F. Moura Pires, S. Abreu (Eds.), Progress in Artificial Intelligence. Proceedings, 2003. XV, 504 pages. 2003. (Subseries LNAI).

Vol. 2903: T.D. Gedeon, L.C.C. Fung (Eds.), AI 2003: Advances in Artificial Intelligence. Proceedings, 2003. XVI, 1075 pages. 2003. (Subseries LNAI).

Vol. 2904: T. Johansson, S. Maitra (Eds.), Progress in Cryptology – INDOCRYPT 2003. Proceedings, 2003. XI, 431 pages. 2003.

Vol. 2905: A. Sanfeliu, J. Ruiz-Shulcloper (Eds.), Progress in Pattern Recognition, Speech and Image Analysis. Proceedings, 2003. XVII, 693 pages. 2003.

Vol. 2906: T. Ibaraki, N. Katoh, H. Ono (Eds.), Algorithms and Computation. Proceedings, 2003. XVII, 748 pages. 2003.

Vol. 2910: M.E. Orlowska, S. Weerawarana, M.P. Papazoglou, J. Yang (Eds.), Service-Oriented Computing – ICSOC 2003. Proceedings, 2003. XIV, 576 pages. 2003.

Vol. 2911: T.M.T. Sembok, H.B. Zaman, H. Chen, S.R. Urs, S.H.Myaeng (Eds.), Digital Libraries: Technology and Management of Indigenous Knowledge for Global Access. Proceedings, 2003. XX, 703 pages. 2003.

Vol. 2913: T.M. Pinkston, V.K. Prasanna (Eds.), High Performance Computing – HiPC 2003. Proceedings, 2003. XX, 512 pages. 2003.

Vol. 2914: P.K. Pandya, J. Radhakrishnan (Eds.), FST TCS 2003: Foundations of Software Technology and Theoretical Computer Science. Proceedings, 2003. XIII, 446 pages. 2003.

Vol. 2916: C. Palamidessi (Ed.), Logic Programming. Proceedings, 2003. XII, 520 pages. 2003.

Vol. 2918: S.R. Das, S.K. Das (Eds.), Distributed Computing – IWDC 2003. Proceedings, 2003. XIV, 394 pages. 2003.

Vol. 2920: H. Karl, A. Willig, A. Wolisz (Eds.), Wireless Sensor Networks. Proceedings, 2004. XIV, 365 pages. 2004.

Vol. 2922: F. Dignum (Ed.), Advances in Agent Communication. Proceedings, 2003. X, 403 pages. 2004. (Subseries LNAI).

Vol. 2923: V. Lifschitz, I. Niemelä (Eds.), Logic Programming and Nonmonotonic Reasoning. Proceedings, 2004. IX, 365 pages. 2004. (Subseries LNAI).

Vol. 2927: D. Hales, B. Edmonds, E. Norling, J. Rouchier (Eds.), Multi-Agent-Based Simulation III. Proceedings, 2003. X, 209 pages. 2003. (Subseries LNAI).

Vol. 2928: R. Battiti, M. Conti, R. Lo Cigno (Eds.), Wireless On-Demand Network Systems. Proceedings, 2004. XIV, 402 pages. 2004.

Vol. 2929: H. de Swart, E. Orlowska, G. Schmidt, M. Roubens (Eds.), Theory and Applications of Relational Structures as Knowledge Instruments. Proceedings. VII, 273 pages. 2003.

Vol. 2932: P. Van Emde Boas, J. Pokorný, M. Bieliková, J. Štuller (Eds.), SOFSEM 2004: Theory and Practice of Computer Science. Proceedings, 2004. XIII, 385 pages. 2004.

Vol. 2937: B. Steffen, G. Levi (Eds.), Verification, Model Checking, and Abstract Interpretation. Proceedings, 2004. XI, 325 pages. 2004.

Vol. 2950: N. Jonoska, G. Păun, G. Rozenberg (Eds.), Aspects of Molecular Computing. XI, 391 pages. 2004.

GPSR Compliance
The European Union's (EU) General Product Safety Regulation (GPSR) is a set of rules that requires consumer products to be safe and our obligations to ensure this.

If you have any concerns about our products, you can contact us on

ProductSafety@springernature.com

In case Publisher is established outside the EU, the EU authorized representative is:

Springer Nature Customer Service Center GmbH
Europaplatz 3
69115 Heidelberg, Germany

www.ingramcontent.com/pod-product-compliance
Ingram Content Group UK Ltd.
Pitfield, Milton Keynes, MK11 3LW, UK
UKHW021858190726
13853UKWH00003B/1326